Environmental Fluid Mechanics

Environmental Fluid Mechanics

Hillel Rubin

Technion–Israel Institute of Technology
Haifa, Israel

Joseph Atkinson

State University of New York at Buffalo
Buffalo, New York

CRC Press
Taylor & Francis Group
Boca Raton   London   New York

CRC Press is an imprint of the
Taylor & Francis Group, an **informa** business

CRC Press
Taylor & Francis Group
6000 Broken Sound Parkway NW, Suite 300
Boca Raton, FL 33487-2742

First issued in paperback 2019

© 2001 by Taylor & Francis Group, LLC
CRC Press is an imprint of Taylor & Francis Group, an Informa business

No claim to original U.S. Government works

ISBN-13: 978-0-8247-8781-3 (hbk)
ISBN-13: 978-0-367-39703-6 (pbk)

**Visit the Taylor & Francis Web site at
http://www.taylorandfrancis.com**

**and the CRC Press Web site at
http://www.crcpress.com**

To our wives, Elana Rubin and Nancy Atkinson

and our families

for their continued love and support

Preface

The purpose of this text is to provide the basis for an upper-level undergraduate or graduate course over one or two semesters, covering basic concepts and examples of fluid mechanics with particular applications in the natural environment. The book is designed to meet a dual purpose, providing an advanced fundamental background in the fluid mechanics of environmental systems and also applying fluid mechanics principles to a variety of environmental issues. Our basic motivation in preparing such a text is to share our experience gained by teaching courses in fluid mechanics, environmental fluid mechanics, and surface- and groundwater quality modeling and to provide a textbook that covers this particular collection of material.

This text presents a contemporary approach to teaching fluid mechanics in disciplines connected with environmental issues. There are many good fluid mechanics texts that overlap with various parts of this text, but they do not directly address themes and applications associated with the environment. On the other hand, there are also several texts that address water quality modeling, calculations of transport phenomena, and other issues of environmental engineering. Generally, such texts do not cover the fundamental topics of fluid mechanics that are relevant when describing fluid motions in the environment. Besides presenting contemporary environmental fluid mechanics topics, this text bridges the gap between those limited to fluid mechanics principles and those addressing the quality of the environment.

The term *environmental fluid mechanics* covers a broad spectrum of subjects. We have adopted the principle that this topic incorporates all issues of small-scale and global fluid flow and contaminant transport in our environment. We have chosen to consider these topics as divided into two general areas, one involving fundamental fluid mechanics principles relevant to the environment and the second concerning various types of applications of these principles to specific environmental flows and issues of water quality modeling. This division is reflected in the organization of the text into two main parts. The intent is to provide flexibility for instructors to choose material best suited for a particular curriculum. A full two-semester course could be developed by following the entire text. However, other options are possible. For example, a one-semester course could concentrate on the advanced fluid mechanics topics of the first part, with perhaps some chapters from the second part added to emphasize the environmental content. The second part by itself can be used in a course concentrating on environmental applications for students with appropriate fluid mechanics backgrounds. Although the book addresses principles of fluid mechanics relevant to the entire environment, the emphasis is mostly on water-related issues.

The material is designed for students who have already taken at least one undergraduate course in fluid mechanics and have an appropriate background in mathematics. Other courses in numerical modeling and environmental studies would be helpful but are not necessary. Because of the breadth of material that could be considered, some subjects have necessarily been omitted or treated only at an introductory level. These topics are left for continuing studies in the student's particular discipline, such as oceanography, meteorology, groundwater hydrology and contaminant transport, surface water quality modeling, etc. References are provided in each chapter so that students can easily get started in pursuing a particular subject in greater detail. Example problems and solutions are included wherever possible, and there is a set of homework problems at the end of each chapter.

We believe it is very important to introduce students to the proper use of physical and numerical models and computational approaches in the framework of analysis and calculation of environmental processes. Therefore, discussion and examples have been included that refer to scaling procedures and to various numerical methods that can be applied to obtain solutions for a given problem. A full discussion of numerical modeling approaches is included.

Both parts of the text are organized to provide (1) a review of introductory material and basic principles, (2) improvement and strengthening of basic knowledge, and (3) presentation of specific topics and applications in environmental fluid mechanics, along with problem-solving approaches. These topics have been chosen to introduce the student to the wide variety of issues addressed within the context of environmental fluid mechanics, regarding fluid motions on the earth's surface, underground, and in the oceans and atmosphere.

We believe that the wide scope of topics in environmental fluid mechanics covered in this text is consistent with present teaching needs in advanced undergraduate and graduate programs in fluid mechanics principles and topics related to the environment. These needs are subject to continuous growth and change due to our increasing interest in the fate of ecological systems and the need for understanding transport phenomena in our environment.

The authors are grateful to the US–Israel Fulbright Foundation for supporting a sabbatical leave for Joseph Atkinson at the Technion–Israel Institute of Technology, without which this text might not have been completed.

Finally, we are indebted to our own teachers, colleagues, and students, who have each made contributions to our understanding of this material and have helped in shaping the presentation of this text. We hope the book will contribute to this legacy.

Hillel Rubin
Joseph Atkinson

Contents

1

Preliminary Concepts

1.1 INTRODUCTION

1.1.1 Historical Perspective

Fluid mechanics and hydraulics have long been major components of civil engineering works and were probably originally associated with problems of water supply in ancient civilizations. One of the first well-documented hydraulic engineers was Archimedes (ca. 287–212 B.C.). His discovery of the basic principles of buoyancy serves today as one of the fundamental building blocks in describing fluid behavior. He also designed simple pumps for agricultural applications, and some of his designs are still in use today. Other early engineers had to deal with moving water over large distances from sources to cities, as with the Roman aqueducts found in many parts of Europe and the Middle East (see Fig. 1.1). These designs needed to incorporate basic aspects of open channel flow, such as finding the proper slope to obtain a desired flow rate. Remains of water storage and conveyance systems have also been found from some of the earliest civilizations known, both in the Near East and in the Far East. Rouse (1957) provides an interesting history of the science and engineering of hydraulics, which is also summarized by Graf (1971), particularly as it relates to open channel flow. In a sense, these were the first kinds of problems that can be associated with the field of environmental fluid mechanics.

An equally important task for early engineers was to design procedures for disposing of wastewater. The simplest means of doing this, which was in use until the relatively recent past, consisted of systems of gutters and drainage ditches, usually with direct discharge into ponds or streams. Septic tanks, with associated leeching fields, are another example of a simple wastewater treatment system, though these can handle only relatively small flow rates. Within the last century the practice of wastewater collection and treatment has evolved considerably, to enable varying degrees of treatment of a waste stream before it is discharged back into the natural environment. This development has been

1

Figure 1.1 Remains of Roman aqueduct, built in northern Israel.

driven by increased demands (both in quantity and in quality) for treating municipal sewage, as well as increased needs for treating industrial wastes. Sanitary engineering, within the general profession of civil engineering, traditionally dealt with designing water and wastewater collection and treatment systems. This has evolved into the contemporary field of environmental engineering, which now encompasses the general area of water quality modeling, for both surface and groundwater systems. This has necessitated the incorporation of other fields of science, such as chemistry and biology, to address the wider range of problems now being faced in treating waste streams with a variety of characteristics and needs.

In addition to treating municipal or industrial wastewater, environmental engineers currently are involved in solving problems of chemical fate and transport in natural environmental systems, including subsurface (groundwater) and surface waters, sediment transport, and atmospheric systems. A knowledge and understanding of fluid flow and transport processes is necessary to describe the transport and dispersion of pollutants in the environment, and chemical and biological processes must be incorporated to describe source and sink terms for contaminants of interest. Typical kinds of problems might involve calculating the expected chemical contaminant concentration at a water supply intake due to an upstream spill, evaluating the spreading of waste heat discharged from power plant condensers, predicting lake or reservoir stratification and associated effects on nutrient and dissolved oxygen distributions, determining the relative importance of contaminated sediments as a continuing source of pollutants to a river or lake system, calculating the expected recovery time of a lake when contaminant loading is discontinued, or evaluating the effectiveness of different remediation options for a contaminated groundwater source. All of these kinds of problems require an understanding of fluid flow phenomena and of biochemical behavior of materials in the environment.

1.1.2 Objectives and Scope

The primary objective of this text is to provide a basis for teaching upper-level fluid mechanics and water quality modeling courses dealing with environmentally related issues and to give a compilation of applications of environmental fluid mechanics seen in contemporary problems. The text also is meant to serve as a reference for further study in the various subjects covered, so references are included for additional reading. It would be impossible to include an exhaustive discussion of all possible subjects in one text, and inclusion of these additional references should provide a good starting point for more in-depth study. Example problems are provided where appropriate, to amplify the discussion or help reinforce certain concepts, and unsolved problems are included at the back of each chapter, to provide exercises that might be included in a course.

Today, the area of environmental fluid mechanics spans a broad range of issues, including open channel hydraulics, sediment transport, stratified flow phenomena, transport and mixing processes, and various issues in water quality and atmospheric modeling. These topics are studied in a variety of ways, such as by theoretical analyses, physical model experiments, field studies, and numerical modeling. This text presents material that might traditionally be included in two separate courses, one in fluid mechanics and the other in water quality modeling. The emphasis here is on aqueous systems, both in surface and subsurface flows, though the basic principles are mostly applicable also for atmospheric studies. A major link between classic hydraulic engineering and water quality studies is in defining the advective and diffusive (or dispersive) transport terms of a water quality model, which are normally estimated from hydrodynamic calculations. Fluid mechanics deals with the study of fluid motion, or the response of a fluid to applied forces, and environmental fluid mechanics refers to the application of fluid mechanics principles to problems involving environmental flows, including purely physical applications (e.g., open channel flow, groundwater flow, sediment transport) and problems of water quality modeling. In the following chapters the analytical bases for the engineering evaluation and solution of these types of problems are developed. Governing equations for fluid motion are derived, as well as the equation expressing mass balance for a dissolved tracer, otherwise known as the advection–diffusion equation. Conservation equations for both mechanical and thermal energy also are developed, and these lead to descriptions of turbulent kinetic energy and temperature, respectively.

The text is divided into two parts. The first part is a discussion of theoretical principles used in describing fluid motion and includes the derivation of the basic mathematical equations governing fluid flow. Chapters 4 through 9 include discussions of potential flow theory, introductions to turbulence and boundary layer theory, groundwater flow, and large-scale motions where the rotation of the earth must be incorporated into the equations of motion. The second part of the text contains material more directly applied to environmental problems. Fundamental transport processes for contaminants are discussed, including advection, diffusion, and dispersion, and applications are described in modeling groundwater flow and contaminant transport, exchange processes between water surfaces and the atmosphere, stratified flows, jets and plumes, sediment transport, and remediation issues. Sections in various chapters are included that discuss associated numerical modeling issues, as we recognize the important role of numerical solutions in many of the problems faced in environmental fluid mechanics. Different solution approaches, boundary conditions, numerical dispersion and scaling considerations are addressed. The intent is that the material contained herein could serve as the basis for a two-semester upper level undergraduate or graduate course, with each part of the text providing a focus for each semester of instruction. Of course, single-semester courses can be developed, based on individual chapters.

The remainder of the present chapter is devoted to a review of fluid properties and mathematical preliminaries.

1.2 PROPERTIES OF FLUIDS

1.2.1 General

Most substances are categorized as existing in one of two states: solid or fluid. Solid elements have a rigid shape that can be modified as a result of stresses. This shape modification is termed *deformation* or *strain*. Different types of solids are identified by different relationships between the shear stress and the strain. A strained solid body is in a state of equilibrium with the stresses applied on that body. When applied stresses vanish, the solid body relaxes to its original shape.

Solid boundaries (i.e., a container) and interfaces with other fluids determine the shape of a fluid body. Unlike solids, even an infinitesimal shear force changes the shape of fluid elements. Differences between different types of fluid are identified by different relationships between the shear stress and the *rate of strain*. When applied stresses vanish, fluid elements do not return to their original shape. In addition, fluids usually do not support tensile stresses, though in many cases they strongly resist normal compressive stresses. In many cases they can be considered as *incompressible* materials or materials subject to *incompressible flow*, meaning that their density is not a function of pressure. In general, fluids may be divided into liquids, for which *compressibility* is generally negligible, and gases, which are *compressible* fluids. In other words, the volume of a liquid mass is almost constant, and it occupies the lowest portion of a container in which it is held. It also has a horizontal free surface in a stationary container. A gas always expands and occupies the entire volume of any container. However, gases like air are usually well described in the atmosphere using *incompressible* flow theory.

1.2.2 Continuum Assumptions

All materials are composed of individual molecules subject to relative movement. However, in the framework of fluid mechanics we consider the fluid as a continuum. We are generally interested in the *macroscopic* behavior of a fluid material, so that the smallest fluid mass of interest usually consists of a *fluid particle* that is much larger than the mean free path of a single molecule. It is therefore possible to ignore the discrete molecular structure of the matter and to refer to it as a *continuum*. The continuum approach is valid if the characteristic length, or size of the flow system (e.g., the diameter of a solid sphere submerged in a flowing fluid) is much larger than the mean free path of the molecules. For example, in a standard atmosphere the molecular free path is of the order of 10^{-8} m, but in the upper altitudes of the atmosphere the molecule mean free path is of the order of 1 m. Therefore, in order to study the dynamics of a rarefied gas in such heights a kinetic theory approach would be necessary, rather than the continuum approach.

1.2.3 Review of Fluid Properties

The *density* ρ of a fluid is a measure of the concentration of matter and is expressed in terms of mass per unit volume. The volume and mass of fluid considered for the calculation of the fluid density should be small, but not so small that variations on a molecular level would become important. Therefore, we define

$$\rho = \lim_{\delta V \to \delta V'} \frac{\delta m}{\delta V} \tag{1.2.1}$$

where δm is an amount of mass contained in a small volume δV, and $\delta V'$ is the volume of the smallest fluid particle that is still much larger than the mean free molecular path. The *specific weight* γ is the force of gravity on the mass contained in a unit volume of the substance,

$$\gamma = \rho g \tag{1.2.2}$$

The density of water is 1000 kg/m^3 (at 4°C) and the acceleration of gravity $g = 9.81$ m/s^2. Therefore, the nominal specific weight of water is

$$\gamma = (1000 \text{ kg/m}^3)(9.81 \text{ m/s}^2) = 9810 \text{ N/m}^3 \tag{1.2.3}$$

The diffusive flux of a dissolved constituent in a fluid is expressed by *Fick's law*, which states that the flux is proportional to the constituent concentration gradient (see also Chap. 10). In a one-dimensional domain this law is expressed as

$$q_m = -k_m \frac{\partial C}{\partial x} \tag{1.2.4}$$

where q_m is the mass flux (kg m^{-2} s^{-1}) of the constituent in the x direction, C is the constituent concentration (kg m^{-3}), and k_m is the mass diffusivity (m^2 s^{-1}), whose value depends on the fluid and on the constituent. The relationship represented by Eq. (1.2.4) is based on empirical evidence and is called a *phenomenological law*. A similar phenomenological law is *Fourier's law of heat diffusion*, which in a one-dimensional domain can be written as

$$q = -k \frac{\partial T}{\partial x} \tag{1.2.5}$$

where q is the heat flux (J m^{-2} s^{-1}), T is the temperature (°C), and k is the thermal conductivity (J m^{-1} s^{-1} °C^{-1}), whose value depends on the fluid.

Another phenomenological law is the law of Newton, expressing proportionality between the strain rate and the shear stress in so-called *Newtonian fluids*. In a one-directional flow with velocity u in the x direction and with the velocity a function of y, the shear stress τ that develops between fluid layers is expressed as

$$\tau = \mu \frac{\partial u}{\partial y} \tag{1.2.6}$$

Here the constant of proportionality μ (Pa s) is the *dynamic viscosity*, whose value depends on the fluid and on temperature. The ratio of dynamic viscosity to density appears often in the equations describing fluid motion and is called the *kinematic viscosity* v (m^2 s^{-1}),

$$v = \frac{\mu}{\rho} \tag{1.2.7}$$

There is some similarity between Eqs. (1.2.4), (1.2.5), and (1.2.6). However, the mass flux given by Eq. (1.2.4) and heat flux given by Eq. (1.2.5) are components of flux vectors, whereas the shear stress given by Eq. (1.2.6) is a component of a tensor. These issues are described further in the following sections of this chapter.

The interface between two immiscible fluids behaves like a stretched membrane, in which tension originates from intermolecular attractive (cohesive) forces. Near an interface, say between the fluid and another fluid or between the fluid and the solid walls of a boundary or container, all the fluid molecules are trying to pull the molecules on the interface inward. The magnitude of the tensile force per unit length of a line on the interface is called *surface tension* σ (N m^{-1}), whose value depends on the pair of fluids and the temperature. If p_1 and p_2 are the fluid pressures on the two sides of an interface, then a simple force balance yields

$$\sigma(2\pi R) = (p_1 - p_2)\pi R^2$$

where R is the radius of curvature of the interfacial surface. This result is also written as

$$\sigma = \frac{(p_1 - p_2)R}{2} \qquad (1.2.8)$$

For a general surface, the radii of curvature along two orthogonal directions R_1 and R_2 are used to specify the curvature. In this case, the relationship between surface tension and pressure is

$$\sigma = \frac{(p_1 - p_2)R_1 R_2}{R_1 + R_2} \qquad (1.2.9)$$

If a fluid and its vapor coexist in equilibrium, the vapor is a *saturated vapor*, and the pressure exerted by this saturated vapor is called the *vapor pressure*, with symbol p_v. The vapor pressure depends on the fluid and the temperature.

The *compressibility* of a fluid is defined in terms of the average modulus of elasticity K (Pa), defined as

$$K = -\frac{dp}{dV/V} = \frac{dp}{d\rho/\rho} \qquad (1.2.10)$$

where dV is the change in volume accompanying a change in pressure dp, and V and ρ are the original volume and density, respectively. The second expression in Eq. (1.2.10) refers to density changes, but the negative sign is dropped since the density changes in the opposite direction to that of volume.

1.3 MATHEMATICAL PRELIMINARIES

1.3.1 Vectors and Tensors

A point in a three dimensional space is defined by its *coordinates*,

$$x^1, x^2, x^3 \tag{1.3.1}$$

A curve is defined as the totality of points given by the equation

$$x^i = f^i(u) \quad (i = 1, 2, 3) \tag{1.3.2}$$

Here, u is an arbitrary parameter and the f^i are three arbitrary functions.

The point given by Eq. (1.3.1) can be represented by a new set of coordinates (x'^1, x'^2, x'^3), where

$$x'^i = f^i(x^1, x^2, x^3) \tag{1.3.3}$$

The Jacobian of the transformation is

$$J' = \left| \frac{\partial x'^i}{\partial x^j} \right| \quad (i, j = 1, 2, 3) \tag{1.3.4}$$

Eq. (1.3.2) also can be represented by another transformation,

$$x^i = g^i(x'^1, x'^2, x'^3) \tag{1.3.5}$$

Differentiation of Eq. (1.3.3) then yields

$$dx'^i = \frac{\partial x'^i}{\partial x^j} dx^j \tag{1.3.6}$$

where index *summation convention* is used. That is, summation is made with regard to the repeating superscript j. Such repeated indices are often referred to as dummy indices. Any such pair may be replaced by any other pair of repeated indices without changing the value of the expression.

For future reference, we introduce the *Kronecker delta*, δ_i^j, defined as

$$\delta_i^j = 1 \quad \text{if} \quad i = j$$
$$\delta_i^j = 0 \quad \text{if} \quad i \neq j \tag{1.3.7}$$

It is evident that

$$\frac{\partial x^i}{\partial x^j} = \delta_j^i \tag{1.3.8}$$

Contravariant Vectors and Tensors, Invariants

Consider a point P with coordinates x^i and a neighboring point Q with coordinates $x^i + dx^i$. These two points define a *vector*, termed the *displacement*, whose components are dx^i. We may still think about the same two points, but apply a different coordinate system $x^{'i}$. In this coordinate system the components of the displacement vector are $dx^{'i}$. Components of the displacement tensor in the two systems of coordinates are related by Eq. (1.3.6).

If we keep the point P fixed, but vary Q in the neighborhood of P, the coefficient $\partial x^{'i}/\partial x^j$ remains constant. Under these conditions, Eq. (1.3.6) is a linear homogeneous (or *affine*) transformation.

The vector has an absolute meaning, but the numbers describing this vector depend on the employed coordinate system. The infinitesimal displacements given by Eq. (1.3.6) satisfy the rule of transformation of *contravariant* vectors. Later we also will refer to *covariant* vectors. A contravariant vector is one in which the vector components comprise a set of quantities A^i associated with a point P that transform, on change of coordinates, according to the equation

$$A^{'i} = A^j \frac{\partial x^{'i}}{\partial x^j} \tag{1.3.9}$$

where the partial derivatives are evaluated at point P. The expression for the infinitesimal displacements given by Eq. (1.3.6) represents a particular example of a contravariant vector.

A set of quantities A^{ij} represents components of a contravariant tensor of the *second order* if they transform according to the equation

$$A^{'ij} = A^{km} \frac{\partial x^{'i}}{\partial x^k} \frac{\partial x^{'j}}{\partial x^m} \tag{1.3.10}$$

The product $A^i \times B^j$ of two contravariant vectors is a contravariant tensor of the second order.

Equation (1.3.10) provides a basic format for the definition of contravariant tensors of the third or higher order. We also can conclude that there is a contravariant tensor of the zero order that is a single component quantity, transformed according to the identity relation

$$A' = A \tag{1.3.11}$$

Such a quantity is called an *invariant*, and its value is independent of the employed coordinate system.

Covariant Vectors and Tensors, Mixed Tensors

If H is an invariant then we may introduce

$$\frac{\partial H}{\partial x^{'i}} = \frac{\partial H}{\partial x^j} \frac{\partial x^j}{\partial x^{'i}} \tag{1.3.12}$$

This transformation is very similar to that of Eq. (1.3.6), but the partial derivative involving the two sets of coordinates is reversed. Equation (1.3.6) indicates that the infinitesimal displacement is the prototype of the contravariant vector. Equation (1.3.12) shows that the partial derivative of an invariant represents a prototype of the general covariant vector. The components of a covariant vector comprise a set of quantities that transform according to

$$A'_i = A_j \frac{\partial x^j}{\partial x'^i} \tag{1.3.13}$$

Suffixes indicating contravariant character are placed as superscripts, and those indicating covariant character are subscripts. This convention means that coordinates should be written x^i rather than x_i, although it is only the differentials of the coordinates, and not the coordinates themselves, that have tensor character.

We may extend Eq. (1.3.13) to define higher order covariant tensors. Following the definitions of contravariant and covariant tensors, *mixed tensors* can be defined. As an example, consider a third-order mixed tensor,

$$A'^i_{jk} = A^m_{np} \frac{\partial x'^i}{\partial x^m} \frac{\partial x^n}{\partial x'^j} \frac{\partial x^p}{\partial x'^k} \tag{1.3.14}$$

It then follows that the Kronecker delta is a second-order mixed tensor represented by the transformation

$$\delta'^i_j = \delta^m_n \frac{\partial x'^i}{\partial x^m} \frac{\partial x^n}{\partial x'^j} \tag{1.3.15}$$

The left-hand side of Eq. (1.3.15) is unity if $i = j$ and zero otherwise. Holding m fixed and summing with respect to n, there is no contribution to the sum unless $n = m$. Therefore the right-hand side of Eq. (1.3.15) reduces to

$$\frac{\partial x'^i}{\partial x^m} \frac{\partial x^m}{\partial x'^j} \tag{1.3.16}$$

and this expression is equal to δ^i_j.

Addition, Multiplication, and Contraction of Tensors

Two tensors of the same order and type can be added together to give another tensor of the same order and type. For example, we can write

$$C'^i_{jk} = A'^i_{jk} + B'^i_{jk} \tag{1.3.17}$$

A second-order tensor is called a *symmetric tensor* if its components satisfy the relationship

$$A_{ij} = A_{ji} \tag{1.3.18}$$

A second-order tensor is *antisymmetric* or *skew-symmetric* if its components satisfy

$$A_{ij} = -A_{ji} \tag{1.3.19}$$

The definitions given by Eqs. (1.3.18) and (1.3.19) can be extended to more complicated tensors. A tensor is symmetric with respect to a pair of suffixes if the value of the components is unchanged on interchanging these suffixes. A tensor is antisymmetric with respect to a pair of suffixes if interchanging these suffixes leads to a change of sign with no change of absolute value.

Any tensor of the second order can be expressed as the sum of a symmetric and an antisymmetric tensor. As an example, we can write

$$A_{ij} = \frac{1}{2}(A_{ij} + A_{ji}) + \frac{1}{2}(A_{ij} - A_{ji}) \tag{1.3.20}$$

The first term on the right-hand side of Eq. (1.3.20) is a symmetric tensor, and the second one is an antisymmetric tensor. This property is useful when discussing stresses in fluid flow (Chap. 2).

Addition or subtraction can be done only with tensors of the same order and type. In multiplication the only restriction is that we never multiply two components with the same literal suffix at the same level in each component. We may take tensors of different types and different literal suffixes. Then the product is a tensor whose order is equal to the sum of orders of the multiplied tensors. As an example,

$$C_{ijk}^{m} = A_{ij}B_{k}^{m} \tag{1.3.21}$$

The product exemplified by Eq. (1.3.21) is called an *outer product*. The *inner product* is associated with *contraction*. It is obtained by multiplication of tensors with lower suffixes identical to lower ones. An example is

$$C_{i}^{m} = A_{ij}B^{jm} \tag{1.3.22}$$

The process of contraction cannot be applied to suffixes at the same level. Indices appearing at lower and upper levels represent summation.

The Metric Tensor and the Line Element

Suppose that y^1, y^2, y^3 are rectangular Cartesian coordinates. Then the square of the distance between adjacent points is

$$ds^2 = (dy^1)^2 + (dy^2)^2 + (dy^3)^2 \tag{1.3.23}$$

Any system of curvilinear coordinates is represented by x^1, x^2, x^3 (e.g., cylindrical or spherical polar). The y^i coordinates are functions of the x^i coordinates, and the dy^i components of the infinitesimal displacement are linear homogeneous functions of the dx^i components. We introduce the relationships of Eq. (1.3.6) to obtain a homogeneous quadratic expression in the dx^i components, which may be written as

$$ds^2 = g_{ij}dx^i dx^j \tag{1.3.24}$$

where the coefficients g_{ij} are functions of the x^i coordinates. As the g_{ij} do not occur separately, but only in the combinations $(g_{ij} + g_{ji})$, there is no loss of generality in taking g_{ij} as a symmetric tensor.

As the distance between two given points is not dependent on the applied coordinates, the value of ds or ds^2 is an invariant. According to Eq. (1.3.6), dx^i is a contravariant vector. Therefore, g_{ij} is a second-order covariant tensor. It is called the *metric tensor*.

By applying Eqs. (1.3.23) and (1.3.24), we obtain

$$g_{ij} = \frac{\partial y^1}{\partial x^i}\frac{\partial y^1}{\partial x^j} + \frac{\partial y^2}{\partial x^i}\frac{\partial y^2}{\partial x^j} + \frac{\partial y^3}{\partial x^i}\frac{\partial y^3}{\partial x^j} \tag{1.3.25}$$

As an example, we consider a cylindrical coordinate system in which $x^1 = r$, $x^2 = \theta$, $x^3 = z$. The relationships between the y^i coordinates and x^i coordinates are $y^1 = x^1 \cos x^2$, $y^2 = x^1 \sin x^2$, and $y^3 = x^3$. By introducing these relationships into Eq. (1.3.25), we obtain for the cylindrical coordinate system

$$g_{ij} = 0 \qquad \text{for} \quad i \neq j$$
$$g_{11} = 1 \qquad g_{22} = r^2 \qquad g_{33} = 1 \tag{1.3.26}$$

The Conjugate Tensor; Lowering and Raising Suffixes

From the covariant metric tensor g_{ij} we can obtain a contravariant tensor g^{ij} given by

$$g^{ij} = \frac{C^{ij}}{g} \tag{1.3.27}$$

where C^{ij} is the *cofactor* of g_{ij} and g is the *determinant* of g_{ij}. The following relationships then hold:

$$g_{ij}C^{ik} = g_{ji}C^{ki} = \delta_j^k \tag{1.3.28}$$

By multiplying both sides of this expression by C^{jm} we obtain

$$g\delta_i^j g^{ik} = \delta_m^k C^{jm} \tag{1.3.29}$$

If $g^{ij} = 0$ for $i \neq j$, then

$$g^{11} = \frac{1}{g_{11}} \qquad g^{22} = \frac{1}{g_{22}} \qquad g^{33} = \frac{1}{g_{33}}$$

$$g^{ij} = 0 \qquad \text{for} \qquad i \neq j \tag{1.3.30}$$

The covariant metric tensor and its contravariant conjugate can be used for *lowering* and *raising* of suffixes. As an example,

$$U_{ijk} = g_{im} V^m_{jk} \tag{1.3.31}$$

Now we may refer to a tensor as a geometrical object that has different representations in different coordinate systems. Until now we could consider that the tensors U^{ij} and U_{ij} were entirely unrelated; one was contravariant and the other covariant, and there was no connection between them. At present we realize that use of the same symbol U for these tensors means that each of them represents the same geometrical object, and internal products with the metric tensors give the relationships between their components.

Geodesics and Christoffel Symbols

A *geodesic* is a curve whose length has a stationary value with respect to arbitrary small variations of the curve while its end points are kept fixed. By using some techniques of variational calculus, it is possible to show that the differential equation of a geodesic is

$$g_{ij} \frac{dp^j}{ds} + [jk, i] p^j p^k = 0 \tag{1.3.32}$$

where s is the arc length along the geodesic and $p^i = dx^i/ds$. The expression given in the square brackets is called the *Christoffel symbol of the first kind*, which is defined by

$$[jk, i] = \frac{1}{2} \left(\frac{\partial g_{ij}}{\partial x^k} + \frac{\partial g_{ik}}{\partial x^j} - \frac{\partial g_{jk}}{\partial x^i} \right) \tag{1.3.33}$$

The Christoffel of the second kind is defined as

$$\sum_{jk}^{i} = g^{im} [jk, m] \tag{1.3.34}$$

If we multiply Eq. (1.3.32) by g^{in}, we obtain another form for the equation of a geodesic,

$$\frac{dp^i}{ds} + \sum_{jk}^{i} p^j p^k = 0 \tag{1.3.35}$$

This expression also can be represented by

$$\frac{d^2x^i}{ds^2} + \sum_{jk} \overset{i}{} \frac{dx^j}{ds}\frac{dx^k}{ds} = 0 \tag{1.3.36}$$

The differential equation of a geodesic in terms of an arbitrary parameter t is identical to Eq. (1.3.36) in which t replaces s.

Derivatives of Tensors

From Eq. (1.3.12) it is shown that the partial derivative of an invariant with respect to a coordinate is a covariant vector. However, as discussed and shown hereinafter, the partial derivative of a tensor is not a tensor.

We refer to a contravariant vector field U^i, defined along a curve $x^i = x^i(t)$. Then the *absolute derivative of* U^i with regard to t is defined as

$$\frac{\delta U^i}{\delta t} = \frac{dU^i}{dt} + \sum_{jk} \overset{i}{} U^j \frac{dx^k}{dt} \tag{1.3.37}$$

This expression is itself a contravariant vector. If the absolute derivative expression of Eq. (1.3.37) vanishes, then the vector U^i is said to be propagated *parallel* along the curve. In the case of a Cartesian coordinate system, the Christoffel symbols vanish and Eq. (1.3.37) yields $dU^i/dt = 0$. In this case the vector passes through a sequence of parallel positions.

The absolute derivative of the vector given by Eq. (1.3.37) means that the vector characteristic is given along a curve. Therefore, Eq. (1.3.37) can be represented by

$$\frac{\delta U^i}{\delta t} = \left(\frac{\partial U^i}{\partial x^k} + \sum_{jk} \overset{i}{} U^j \right) \frac{dx^k}{dt} \tag{1.3.38}$$

The left-hand side of Eq. (1.3.38) represents a contravariant vector. The term dx^k/dt also is a contravariant vector. Therefore, the expression between parentheses of Eq. (1.3.38) is a second-order mixed tensor. We call it the *covariant derivative of a contravariant vector*. It is represented as

$$U^i_{,k} = \frac{\partial U^i}{\partial x^k} + \sum_{jk} \overset{i}{} U^j \tag{1.3.39}$$

The same method can be applied to obtain the covariant derivative of any tensor from the absolute derivative. In the following equations we provide expressions for the covariant derivative of various types of tensors:

$$U_{i,k} = \frac{\partial U_i}{\partial x^k} - \sum_{ik}^{j} U_j \tag{1.3.40}$$

$$U^{ij}_{,k} = \frac{\partial U^{ij}}{\partial x^k} + \sum_{mk}^{i} U^{mj} + \sum_{mk}^{j} U^{im} \tag{1.3.41}$$

$$U_{ij,k} = \frac{\partial U_{ij}}{\partial x^k} - \sum_{ik}^{m} U_{mj} - \sum_{jk}^{m} U_{im} \tag{1.3.42}$$

$$U^{i}_{j,k} = \frac{\partial U^{i}_{j}}{\partial x^k} + \sum_{mk}^{i} U^{m}_{j} - \sum_{jk}^{m} U^{i}_{m} \tag{1.3.43}$$

Cartesian Tensors

If we refer to two Cartesian coordinate systems z^i and $z^{\cdot i}$, then for a contra-variant tensor of the second order we may write the following law of transformation:

$$U'^{ij} = U^{mn} \frac{\partial z'^{i}}{\partial z^m} \frac{\partial z'^{j}}{\partial z^n} \tag{1.3.44}$$

However, the partial derivatives of Eq. (1.3.44) represent the cosine between the relevant axes of the two Cartesian coordinate systems. Therefore we may write

$$\frac{\partial z'^{i}}{\partial z^m} = \frac{\partial z^m}{\partial z'^{i}} = \cos(z'^{i} z^m) \tag{1.3.45}$$

By introducing the relationships of Eq. (1.3.45) into Eq. (1.3.44), we obtain

$$U'^{ij} = U^{mn} \frac{\partial z^m}{\partial z'^{i}} \frac{\partial z^n}{\partial z'^{j}} \tag{1.3.46}$$

This expression is identical to the transformation of a covariant tensor. We may conclude that in every case of Cartesian tensors, the law of transformation remains unchanged when a subscript is raised or a superscript is lowered. Therefore, when dealing with Cartesian tensors, it is common to apply subscripts exclusively. Also, coordinates are represented with subscripts in such cases. The Kronecker delta is identical to the metric tensor and is written as δ_{ij}, which also is identical to the unit matrix.

The *permutation tensor* ε_{ijk} is defined as

$\varepsilon_{ijk} = 0$ if two of the suffixes are equal
$\varepsilon_{ijk} = 1$ if the sequence of numbers ijk is the sequence of 1-2-3,
      or an even permutation of the sequence
$\varepsilon_{ijk} = -1$ if the sequence of numbers ijk is an odd permutation of
      the sequence 1-2-3

Examples of the application of these rules are

$$\varepsilon_{123} = \varepsilon_{231} = \varepsilon_{312} = 1 \qquad \varepsilon_{132} = \varepsilon_{213} = \varepsilon_{321} = -1 \qquad (1.3.47)$$

Using the permutation tensor, the *vector product* C_i of two vectors A_j and B_k is given by

$$C_i = \varepsilon_{ijk}A_jB_k \qquad (1.3.48)$$

In addition, the *curl* operator is given by

$$C_i = \varepsilon_{ijk}A_{k,j} \qquad (1.3.49)$$

The following useful relation is the *epsilon delta relation*

$$\varepsilon_{ijk}\varepsilon_{kmn} = \delta_{im}\delta_{jn} - \delta_{in}\delta_{jm} \qquad (1.3.50)$$

Physical Components of Tensors

Consider a vector whose components in a Cartesian coordinate system z_i are represented by Z_i. As the coordinate system is a Cartesian one, covariant and contravariant components are identical. The quantities Z_i also are called the *physical components of the vector along the coordinate axes.*

    If we introduce curvilinear coordinates x^j, the definition of *contravariant* and *covariant components* X^j and X_j, respectively, of the vector for the coordinate system x^j is given by

$$X^j = Z_i\frac{\partial x^j}{\partial z_i} \qquad X_j = Z_i\frac{\partial z_i}{\partial x^j} \qquad (1.3.51)$$

In connection with these components, we do not use the word physical, since in general such components have no direct physical meaning. They may even have physical dimensions different from those of the physical components Z_i.

    Let x^j be a curvilinear coordinate system with metric tensor g_{ij}, and let X^j be contravariant components of a vector. We define *the physical components of the vector X^j in the direction λ^j* as the invariant

$$g_{ij}X^i\lambda^j = X^i\lambda_i = X_i\lambda^i \qquad (1.3.52)$$

If the curvilinear coordinates x^j are orthogonal coordinates, then the line element is given by

$$ds^2 = (h_1 dx^1)^2 + (h_2 dx^2)^2 + (h_3 dx^3)^2 \tag{1.3.53}$$

where h_1, h_2, and h_3 are the *geometrical scales* associated with the respective coordinates. We take a unit vector λ^i in the direction of x^1. Therefore the three components of λ^i are

$$\lambda^1 = \frac{dx^1}{ds} \qquad \lambda^2 = 0 \qquad \lambda^3 = 0 \tag{1.3.54}$$

Since λ^i is a unit vector, we have

$$g_{ij}\lambda^i\lambda^j = h_1^2(\lambda^1)^2 \qquad \lambda^1 = \frac{1}{h_1} \tag{1.3.55}$$

By multiplying by the metric tensor, we lower superscripts and obtain

$$\lambda_1 = h_1 \qquad \lambda_2 = \lambda_3 = 0 \tag{1.3.56}$$

Equations (1.3.52)–(1.3.56) imply that the physical components of the vector X^j along the parametric line of x^1 are X_1/h_1 or $h_1 X^1$. Considering all geometrical scales of the coordinate system we obtain the following expressions for the physical components of the vector:

$$\frac{X_1}{h_1} \qquad \frac{X_2}{h_2} \qquad \frac{X_3}{h_3} \qquad \text{or} \qquad h_1 X^1 \qquad h_2 X^2 \qquad h_3 X^3 \tag{1.3.57}$$

In order to define the physical components of a second order tensor we apply two unit vectors in the directions of two parametric lines of two coordinates. Such an operation leads to the following expressions for the physical components of the second order tensor, in terms of its covariant components:

$$\begin{matrix} \dfrac{X_{11}}{h_1^2} & \dfrac{X_{12}}{h_1 h_2} & \dfrac{X_{13}}{h_1 h_3} \\[2ex] \dfrac{X_{21}}{h_2 h_1} & \dfrac{X_{22}}{h_2^2} & \dfrac{X_{23}}{h_2 h_3} \\[2ex] \dfrac{X_{31}}{h_3 h_1} & \dfrac{X_{32}}{h_3 h_2} & \dfrac{X_{33}}{h_3^2} \end{matrix} \tag{1.3.58}$$

In terms of the contravariant components of the second order tensor, the physical components of Eq. (1.3.58) are given by

$$\begin{matrix} X^{11} h_1^2 & X^{12} h_1 h_2 & X^{13} h_1 h_3 \\[1ex] X^{21} h_2 h_1 & X^{22} h_2^2 & X^{23} h_2 h_3 \\[1ex] X^{31} h_3 h_1 & X^{32} h_3 h_2 & X^{33} h_3^2 \end{matrix} \tag{1.3.59}$$

As an example, we calculate the relationships between the Cartesian components of the velocity vector and its contravariant, covariant, and physical components in spherical polar coordinates. The spherical polar coordinates are r, θ, and ϕ, which are referred to, respectively, as x^1, x^2, x^3. These coordinates are related to the Cartesian coordinates z_1, z_2, z_3, by

$$z_1 = x^1 \sin x^2 \cos x^3 \qquad z_2 = x^1 \sin x^2 \sin x^3 \qquad z_3 = x^1 \cos x^2 \quad (1.3.60)$$

Components of the velocity vector in the Cartesian and spherical coordinate systems are given, respectively, by

$$V_i = \frac{dz_i}{dt} \qquad v^i = \frac{dx^i}{dt} \tag{1.3.61}$$

The relationships between the contravariant, contravariant spherical coordinate components and Cartesian components are given by

$$v^i = V_j \frac{\partial x^i}{\partial z_j} \qquad v_i = V_j \frac{\partial z_j}{\partial x^i} \tag{1.3.62}$$

By applying Eq. (1.3.61), we calculate the partial derivatives required by Eq. (1.3.62) and define the relationships between the Cartesian and spherical coordinate components of the velocity vector.

The line element in spherical coordinates is given by

$$ds^2 = dr^2 + r^2\, d\theta^2 + r^2 \sin^2 \theta d\phi^2 = (dx^1)^2 + (x^1 dx^2)^2 + (x^1 \sin x^2 dx^3)^2 \tag{1.3.63}$$

This expression indicates that the metric tensor components are

$$g_{11} = 1 \quad g_{22} = r^2 \quad g_{33} = r^2 \sin^2 \theta \quad g_{ij} = 0 \quad \text{for} \quad i \neq j \tag{1.3.64}$$

Equation (1.3.61) specifies the various contravariant components of the velocity vector. By multiplying the contravariant velocity vector by the metric tensor we obtain the covariant components of the velocity vector in the spherical coordinate system. The contravariant and covariant components of this vector are given, respectively, by

$$v^1 = \frac{dr}{dt} \qquad v^2 = \frac{d\theta}{dt} \qquad v^3 = \frac{d\phi}{dt}$$

$$v_1 = \frac{dr}{dt} \qquad v_2 = r^2 \frac{d\theta}{dt} \qquad v_3 = r^2 \sin^2 \theta \frac{d\phi}{dt} \tag{1.3.65}$$

According to Eq. (1.3.64) the geometric scales of the spherical coordinate system are

$$h_1 = 1 \qquad h_2 = r \qquad h_3 = r \sin \theta \tag{1.3.66}$$

By applying Eqs. (1.3.61) and (1.3.65) we obtain the following expressions for the physical components of the velocity vector in the spherical coordinate system:

$$v_r = \frac{dr}{dt} \qquad v_\theta = r\frac{d\theta}{dt} \qquad v_\phi = r\sin\theta\frac{d\phi}{dt} \qquad (1.3.67)$$

We can identify the principal components of a symmetric tensor, and its principal axes. The symmetric tensor has only diagonal components in a coordinate system comprising its principal axes. These components are called principal components. Basically, the principal components are eigenvalues of the matrix representing the symmetric tensor. The principal axes are represented by a set of unit mutually orthogonal vectors called eigenvectors. The principal components λ_i of the symmetric tensor B_{ij} satisfy the equation

$$\det|B_{ij} - \lambda\delta_{ij}| = 0 \qquad (1.3.68)$$

This expression represents a third-order equation whose solution provides values of the principal components λ_1, λ_2, and λ_3.

Each of the three eigenvectors is found by solving the following set of equations:

$$(B_{ij} - \lambda\delta_{ij})b_j = 0 \qquad (1.3.69)$$

According to Eq. (1.3.69), each of the principal components λ_k is associated with three components of the relevant eigenvector b^k. If the coordinate system is rotated to coincide with the eigenvectors, then the second-order symmetric tensor B_{ij} is transformed to a diagonal matrix with elements λ_1, λ_2, and λ_3. Available computing libraries that include matrix calculation and linear algebra usually include programs aimed at the identification of eigenvalues and eigenvectors of matrices. Such computer codes can be used to identify the principal components and axes of symmetric tensors.

1.3.2 Complex Variables

Complex Numbers

A *complex number* incorporates a *real* and an *imaginary* part. The Cartesian representation of the complex variable z is

$$z = x + iy \qquad (1.3.70)$$

Here, x is the real part and y is the imaginary part. The symbol i is given by

$$i = \sqrt{-1} \qquad i^2 = -1 \qquad (1.3.71)$$

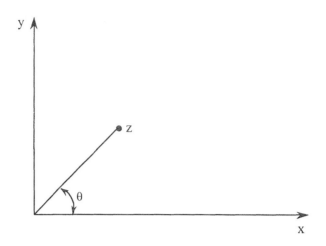

Figure 1.2   Representation of a general complex number, z.

The *Argand diagram* shown in Fig. 1.2 provides a geometric presentation of complex numbers. The length between the coordinate origin and the point represented by z is the *modulus* of the complex variable. It can be represented by r or $|z|$. As shown in Fig. 1.2,

$$|z| = \sqrt{x^2 + y^2} \tag{1.3.72}$$

Also, it is seen that $x = r \cos\theta$ and $y = r \sin\theta$. Therefore the complex variable z can be given by its trigonometric representation as

$$z = r(\cos\theta + i\sin\theta) \tag{1.3.73}$$

Complex variables z_1 and z_2 are added like vectors, i.e., the real part of z_1 is added to the real part of z_2, and the imaginary part of z_1 is added to the imaginary part of z_2. Thus

$$z = z_1 + z_2 = x_1 + x_2 + i(y_1 + y_2) \tag{1.3.74}$$

The factor i is an operator that upon multiplication rotates a complex number through $90°$. Powers of i are as follows:

$$i^2 = -1 \qquad i^3 = -i \qquad i^4 = 1 \tag{1.3.75}$$

Also, the product of two complex variables z_1 and z_2 is

$$z_1 z_2 = (x_1 + i y_1)(x_2 + i y_2) = x_1 x_2 - y_1 y_2 + i(x_1 y_2 + x_2 y_1) \tag{1.3.76}$$

A complex number also can be expressed in an exponential form. It is based on an infinite series expansion of the exponential and trigonometric functions.

For example, the Maclaurin series expansions for e^x, $\sin x$, and $\cos x$ are given, respectively, by

$$e^x = 1 + x + \frac{x^2}{2!} + \frac{x^3}{3!} + \frac{x^4}{4!} + \cdots \tag{1.3.77}$$

$$\sin x = x - \frac{x^3}{3!} + \frac{x^5}{5!} - \frac{x^7}{7!} + \cdots \tag{1.3.78}$$

$$\cos x = 1 - \frac{x^2}{2!} + \frac{x^4}{4!} - \frac{x^6}{6!} + \cdots \tag{1.3.79}$$

All these series are convergent for all values of x. Replacing x by $i\theta$ in Eq. (1.3.77) and using Eq. (1.3.75), we obtain

$$e^{i\theta} = 1 + i\theta - \frac{\theta^2}{2!} - i\frac{\theta^3}{3!} + \frac{\theta^4}{4!} + i\frac{\theta^5}{5!} - \frac{\theta^6}{6!} - \cdots \tag{1.3.80}$$

Or, upon rearranging,

$$e^{i\theta} = 1 - \frac{\theta^2}{2!} + \frac{\theta^4}{4!} - \frac{\theta^6}{6!} + \cdots + i\left(\theta - \frac{\theta^3}{3!} + \frac{\theta^5}{5!} - \frac{\theta^7}{7!} + \cdots\right) \tag{1.3.81}$$

By applying Eqs. (1.3.78) and (1.3.79), Eq. (1.3.81) becomes

$$e^{i\theta} = \cos\theta + i\sin\theta \tag{1.3.82}$$

All three forms of a complex number are then

$$z = x + iy = r(\cos\theta + i\sin\theta) = re^{i\theta} \tag{1.3.83}$$

Following these definitions, the nth power of a complex number is given by

$$z^n = r^n e^{in\theta} = r^n(\cos n\theta + i\sin n\theta) \tag{1.3.84}$$

The product of two complex numbers is

$$z_1 z_2 = r_1 r_2 e^{i(\theta_1 + \theta_2)} \tag{1.3.85}$$

and the division of two complex numbers yields

$$\frac{z_1}{z_2} = \frac{r_1 e^{i\theta_1}}{r_2 e^{i\theta_2}} = \frac{r_1}{r_2} e^{i(\theta_1 - \theta_2)} \tag{1.3.86}$$

Alternatively, the division of two complex variables can be represented by

$$\frac{z_1}{z_2} = \frac{x_1 + iy_1}{x_2 + iy_2} = \frac{(x_1 + iy_1)(x_2 - iy_2)}{(x_2 + iy_2)(x_2 - iy_2)} = \frac{x_1 x_2 + y_1 y_2}{x_2^2 + y_2^2} + i\frac{y_1 x_2 - x_1 y_2}{x_2^2 + y_2^2} \tag{1.3.87}$$

In order to avoid the presence of imaginary terms in the denominator of Eq. (1.3.87), its numerator and denominator have been multiplied by the *complex conjugate* of z_2. The complex conjugate of a complex variable is defined by replacing i by $-i$. Finally, the logarithm of a complex number can be written

$$\ln z = \ln(re^{i\theta}) = \ln r + i\theta = \frac{1}{2}\ln(x^2 + y^2) + i\tan^{-1}\frac{y}{x} \tag{1.3.88}$$

A function of a complex variable is defined as

$$w = f(z) = f(x + iy) \tag{1.3.89}$$

The complex function w can be separated into real and imaginary parts, called ϕ and ψ, respectively,

$$w = \phi(x, y) + i\psi(x, y) \tag{1.3.90}$$

where ϕ and ψ are both real functions of x and y. The function w is called *holomorphic, regular*, or *analytic* in a region, provided that within this region (1) there is one and only one value of w for each value of z and that value is finite, and (2) w has a single-valued derivative at each point of the region.

The derivative of $f(z)$ is also a complex function, given by

$$\lim_{\delta z \to 0} \frac{f(z + \delta z) - f(z)}{\delta z} \tag{1.3.91}$$

where the infinitesimal value δz is given by

$$\delta z = \delta x + i\delta y \tag{1.3.92}$$

There is no limitation on the relationship between δx and δy. We may choose paths of $\delta z \to 0$ in which $\delta x = 0$ or $\delta y = 0$, for example. These options imply

$$\lim_{\delta x \to 0;\ \delta y = 0} \frac{f(z + \delta z) - f(z)}{\delta x + i\delta y} = \lim_{\delta x \to 0} \frac{f(z + \delta x) - f(z)}{\delta x} = \frac{\delta f}{\delta x} \tag{1.3.93}$$

$$\lim_{\delta x = 0;\ \delta y \to 0} \frac{f(z + \delta z) - f(z)}{\delta x + i\delta y} = \lim_{\delta x \to 0} \frac{f(z + i\delta y) - f(z)}{i\delta y}$$

$$= \frac{1}{i}\frac{\partial f}{\partial y} = -i\frac{\partial f}{\partial y} \tag{1.3.94}$$

As the derivative of the analytic function does not depend on the path of $\delta z \to 0$, Eqs. (1.3.93) and (1.3.94) imply

$$\frac{\partial f}{\partial x} = -i\frac{\partial f}{\partial y} \tag{1.3.95}$$

The derivative of f comprises real and imaginary parts given by

$$\frac{\partial f}{\partial x} = \frac{\partial \phi}{\partial x} + i\frac{\partial \psi}{\partial x} \qquad \frac{\partial f}{\partial y} = \frac{\partial \phi}{\partial y} + i\frac{\partial \psi}{\partial y} \tag{1.3.96}$$

Introducing Eqs. (1.3.90) and (1.3.92) into Eq. (1.3.96), we obtain

$$\frac{\partial \phi}{\partial x} = \frac{\partial \psi}{\partial y} \qquad \frac{\partial \phi}{\partial y} = -\frac{\partial \psi}{\partial x} \tag{1.3.97}$$

These relations are called the *Cauchy–Riemann* relations.

Differentiating the first of Eq. (1.3.97) with respect to x and the second with respect to y and adding, and differentiating the first of Eq. (1.3.97) with respect to y and the second with respect to x and subtracting one from the other, we obtain, respectively,

$$\frac{\partial^2 \phi}{\partial x^2} + \frac{\partial^2 \phi}{\partial y^2} = 0 \qquad \frac{\partial^2 \psi}{\partial x^2} + \frac{\partial^2 \psi}{\partial y^2} = 0 \tag{1.3.98}$$

These expressions indicate that both functions ϕ and ψ satisfy the *Laplace* equation in two-dimensional Cartesian coordinates.

1.3.3 Partial Differential Equations

All basic processes typical of environmental fluid mechanics can be formulated as partial differential equations (PDEs). Partial differential equations arise because the functions for which solutions are sought (e.g., concentrations, velocities, temperature, etc.) tend to depend on one or more spatial coordinates as well as time. As will be seen in subsequent chapters, most equations of interest contain diffusion processes, which involve second-order spatial derivatives. The solution of the relevant differential equation(s) subject to appropriate initial and boundary conditions provides the basis for mathematical simulation of the physical problem. In the following paragraphs, we review the basic types of partial differential equations encountered with environmental fluid mechanics issues.

Identification of the partial differential equation connected with the particular problem of interest is of major importance. Different criteria of convergence and stability are typical of each type of partial differential equation, as described below. The equation provides the basic guideline for the development of a mathematical model that can be applied to the solution of that problem. In cases of numerical simulations, particular rules for the development of the numerical scheme are used for the particular differential equation that is associated with a given problem. Problems of environmental fluid mechanics can be classified into two general categories: problems of equilibrium and problems of propagation.

The general format of a second-order linear PDE in a two dimensional domain is given by

$$a\frac{\partial^2 \varphi}{\partial x^2} + b\frac{\partial^2 \varphi}{\partial x \partial y} + c\frac{\partial^2 \varphi}{\partial y^2} = f \qquad (1.3.99)$$

where a, b, and c are constant coefficients and f represents a linear combination of coefficients multiplied by lower order derivatives of the dependent variable φ.

The method and form of the solution of a PDE subject to initial and boundary conditions depends on the type of the PDE. It is common to classify PDEs according to the relationships between the coefficients of Eq. (1.3.99) as follows:

If $b^2 - 4ac > 0$ then the PDE is *hyperbolic*   (1.3.100a)

If $b^2 - 4ac = 0$ then the PDE is *parabolic*   (1.3.100b)

If $b^2 - 4ac < 0$ then the PDE is *elliptic*   (1.3.100c)

The terms hyperbolic, parabolic, and elliptic chosen to classify partial differential equations stems from the analogy between the form of the discriminant ($b^2 - 4ac$) for partial differential equations and the form of the discriminant that classifies conic sections. There is no other significance to this terminology. If the PDE refers to a domain with n dimensions, then the characteristics, if real characteristics exist, are surfaces of $(n - 1)$ dimensions, along which signals, or information, propagate. If no real characteristics exist, then there are no preferred paths of information propagation. Therefore the existence or absence of characteristics has a significant impact on the solution of the partial differential equation.

First-order partial differential equations refer to advection or convection of a property φ, such as solute concentration or heat. The general form of such an equation in the (x, t) domain, where x refers to a spatial coordinate and t refers to time, is given by

$$\frac{\partial \varphi}{\partial t} + u\frac{\partial \varphi}{\partial x} = 0 \qquad (1.3.101)$$

where u is the advection velocity. If φ refers to dissolved mass of a solute, then the second term in Eq. (1.3.101) incorporates the process of solute mass being carried (advected) by a fluid particle as it moves through the domain. The location of any fluid particle is related to its velocity u by a simple relationship representing the differential equation of the particle pathline:

$$\frac{dx}{dt} = u \qquad (1.3.102)$$

Thus the pathline of a fluid particle is given by

$$x = x_0 + \int_{t_0}^{t} u \, dt \tag{1.3.103}$$

Along the pathline of the fluid particle the advection equation can be written as

$$\frac{\partial \varphi}{\partial t} + u \frac{\partial \varphi}{\partial x} = \frac{\partial \varphi}{\partial t} + \frac{dx}{dt} \frac{\partial \varphi}{\partial x} = \frac{d\varphi}{dt} = 0 \tag{1.3.104}$$

The last part of this equation shows that φ is constant along the pathline of the fluid particle. This pathline is the characteristic path associated with the advection equation. The first-order differential equation of the form given by Eq. (1.3.101) is termed a first-order hyperbolic partial differential equation, and it has a single family of characteristic curves, along which the information propagates in the domain. A single first-order partial differential equation is always hyperbolic. In second-order hyperbolic partial differential equations there are two families of characteristic curves, along which the information propagates.

Parabolic and hyperbolic differential equations are typical of *propagation problems*. The propagation is in time and space. This means that parabolic and hyperbolic differential equations usually refer to problems of a property propagating in the domain. The features of the propagation of the property in cases of parabolic differential equations are different from those of hyperbolic differential equations. Elliptic partial differential equations generally concern equilibrium problems, i.e., ones that do not involve time derivatives.

A typical parabolic equation associated with environmental fluid mechanics is the equation of diffusion. In the (x, t) domain, the form of this equation is given by

$$\frac{\partial \varphi}{\partial t} = \alpha \frac{\partial^2 \varphi}{\partial x^2} \tag{1.3.105}$$

where α is the *diffusion coefficient*, or *diffusivity*. In many applications an advective term is added, forming an advection–diffusion equation (see Chap. 10).

A typical hyperbolic equation associated with environmental fluid mechanics is the *wave equation*. In the (x, t) domain, the form of this equation is given by

$$\frac{\partial^2 \varphi}{\partial t^2} = c^2 \frac{\partial^2 \varphi}{\partial x^2} \tag{1.3.106}$$

where c is the propagation speed of the wave.

A typical elliptic equation, associated with environmental fluid mechanics is the Laplace equation. In the (x, y) domain, the form of this equation is given by

$$\frac{\partial^2 \varphi}{\partial x^2} + \frac{\partial^2 \varphi}{\partial y^2} = 0 \qquad (1.3.107)$$

The solution of a parabolic or hyperbolic partial differential equation, of the types given by Eqs. (1.3.105) and (1.3.106), can be obtained, provided that adequate initial and boundary conditions are given. Initial conditions refer to values of the unknown variables and possibly their space derivatives at a time of reference. Boundary conditions refer to values of the unknown variables and their space derivatives at the boundaries or other specific locations of the domain. The solution of an elliptic partial differential equation of the type given by Eq. (1.3.107) can be obtained, provided that adequate boundary conditions of the domain are given. For elliptic partial differential equations there are no initial conditions, since time derivatives are not involved.

There are three types of linear boundary conditions that can be applied to the solution of partial differential equations:

1. All values of the dependent variable, φ, are specified on the boundaries of the domain:

$$\varphi = f(x, y) \qquad \text{where} \qquad (x, y) \in G \qquad (1.3.108)$$

where G is the surface of the domain. Boundary conditions of this type are referred to as *Dirichlet* boundary conditions.

2. All values of the gradient of the dependent variable, φ, are specified on the boundaries of the domain:

$$\frac{\partial \varphi}{\partial n} = f(x, y) \qquad \text{where} \qquad f(x, y) \in G \qquad (1.3.109)$$

where n represents a coordinate normal to the boundary G. Boundary conditions of this type are referred to as *Neumann* boundary conditions.

3. A general linear combination of Dirichlet and Neumann boundary conditions:

$$a\varphi + b\frac{\partial \varphi}{\partial n} = c \qquad (1.3.110)$$

where a, b, and c are functions of (x, y). This type of boundary condition can be used to specify total flux, as will be described in later chapters.

It should be noted that besides linear boundary conditions, the domain may be subject to nonlinear boundary conditions. An example is application of boundary conditions at a water-free surface, which may be part of the solution of the problem. Application of such conditions is generally very complicated.

1.4 DIMENSIONAL REASONING

1.4.1 Uses of Dimensional Analysis

Dimensional analysis provides a powerful tool to evaluate relationships between various parameters of a problem when the governing equation is not known from some other source, such as a theoretical result. The basic premise underlying any dimensional reasoning is that all physically realistic expressions must be dimensionally consistent. In fact, the *Buckingham π theorem*, introduced in the following section, can be seen as a formal statement of a relationship between variables based simply on their dimensional units. Following this idea, any physical equation that is dimensionally balanced (that is, the dimensional units are the same for each of the terms in the equation) can be written in nondimensional form. The easiest way to see this is to divide all the terms of the equation by one of the terms. Done properly, this usually results in equations expressed in terms of common *dimensionless parameters*. Since all the terms have the same physical dimensions, the result of this process is a relationship between these dimensionless variables, which can be used to evaluate the relative importance of different terms in any given equation. For example, it would be possible to gain some understanding of the relative importance of different forces in a particular flow field by looking at the values of the parameters in dimensionless forms of the momentum equations. This process sometimes allows simplification of a general governing equation, by eliminating terms that are seen as being of lesser importance, compared with others.

A common example of a dimensionless number is the *Reynolds number*, defined as

$$\text{Re} = \frac{UL}{\nu} \tag{1.4.1}$$

where U is a characteristic velocity and L is a characteristic length of the problem being studied, and ν is kinematic viscosity of the fluid. Re represents the relative importance of inertia to viscous forces. For example, a high value of Re indicates that viscous forces are not very important. (As will be seen later, a high Re is associated with *turbulent* flow.)

The result of dimensional analysis is a definition of a relationship between the appropriate dimensionless variables resulting from grouping the parameters of the problem. The specific form of the relationship is not revealed using dimensional analysis — physical experiments must be performed to provide additional information. For example, dimensional analysis can be used to show that a dimensionless group incorporating the drag on a sphere moving at constant velocity through a fluid should depend on Re. However, the actual form of the relationship is determined from experimental results.

One other important application of dimensional analysis is in providing a means of scaling the results of a model study to prototype conditions. This is necessary, for instance, in extrapolating results from laboratory physical modeling studies to field conditions. In order to do this, conditions of *similarity* must be satisfied. There are three kinds of similarity. Intuitively, a model or experiment should be *geometrically* similar to the field situation, which means that the ratio of all length scales is the same between the model and the prototype. *Kinematic* similarity incorporates similarity of length and time quantities. *Dynamic* similarity also must be satisfied in order to properly scale results concerning forces and stresses. Kinematic and dynamic similarity are obtained when appropriate dimensionless parameters are the same in the model and in the prototype. Dynamic similarity is equivalent to saying the ratios of relevant forces are the same.

For example, consider an open channel flow, as sketched in Fig. 1.3. For simplicity, we assume a rectangular cross section of width b and flow depth h. Geometric similarity implies

$$L_{\mathrm{r}} = \frac{L_1}{L_2} \tag{1.4.2}$$

where L_{r} is the length scale ratio and L represents any length for the problem, in this case either b or h. Subscripts 1 and 2 refer to the two systems (prototype and model — expressing the ratio in this way avoids very small values for L_{r}). Thus $h_1/h_2 = b_1/b_2$ and $h_1/b_1 = h_2/b_2$ (i.e., the flow aspect ratio is the same in the two systems). In some cases *distorted scale* models are necessary,

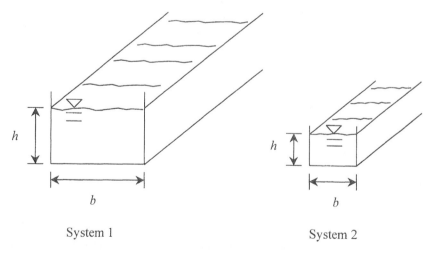

System 1   System 2

Figure 1.3 Open channel flow in two geometrically similar rectangular channels.

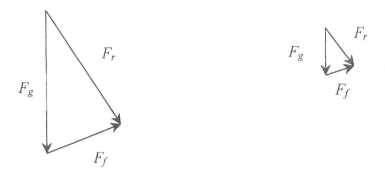

Figure 1.4 Force diagrams for the two systems shown in Fig. 1.3; F_g is the gravity force, F_f is friction, and F_r is the resultant force.

such as when a physical model of a large lake is used. In this case, because the horizontal dimensions are generally much greater than the vertical scale, the horizontal scale ratio is chosen to be much larger than the vertical ratio. This is to avoid models with extremely shallow water layers. Scaling is otherwise similar to that in nondistorted models; one should be careful to maintain common values of the relevant dimensionless parameters between the model and prototype.

If we consider a small fluid element in either system and assume that the important forces for this problem are gravity and friction, the resultant forces on the fluid element can be calculated, and we obtain force diagrams like those in Fig. 1.4. The shapes of these force diagrams must be similar for the two systems if there is dynamic similarity. As shown in the following section, this condition is satisfied when the corresponding values of properly defined dimensionless variables are the same.

1.4.2 Dimensionless Parameters

Buckingham π Theorem

The Buckingham π theorem states that a group of physical variables defined for a given problem may be combined in such a way as to form a non-dimensional representation of the same problem. Moreover, since the original variables are functionally related, i.e.,

$$f(x_1, x_2, \ldots, x_n) = 0 \tag{1.4.3}$$

where the $x_1, x_2, \ldots, x_n$ represent the n physical variables of a problem, then the nondimensional variables also are functionally related. If there are k physical dimensional units involved with the n variables, then $(n\text{-}k)$ dimensionless parameters should be formed, and

$$f(\pi_1, \pi_2, \ldots, \pi_{(n-k)}) = 0 \tag{1.4.4}$$

where the $\pi_1, \pi_2, \ldots, \pi_{(n-k)}$ are the dimensionless groupings.

Consider the force diagrams indicated in Fig. 1.4. The shapes will be the same when the ratios of any two of the forces are the same in each system. The resultant or inertial force is represented by $(\rho \forall U/t)$, where ρ is the fluid density, $\forall$ is the volume of a fluid element, U is its velocity, and t is some appropriate time scale. Here, (U/t) has been used to approximate acceleration, and t will be estimated as $t = L/U$, where L is a characteristic length scale. The viscous force, acting on area A, is approximately $(\mu U/L)(A)$, where (U/L) has been used to estimate the velocity gradient. Substituting L^3 for $\forall$ and L^2 for A, the ratio of inertial to viscous force is then

$$\frac{\rho(L^3)U^2/L}{\mu(U/L)L^2} = \frac{\rho L U}{\mu} = \text{Re} \tag{1.4.5}$$

In other words, this is the Reynolds number as defined in Eq. (1.4.1). By going through a similar procedure for the ratio of inertial to gravity force, where the force of gravity F_g is approximated by $\rho \forall g$, we obtain

$$\frac{\rho L^2 U^2}{\rho L^3 g} = \text{Fr} \tag{1.4.6}$$

which defines Fr as a second dimensionless parameter, the *Froude number.*

Thus by insuring that the Reynolds numbers and the Froude numbers are the same for both systems, the shape of the resulting force diagrams will be the same, and dynamic similarity will be achieved. This type of reasoning may be applied to problems with a greater number of relevant forces, with the result that additional dimensionless parameters would need to be defined. Many different dimensionless parameters have been defined for various problems in fluid mechanics. Rather than attempting to list them all here, we shall introduce them as needed within the context of a given problem or derivation.

In order to illustrate the application of the Buckingham π theorem, let us consider the problem of finding the drag on a smooth sphere fully immersed and moving at constant velocity through a fluid. It is assumed that the drag is a function of the velocity and diameter of the sphere, and the density and viscosity of the fluid. Note that one limitation of the Buckingham π theorem is that it does not provide specific guidance on which parameters should be chosen for a given problem. These must be chosen on the basis of experience and physical intuition, with perhaps some trial and error to be expected in some cases. Usually, it will be clear when the wrong set of parameters is chosen, since it will be difficult to perform experimental tests to obtain a clear relationship between the dimensionless parameters defined. For the present problem, a functional relationship is defined by

$$f(D, U, d, \rho, \mu) = 0 \tag{1.4.7}$$

where D is drag, U is velocity, d is the sphere diameter, and ρ and μ are fluid density and viscosity, respectively. There are five variables and three physical dimensional units, mass (M), length (L), and time (T), so two π's will be defined. First, a subset of variables is defined, called the *basis set*, with the following characteristics:

> The number of variables in the basis set is equal to the number of physical dimensions.
> All the dimensions of the problem are represented by the variables, as simply as possible.
> Variables are chosen so that recognizable dimensionless groupings are found.
> The main parameter of interest (the dependent variable) is not chosen for the basis.

The third of these conditions is not absolutely necessary, but it usually helps to interpret results, particularly in view of the above interpretation of many of these dimensionless groups as force ratios. In many cases there is not a unique basis set, and different basis sets will result in definitions of different sets of dimensionless numbers. This is acceptable, from a purely dimensional analysis point of view, but is it preferable to form common dimensionless groupings whenever possible.

For the current example, U, d, and ρ are chosen as the basis variables. These are combined with D and μ, in turn, to form two π groups. The first of these is found from

$$\pi_1 = D(U)^a(d)^b(\rho)^c = \left(\frac{ML}{T^2}\right)\left(\frac{L}{T}\right)^a (L)^b \left(\frac{M}{L^3}\right)^c$$

Separate equations are then formed for each of the dimensional units, to find a, b, and c so that π_1 is dimensionless. For mass M, $(1 + c = 0)$ gives $c = -1$. The equation for time T is $(-2 - a = 0)$, or $a = -2$. The last equation for length L gives $(1 + a + b - 3c = 0)$, or $(a + b = -4)$. Then $b = -2$ and

$$\pi_1 = \frac{D}{\rho U^2 d^2}$$

This is a dimensionless drag and is commonly referred to as a *drag coefficient*, C_D. Following a similar procedure using μ, it is easily shown that a Reynolds number results for π_2. The dimensionless result analogous to

Eq. (1.4.7) is then

$$f(C_D, \text{Re}) = 0 \tag{1.4.8}$$

and experimental results are needed to describe the specific form of the functional relationship. It should be noted that an equally valid dimensional analysis result would be obtained by using one of the π's raised to some power, or using the inverse. However, use of common dimensionless parameters is preferred, as noted above.

1.4.3 Scales of Motion

It is evident from the above discussion that it is necessary to define certain parameters of a problem in order to carry out dimensional analysis. A similar requirement is to define certain characteristic scales to represent a problem, in order to use dimensional reasoning to carry out scaling analyses. The usual scales of interest for kinematic problems are those for length, velocity, or time, though other types of parameters are sometimes needed. Often the choices for these scales are obvious. A simple example is with open channel flow, where the flow mean velocity is usually chosen as the velocity scale, and depth (or hydraulic radius) is chosen as the characteristic length scale. The choice for these scales determines values for the nondimensional variables discussed in the previous section, so some care must be taken. As shown in Sec. 2.7, one of the principal applications of scaling analysis is in developing an understanding of the relative importance of the various terms of a relationship, with a view to simplifying the equation whenever possible. In addition to possibly simplifying the equation, the main advantage of developing nondimensional forms of equations is that the actual scale becomes secondary — it is only the dimensionless groups that are important. Nondimensional equations and parameters apply equally to systems with very different scales (e.g., values of L and U), as long as the values of the dimensionless groupings are similar. This idea forms the basis for physical modeling tests and provides the means for scaling model results to estimate prototype conditions.

PROBLEMS

Solved Problems

Problem 1.1 The *material* or *substantial derivative* of the velocity vector represents the acceleration of the fluid particles. In Cartesian coordinates the acceleration is expressed by

$$a_i = \frac{\partial u_i}{\partial t} + u_j \frac{\partial u_i}{\partial x_j}$$

where u_i are components of the velocity vector. What is the expression for the acceleration in a general coordinate system? What are the expressions for the contravariant, covariant, and physical components of the acceleration in a cylindrical coordinate system?

Solution

The expression for the contravariant acceleration vector is

$$a^i = \frac{\partial u^i}{\partial t} + u^j u^i{}_{,j}$$

Multiplying this expression by the metric tensor and replacing indices, we obtain the following expression for the covariant acceleration vector:

$$a_i = \frac{\partial u_i}{\partial t} + u^j u_{i,j}$$

In a cylindrical coordinate system the physical components of the velocity and acceleration vectors are given by the following symbols, respectively:

u, v, w (physical components in directions r, θ, z, respectively)

a_r, a_θ, a_z

The line element in a cylindrical coordinate system is given by

$$ds^2 = (dr)^2 + (r\,d\theta)^2 + (dz)^2$$

Therefore components of the metric tensor and geometrical scales in the r, θ, and z directions are given by

$$g_{11} = 1 \qquad g_{22} = r^2 \qquad g_{33} = 1$$

$$h_1 = 1 \qquad h_2 = r \qquad h_3 = 1$$

By applying Eq. (1.3.57), the following relationships between physical, covariant, and covariant components of the velocity and acceleration vectors are obtained:

$$u = u_1 = u^1 \qquad v = \frac{u_2}{r} = ru^2 \qquad w = u_3 = u^3$$

$$a_r = a_1 = a^1 \qquad a_\theta = \frac{a_1}{r} = ra^1 \qquad a_z = a_3 = a^3$$

By applying the general expressions for Christoffel symbols given by Eqs. (1.3.33) and (1.3.34), we obtain values of the second symbols of Christoffel. The only nonzero symbols in a cylindrical coordinate system are

$$\sum_{22}^{1} = -r \qquad \sum_{12}^{2} = \sum_{21}^{2} = \frac{1}{r}$$

We apply the general expression for the contravariant acceleration vector with these expressions for the second symbols of Christoffel to obtain

$$a^1 = \frac{\partial u^1}{\partial t} + u^1 \frac{\partial u^1}{\partial r} + u^2 \left(\frac{\partial u^1}{\partial \theta} - r u^2 \right) + u^3 \frac{\partial u^1}{\partial z}$$

$$a^2 = \frac{\partial u^2}{\partial t} + u^1 \left(\frac{\partial u^2}{\partial r} + \frac{u^2}{r} \right) + u^2 \left(\frac{\partial u^2}{\partial \theta} + \frac{u^1}{r} \right) + u^3 \frac{\partial u^2}{\partial z}$$

$$a^3 = \frac{\partial u^3}{\partial t} + u^1 \frac{\partial u^3}{\partial r} + u^2 \frac{\partial u^3}{\partial \theta} + u^3 \frac{\partial u^3}{\partial z}$$

By introducing the physical components of the velocity and acceleration vectors into these expressions we obtain

$$a_r = \frac{\partial u}{\partial t} + u \frac{\partial u}{\partial r} + \frac{v}{r} \frac{\partial u}{\partial \theta} - \frac{v^2}{r} + w \frac{\partial u}{\partial z}$$

$$a_\theta = \frac{\partial v}{\partial t} + u \frac{\partial v}{\partial r} + \frac{v}{r} \frac{\partial v}{\partial \theta} + \frac{uv}{r} + w \frac{\partial v}{\partial z}$$

$$a_z = \frac{\partial w}{\partial t} + u \frac{\partial w}{\partial r} + \frac{v}{r} \frac{\partial w}{\partial \theta} + w \frac{\partial w}{\partial z}$$

Problem 1.2   Develop the expression for div $\overrightarrow{V}$ in cylindrical coordinates by applying the contravariant as well as covariant components of the velocity vector $\overrightarrow{V}$.

Solution

The general required expressions for div $\overrightarrow{V}$ are

$$\nabla \cdot \overrightarrow{V} = u^i_{,i} = g^{ij} u_{i,j}$$

The expression with contravariant components of the velocity vector is

$$u^i_{,i} = \frac{\partial u^i}{\partial x^i} + \sum_{ji} u^j = \frac{\partial u^1}{\partial r} + \frac{\partial u^2}{\partial \theta} + \frac{u^1}{r} + \frac{\partial u^3}{\partial z} = \frac{\partial u}{\partial r} + \frac{1}{r} \frac{\partial v}{\partial \theta} + \frac{u}{r} + \frac{\partial w}{\partial z}$$

The expression with covariant components of the velocity vector is

$$g^{ij} u_{i,j} = g^{ij} \frac{\partial u_i}{\partial x^j} - g^{ij} \sum_{ij}^{k} u_k = \frac{\partial u_1}{\partial r} + \frac{1}{r^2} \frac{\partial u_2}{\partial \theta} + \frac{u_1 r}{r^2} + \frac{\partial u_3}{\partial z}$$

$$= \frac{\partial u}{\partial r} + \frac{u}{r} + \frac{1}{r} \frac{\partial v}{\partial \theta} + \frac{\partial w}{\partial z}$$

Problem 1.3 Prove the following relationships:

$$\sin x = \frac{e^{ix} - e^{-ix}}{2i} \qquad \cos x = \frac{e^{ix} + e^{-ix}}{2} \qquad \sinh ix = i \sin x$$

Solution

We apply the Euler relationships,

$$e^{ix} = \cos x + i \sin x \qquad e^{-ix} = \cos x - i \sin x$$

Introducing these expressions into the expressions for $\sin x$ and $\cos x$, we obtain the relationships written above. Introducing the explicit expression for $\sinh ix$, we obtain the last identity.

Problem 1.4 Find the complex numbers given by

(a) $(2 + i)(3 - 2i)$ (b) $\dfrac{1 + 3i}{1 - i}$ (c) $\ln(3 + 4i)$

Solution

(a) $(2 + i)(3 - 2i) = 6 + 3i - 4i + 2 = 8 - i$

(b) $\dfrac{1 + 3i}{1 - i} = \dfrac{(1 + 3i)(1 + i)}{(1 - i)(1 + i)} = \dfrac{1 + 3i + i - 3}{1 - i + i + 1} = \dfrac{-2 + 4i}{2} = -1 + 2i$

(c) $\ln(3 + 4i) = \dfrac{1}{2} \ln(3^2 + 4^2) + i \tan^{-1} \dfrac{4}{3} = 1.61 + i0.93$

Problem 1.5 Separate the following functions of z into their real and imaginary parts ϕ and ψ:

(a) $\dfrac{1}{z}$ (b) $\ln z^2$ (c) e^{iz}

Solution

(a) $\dfrac{1}{z} = \dfrac{1}{x + iy} = \dfrac{x - iy}{(x + iy)(x - iy)} = \dfrac{x - iy}{x^2 + y^2} = \dfrac{x}{x^2 + y^2} - i\dfrac{y}{x^2 + y^2}$

Therefore

$$\phi = \frac{x}{x^2 + y^2}; \ \psi = \frac{y}{x^2 + y^2}$$

(b) $\ln z^2 = \ln[(x + iy)(x + iy)] = \ln(x^2 + y^2 + i2xy)$

$$= \frac{1}{2} \ln[(x^2 + y^2)^2 + 4x^2 y^2] + i \tan^{-1} \frac{2xy}{x^2 + y^2}$$

Therefore

$$\phi = \frac{1}{2}\ln[(x^2 + y^2)^2 + 4x^2y^2]; \; \psi = \tan^{-1}\frac{2xy}{x^2 + y^2}$$

(c) $e^{iz} = \exp[i(x + iy)] = \exp(ix - y) = e^{-y}e^{ix} = e^{-y}(\cos x + i\sin x)$

Problem 1.6 Consider the problem of dumping sewage from a barge into a linearly stratified ocean, as illustrated in Fig. 1.5.

A volume V_s of sludge of density ρ_s is released suddenly from the barge into water of density ρ_0 and density gradient $(-d\rho_a/dz)$. Find the maximum depth of penetration, d_{max}, the minimum dilution at that depth, and the time of descent. (Note that the sludge cloud seeks its density equilibrium position, which also depends on entrainment.)

Solution

First, we define

$S =$ (total sample volume)/(volume of effluent in sample)

$P = 1/S =$ volume fraction of effluent($=$ relative concentration)

A definition of dilution is

$$D = \frac{1 - P}{P}$$

where $D =$ (volume ambient water in sample)/(volume effluent in sample)$=$ $S - 1$.

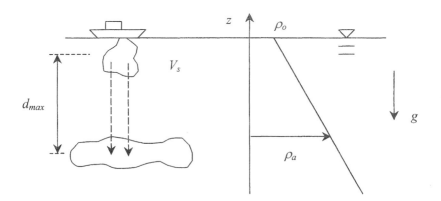

Figure 1.5 Definition sketch, Problem 1.6.

The variables of the problem are

$$d_{max} \qquad \forall s \qquad g \qquad \rho_0 \qquad \rho_s \qquad -d\rho_a/dz$$

with corresponding units

$$(L) \qquad (L^3) \qquad (L/T^2) \qquad (M/L^3) \qquad (M/L^3) \qquad (M/L^4)$$

According to the Buckingham π theorem, with six variables and three dimensional quantities, there should be three dimensionless groupings. However, it is usually more convenient to work with one or two dimensionless groups, for ease of analysis. Therefore, define

$$\Delta\rho = \rho_s - \rho_a$$

and

$$V_s g \Delta\rho = \text{submerged weight of sludge}$$

It also will be convenient to combine g and $(-d\rho_a/dz)$. Then we have for variables

$$d_{max} \qquad \forall_s g \Delta\rho \qquad -gd\rho_a/dz \qquad \rho_0$$

the corresponding units

$$(L) \qquad (ML/T^2) \qquad (M/L^3T^2) \qquad (M/L^3)$$

There are now four variables and three dimensions, so only one dimensionless grouping (π) is needed. If we now set

$$\pi = (d_{max})^a (\forall_s g \Delta\rho)^b \left(-g\frac{d\rho_a}{dz}\right)^c (\rho_0)^d$$

$$= (L)^a \left(\frac{ML}{T^2}\right)^b \left(\frac{M}{L^3T^2}\right)^c \left(\frac{M}{L^3}\right)^d$$

and solve individually for each of the power coefficients,

(M): $b = -c - d$

(L): $a + b - 3c - 3d = 0$

(T): $b = -c$

then we can solve for the power coefficients to define π. However, we have four power coefficients and only three equations. Therefore it is necessary to set the value for one of the power coefficients arbitrarily. Anticipating the desired result, we set $c = 1$ and solve for the remaining values based on this assumption. Note that an equally valid result could be obtained starting with other values for c. Also, from examination of the equations for M and T, it is seen that the only solution that can satisfy both is when $d = 0$; however, again anticipating the result, we include ρ_o, twice, so that it cancels. This is because of the desire to develop a solution that has recognizable parameters. The final result is

$$\pi = (d_{max})^4 \frac{\left(-\frac{g}{\rho_0} \frac{d\rho_a}{dz} \right)}{\left(V_{sg} \frac{\Delta\rho}{\rho_0} \right)}$$

Note that ρ_0 cancels in the numerator and denominator, so that effectively $d = 0$. It is also evident that g cancels, but it, too, is kept because of conventional definitions. For example, the square root of the term in parentheses in the numerator is called the *buoyancy frequency*, N, and the term in the denominator is the buoyancy force per unit mass acting on the submerged sludge. If we further define $g' = g\Delta\rho/\rho_0 = reduced\ gravity$, and note that, since there is only one π for this problem, then it must equal a constant (say A^4), then the final result is

$$d_{max} = A \left(\frac{V_s g'}{N^2} \right)^{1/4}$$

Now, if experiments are done, measuring d_{max} while varying the other parameters in this expression, then a plot of d_{max} versus $(V_s g'/gN^2)^{1/4}$ should result in a straight line with slope corresponding to the value for A, such as is illustrated in Fig. 1.6.

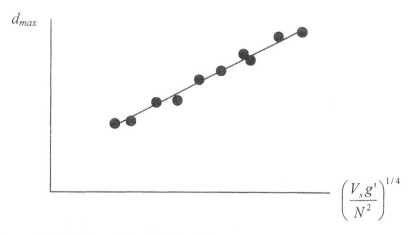

Figure 1.6   Variation of d_{max}, Problem 1.6.

From experimental studies, the value of A is found to be approximately 2.66. The remainder of this problem solution is left as an exercise for the student.

Unsolved Problems

Problem 1.7   The expression for the vorticity (ω_{ij}) tensor is

$$\omega_{ij} = \frac{1}{2}(u_{i,j} - u_{j,i})$$

where the u_i are components of the velocity vector. Find the components of the vorticity tensor in cylindrical and spherical coordinates.

Problem 1.8   The divergence of a second-order tensor is a vector expressed as

$$(\nabla \cdot \widetilde{\widetilde{B}})_i = B^j_{i,j}$$

Find the expressions for the components of div$B$ in Cartesian, cylindrical, and spherical coordinates.

Problem 1.9   The stress tensor for a Newtonian incompressible fluid is given by

$$\tau_{ij} = -pg_{ij} + \mu(u_{i,j} + u_{j,i})$$

where p is the pressure, μ is the fluid viscosity, and the u_i are components of the velocity vector. Find expressions for components of τ_{ij} and div τ in Cartesian and cylindrical coordinates.

Problem 1.10   Prove the following expressions by using indicial notation:

$$\vec{a} \times (\vec{b} \times \vec{c}) = (\vec{a} \cdot \vec{c})\vec{b} - (\vec{a} \cdot \vec{b})$$
$$\vec{V} \cdot \nabla \vec{V} = \nabla \frac{|V|^2}{2} - \vec{V} \times \nabla \times \vec{V}$$

Problem 1.11   Prove the vector identity

$$\nabla^2 \vec{V} = \vec{\nabla}(\vec{\nabla} \cdot \vec{V}) - \vec{\nabla} \times (\vec{\nabla} \times \vec{V})$$

Problem 1.12   How many separate quantities are represented by each of the following expressions?

(a) $\varepsilon_{ijk}\dfrac{\partial u_k}{\partial x_j}$ (b) $\dfrac{\partial u_j}{\partial x_j}$

(c) $\dfrac{\partial^2 u_i}{\partial x_j^2} + \dfrac{1}{3}\dfrac{\partial^2 u_j}{\partial x_i \partial x_j}$ (d) $\dfrac{1}{2}\left(\dfrac{\partial u_i}{\partial x_j} + \dfrac{\partial u_j}{\partial x_i}\right)$

Problem 1.13   Use the properties of the alternating tensor ε_{ijk} to prove the vector identity

$$\vec{\nabla} \cdot (\vec{\nabla} \times \vec{V}) = 0 \qquad \text{for } \vec{V} = \text{any vector}$$

Problem 1.14   Find the complex numbers given by

   (a) $5e^i + 3e^{1.5i}$ (b) $(1+i)(2-i)(1+3i)$

   (c) $\ln(i)$ (d) $\ln(-1)$

Problem 1.15   Separate the following functions into their real and imaginary parts:

   (a) $\dfrac{z}{\tilde{z}}$ where $\tilde{z} = x - iy$ (b) $\dfrac{z - \tilde{z}}{z + \tilde{z}}$

   (c) $\tilde{z} + \dfrac{1}{z}$ (d) $\ln\left(\dfrac{1}{z}\right)$ (e) $\tilde{z}^2 z$

Problem 1.16   Show that Cauchy–Riemann relations in two-dimensional cylindrical coordinates are

$$\dfrac{\partial \phi}{\partial r} = \dfrac{1}{r}\dfrac{\partial \psi}{\partial \theta} \qquad \dfrac{1}{r}\dfrac{\partial \phi}{\partial \theta} = -\dfrac{\partial \psi}{\partial r}$$

Problem 1.17   Which of the following functions are analytic functions?:

   (a) $r\cos\dfrac{\theta}{2} + ir\sin\dfrac{\theta}{2}$ (b) $\sqrt{r}\cos\dfrac{\theta}{2} + i\sqrt{r}\sin\dfrac{\theta}{2}$

   (c) $\dfrac{1}{x^2} + i\dfrac{1}{y^2}$ (d) $\dfrac{x^2 + y^2}{x - iy}$

   (e) $\dfrac{x}{x^2 + y^2} - i\dfrac{y}{x^2 + y^2}$ (f) $x^2 - y^2 - x + i(2xy - y)$

Problem 1.18   Determine the derivatives of the following analytic functions and separate the derivatives into their real and imaginary parts:

(a) $w = (1 + i) \ln z$ (b) $z = \ln w$

(c) $w = \dfrac{i}{z} + z$ (d) $w = z^2 + iz$

(e) $w = \sqrt{z}$ (f) $w = \ln z + z$

Problem 1.19   Prove the following identities:

(a) $\cosh ix = \cos x$

(b) $\sinh z = \sinh x \cos y + i \sinh x \sin y$

(c) $\sin z = \sin x \cosh y + i \cos x \sinh y$

(d) $\cos z = \cos x \cosh y - i \sin x \sinh y$

Problem 1.20   Assuming that the drag (D) experienced by an object moving through a fluid is a function of its projected area (A) in the direction of motion, its velocity (V), and the density (ρ) and viscosity (μ) of the fluid, develop a dimensionless relationship to show how the drag should be related to the other variables of the problem.

Problem 1.21   Use dimensional analysis to develop an expression for the vertical velocity (w) produced in a container of a fluid of depth h, when heated from below with input power P $(= \text{energy input per unit time}, ML^2/T^3)$. Assume that w is a function of h and P, as well as fluid density $(\rho, M/L^3)$, thermal expansion coefficient $(\alpha, 1/T)$, and specific heat $(c, \text{energy per unit mass, per unit temperature}, L^2/T^2\theta)$. To simplify, combine α, ρ, and c as $(\alpha/\rho c)$.

Problem 1.22   It is desired to formulate an expression to predict the mixing generated by wind blowing over a stratified water body, as shown in Fig. 1.7. Specifically, the wind transfers energy into the water by a surface shear stress, which may be characterized by the friction velocity, $u_* = (\tau/\rho_0)^{1/2}$. Part of this energy is used to mix fluid across the density interface, resulting in a deepening of the upper layer. Formulate a nondimensional expression that could be used to relate the entrainment velocity, $u_e = dh/dt$, to other variables of the problem (remember to include g). The result should be written in terms of the *bulk Richardson number*,

$$\text{Ri} = \frac{g'h}{u_*^2}$$

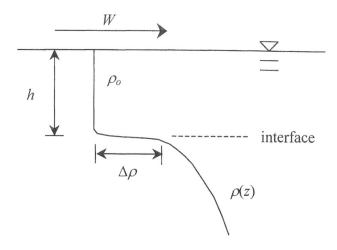

Figure 1.7   Mixed layer structure, Problem 1.22.

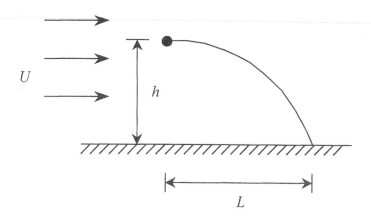

Figure 1.8   Definition sketch, Problem 1.23.

Problem 1.23   A $1:10$ scale model is to be used to test the distance (L) a sphere of diameter (d) will travel when released at a height (H) in a fluid stream moving at velocity (U) (see Fig. 1.8). It is assumed that L is a function of these other variables, as well as the fluid viscosity and specific weight, i.e., $L = f(H, d, U, \gamma, \mu)$. The model and prototype viscosities are the same, but the model specific weight is nine times the specific weight of the prototype.

   (a)   Determine an appropriate set of dimensionless parameters to characterize this problem.

(b) If the prototype velocity is 50 mph, what should be the model velocity?

(c) If L is measured for a particular test with the model to be 0.1 m, what would the corresponding L be for the prototype?

SUPPLEMENTAL READING

Aris, R., 1962. *Vectors, Tensors and the Basic Equations of Fluid Mechanics.* Prentice-Hall, Englewood Cliffs, New Jersey. (This book provides an easy treatment of tensors and their application to fluid mechanics.)

Graf, W., 1971. *Hydraulics of Sediment Transport,* McGraw-Hill, New York.

Prager, W., 1961. *Introduction to Mechanics of Continua.* Dover, New York. (Contains coverage of Cartesian tensors.)

Rouse, H., and Ince, S., 1957. *History of Hydraulics.* Iowa Institute of Hydraulic Research, State University of Iowa.

Sommerfeld, A., 1964. *Mechanics of Deformable Bodies.* Academic Press, New York. (Contains coverage of Cartesian tensors.)

Streeter, V. L., 1948. *Fluid Dynamics.* McGraw-Hill, New York. (Contains a fundamental treatment of complex variable applications in fluid mechanics.)

Synge, J. L., and Schild, A., 1978. *Tensor Calculus.* Dover, New York. (Gives a clear and comparatively easy treatment of all kinds of tensors and their application in mechanics.)

2
Fundamental Equations

2.1 INTRODUCTION

The basic equations of fluid mechanics are derived by considering conservation statements (i.e., of mass, momentum, energy, etc.) applied to a finite volume of fluid continuum which is called a *system* or *material volume* and consists of a collection of infinitesimal fluid particles. Quantities involving space and time only are associated with the *kinematics* of the fluid particles. Examples of variables related to the kinematics of the fluid particles are displacement, velocity, acceleration, rate of strain, and rotation. Such variables represent the motion of the fluid particles, in response to applied *forces*. All variables connected with these forces involve space, time, and mass dimensions. These are related to the *dynamics* of the fluid particles.

In the following sections of this chapter we provide information concerning the basic representation of kinematic and dynamic variables and concepts associated with fluid particles and fluid systems.

2.2 FLUID VELOCITY, PATHLINES, STREAMLINES, AND STREAKLINES

A *pathline* represents the trajectory of a fluid particle. At a time of reference t_0, consider a fluid particle to be at position $\vec{r}_0$. In Cartesian coordinates this location is represented by (x_0, y_0, z_0). Due to its motion, the fluid particle is at position $\vec{r}$ at time t, and this new position is represented by coordinates (x, y, z). The functional representation of the pathline is given by

$$\vec{r} = \vec{r}(\vec{r}_0, t) \qquad \text{or} \qquad \vec{x} = \vec{x}(\vec{x}_0, t) \tag{2.2.1}$$

The vector $\vec{r}_0$ (or $\vec{x}_0$) represents the *label* of the particular fluid particle. The concept of pathline is a basic feature of the *Lagrangian* approach, which is explained in greater detail in Sec. 2.4.

As an example of the pathline concept, consider the following description of pathlines in a two-dimensional flow field:

$$x = x_0 e^{-at} \qquad y = y_0 e^{at} \tag{2.2.2}$$

It is possible to eliminate t from these expressions and obtain an equation describing the shape of the pathline in the $x-y$ plane, as

$$xy = x_0 y_0 \tag{2.2.3}$$

This expression shows that pathlines are hyperbolas whose asymptotes are the coordinate axes.

By differentiating the equation of the pathline with regard to time we obtain the Lagrangian expressions for the velocity components. By further differentiating the latter expressions with regard to time, we obtain the Lagrangian expressions for the acceleration components:

$$\vec{V} = \vec{V}(\vec{r}_0, t) = \frac{\partial \vec{r}}{\partial t} \qquad \vec{a} = a(\vec{r}_0, t) = \frac{\partial^2 \vec{r}}{\partial t^2} \tag{2.2.4}$$

For the example pathlines of Eq. (2.2.2), the Lagrangian velocity components are

$$u(x_0, y_0, t) = -ax_0 e^{-at} \qquad v(x_0, y_0, t) = ay_0 e^{at} \tag{2.2.5}$$

By eliminating x_0 and y_0 from Eq. (2.2.5), we obtain the *Eulerian* presentation (which will be discussed hereinafter) of the velocity components,

$$u(x, y, t) = -ax \qquad v(x, y, t) = ay \tag{2.2.6}$$

The Eulerian presentation is the most common way of describing a flow field, where a spatial distribution of velocity values is given (note that velocities do not depend on an initial position in this presentation). It should be further noted that the pathline equation given by Eq. (2.2.2) can be obtained by direct integration of Eq. (2.2.5) or integration of Eq. (2.2.6), while considering that $x = x(x_0, y_0, t)$; $y = y(x_0, y_0, t)$.

By differentiation of Eq. (2.2.5) with regard to time, we obtain the Lagrangian presentation of the acceleration component,

$$a_x(x_0, y_0, t) = a^2 x_0 e^{-at} \qquad a_y(x_0, y_0, t) = a^2 y_0 e^{at} \tag{2.2.7}$$

Again, by eliminating x_0 and y_0 from Eq. (2.2.7), the Eulerian presentation of the acceleration components is

$$a_x(x, y, t) = a^2 x \qquad a_y(x, y, t) = a^2 y \tag{2.2.8}$$

Flow fields are often depicted using *streamlines*. Streamlines are curves that are everywhere tangent to the velocity vector, as shown in Fig. 2.1. A

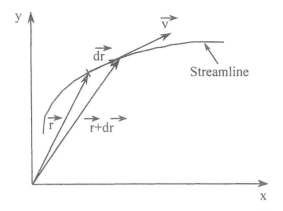

Figure 2.1 Example of streamline.

streamline is associated with a particular time and may be considered as an instantaneous "photograph" of the velocity vector directions for the entire flow field.

As implied in Fig. 2.1 (since the streamlines are tangent to the velocity), a streamline may be described by

$$\vec{V} \times d\vec{r} = 0 \qquad \text{where} \qquad \vec{V} = \vec{V}(\vec{x}, t) \tag{2.2.9}$$

where $\vec{V}$ is the velocity vector, $d\vec{r}$ is an infinitesimal element along the streamline, and $\vec{x}$ is the coordinate vector. In a Cartesian coordinate system, Eq. (2.2.9) yields

$$\frac{dx}{u} = \frac{dy}{v} = \frac{dz}{w} \tag{2.2.10}$$

where u, v, and w are the velocity components in the x, y, and z directions, respectively.

According to Eq. (2.2.10), the shape of the streamlines is constant if the velocity vector can be expressed as a product of a spatial function and a temporal function. Such a case is represented by either one of the following conditions:

$$\vec{V}(\vec{x}, t) = \vec{U}(\vec{x}) f(t) \qquad \frac{\vec{V}}{|\vec{V}|} \neq f(t) \tag{2.2.11}$$

If $\vec{V}$ is solely a spatial function [i.e., $f(t)$ is a constant], then the flow field is subject to *steady state conditions* and the shape of the streamlines is identical to that of the pathlines. As an example, consider the velocity vector represented

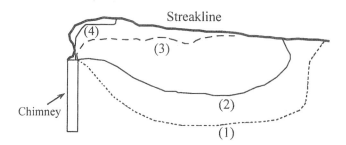

Figure 2.2 Four pathlines and a streakline at a chimney.

by Eq. (2.2.6). The differential equation of the streamlines is

$$-\frac{dx}{x} = \frac{dy}{y} \qquad (2.2.12)$$

Direct integration of this equation yields

$$xy = C \qquad (2.2.13)$$

where C is a constant of the particular streamline. Since Eq. (2.2.6) refers to steady state conditions, the shape of the streamlines represented by Eq. (2.2.13) is identical to that of the pathlines, which is given by Eq. (2.2.3).

A *streakline* is defined as a line connecting a series of fluid particles with their point source. An example of pathlines and a streakline that might be produced by smoke particles is presented in Fig. 2.2. In this figure the pathlines are enumerated. Pathline (1) refers to the first particle that left the chimney outlet. Pathline (2) refers to the second particle, etc.

2.3   RATE OF STRAIN, VORTICITY, AND CIRCULATION

In this section we discuss variables characterizing the kinematics of the flow field, which are associated with the velocity vector distribution in the domain. All such variables originate from the Eulerian presentation of the velocity vector.

In Fig. 2.3 are described two points in a flow field, A and B. The rates of change of the coordinate intervals between these points are represented by the following expressions given in Cartesian indicial format:

$$\frac{d}{dt}(\Delta x_i) = \Delta u_i = \frac{\partial u_i}{\partial x_j} dx_j \qquad (2.3.1)$$

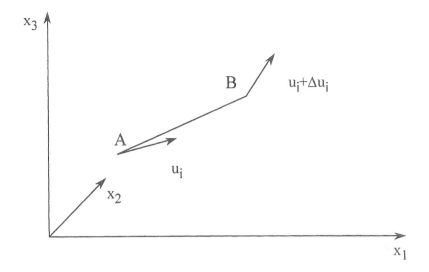

Figure 2.3 Rate of change of distance between two points.

Applying this expression, we obtain a second-order tensor that describes the rate of change of the coordinate intervals per unit length. This second-order tensor can be separated into symmetric and asymmetric tensors,

$$\frac{\partial u_i}{\partial x_j} = \frac{1}{2}\left(\frac{\partial u_i}{\partial x_j} + \frac{\partial u_j}{\partial x_i}\right) + \frac{1}{2}\left(\frac{\partial u_i}{\partial x_j} - \frac{\partial u_j}{\partial x_i}\right) \tag{2.3.2}$$

The first tensor on the right-hand side of Eq. (2.3.2) is the symmetric tensor, called the *rate of strain tensor*. The second tensor is the asymmetric one, called the *vorticity tensor*. Each of these tensors has a distinct physical meaning, as described below.

The rate of strain tensor is represented by

$$e_{ij} = \frac{1}{2}\left(\frac{\partial u_i}{\partial x_j} + \frac{\partial u_j}{\partial x_i}\right) \tag{2.3.3}$$

In Fig. 2.4 the rate of elongation of an elementary fluid volume in a two-dimensional flow field is illustrated. The rate of elongation per unit length of that elementary volume in the x_i direction is called the *linear* or *normal strain rate*. It is represented by

$$\frac{u_1 + \Delta u_1 - u_1}{\Delta x_1} = \frac{(\partial u_1/\partial x_1)\Delta x_1}{\Delta x_1} = \frac{\partial u_1}{\partial x_1} \tag{2.3.4}$$

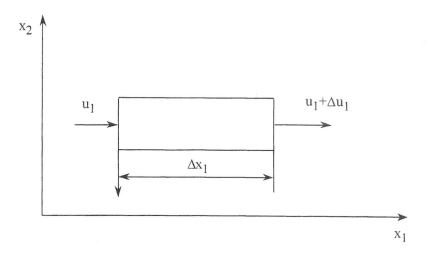

Figure 2.4 Elongation of an elementary fluid volume.

This expression gives the component e_{11} of the strain rate tensor. The components e_{22} and e_{33} represent the linear strain in the x_2 and x_3 directions. They are given, respectively, by

$$e_{22} = \frac{\partial u_2}{\partial x_2} \qquad e_{33} = \frac{\partial u_3}{\partial x_3} \tag{2.3.5}$$

Thus it is seen that diagonal components of the rate of strain tensor describe the linear rate of strain. The *volumetric strain rate* of an elementary volume is given by the *trace* of the strain rate tensor, i.e., the sum of the diagonal components, since

$$\frac{1}{\Delta x_1 \Delta y_1 \Delta z_1} \frac{d}{dt} (\Delta x_1 \Delta y_1 \Delta z_1)$$

$$= \frac{1}{\Delta x_1} \frac{d}{dt} (\Delta x_1) + \frac{1}{\Delta x_2} \frac{d}{dt} (\Delta x_2) + \frac{1}{\Delta x_3} \frac{d}{dt} (\Delta x_3)$$

$$= \frac{\partial u_1}{\partial x_1} + \frac{\partial u_2}{\partial x_2} + \frac{\partial u_3}{\partial x_3} = e_{11} + e_{22} + e_{33} \tag{2.3.6}$$

With regard to components of the rate of strain tensor that are not on the diagonal, we consider in Fig. 2.5 the rate of change of the angle of the elementary rectangle, which is called the *shear strain rate*. The expression for the shear strain rate is

$$\frac{u_1 + \Delta u_1 - u_1}{\Delta x_2} + \frac{u_2 + \Delta u_2 - u_2}{\Delta x_1}$$

$$= \frac{(\partial u_1/\partial x_2)\Delta x_2}{\Delta x_2} + \frac{(\partial u_2/\partial x_1)\Delta x_1}{\Delta x_1} = \frac{\partial u_1}{\partial x_2} + \frac{\partial u_2}{\partial x_1} \tag{2.3.7}$$

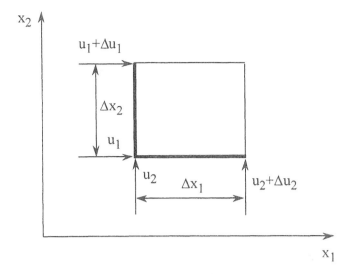

Figure 2.5 Elementary fluid volume subject to shear strain.

This expression is proportional to e_{12}, where

$$e_{12} = \frac{1}{2}\left(\frac{\partial u_1}{\partial x_2} + \frac{\partial u_2}{\partial x_1}\right) \tag{2.3.8}$$

Components of the strain rate tensor that are off the main diagonal thus represent deformation of shape. They are equal to half of the corresponding shear rate.

The *vorticity tensor* is an asymmetric tensor given in Cartesian coordinates by

$$\omega_{ij} = \left(\frac{\partial u_i}{\partial x_j} - \frac{\partial u_j}{\partial x_i}\right) \tag{2.3.9}$$

By considering Fig. 2.5, it is possible to visualize the physical meaning of the vorticity tensor. In this figure the velocity components that lead to rotation of an elementary fluid volume in a two-dimensional flow field are shown. The average angular velocity of that volume in the counterclockwise direction is given by

$$\frac{1}{2}\left(\frac{u_2 + \Delta u_2 - u_2}{\Delta x_1} - \frac{u_1 + \Delta u_1 - u_1}{\Delta x_2}\right)$$

$$= \frac{1}{2}\left(\frac{(\partial u_2/\partial x_1)\Delta x_1}{\Delta x_1} - \frac{(\partial u_1/\partial x_2)\Delta x_2}{\Delta x_2}\right)$$

$$= \frac{1}{2}\left(\frac{\partial u_2}{\partial x_1} - \frac{\partial u_1}{\partial x_2}\right) = \omega_{21} = -\omega_{12} \tag{2.3.10}$$

This expression indicates that the vorticity tensor is associated with rotation of the fluid particles.

In general, a second-order asymmetric tensor has three pairs of nonzero components. Each pair of components has identical magnitudes but opposite signs. Such a tensor also can be represented by a vector that has three components. Components of the vorticity tensor are proportional to components of the vorticity vector, which is the *curl* of the velocity vector,

$$\vec{\omega} = \nabla \times \vec{V} \qquad \text{or} \qquad \omega_i = \varepsilon_{ijk} \frac{\partial u_k}{\partial x_j} \tag{2.3.11}$$

According to this expression, components of the vorticity vector are given by

$$\omega_1 = \frac{\partial u_3}{\partial x_2} - \frac{\partial u_2}{\partial x_3} \qquad \omega_2 = \frac{\partial u_1}{\partial x_3} - \frac{\partial u_3}{\partial x_1} \qquad \omega_3 = \frac{\partial u_2}{\partial x_1} - \frac{\partial u_1}{\partial x_2} \tag{2.3.12}$$

Irrotational flow is a flow in which all components of the vorticity vector are equal to zero. In such a flow the velocity vector originates from a potential function, namely

$$\vec{V} = \nabla \Phi \qquad \text{or} \qquad u_i = \frac{\partial \Phi}{\partial x_i} \tag{2.3.13}$$

Potential flows are discussed in greater detail in Chap. 4.

The *circulation* is defined as the line integral of the tangential component of velocity. It is given by

$$\Gamma = \oint_c \vec{V} \cdot d\vec{s} \qquad \text{or} \qquad \Gamma = \oint_c u_i \, ds_i \tag{2.3.14}$$

By applying the Stokes theorem, the line integral of Eq. (2.3.14) is converted to an area integral,

$$\oint_c \vec{V} \cdot d\vec{s} = \int_A (\nabla \times \vec{V}) \cdot d\vec{A} \quad \text{or} \quad \oint_c u_i \, ds_i = \int_A \varepsilon_{ijk} \frac{\partial u_k}{\partial x_j} \, dA_i \tag{2.3.15}$$

This form of the equation is sometimes more useful.

2.4   LAGRANGIAN AND EULERIAN APPROACHES

2.4.1   General Presentation of the Approaches

Some basic concepts of the *Lagrangian* and *Eulerian* approaches have already been represented in the previous section. In the present section we expand on those concepts and describe some derivations of the basic conceptual approaches.

In the Lagrangian approach interest is directed at fluid particles and changes of properties of those particles. The Eulerian approach refers to spatial and temporal distributions of properties in the domain occupied by the fluid. Whereas the Lagrangian approach represents properties of individual fluid particles according to their initial location and time, the Eulerian approach represents the distribution of such properties in the domain with no reference to the history of the fluid particles. The concept of pathlines originates from the Lagrangian approach, while the concept of streamlines is associated with the Eulerian approach.

Every property F of an individual fluid particle can be represented in the Lagrangian approach by

$$F = F(\vec{x}_0, t) \tag{2.4.1}$$

where $\vec{x}_0$ is the location of the fluid particle at time t_0 and t is the time. The property F, according to the Eulerian approach, is distributed in the domain occupied by the fluid. Therefore its functional presentation is given by

$$F = F(\vec{x}, t) \tag{2.4.2}$$

where $\vec{x}$ and t are the spatial coordinates and time, respectively.

According to the Lagrangian approach, the rate of change of the property F of the fluid particle is given by

$$\frac{\partial F(\vec{x}_0, t)}{\partial t} \tag{2.4.3}$$

Therefore the velocity and acceleration of the fluid particle are given by

$$u_i(\vec{x}_0, t) = \frac{\partial x_i(\vec{x}_0, t)}{\partial t} \qquad a_i(\vec{x}_0, t) = \frac{\partial u_i(\vec{x}_0, t)}{\partial t} = \frac{\partial^2 x_i(\vec{x}_0, t)}{\partial t^2} \tag{2.4.4}$$

For example, consider the flow field defined by the pathlines given in Eq. (2.2.2). The Lagrangian velocity components are given by Eq. (2.2.5), and the Lagrangian acceleration components are given by Eq. (2.2.7).

The rate of change of the property F of the fluid particles, according to the Eulerian approach, can be expressed through use of the *material* or *absolute derivative*. This derivative expresses the rate of change of the property F by an observer moving with the fluid particle. The expression of the material derivative is given by

$$\frac{DF[\vec{x}(t), t]}{Dt} = \frac{\partial F}{\partial t} + (\nabla F)\frac{d\vec{x}}{dt} = \frac{\partial F}{\partial t} + \frac{\partial F}{\partial x_i}\frac{dx_i}{dt} \tag{2.4.5}$$

Therefore the velocity and acceleration distributions in the flow field, according to the Eulerian approach, are given, respectively, by

$$\vec{V} = \frac{d\vec{x}}{dt} \qquad \vec{a} = \frac{\partial \vec{V}}{\partial t} + \vec{V} \cdot \nabla \vec{V}$$

$$\text{or} \qquad u_i = \frac{dx_i}{dt} \qquad a_i = \frac{\partial u_i}{\partial t} + u_k \frac{\partial u_i}{\partial x_k}$$

$$(2.4.6)$$

As an example, consider the Eulerian velocity distribution given by Eq. (2.2.6). By introducing the expressions of Eq. (2.2.6) into Eq. (2.4.6) we obtain the Eulerian acceleration distribution given by Eq. (2.2.8).

2.4.2 System and Control Volume

The previous paragraphs refer to individual fluid particles and their properties. Presently we will refer to aggregates of fluid particles comprising a finite fluid volume. A finite volume of fluid incorporating a constant quantity of fluid particles (or matter) is called a *system* or *material volume*. A system may change shape, position, thermal condition, etc., but it always incorporates the same matter. In contrast, a *control volume* is an arbitrary volume designated in space. A control volume may possess a variable shape, but in most cases it is convenient to consider control volumes of constant shape. Therefore fluid particles may pass into or out of the fixed control volume across its surface.

Figure 2.6 shows an arbitrary flow field. Several streamlines describing the flow direction at time t are depicted. The figure shows a system at time t. A control volume (CV) identical to the system at time t also is shown. At time $t + \Delta t$ the system has a shape different from its shape at time t, but the control volume has its original fixed shape from time t. We may identify three partial volumes, as indicated by Fig. 2.6: volume I represents the portion of the control volume evacuated by particles of the system during the time interval Δt; volume II is the portion of the control volume occupied by particles of the system at time $t + \Delta t$; volume III is the space to which particles of the system have moved during the time interval Δt. Particles of the system also convey properties of the flow. In the following paragraphs we consider the presentation of the rate of change of an arbitrary property η in the system by reference to a control volume.

2.4.3 Reynolds Transport Theorem

The Reynolds transport theorem represents the use of a control volume to calculate the rate of change of a property of a material volume. The rate of

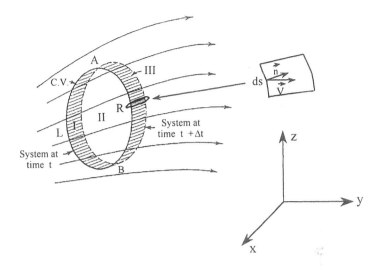

Figure 2.6 System (material volume) and control volume.

change of a property, η, of a material volume is represented by

$$\frac{D}{Dt} \int_{M.V.} \eta \, dU \tag{2.4.7}$$

where $M.V.$ represents material volume and dU is an elementary volume element. In Fig. 2.6, the integral of Eq. (2.4.7) incorporates two parts. One part consists of the control volume, CV, namely volume I and the material volume of Fig. 2.6, and the second part incorporates volumes I and III. An elementary volume ΔU of volumes I and III, as shown in Fig. 2.6, is represented by $\Delta U = (\vec{V} \cdot \vec{n} \, ds)\Delta t$, where $\vec{n}$ is a unit vector normal to the surface of the control volume (by convention, the direction of this vector is outward of the control volume) and ds is an elementary surface element. Summation of all elementary volumes ΔU leads to a surface integral, which is taken over the surface of the control volume, also known as the control surface (S). Therefore the rate of change of the material volume property, η, which is expressed by Eq. (2.4.7), can be given, by reference to the control volume, as

$$\frac{D}{Dt} \int_{M.V.} \eta \, dU = \frac{\partial}{\partial t} \int_{U} \eta \, dU + \int_{S} \eta (\vec{V} \cdot \vec{n}) \, ds \tag{2.4.8}$$

where U is the volume of the control volume. If a fixed control volume is considered, then the partial derivative of the first term of the RHS of Eq. (2.4.8) can be moved inside the volume integral of that expression. It should be noted that the property η can be a scalar as well as a vector quantity. This is illustrated in the following sections.

2.5 CONSERVATION OF MASS

2.5.1 The Finite Control Volume Approach

By definition, the total mass of a material volume or system is constant. Therefore,

$$\frac{D}{Dt} \int_{M.V.} \rho \, dU = 0 \tag{2.5.1}$$

Comparison of this expression with Eq. (2.4.7) indicates that the property η of Eq. (2.4.7) was replaced by the density ρ in Eq. (2.4.8). We may, therefore, apply the transport theorem of Reynolds, namely Eq. (2.4.8), to obtain

$$\frac{\partial}{\partial t} \int_U \rho \, dU + \int_S \rho(\vec{V} \cdot \vec{n}) \, ds = 0 \qquad \text{or}$$

$$\frac{\partial}{\partial t} \int_U \rho \, dU + \int_S \rho(u_i n_i) \, ds = 0 \tag{2.5.2}$$

Here, the first term represents the rate of change of mass included in the control volume. The second term represents the mass flux flowing through the surface of the control volume. Equation (2.5.2) represents the integral expression for the conservation of mass.

If we refer to a fixed control volume, and the density ρ of the fluid is constant, then the first term of Eq. (2.5.2) vanishes, and

$$\int_S (\vec{V} \cdot \vec{n}) ds = 0 \qquad \text{or} \qquad \int_S u_i n_i \, ds = 0 \tag{2.5.3}$$

This equation represents the integral expression for continuity. It indicates that if the fluid density is constant, then the total mass flux entering the control volume is identical to the total mass flux flowing out of the control volume (for a fixed volume). When applied to a control volume of a stream tube, as shown in Fig. 2.7, Eq. (2.5.3) leads to

$$\vec{V} \cdot \vec{n} A = \text{const} \tag{2.5.4}$$

2.5.2 The Differential Approach

Consider again a fixed control volume. We transform the surface integral of the second term on the RHS of Eq. (2.5.2) to a volume integral by the divergence theorem and obtain

$$\int_U \left[\frac{\partial \rho}{\partial t} + \nabla \cdot (\rho \vec{V}) \right] dU = 0 \tag{2.5.5}$$

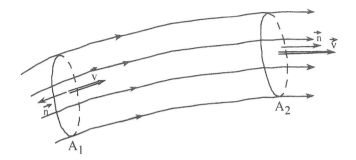

Figure 2.7 The integral continuity expression for a stream tube.

If the control volume is an arbitrarily small elementary volume, then Eq. (2.5.5) yields

$$\frac{\partial \rho}{\partial t} + \nabla \cdot (\rho \vec{V}) = 0 \qquad \text{or}$$

$$\frac{\partial \rho}{\partial t} + \frac{\partial (\rho u_i)}{\partial x_i} = 0 \qquad \text{or} \qquad (2.5.6)$$

$$\frac{D\rho}{Dt} + \rho (\nabla \cdot \vec{V}) = 0$$

This expression represents the differential equation of mass conservation. If the density of the fluid is fixed (i.e., $D\rho/Dt = 0$), then the flow is called *incompressible flow*, and Eq. (2.5.6) gives

$$\nabla \cdot \vec{V} = 0 \qquad \text{or} \qquad \frac{\partial u_i}{\partial x_i} = 0 \qquad (2.5.7)$$

This expression represents the differential *continuity equation*.

2.5.3 The Stream Function

If the flow field is two dimensional, and a Cartesian coordinate system is assumed, then Eq. (2.5.7) implies

$$\frac{\partial u}{\partial x} + \frac{\partial v}{\partial y} = 0 \qquad (2.5.8)$$

Then a *stream function* Ψ may be defined that satisfies Eq. (2.5.8),

$$u = \frac{\partial \Psi}{\partial y} \qquad v = \frac{\partial \Psi}{\partial x} \qquad (2.5.9)$$

Then, introducing Eq. (2.5.9) into Eq. (2.2.10), it is seen that streamlines are defined by

$$\frac{\partial \Psi}{\partial x} dx + \frac{\partial \Psi}{\partial y} dy = 0 \tag{2.5.10}$$

This expression indicates that the differential of the stream function vanishes on the streamlines. Therefore the stream function has a constant value on a streamline, and the value of the stream function can be used for the identification of particular streamlines in the flow field.

Figure 2.8 shows two streamlines, which are identified by Ψ_A and Ψ_B. The discharge per unit width flowing through the stream tube bounded by the streamlines Ψ_A and Ψ_B is given by

$$q = \int_A^B (u\, dy - v\, dx) = \int_A^B \left(\frac{\partial \Psi}{\partial y} dy + \frac{\partial \Psi}{\partial x} dx \right)$$

$$= \int_A^B d\Psi = \Psi_B - \Psi_A \tag{2.5.11}$$

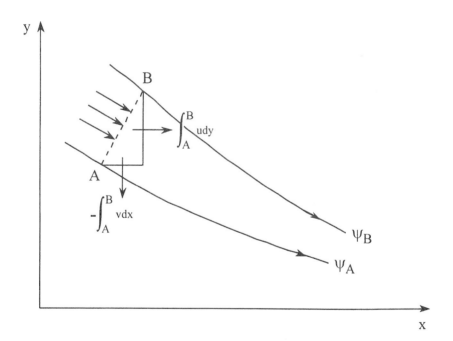

Figure 2.8   Illustration of volumetric flux between two streamlines.

Thus the difference between values of the stream function for two streamlines represents the discharge flowing between those streamlines.

If the flow field is represented by a cylindrical coordinate system, then the employment of the covariant derivative and the relevant scale yield the following expression for the differential continuity equation:

$$\nabla \cdot \vec{V} = \frac{\partial u_r}{\partial r} + \frac{u_r}{r} + \frac{1}{r}\frac{\partial v_\theta}{\partial \theta} + \frac{\partial w_z}{\partial z}$$
$$= \frac{1}{r}\frac{\partial(ru_r)}{\partial r} + \frac{1}{r}\frac{\partial v_\theta}{\partial \theta} + \frac{1}{r}\frac{\partial(rw_z)}{\partial z} = 0 \qquad (2.5.12)$$

where u_r, v_θ, and w_z are physical components of the velocity vector in the r, θ, and z directions, respectively. We may use the concept of stream function in cylindrical coordinates for two types of flow field. One type is a two-dimensional flow field expressed by reference to coordinates r and θ. The other type is an axisymmetric flow field expressed by coordinates r and z.

In the case of two-dimensional flow, there is no flow in the z-direction, and velocity components do not depend on the z coordinate. Therefore the term referring to z and w_z of Eq. (2.5.12) vanishes, and the expressions for u_r and v_θ are given by the stream function as

$$u_r = \frac{1}{r}\frac{\partial \Psi}{\partial \theta} \qquad v_\theta = -\frac{\partial \Psi}{\partial r} \qquad (2.5.13)$$

In cases of axisymmetric flow, there is no flow in the θ-direction, and velocity components do not depend on the θ coordinate. Then the presentation of u_r and w_z by the stream function is given as

$$u_r = \frac{1}{r}\frac{\partial \Psi}{\partial z} \qquad w_z = -\frac{1}{r}\frac{\partial \Psi}{\partial r} \qquad (2.5.14)$$

Note that the stream function of Eq. (2.5.13) has dimensions of discharge per unit width, whereas the stream function of Eq. (2.5.14) has dimensions of volumetric discharge.

2.5.4 Stratified Flow

In cases of *stratified flow*, where the density field is not constant, the differential equation of mass conservation, namely Eq. (2.5.6), is still

$$\frac{\partial \rho}{\partial t} + \vec{V} \cdot \nabla \rho + \rho \nabla \cdot \vec{V} = 0 \quad \text{or} \quad \frac{\partial \rho}{\partial t} + u_i \frac{\partial \rho}{\partial x_i} + \rho \frac{\partial u_i}{\partial x_i} = 0 \qquad (2.5.15)$$

(Recall that there were no constraints placed on density in deriving the mass conservation expression.) In particular, consider the second of these expressions, which is rewritten as

$$\frac{D\rho}{Dt} + \rho\vec{\nabla} \cdot \vec{V} = 0 \quad \text{or} \quad \frac{D\rho}{Dt} + \rho\frac{\partial u_i}{\partial x_i} = 0 \tag{2.5.16}$$

This expression indicates that incompressible flow is identified by the vanishing material derivative of the density. In other words, density is constant, *following a fluid particle*. In cases of steady stratified flow, the temporal derivative of the density is zero. If the flow is also incompressible, namely $\vec{\nabla} \cdot \vec{V} = 0$ [Eq. (2.5.7)], then according to Eq. (2.5.15), the velocity vector is perpendicular to the density gradient.

In cases of steady two-dimensional flow, Eq. (2.5.6) yields

$$\frac{\partial(\rho u)}{\partial x} + \frac{\partial(\rho v)}{\partial y} = 0 \tag{2.5.17}$$

This equation can be identically satisfied by a stream function defined by

$$\rho u = \frac{\partial \Psi}{\partial y} \quad \rho v = -\frac{\partial \Psi}{\partial x} \tag{2.5.18}$$

This stream function has dimensions of mass flux per unit width.

2.6 CONSERVATION OF MOMENTUM

The property $\rho\vec{V}$ represents the momentum of a unit volume of the fluid. The rate of change of momentum of a fluid material volume is equal to the sum of forces acting on that material volume. Using the Reynolds transport theorem, Eq. (2.4.8) applied to $\rho\vec{V}$ yields

$$\frac{\partial}{\partial t}\int_U \rho\vec{V}\,dU + \int_S \rho\vec{V}(\vec{V} \cdot \vec{n})\,ds$$

$$= \int_U \rho\vec{g}\,dU + \int_S \tilde{S} \cdot \vec{n}\,ds + \vec{F}_s \tag{2.6.1a}$$

$$\text{or} \quad \frac{\partial}{\partial t}\int_U \rho u_i\,dU + \int_S \rho u_i(u_k n_k)\,ds$$

$$= \int_U \rho g_i\,dU + \int_S S_{ik}n_k\,ds + F_{si} \tag{2.6.1b}$$

where $\tilde{S}$ is the stress tensor, which refers to forces acting on the fluid surface of the control volume, and $\vec{F}_s$ represents forces acting on solid surfaces comprising portions of the surface of the control volume.

The first RHS term of Eq. (2.6.1) represents body forces originating from gravity. The gravitational acceleration vector, $\vec{g}$, is equal to the gravity,

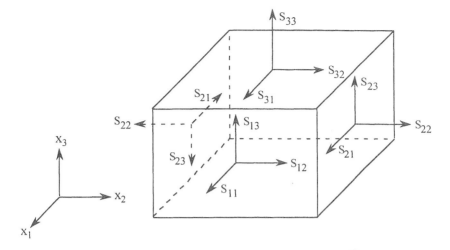

Figure 2.9 Components of the stress tensor acting on a small rectangle.

g, multiplied by a unit vector in the negative direction of the normal to the earth's surface. The second RHS term represents surface forces.

The stress tensor at each point of the surface of the control volume can be completely defined by the nine components of the stress tensor, $\tilde{S}$. Figure 2.9 shows an infinitesimal rectangular parallelepiped with faces having normal unit vectors parallel to the coordinate axes. The force per unit area acting on each face of the parallelepiped is divided into a normal component and two shear components (shear stresses) that are perpendicular to the normal component. Figure 2.9 exemplifies the decomposition of the force per unit area over four different faces. Directions of the stress tensor components shown in Fig. 2.9 are considered positive, by convention. The first subscript of the stress component represents the direction of the normal of the particular face on which the stress acts. The second subscript represents the direction of the component of the stress.

In Fig. 2.10 are shown components of the shear stress creating torque, which may lead to rotation of the elementary rectangle around its center of gravity, G. The total torque is expressed by

$$\text{Torque} = \left(S_{12} + \frac{1}{2}\frac{\partial S_{12}}{\partial x_1} dx_1 \right) dx_2 \frac{dx_1}{2} + \left(S_{12} - \frac{1}{2}\frac{\partial S_{12}}{\partial x_1} dx_1 \right) dx_2 \frac{dx_1}{2}$$
$$- \left(S_{21} + \frac{1}{2}\frac{\partial S_{21}}{\partial x_2} dx_2 \right) dx_1 \frac{dx_2}{2} - \left(S_{21} - \frac{1}{2}\frac{\partial S_{21}}{\partial x_2} dx_2 \right) dx_1 \frac{dx_2}{2}$$

$$(2.6.2)$$

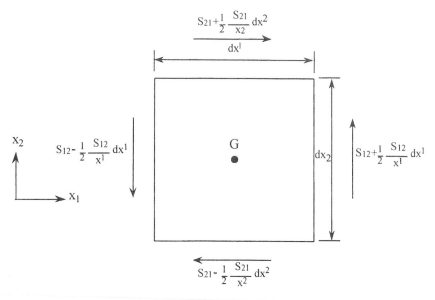

Figure 2.10   Torque applied on an elementary rectangle of fluid.

Also the total torque is equal to the moment of inertia multiplied by the angular acceleration. Therefore, Eq. (2.6.2) yields

$$(S_{12} - S_{21})\, dx_1\, dx_2 = \frac{\rho}{12}\, dx_1\, dx_2 \left[(dx_1)^2 + (dx_2)^2\right] \alpha \tag{2.6.3}$$

where α is the angular acceleration.

Upon dividing Eq. (2.6.3) by the area of the elementary rectangle and allowing dx_1 and dx_2 to approach zero, the RHS of Eq. (2.6.3) vanishes. This result indicates that the stress tensor is a symmetric tensor, namely

$$S_{ij} = S_{ji} \tag{2.6.4}$$

The stress tensor can be decomposed into two tensors, as

$$\tilde{S} = -p\tilde{I} + \tilde{\tau} \quad \text{or} \quad S_{ij} = -p\delta_{ij} + \tau_{ij} \tag{2.6.5}$$

where $\tilde{I}$ is a unit matrix, which also can be represented by δ_{ij}, p is the pressure, and $\tilde{\tau}$ is the *deviator stress tensor*, related to shear stresses (see below).

The first term on the RHS of Eq. (2.6.5) is an isotropic tensor, namely a tensor that has components only on its diagonal, and all diagonal components are identical, provided that we apply a Cartesian coordinate system. Components of the isotropic tensor are not modified by rotation of the coordinate system. The pressure, p, is equal to the negative one-third of the *trace* of the stress tensor,

$$p = -\frac{1}{3}(S_{11} + S_{22} + S_{33}) \tag{2.6.6}$$

where the trace of a tensor is defined as the sum of its diagonal components. Note that the trace of the deviator stress tensor is zero. Positive normal stress means tension. However, fluids can only resist and convey negative normal stresses. The definition of Eq. (2.6.6) yields a positive value for the pressure.

Incorporating the definitions and expressions developed in the preceding paragraphs, Eq. (2.6.1) is rewritten to express conservation of momentum in a fluid material volume:

$$\frac{\partial}{\partial t} \int_U \rho u_i \, dU + \int_s \rho u_i(u_k n_k) \, ds$$
$$= -\int_s p n_i \, ds + \int_s \tau_{ik} n_k \, ds - \int_U \rho g k_i \, dU + F_{Si} \tag{2.6.7}$$

where k_i represents the component of a unit vector perpendicular to the earth, directed toward the atmosphere. For a fixed control volume, the derivative of the first term on the LHS of Eq. (2.6.7) can be moved into the integral of that term.

When Eq. (2.6.7) is applied to an elementary volume of fluid, the last term vanishes since there are no solid surfaces. Then, using the divergence theorem to convert surface integrals to volume integrals, we have

$$\int_{\Delta U} \left[\frac{\partial(\rho u_i)}{\partial t} + \frac{\partial(\rho u_i u_k)}{\partial x_k} + \frac{\partial p}{\partial x_i} - \frac{\partial \tau_{ik}}{\partial x_k} + \rho g k_i \right] = 0 \tag{2.6.8}$$

By introducing the conservation of mass, expressed by Eq. (2.5.6), into Eq. (2.6.8), and considering that ΔU is small but different from zero,

$$\rho \left[\frac{\partial u_i}{\partial t} + u_k \frac{\partial u_i}{\partial x_k} \right] = -\frac{\partial p}{\partial x_i} + \frac{\partial \tau_{ik}}{\partial x_k} - \rho g k_i \tag{2.6.9a}$$

$$\text{or} \quad \rho \left[\frac{\partial \vec{V}}{\partial t} + (\vec{V} \cdot \nabla)\vec{V} \right] = -\nabla(p + \rho g Z) + \nabla \cdot \tilde{\tau} \tag{2.6.9b}$$

where Z is the elevation with regard to an arbitrary level of reference. Equation (2.6.9) is the *equation of motion*, or the differential equation of conservation of momentum.

The *Bernoulli equation* can be derived by direct integration of Eq. (2.6.9). First, note that the nonlinear term of the LHS of Eq. (2.6.9) can be expressed as

$$(\vec{V} \cdot \nabla)\vec{V} = \nabla \frac{V^2}{2} - \vec{V} \times (\nabla \times \vec{V}) \tag{2.6.10}$$

If the velocity vector is derived from a potential function, then shear stresses also are negligible, and $\nabla \times \vec{V} = 0$. Therefore, in such a case Eqs. (2.6.9) and (2.6.10) yield

$$\rho \left[\frac{\partial}{\partial t}(\nabla \Phi) + \nabla \frac{V^2}{2} \right] = -\nabla(p + \rho g Z) \qquad (2.6.11)$$

where Φ is the potential function, defined in Eq. (2.3.13). For steady state cases, direct integration of Eq. (2.6.11) and division by the specific weight of the fluid yield

$$\frac{V^2}{2g} + \frac{p}{\gamma} + Z = const \qquad (2.6.12)$$

where $\gamma = \rho g$ is the specific weight of the fluid. This is called the *Bernoulli equation*. The sum of the terms on the LHS of this equation is called the total head, which incorporates the velocity head, the pressure head, and the elevation (or elevation head). The sum of pressure head and elevation is called the *piezometric head*. According to Eq. (2.6.12) the total head is constant in a domain of steady potential flow.

In cases of steady flow with negligible effect of the shear stresses, consider a natural coordinate system that incorporates a coordinate, s, tangential to the streamline, and a coordinate, n, perpendicular to the streamline. The velocity vector has only a component tangential to the streamline. Therefore, Eq. (2.6.9) yields for the tangential direction,

$$\rho \left[V \frac{\partial V}{\partial s} \right] = -\frac{\partial}{\partial s}(p + \rho g Z) \qquad (2.6.13)$$

Direct integration of this expression indicates that the total head is constant along the streamline even if the flow is nonpotential flow, provided that the effect of shear stresses is negligible.

A *moving coordinate system* is sometimes applied to calculate momentum conservation. All basic equations applicable to a stationary coordinate system also can be applied to cases in which the coordinate system moves with a constant velocity. It should be noted that the Bernoulli equation, represented by Eq. (2.6.12), is applicable only in cases of steady state. The application of a moving coordinate system may sometimes enable use of Bernoulli's equation in cases of unsteady state conditions.

A *noninertial coordinate system* is one that is subject to acceleration. All momentum quantities in the conservation of momentum equation must be written with respect to an inertial coordinate system. If a noninertial system is used, then the acceleration measured by a fixed observer, $\vec{a}_{F.O.}$, is given by

$$\vec{a}_{F.O.} = \vec{a}_{M.O.} + \vec{a}_t + 2\vec{\omega} \times \vec{V}_{M.O} + \frac{d\vec{\omega}}{dt}$$
$$\times \vec{r}_{M.O.} + \vec{\omega} \times (\vec{\omega} \times \vec{r}_{M.O.}) \qquad (2.6.14)$$

where subscript F.O. refers to a fixed observer, M.O. refers to an observer moving with the coordinate system, a_t is the translational acceleration of the moving coordinate system, ω is the angular velocity of the moving coordinate system, $V_{M.O.}$ is the velocity of the fluid particle measured by the moving observer, and $r_{M.O.}$ is the position of the fluid particle measured by the moving observer. The momentum conservation Eq. (2.6.7) can be applied, with minor modification, to cases in which noninertial coordinate systems are used. In such cases, the integral equation of momentum conservation is given by

$$\frac{\partial}{\partial t} \int_U \rho \vec{V} \, dU + \int_s \rho \vec{V} (\vec{V} \cdot \vec{n}) \, ds$$

$$= -\int_s p\vec{n} \, ds + \int_s \vec{\tau} \cdot \vec{n} \, ds - \int_U \rho g \vec{k} \, dU + \vec{F}_s$$

$$- \int_U \left[\vec{a}_t + 2\vec{\omega} \times \vec{V} + \frac{d\vec{\omega}}{dt} \times \vec{r} + \vec{\omega} \times (\vec{\omega} \times \vec{r}) \right] \rho \, dU \qquad (2.6.15)$$

The following section provides further discussion of coordinate systems subject to rotational velocity originating from the earth's rotation. This is also described in further detail, using a dimensional scaling approach, in Sec. 2.9.3.

2.7 THE EQUATIONS OF MOTION AND CONSTITUTIVE EQUATIONS

In the preceding section it was shown that the equations of motion represent the conservation of momentum in an elementary fluid volume. The general form of the equations of motion is represented by Eq. (2.6.9), which is again given as

$$\rho \left[\frac{\partial u_i}{\partial t} + u_k \frac{\partial u_i}{\partial x_k} \right] = -\frac{\partial p}{\partial x_i} + \frac{\partial \tau_{ik}}{\partial x_k} - \rho g k_i \qquad (2.7.1a)$$

or $$\rho \left[\frac{\partial \overset{r}{V}}{\partial t} + (\overset{r}{V} \cdot \nabla)\overset{r}{V} \right] = -\nabla(p + \rho gZ) + \nabla \cdot \tilde{\tau} \qquad (2.7.1b)$$

Different types of fluids are identified by their *constitutive equations*, which provide the relationships between the deviatoric stress tensor, τ_{ij}, and kinematic parameters. For a *Newtonian fluid* the shear stress is assumed to be proportional to the rate of strain, and the constitutive equation for such a fluid is

$$\tau_{ij} = -\left(p + \frac{1}{3}\mu \frac{\partial u_k}{\partial x_k} \right) \delta_{ij} + 2\mu e_{ij} \qquad (2.7.2)$$

where e_{ij} is the rate of strain tensor,

$$e_{ij} = \frac{1}{2}\left(\frac{\partial u_i}{\partial x_j} + \frac{\partial u_j}{\partial x_i}\right) \tag{2.7.3}$$

By introducing Eq. (2.7.2) into Eq. (2.7.1), the general form of the *Navier–Stokes equations* is obtained,

$$\rho\frac{Du_i}{Dt} = -\frac{\partial p}{\partial x_i} - \rho g k_i + 2\mu\left[\frac{\partial e_{ij}}{\partial x_j} - \frac{1}{3}\frac{\partial^2 u_i}{\partial x_i \partial x_j}\right]$$

$$= -\frac{\partial p}{\partial x_i} - \rho g k_i + \mu\left[\frac{\partial^2 u_i}{\partial x_j^2} + \frac{1}{3}\frac{\partial^2 u_i}{\partial x_i \partial x_j}\right] \tag{2.7.4}$$

For incompressible flow, Eq. (2.7.4) reduces to

$$\rho\frac{D\vec{V}}{Dt} = -\nabla(p + \rho g Z) + \mu\nabla^2\vec{V} \tag{2.7.5a}$$

or $\quad \rho\dfrac{Du_i}{Dt} = -\dfrac{\partial}{\partial x_i}(p + \rho g Z) + \mu\dfrac{\partial^2 u_i}{\partial x_j^2} \tag{2.7.5b}$

Non-Newtonian fluids are characterized by constitutive equations different from Eq. (2.7.2). These types of fluids are not considered here.

The equations of motion given in the preceding paragraphs are valid in an inertial or fixed frame of reference. In comparatively small hydraulic systems, it is possible to refer to such equations of motion, while considering that the frame of reference, namely the earth, is stationary. In geophysical applications the rotation of the earth must be considered.

Figure 2.11 shows two coordinate systems: coordinate system (X_1, X_2, X_3), which is stationary, and coordinate system (x_1, x_2, x_3), which rotates at angular velocity Ω with regard to the fixed coordinate system. Any vector associated with the point G has three components in each of the coordinate systems. As an example, the decomposition of the vector $\vec{r}$ into three components of the rotating coordinate system is shown. A general vector $\vec{R}$ is represented in the rotating coordinate system by

$$\vec{R} = R_1\vec{i}_1 + R_2\vec{i}_2 + R_3\vec{i}_3 \tag{2.7.6}$$

A fixed observer, F.O., observes the rate of change of the vector $\vec{R}$ as

$$\left(\frac{d\vec{R}}{dt}\right)_{\text{F.O.}} = \frac{d}{dt}(R_1\vec{i}_1 + R_2\vec{i}_2 + R_3\vec{i}_3)$$

$$= \vec{i}_1\frac{dR_1}{dt} + \vec{i}_2\frac{dR_2}{dt} + \vec{i}_3\frac{dR_3}{dt} + R_1\frac{d\vec{i}_1}{dt} + R_2\frac{d\vec{i}_2}{dt} + R_3\frac{d\vec{i}_3}{dt} \tag{2.7.7}$$

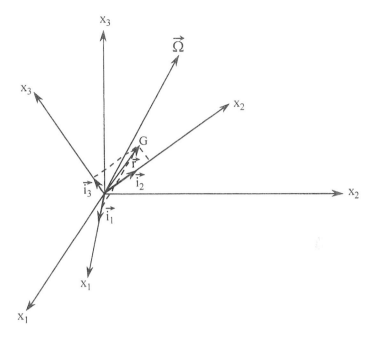

Figure 2.11 Coordinate system x_1, x_2, x_3 rotates with angular velocity Ω with regard to the stationary coordinate system X_1, X_2, X_3.

The first three terms on the RHS represent the rate of change of the vector, as observed by an observer, R.O., rotating with the rotating coordinate system. The second group of three terms represents the rate of change of the vector, originating from rotation of the coordinate system. Therefore Eq. (2.7.7) can be expressed as

$$\left(\frac{d\vec{R}}{dt}\right)_{F.O.} = \left(\frac{d\vec{R}}{dt}\right)_{R.O.} + R_1 \frac{d\vec{i}_1}{dt} + R_2 \frac{d\vec{i}_2}{dt} + R_3 \frac{d\vec{i}_3}{dt} \qquad (2.7.8)$$

Due to its rotation around the axis, $\vec{\Omega}$, each unit vector $\vec{i}$ traces a cone as shown in Fig. 2.12. The rate of change of this vector is given by

$$\left|\frac{d\vec{i}}{dt}\right| = \sin \beta \left(\frac{d\theta}{dt}\right) = \Omega \sin \beta \qquad (2.7.9)$$

The direction of the rate of change of the vector $\vec{i}$ is perpendicular to the plane made by the vectors $\vec{i}$ and $\vec{\Omega}$. Therefore

$$\frac{d\vec{i}}{dt} = \vec{\Omega} \times \vec{i} \qquad (2.7.10)$$

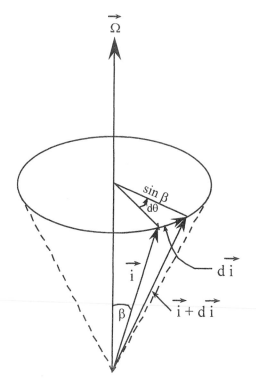

Figure 2.12   Cone of rotation of a unit vector.

The sum of the last three terms of Eq. (2.7.8) is given by

$$R_1 \vec{\Omega} \times \vec{i}_1 + R_2 \vec{\Omega} \times \vec{i}_2 + R_3 \vec{\Omega} \times \vec{i}_3 = \vec{\Omega} \times \vec{R} \qquad (2.7.11)$$

Introducing Eq. (2.7.11) into Eq. (2.7.8), we obtain

$$\left(\frac{d\vec{R}}{dt} \right)_{\text{F.O.}} = \left(\frac{d\vec{R}}{dt} \right)_{\text{R.O.}} + \vec{\Omega} \times \vec{R} \qquad (2.7.12)$$

This expression gives the relationship between the velocity vector measured by the fixed and rotating observers as

$$\vec{V}_{\text{F.O.}} = \vec{V}_{\text{R.O.}} + \vec{\Omega} \times \vec{r} \qquad (2.7.13)$$

Equation (2.7.12) also implies that acceleration can be expressed as

$$\left(\frac{d\vec{V}_{\text{F.O.}}}{dt} \right)_{\text{F.O.}} = \left(\frac{d\vec{V}_{\text{F.O.}}}{dt} \right)_{\text{R.O.}} + \vec{\Omega} \times \vec{V}_{\text{F.O.}} \qquad (2.7.14)$$

By introducing Eq. (2.7.13) into Eq. (2.7.14), we obtain

$$\frac{d\vec{V}_{F.O.}}{dt} = \frac{d}{dt}[\vec{V}_{R.O.} + \vec{\Omega} \times \vec{r}]_{R.O.} + \vec{\Omega} \times (\vec{V}_{R.O.} + \vec{\Omega} \times \vec{r})$$

$$= \left(\frac{d\vec{V}_{R.O.}}{dt}\right)_{R.O.} + \vec{\Omega} \times \left(\frac{d\vec{r}}{dt}\right)_{R.O.} + \vec{\Omega} \times \vec{V}_{R.O.} + \vec{\Omega} \times (\vec{\Omega} \times \vec{r})$$

$$(2.7.15)$$

Thus the relationship between the acceleration in the two coordinate systems is

$$\vec{a}_{F.O.} = \vec{a}_{R.O.} + 2\vec{\Omega} \times \vec{V}_{R.O.} + \vec{\Omega} \times (\vec{\Omega} \times \vec{r}) \tag{2.7.16}$$

Upon introducing the vector $\vec{R}$, which is perpendicular to the axis of rotation represented by the vector $\vec{\Omega}$ (also refer to Fig. 2.13), we find

$$\vec{\Omega} \times \vec{r} = \vec{\Omega} \times \vec{R} \tag{2.7.17}$$

Also, using the vector identity,

$$\vec{\Omega} \times (\vec{\Omega} \times \vec{R}) = (\vec{\Omega} \cdot \vec{R})\vec{\Omega} - (\vec{\Omega} \cdot \vec{\Omega})\vec{R} = -(\vec{\Omega} \cdot \vec{\Omega})\vec{R} = -\Omega^2 \vec{R} \tag{2.7.18}$$

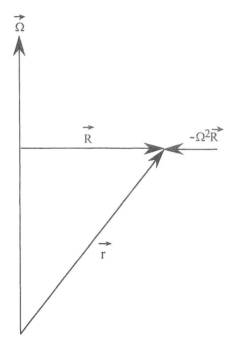

Figure 2.13 Relationships between vectors r, R and the centripetal acceleration.

with Eq. (2.7.16), we obtain

$$\vec{a}_{\text{F.O.}} = \vec{a} + 2\vec{\Omega} \times \vec{V} - \Omega^2 \vec{R} \qquad (2.7.19)$$

where $\vec{V}$ and $\vec{a}$ are the velocity and acceleration vectors, respectively, in the rotating coordinate system. The second term on the RHS of this last result represents the *Coriolis acceleration*. The last term on the RHS of this equation represents *centripetal acceleration*.

The preceding paragraphs indicate that the equations of motion for *geostrophic* (or, "earth-turned") *scales* should incorporate terms originating from the rotation of earth. Introducing Eq. (2.7.17) into Eq. (2.7.5) yields

$$\frac{D\vec{V}}{Dt} = -\frac{1}{\rho}\nabla(p + \rho g Z) + v\nabla^2 \vec{V}^2 + \Omega^2 \vec{R} - 2\vec{\Omega} \times \vec{V} \qquad (2.7.20)$$

Normally, the centrifugal acceleration term is considered as a minor adjustment to Newtonian gravity, with the sum of these two terms referred to as *effective gravitational acceleration*, $\vec{g}_{\text{eff}}$,

$$\vec{g}_{\text{eff}} = \nabla(-gZ) + \Omega^2 \vec{R} \qquad (2.7.21)$$

The relationships between the vectors $\vec{\Omega}$, $\vec{R}$, $\vec{g}$, $\Omega^2\vec{R}$, and $\vec{g}_{\text{eff}}$ in the northern hemisphere are shown in Fig. 2.14.

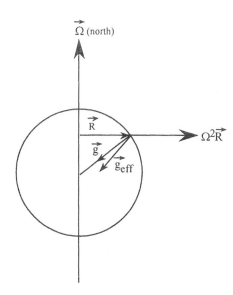

Figure 2.14   Relationships between the vectors Ω, R, g, Ω^2R, and g_{eff}.

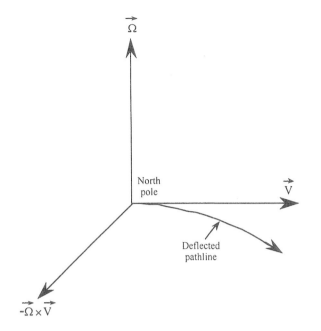

Figure 2.15 Relationships between the vectors Ω, V, and $-\Omega \times V$.

In Fig. 2.15 we show the relationships between the vectors $\vec{\Omega}$, $\vec{V}$, and $-\vec{\Omega} \times \vec{V}$. This figure indicates that Coriolis force induces a deflection of pathlines of the fluid particles to the right of their direction in the Northern Hemisphere.

The equation of motion represented by Eq. (2.7.20) is applicable in cases of *geostrophic flows*, in which the effect of the centrifugal acceleration and Coriolis force are significant. For small-scale flows, in small hydraulic systems, such effects are usually negligible. It is usually possible to determine the relative importance of different terms in the equations of motion by scaling analysis, as demonstrated in Sec. 2.9.

2.8 CONSERVATION OF ENERGY

Consider the material volume shown in Fig. 2.16. In general, this material volume may be subject to movement and deformation. The net heat added to the material volume during a short time period dt is dQ. During that time interval, the material volume exerts work dW on its surroundings. According to the first law of thermodynamics,

$$dE = dQ - dW \tag{2.8.1}$$

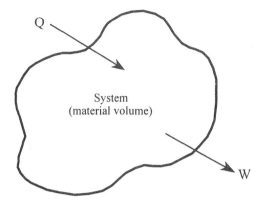

Figure 2.16 Heat Q added to a material volume and work W done by this volume.

where E is the total energy stored within the material volume. This variable incorporates the kinetic, potential, and internal energy [see Eq. (2.8.4) below]. Note that the normal convention is used to express work as a positive quantity when the material volume does work on its surroundings.

The variables Q and W are not point functions, whereas the variable E is a point function distributed within the material volume. Therefore the relationship between the rates of change of the variables given in Eq. (2.8.1) is represented by

$$\frac{DE}{Dt} = \frac{dQ}{dt} - \frac{dW}{dt} \tag{2.8.2}$$

By applying the Reynolds transport theorem, written for energy, we obtain

$$\frac{DE}{Dt} = \frac{\partial}{\partial t} \int_U \rho e \, dU + \int_S \rho e (\vec{V} \cdot \vec{n}) \, dS \tag{2.8.3}$$

where e is the stored energy per unit mass, given by

$$e = \frac{V^2}{2} + gz + u \tag{2.8.4}$$

The first term on the RHS of this equation represents kinetic energy, the second term represents potential energy, and the third term represents internal energy.

The work W done by the control volume on its surroundings incorporates flow work W_f, which is associated with stresses acting at the surface of the control volume, and shaft work, which is transferred from the control volume, for instance by turbomachines. The rate of change of the flow work can be represented by

$$\frac{dW_f}{dt} = -\int_S \tilde{S} \cdot \vec{n} \cdot \vec{V}\, dS = \int_S p\vec{V} \cdot \vec{n}\, dS - \int_S \tilde{\tau} \cdot \vec{n} \cdot \vec{V}\, dS \qquad (2.8.5)$$

where $\tilde{S}$ is the stress tensor, p is the pressure, and $\tilde{\tau}$ is the deviator stress tensor. It should be noted that the product $\tilde{\tau} \cdot \vec{n}$ represents stresses normal to the control volume surface. The velocity vector of viscous flow vanishes at solid surfaces, and has no component perpendicular to a solid surface. Therefore, the last term of Eq. (2.8.5) almost vanishes. The only contribution of this term is due to diagonal components of the deviator stress tensor at fluid surfaces subject to flow. In the following development, the last term of Eq. (2.8.5) is neglected.

Introducing Eqs. (2.8.3)–(2.8.5) into Eq. (2.8.2), we obtain

$$\frac{dQ}{dt} - \frac{dW_s}{dt} - \int_S p\vec{V} \cdot \vec{n}\, dS$$

$$= \frac{\partial}{\partial t} \int_U \rho e\, dU + \int_S \left(\frac{V^2}{2} + gz + u\right)(\rho\vec{V} \cdot \vec{n}\, dS) \qquad (2.8.6)$$

Using the divergence theorem to rewrite the last term on the LHS of Eq. (2.8.6), an integral expression for conservation of energy is obtained as

$$\frac{dQ}{dt} - \frac{dW_s}{dt} = \frac{\partial}{\partial t} \int_U \rho e\, dU + \int_S \left(\frac{V^2}{2} + gz + u + \frac{p}{\rho}\right)(\rho\vec{V} \cdot \vec{n}\, dS) \qquad (2.8.7)$$

Application of this equation is illustrated by considering Fig. 2.17, which shows a control volume with two openings. The fluid enters the control volume through one of the openings, of cross-sectional area A_1, with velocity V_1, pressure p_1, and temperature T_1. The fluid flows out of the control volume through the second opening, of cross-sectional area A_2, with velocity V_2, pressure p_2, and temperature T_2.

Referring to this control volume, under steady state conditions Eq. (2.8.7) yields

$$\frac{dQ}{dt} - \frac{dW_s}{dt} = -\left[\frac{V_1^2}{2} + g(z_c)_1 + h_1\right]\rho_1 V_1 A_1$$

$$+ \left[\frac{V_2^2}{2} + g(z_c)_2 + h_2\right]\rho_2 V_2 A_2 \qquad (2.8.8)$$

where z_c is the elevation of the center of gravity of the cross-sectional area, and h is the specific enthalpy, which is defined by

$$h = u + \frac{p}{\rho} = C_pT = C_vT + \frac{p}{\rho} \qquad (2.8.9)$$

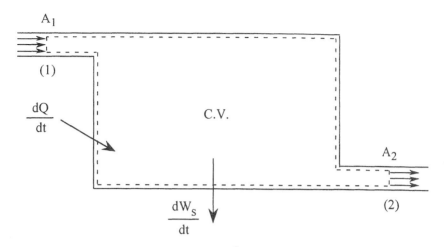

Figure 2.17   Energy conservation in a control volume (C.V.) with a single entrance and a single exit.

where C_p and C_v are the specific heats for constant pressure and constant volume, respectively.

Due to conservation of mass, $\rho_1 V_1 A_1 = \rho_2 V_2 A_2 = dm/dt$, where dm/dt is the mass flow rate which enters and leaves the control volume of Fig. 2.17. Dividing Eq. (2.8.8) by the mass flow rate and rearranging terms,

$$\left[\frac{V_1^2}{2} + g(z_c)_1 + h_1\right] + \frac{dQ/dt}{dm/dt} = \left[\frac{V_2^2}{2} + g(z_c)_2 + h_2\right] + \frac{dW_s/dt}{dm/dt} \quad (2.8.10)$$

The second term on the LHS of this equation represents the ratio between the heat flux into the control volume and the mass flow rate through the control volume. It also can be represented by dQ/dm, namely the net heat added to the control volume per unit mass of flow. The last term of Eq. (2.8.10) can be represented by dW_s/dm, namely the net work done by the control volume per unit mass of flow through the control volume. In the case of incompressible fluid, if the control volume is insulated and does not perform work on its surrounding, then Eq. (2.8.10) indicates

$$\left[\frac{V_1^2}{2} + g(z_c)_1 + \frac{p_1}{\rho}\right] - \left[\frac{V_2^2}{2} + g(z_c)_2 + \frac{p_2}{\rho}\right] = C(T_2 - T_1) \quad (2.8.11)$$

where C is the specific heat of the incompressible fluid. For both Eq. (2.8.10) and Eq. (2.8.11), terms within the square brackets represent the total head in the entrance and exit cross sections, respectively.

Equation (2.8.11) indicates that the difference in total head between cross section 1 and cross section 2, in an insulated control volume, is represented by a raise in temperature multiplied by the specific heat of the fluid. On the other hand, if the control volume is kept at constant temperature, namely *isothermal* conditions, then Eq. (2.8.10) yields

$$\left[\frac{V_1^2}{2} + g(z_c)_1 + \frac{p_1}{\rho} \right] - \left[\frac{V_2^2}{2} + g(z_c)_2 + \frac{p_2}{\rho} \right] = -\frac{dQ}{dm} \qquad (2.8.12)$$

This expression shows that for an isothermal control volume of incompressible fluid, the head difference between the entrance and exit represents the net heat per unit mass of flow that is transferred from the control volume into its surrounding. The heat transferred from the control volume into the surrounding is created in the control volume due to friction (viscous) forces.

Equations (2.8.11) and (2.8.12) indicate that Bernoulli's equation is approximately satisfied if the control volume does not perform any work on its surrounding and if heat transfer between the control volume and the surroundings is negligible. These equations also show that the conservation of energy with some approximation leads to Bernoulli's equation. Section 2.9.3 extends this discussion with the basic issues of thermal energy sources and transport in the environment.

2.9   SCALING ANALYSES FOR GOVERNING EQUATIONS

As described in Sec. 1.4, it is possible to apply dimensional reasoning to the general governing equations in order to simplify them for most ordinary applications. This process requires that characteristic values for various quantities must be defined *(characteristic scales)* and that the analysis be based on developing order-of-magnitude estimates for different terms in the equation. For now, we define the following characteristic scales for a fluid flow problem:

L = length (for some problems both vertical and horizontal length
  scales are needed)

U = velocity

Δp_0 = pressure difference

T = time

ρ_0 = density

$\Delta \rho_0$ = density difference

$\Delta \theta_0$ = temperature difference

ΔC_0 = dissolved solids concentration difference

Ω_0 = rotation rate

These scales will be used in the following discussion to estimate the typical order of magnitude for various terms in each of the basic equations discussed in the preceding sections of this chapter. To some extent, the material is parallel to the previous discussions, though the emphasis here is on relative orders of magnitude of different terms in the equations. First, we consider the mass conservation, or continuity equation.

2.9.1 Mass Conservation

The general statement of continuity, or mass conservation, is given by Eq. (2.5.6),

$$\frac{\partial \rho}{\partial t} + \vec{V} \cdot \vec{\nabla}\rho + \rho\vec{\nabla} \cdot \vec{V} = 0$$

or, dividing by ρ,

$$\frac{1}{\rho}\frac{\partial \rho}{\partial t} + \frac{1}{\rho}\vec{V} \cdot \vec{\nabla}\rho + \vec{\nabla} \cdot \vec{V} = 0 \qquad (2.9.1)$$

The scaling quantities defined above are then substituted to estimate the relative magnitudes for each of the terms and, to provide a simpler means of comparison, we divide all the terms in Eq. (2.9.1) by the divergence term, so that the first and second terms will be compared with 1. The respective relative magnitudes for each of the terms are then

$$\left[\frac{1}{T}\frac{\Delta\rho_0}{\rho_0}\right] + \left[\frac{U}{L}\frac{\Delta\rho_0}{\rho_0}\right] + \left[\frac{U}{L}\right] \approx 0$$

$$\Rightarrow \left[\frac{L}{UT}\frac{\Delta\rho_0}{\rho_0}\right] + \left[\frac{\Delta\rho_0}{\rho_0}\right] + [1] \approx 0 \qquad (2.9.2)$$

The procedure is then to compare the probable magnitudes of the first two terms in brackets with [1]. Except in certain cases, where compressible effects become important, the controlling factor is the possible relative change in density that may exist in a flow. Thus it is necessary to estimate the expected changes in density resulting from changes in environmental conditions.

In general, the density of natural water depends on its temperature, salinity and, to a much lesser extent, pressure. Other dissolved solids may affect water density, but the largest variations are due to salt. The rate of change of density with temperature is given by the thermal expansion coefficient,

$$\alpha = -\frac{1}{\rho}\frac{\partial \rho}{\partial \theta} \qquad (2.9.3)$$

where the negative sign indicates that density decreases with increasing temperature. (It should be noted that this is true only when temperature is above the temperature of maximum density, which for pure water is 4°C, so there is the potential that α changes sign for certain problems.) In terms of the scaling quantities defined above, the magnitude of the relative change in density is

$$\left[\frac{\Delta \rho_0}{\rho_0}\right] \approx [\alpha \Delta \theta_0] \tag{2.9.4}$$

In water, α is generally a function of temperature (water density is a parabolic function of temperature, at least over a range of normal environmental temperatures), with magnitude approximately $10^{-4}\,°C^{-1}$. A typical large temperature variation might be of order 10°C so, using Eq. (2.9.4), the expected magnitude of relative density variations is of order 0.001 (0.1%), which is insignificant compared with 1. Even temperature changes as high as 30–50°C would produce only a relatively negligible change in density for water.

As with temperature, a salinity expansion coefficient can be defined by

$$\beta = \frac{1}{\rho}\frac{\partial \rho}{\partial C} \tag{2.9.5}$$

and

$$\left[\frac{\Delta \rho_0}{\rho_0}\right] \approx [\beta \Delta C_0] \tag{2.9.6}$$

where C indicates the concentration of dissolved solids, primarily salts. Relatively sophisticated expressions have been developed to calculate density in the ocean as a function of temperature and salinity, and a typical value for β is about 8×10^{-4} ppt^{-1}. Density is approximately linearly related to salinity except when concentrations start to approach saturation, but that is not a condition of major interest for most environmental applications. Typical ocean salinity is approximately 30 ppt (parts per thousand) ($C = 0.03$), so the relative density variation is estimated according to Eq. (2.9.6) as 0.024, or 2.4%. Hypersaline lakes exist in some parts of the world, where C may be as high as 200 or 250 ppt. This would result in $(\Delta \rho_0/\rho_0)$ being of order 20%, but for most natural conditions this result is much less than 1 and may be ignored.

The possible effect of pressure is somewhat more complicated. First, we note that the definition of *sonic velocity*,

$$c_0 = \sqrt{\frac{\partial p}{\partial \rho}} \tag{2.9.7}$$

can be rearranged to obtain

$$\left[\frac{\Delta p_0}{\Delta \rho_0 c_0^2}\right] \approx 1 \Rightarrow \left[\frac{\Delta \rho_0}{\rho_0}\right] \approx \left[\frac{\Delta p_0}{\rho_0 c_0^2}\right] \qquad (2.9.8)$$

The value for c_0 is approximately 1,500 m/s in water and with $\rho_0 = 1,000$ kg/m^3, a pressure difference of order 2.25×10^6 kPa is needed before $(\delta \rho_0 / \rho_0)$ becomes of order 1. This is equivalent to the pressure at a depth of 225 km under water, which is clearly unreasonable. This result is, however, consistent with the assumption of incompressible flow that is normally applied for water. Further estimates for δp_0 or $(\delta \rho_0 / \rho_0)$ can be obtained under special conditions by looking at possible balances between terms in scaling analyses of the momentum equation. Results from such an exercise show that pressure effects can be neglected for normal environmental conditions in water. In fact, the only circumstances under which this term becomes important are with high-speed flows, when U approaches c_0, with very high frequency oscillatory flow, or with large-scale atmospheric motions or temperature changes.

Thus it may be concluded that $(\delta \rho_0 / \rho_0)$ is small for normal environmental conditions. Also, the factor (LU/T) appears in Eq. (2.9.2), but this ratio is usually of order 1, and when it is multiplied by $(\delta \rho_0 / \rho_0)$, it becomes very small and may be neglected. Since both the first two terms in Eq. (2.9.2) are negligibly small, and the right-hand side is zero, the only way to balance the equation is to have the third term also equal 0, i.e.,

$$\vec{\nabla} \cdot \vec{V} = 0 \qquad (2.9.9)$$

which is the continuity equation for an incompressible fluid, as defined previously in Eq. (2.5.7). Equivalently, referring back to Eq. (2.9.1), we may conclude that $D\rho/Dt = 0$, i.e., the density "following a fluid particle" remains constant. This is consistent with the conclusion found in Sec. 2.5.4.

2.9.2   Momentum Conservation

In vector notation, the general momentum equation is (refer to Sec. 2.7)

$$\frac{D\vec{V}}{Dt} + 2\vec{\Omega} \times \frac{D\vec{r}}{Dt} + \frac{D\vec{\Omega}}{Dt} \times \vec{r} + \vec{\Omega} \times (\vec{\Omega} \times \vec{r})$$

$$= \vec{g} - \frac{1}{\rho}\vec{\nabla}p + \frac{\mu}{\rho}\left[\nabla^2\vec{V} + \frac{1}{3}\vec{\nabla} \cdot (\vec{\nabla} \cdot \vec{V})\right] \qquad (2.9.10)$$

In general, this equation would have a term added to the LHS, $D^2\vec{R}/Dt^2$, to account for translational acceleration of the coordinate system, but for problems of practical interest this term can be neglected. The time derivative term for position also can be replaced by $D\vec{r}/Dt = \vec{V}$, and incompressible fluid will be assumed, as shown above. With these assumptions, Eq. (2.9.10) reduces to

$$\frac{D\vec{V}}{Dt} + 2\vec{\Omega} \times \vec{V} + \frac{D\vec{\Omega}}{Dt} \times \vec{r} + \vec{\Omega} \times (\vec{\Omega} \times \vec{r}) = \vec{g} - \frac{1}{\rho}\vec{\nabla}p + v\nabla^2\vec{V}$$
(2.9.11)

For problems in environmental fluid mechanics, the frame of reference is the earth's surface, so that $\vec{\Omega}$ represents the rotation of the earth. The earth rotates at a nearly constant rate, so the time derivative term for $\vec{\Omega}$ vanishes. The resulting equation is then similar to Eq. (2.7.20). We now consider the remaining terms.

Figure 2.18 shows a cross section of the earth along a north–south axis, along with the centripetal acceleration vector. The total magnitude of this term is $(\Omega^2 R \cos\theta)$, where θ is the latitude. The components, normal (pointing towards the earth's center) and tangential to the earth's surface, are $(\Omega^2 R \cos^2\theta)$ and $(\Omega^2 R \cos\theta \sin\theta)$, respectively. Similarly, Fig. 2.19 illustrates the components of the Coriolis term, $\vec{\Omega} \times \vec{V}$. The normal and

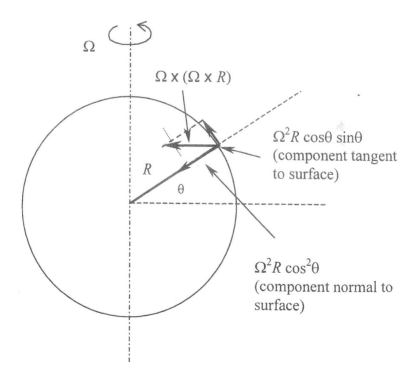

Figure 2.18 Cross section of earth, showing centripetal acceleration term.

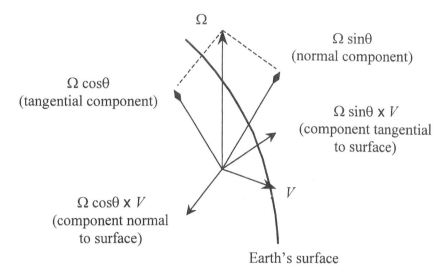

Figure 2.19 Components of Coriolis acceleration, for velocity tangent to surface (note: Coriolis term is $\vec{\Omega} \times \vec{V}$).

tangential components of this term are $((\vec{\Omega} \cos \theta) \times \vec{V})$ and $((\vec{\Omega} \sin \theta) \times \vec{V})$, respectively.

We first compare the normal components with gravity, using values for Ω and R appropriate for rotation of the earth: $\Omega = 2\pi(\text{rad/day}) \cong 7 \times 10^{-5} \ (\text{s}^{-1})$ and $R \cong 6 \times 10^6$ m. The magnitude of the centripetal term is then $(\Omega^2 R) \cong 0.03$ m/s^2, which is much less than g ($\cong 10$ m/s^2). Also, in order for the normal Coriolis term to be comparable to g, the velocity magnitude would have to be of order $O(10^5$ m/s), which is obviously too large for practical consideration.

For the tangential components, first note that the centripetal term is a constant, while the Coriolis term depends on the magnitude of $\vec{V}$. The centripetal term is usually considered as a minor adjustment to gravity, as previously noted (see Eq. 2.7.21) and as shown in Fig. 2.20 (see also Fig. 2.14). For now, we retain the Coriolis term and show in the following discussion under what circumstances it needs to be included. A simplified version of Eq. (2.9.11) is thus

$$\frac{\partial \vec{V}}{\partial t} + \vec{V} \cdot \nabla \vec{V} + 2\vec{\Omega} \times \vec{V} = \vec{g} - \frac{1}{\rho}\nabla p + v\nabla^2 \vec{V} \qquad (2.9.12)$$

Note that this is essentially the same result as Eq. (2.7.20), with Eq. (2.7.21) substituted for $\vec{g}_{\text{eff}}$ (note also that $\vec{g}_{\text{eff}} \cong \vec{g}$).

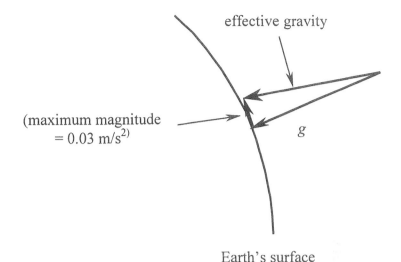

Figure 2.20 Relative importance of the effect of centripetal acceleration as an adjustment to gravity.

This analysis can be extended by considering the pressure term as consisting of hydrostatic and dynamic contributions. Referring to Fig. 2.21, hydrostatic pressure is defined by

$$p_z = p_r - \int_{z_r}^{z} \rho g \, dz \tag{2.9.13}$$

where p_r is a reference value.

The total pressure is the sum of p_z and $p_d =$ dynamic pressure, so the pressure term in Eq. (2.9.12) can be written as

$$\frac{1}{\rho}\vec{\nabla}p = \frac{1}{\rho}\vec{\nabla}p_r - \frac{g}{\rho}\vec{\nabla}\int_{z_r}^{z}\rho \, dz + \frac{1}{\rho}\vec{\nabla}p_d$$

$$= \frac{1}{\rho}\vec{\nabla}p_r - \frac{g}{\rho}\int_{z_r}^{z}\vec{\nabla}\rho \, dz - g\vec{\nabla}z + g\frac{\rho_r}{\rho}\vec{\nabla}z_r + \frac{1}{\rho}\vec{\nabla}p_d \tag{2.9.14}$$

where this last result is obtained using the fact that $\rho = \rho_r$ at $z = z_r$. Then, substituting Eq. (2.9.14) into Eq. (2.9.12), we obtain

$$\frac{\partial \vec{V}}{\partial t} + \vec{V}\cdot\nabla\vec{V} + 2\vec{\Omega}\times\vec{V}$$

$$= -\frac{1}{\rho}\vec{\nabla}p_r + \frac{g}{\rho}\int_{z_r}^{z}\vec{\nabla}\rho \, dh - \frac{\rho_r}{\rho}g\vec{\nabla}z_r - \frac{1}{\rho}\vec{\nabla}p_d + v\nabla^2\vec{V} \tag{2.9.15}$$

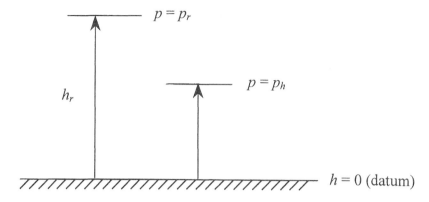

Figure 2.21 Illustration of hydrostatic pressure variations.

If we now let $\rho = \rho_0 + \delta\rho$, where ρ_0 is the constant base, or characteristic value previously defined for density, then

$$\frac{1}{\rho} = \frac{1}{\rho_0 + \Delta\rho} = \frac{1}{\rho_0}\left(\frac{1}{1 + \dfrac{\Delta\rho}{\rho_0}}\right) \qquad \text{and} \qquad \frac{\Delta\rho}{\rho_0} \ll 1$$

(from previous scaling of mass conservation equation), so

$$\frac{1}{\rho} \cong \frac{1}{\rho_0} \qquad \text{and} \qquad \rho_r \cong \rho_0 \tag{2.9.16}$$

This last result is a statement of the *Boussinesq approximation*, which says
density variations are negligible except in the buoyancy terms, as will be
shown below. Eq. (2.9.15) is thus written as

$$\frac{\partial \vec{V}}{\partial t} + \vec{V} \cdot \nabla\vec{V} + 2\vec{\Omega} \times \vec{V}$$

$$= -\frac{1}{\rho_0}\vec{\nabla}(p_r + p_d) + \frac{g}{\rho_0}\int_{Z_r}^{Z}\vec{\nabla}\rho\,dz - g\vec{\nabla}z_r + \upsilon\nabla^2\vec{V} \tag{2.9.17}$$

The first term on the RHS of Eq. (2.9.17) is the net force due to pressure
gradients, the second term is the effect of density variations (important for
stratified fluids), and the third term is the effect of reference surface gradients
(such as waves).

Using the same characteristic scaling variables as in Sec. 2.9.1, the
magnitudes of the terms in Eq. (2.9.17) may be compared under different
scenarios. Dividing by the convective term ($\vec{V} \cdot \vec{\nabla}\vec{V}$), which has characteristic

magnitude (U^2/L), results in relative magnitudes as

$$\left[\frac{L}{UT}\right] + [1] + \left[\frac{\Omega_0 L}{U}\right] \approx \left[\frac{\Delta p_0}{\rho_0 U^2}\right] + \left[\frac{g(\Delta \rho/\rho_0)L}{U^2}\right] - \left[\frac{gL}{U^2}\right] + \left[\frac{v}{UL}\right]$$

(2.9.18)

where

$$\left[\frac{U}{\Omega_0 L}\right] = \text{Rossby number, Ro}$$

$$\left[\frac{\Delta p_0}{\rho_0 U^2}\right] = \text{Euler number, Eu}$$

$$\left[\frac{U}{\sqrt{g(\Delta \rho_0/\rho_0)L}}\right] = \text{densimetric Froude number, Fr}_d$$

$$\left[\frac{U}{\sqrt{gL}}\right] = \text{Froude number, Fr}$$

$$\left[\frac{UL}{v}\right] = \text{Reynolds number, Re}$$

Thus, for example, if Ro is large, Coriolis effects should be negligible in the momentum equation. Similarly, pressure effects are small if Eu is small, density effects are negligible if Fr_d is very large, changes in surface elevation may be neglected if Fr is very large, and viscous effects are small when Re is large.

The time-dependent term $[L/UT]$ is the *Strouhal number*, and it should be clear that a problem may be treated as being steady for large times, $T \to \infty$, when this ratio is small. The values of Ro (order of magnitude) for several representative situations are listed in Table 2.1 using $\Omega_0 \approx 10^{-4}\text{s}^{-1}$, which is valid for mid-latitudes. It is clear from these examples that the Coriolis effect is expected to be important only in systems with larger L (estuaries, large lakes, and ocean currents), depending also on U and Ω_0.

Table 2.1 Estimates of Ro for Different Environmental Systems

	L(m)	U(m/s)	Ro
Stream	1–10	0.1–1	10^2–10^4
Pond	10–100	0.1	10–10^2
River	100	0.1–1	10–10^2
Estuary	10^3–10^4	1	1–10
Large lake	10^3–10^5	0.1	10^{-2}–1
Ocean current	10^5–10^6	0.01–0.1	10^{-4}–10^{-2}

The relative importance of density gradients can be estimated by assuming $Fr_d \approx 1$. This is easily shown to be equivalent to assuming that the convective term is balanced by the buoyancy term in Eq. (2.9.18), i.e.,

$$\vec{V} \cdot \vec{\nabla}\vec{V} \approx \frac{g}{\rho_0} \int_{z_r}^{z} \vec{\nabla}\rho \, dz \Rightarrow U \approx \left[g\frac{\Delta\rho}{\rho_0}L \right]^{1/2} \qquad (2.9.19)$$

Then, for a typical value of $\Delta\rho/\rho_0 \cong 10^{-3}$ (corresponding to a temperature difference of about 10°C), and $g \cong 10$ m/s^2, $L = (1-100$ m), gives $U \cong (0.1-1$ m/s). Thus, at least in the buoyancy term, even a small density difference can generate an appreciable velocity. The Boussinesq approach of neglecting density variations does not apply to the buoyancy term, unless very small characteristic lengths (L) are involved.

As a special case of the general result shown in Eq. (2.9.17), consider a situation of steady, constant density flow, with $\Omega = \nabla z_r = 0$. Then

$$\vec{V} \cdot \vec{\nabla}\vec{V} = -\vec{\nabla}\left(\frac{p}{\rho_0} + gz \right) + v\nabla^2\vec{V} \qquad (2.9.20)$$

where (p/ρ_0) represents dynamic pressure and (gz) is the hydrostatic pressure. This equation is then multiplied by (i.e., take dot product with) $\vec{V}$, to obtain a mechanical energy equation,

$$\vec{V} \cdot \vec{\nabla}\left(\frac{1}{2}V^2 \right) + \frac{1}{\rho_0}(\vec{V} \cdot \vec{\nabla}p) + \vec{V} \cdot \vec{\nabla}(gz) = v(\vec{\nabla}V)^2 = -\varepsilon \qquad (2.9.20)$$

where ε is the viscous dissipation rate for mechanical energy. If we now define total head as

$$H = \frac{V^2}{2g} + \frac{p}{\rho_0 g} + z \qquad (2.9.21)$$

(refer to Eq. 2.8.11), then Eq. (2.9.20) becomes

$$\vec{V} \cdot \vec{\nabla}(gH) = -\varepsilon \qquad (2.9.22)$$

If $\varepsilon \cong 0$, then this is the *Bernoulli equation*, also derived in Sec. 2.8.

Note that for steady flow, the left-hand-side of Eq. (2.9.22) is the same as the material derivative, (gDH/Dt), and if inviscid conditions are assumed ($\varepsilon = 0$), then

$$\frac{DH}{Dt} = 0 \Rightarrow \frac{V^2}{2g} + \frac{p}{\rho_0 g} + z = K \text{ (a constant)} \qquad (2.9.23)$$

which is the usual form of the Bernoulli equation used in many introductory textbooks. In general, this result holds along a streamline (i.e., following a fluid particle). If, however, the flow is also irrotational ($\vec{\omega} = \vec{0}$), the vector identity

$$\nabla^2 \vec{V} = \vec{\nabla}(\vec{\nabla} \cdot \vec{V}) - \vec{\nabla} \times (\vec{\nabla} \times \vec{V}) \tag{2.9.24}$$

can be used to show that the Bernoulli result (2.9.23) is valid everywhere in the flow field. This is because the RHS of Eq. (2.9.24) is 0, due to continuity for the first term and irrotationality for the second.

It is interesting to note that $\nabla^2 \vec{V} = 0$ for an irrotational flow field, independent of the value of Re. However, the value of Re controls the rate at which vorticity grows outward from solid boundaries, which may be important for boundary layer analysis (see Chap. 6).

Another special case of interest is when the velocity vanishes, so Eq. (2.9.17) becomes

$$\vec{0} = -\frac{1}{\rho_0}\vec{\nabla}p + g\vec{\nabla}z \Rightarrow 0 = -\frac{1}{\rho_0}\frac{\partial p}{\partial z} - g \tag{2.9.25}$$

which gives the hydrostatic pressure field (assuming boundary conditions are known).

One additional case of interest is that of *geostrophic flow*. For this case, there is a balance in the momentum equation between the Coriolis and pressure terms, so

$$2\vec{\Omega} \times \vec{V} \approx -\frac{1}{\rho_0}\vec{\nabla}_p \tag{2.9.26}$$

This balance has many applications in meteorology and in the oceans. When this balance occurs, large-scale pressure differences (gradients), for example, can be related to corresponding characteristic velocities by

$$\frac{\Delta p_0}{L} \approx \rho_0 \Omega_0 U \tag{2.9.27}$$

These flows are discussed further in Chap. 9.

2.9.3 Thermal Energy Equation

The thermal energy equation is derived from the general conservation of energy equation and may be written as (see Sec. 12.3.1 for further discussion)

$$\frac{D\theta}{Dt} = \frac{\partial \theta}{\partial t} + \vec{V} \cdot \vec{\nabla}\theta = \frac{\kappa}{\rho c}\nabla^2\theta - \frac{1}{\rho c}\vec{\nabla} \cdot \vec{\varphi}_r + \frac{1}{\rho c}(\rho_0 \varepsilon) - \frac{\alpha c_0^2 \theta}{c}(\vec{\nabla} \cdot \vec{V}) \tag{2.9.28}$$

where θ is temperature, c is *specific heat*, κ is *thermal conductivity*, φ_r is radiation heat flux, ε is the kinematic viscous dissipation rate of mechanical energy, α is the *thermal expansion coefficient*, and c_0 is the *sonic velocity*. The terms on the RHS of this equation relate to conduction (diffusion), radiative heat transfer, viscous heating, and compression or expansion heating, respectively. Following a similar procedure as in the preceding sections, we introduce characteristic values for this equation to derive

$$\left[\frac{\delta\theta_0}{T}\right] + \left[\frac{U\Delta\theta_0}{L}\right] \approx \left[\frac{\kappa\Delta\theta_0}{\rho cL^2}\right] - \left[\frac{\varphi_0}{\rho cL}\right] + \left[\frac{\varepsilon}{c}\right] - \left[\frac{\alpha c_0^2\theta_0}{c}\frac{U}{L}\frac{\Delta\rho_0}{\rho_0}\right] \quad (2.9.29)$$

where the compression/expansion term is scaled as in Sec. 2.9.1, to substitute $(\delta\rho_0/\rho_0)$ for $(\vec{\nabla}\cdot\vec{V})$. To nondimensionalize the equation, each term is divided by the advection term, so

$$\left[\frac{L}{UT}\right] + [1] \approx \left[\frac{\kappa/\rho c}{UL}\right] - \left[\frac{\varphi_0}{\rho_0 cU\Delta\theta_0}\right] + \left[\frac{U^2}{c\Delta\theta_0}\right] + \left[\frac{\alpha^2 c_0^2\theta_0}{c}\frac{1}{\alpha\Delta\theta_0}\frac{\Delta\rho_0}{\rho_0}\right]$$
$$(2.9.30)$$

where $\varepsilon \approx U^3/L$ has been substituted for the dissipation rate (see Chap. 5).

Typical magnitudes for the terms on the right-hand side are estimated as follows:

Heat conduction: First, note that a *thermal diffusivity* may be defined as

$$k_t = \frac{\kappa}{\rho c} \quad (2.9.31)$$

and the conduction term may be rewritten as

$$\left[\frac{k_t}{UL}\right] = \left[\frac{k_t}{v}\right]\left[\frac{v}{UL}\right] = \left(\frac{1}{\text{Pr}}\right)\left(\frac{1}{\text{Re}}\right) \quad (2.9.32)$$

where Pr is the Prandtl number and signifies the ratio of heat transport to momentum transport. Re is the Reynolds number as defined previously. In water, Pr has a value of about 7 (a fixed value), so conductive heat transfer depends on Re.

Radiative heating: The prime heating source by radiation is the sun, and a typical value for φ_0 in temperate latitudes is about 200–250 W/m². The amount of heating that takes place for any given radiative input depends on the length of time over which the heating takes place and, of course, the depth (or volume) of the water body under consideration.

Viscous heating: In water the specific heat is $c \cong 1\text{J/g}°\text{C}$. If $(U^2/c\Delta\theta_0)$ is to be of order 1 (i.e., the magnitude of the viscous heating term would be sufficient to require it to be included in the temperature equation), then estimates for $\Delta\theta_0$ may be obtained based on U. At the upper range of environmental flow conditions, water velocities may be of order $1-10$ m/s. The characteristic temperature change associated with this range of values is 2.5×10^{-4} to 2.5×10^{-2}°C, which may be ignored under most circumstances.

Compression heating: Because of its dependence on fluid compressibility, this term is generally important only for atmospheric studies, or possibly in the deep oceans. Otherwise it can be neglected.

Thus the final usual form of the temperature equation is

$$\frac{\partial\theta}{\partial t} + \vec{V}\cdot\nabla\theta = k_t\nabla^2\theta - \frac{1}{\rho c}\nabla\cdot\vec{\varphi}_r \qquad (2.9.33)$$

and this is examined further in Chap. 12.

PROBLEMS

Solved Problems

Problem 2.1   A two-dimensional flow field is given by the following velocity components:

$$u = V\cos(\omega t) \qquad v = V\sin(\omega t)$$

where u and v represent the velocity in the x and y directions, respectively; V and ω are constant coefficients. Provide expressions for the streamlines and pathlines.

Solution

As velocity components are time dependent, the flow is unsteady. The differential equation for the streamlines is

$$\frac{dx}{V\cos(\omega t)} = \frac{dy}{V\sin(\omega t)}$$

By rearranging this expression to solve for dy, we obtain

$$dy = \tan(\omega t)dx$$

Direct integration of this expression then gives the equation of the streamlines as

$$y = \tan(\omega t)x + C$$

where C is an integration constant. This expression indicates that streamlines are straight lines whose slope is time dependent. The differential equations of the streamlines are

$$\frac{dx}{dt} = V\cos(\omega t) \qquad \frac{dy}{dt} = V\sin(\omega t)$$

Direct integration of these expressions, and considering that at time $t = 0$ the fluid particle is located at $x = x_0$ and $y = y_0$, yields

$$x = x_0 + \frac{V}{\omega}\sin(\omega t) \qquad y = y_0 + \frac{V}{\omega} - \frac{V}{\omega}\cos(\omega t)$$

Eliminating time from these expressions, we obtain

$$(x - x_0)^2 + \left(y - y_0 - \frac{V}{\omega}\right)^2 = \left(\frac{V}{\omega}\right)^2$$

This expression indicates that the pathlines are circles with radius V/ω and that the center of each pathline is located at $x = x_0$ and $y = y_0 + V/\omega$.

Problem 2.2 A two-dimensional flow field is given by the following velocity components:

$$u = \alpha y \qquad v = \alpha x$$

where u and v represent the velocity in the x and y directions, respectively, and α is a constant. Provide expressions for the streamlines and pathlines.

Solution

As velocity components are not time dependent, the flow is steady. Therefore the shape of the streamlines does not change with time, and that shape is identical to that of the pathlines. The differential equation for the streamlines is

$$\frac{dx}{\alpha y} = \frac{dy}{\alpha x}$$

and upon rearranging,

$$y\,dy = x\,dx$$

Direct integration of this expression yields the following equation of the streamlines:

$$x^2 - y^2 = C$$

where C is an integration constant. This equation indicates that streamlines are hyperbolas whose asymptotes have a slope of $45°$. As expected, the shape of the streamlines does not change with time (since the flow is steady).

The differential equations of the pathlines are

$$\frac{dx}{dt} = \alpha y \qquad \frac{dy}{dt} = \alpha x$$

Differentiating the first expression with regard to time, we obtain

$$\frac{d^2x}{dt^2} = \alpha \frac{dy}{dt}$$

Introducing the first two expressions into the last one then gives

$$\frac{d^2x}{dt^2} - \alpha^2 x = 0$$

The solution of this differential equation is

$$x = C_1 \exp(\alpha t) + C_2 \exp(-\alpha t)$$

where $C_1 + C_2 = x_0$.
Introducing this expression into the basic equation of $dy/dt = \alpha x$ and integrating, we obtain

$$y = C_1 \exp(\alpha t) - C_2 \exp(-\alpha t)$$

where $C_1 - C_2 = y_0$.
We may eliminate time from the expressions of x and y and obtain

$$x^2 - y^2 = 4C_1 C_2$$

This expression indicates that pathlines and streamlines have identical shapes, as found previously.

Problem 2.3 A two-dimensional flow field is given by the following velocity components:

$$u = \alpha y t \qquad v = \alpha x t$$

where u and v represent the velocity in the x and y directions, respectively; t is the time and α is a constant. Provide expressions for the streamlines.

Solution

As velocity components are time dependent, the flow is unsteady. However, the velocity vector can be expressed as a product of a space vector with a time function. Therefore the shape of the streamlines does not change with time, and that shape is identical to that of the pathlines, as shown below. The differential equation for the streamlines is

$$\frac{dx}{\alpha y t} = \frac{dy}{\alpha x t}$$

Upon rearranging, this gives

$$y \, dy = x \, dx$$

Direct integration of this expression yields the equation of the streamlines as

$$x^2 - y^2 = C$$

where C is an integration constant. This equation indicates that streamlines are hyperbolas whose asymptotes have a slope of $45°$. As previously noted, the shape of the streamlines does not change with time.

Problem 2.4   For each of the following flow fields, calculate components of the rate of strain, vorticity tensor and vector, and the circulation on the sides of a small square with sides of length $2b$ centered on the origin.

(a) $u = ax$ $v = -ay$
(b) $u = ay$ $v = ax$
(c) $u = ay$ $v = -ax$

Solution

Components of the rate of strain tensor:

$$\text{(a)} \quad e_{11} = \frac{1}{2}\left(\frac{\partial u}{\partial x} + \frac{\partial u}{\partial x}\right) = a \qquad e_{12} = e_{21} = \frac{1}{2}\left(\frac{\partial u}{\partial y} + \frac{\partial v}{\partial x}\right) = 0$$

$$e_{22} = \frac{1}{2}\left(\frac{\partial v}{\partial y} + \frac{\partial v}{\partial y}\right) = -a$$

(b) $e_{11} = 0$ $e_{12} = e_{21} = a$ $e_{22} = 0$
(c) $e_{11} = 0$ $e_{12} = e_{21} = 0$ $e_{22} = 0$

Components of the vorticity tensor:

$$\text{(a)} \quad \zeta_{11} = \zeta_{22} = 0 \qquad \zeta_{12} = -\zeta_{21} = \frac{1}{2}\left(\frac{\partial u}{\partial y} - \frac{\partial v}{\partial x}\right) = 0$$

$$\text{(b)} \quad \zeta_{11} = \zeta_{22} = 0 \qquad \zeta_{12} = -\zeta_{21} = \frac{1}{2}\left(\frac{\partial u}{\partial y} - \frac{\partial v}{\partial x}\right) = 0$$

$$\text{(c)} \quad \zeta_{11} = \zeta_{22} = 0 \qquad \zeta_{12} = -\zeta_{21} = \frac{1}{2}\left(\frac{\partial u}{\partial y} - \frac{\partial v}{\partial x}\right) = a$$

Components of the vorticity vector: Only case (c) is relevant, as the flow is two-dimensional [no vorticity for cases (a) or (b)]. Thus

$$\nabla \times \vec{V} = \left(\frac{\partial v}{\partial x} - \frac{\partial u}{\partial y}\right)\vec{k} = -2a\vec{k}$$

where $\vec{k}$ is a unit vector in the z-direction.

Note that the component ζ_{21} is equal to half of the corresponding vorticity component.

Circulation values: First, note that the circulation is defined by

$$\Gamma = \oint_c \vec{V} \cdot \vec{dl}$$

where C is a closed curve and dl is a line element. As required, the closed line integral should be performed in the counterclockwise direction along the four sides of the small square as shown in Fig. 2.22.

For flow fields (a) and (b) the circulation vanishes. In case (c), we obtain

$$\Gamma = \left[\int_{-b}^{b} v\,dy\right]_{x=b} + \left[\int_{b}^{-b} u\,dx\right]_{y=b} + \left[\int_{b}^{-b} v\,dy\right]_{x=-b} + \left[\int_{-b}^{b} u\,dx\right]_{y=-b}$$

$$= \int_{-b}^{b} -ab\,dy + \int_{b}^{-b} ab\,dx + \int_{b}^{-b} ab\,dy + \int_{-b}^{b} -ab\,dy = 8ab^2$$

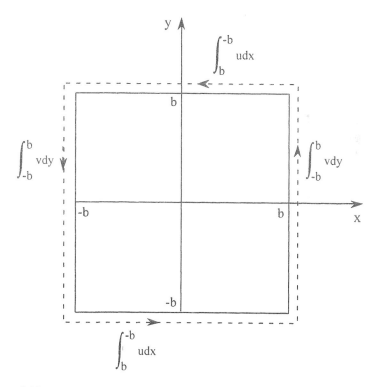

Figure 2.22 Line integration around square element, Problem 2.4.

Problem 2.5   A fluid flow is given by the following pathlines:

$$x = x_0(1 + \alpha t) \qquad y = y_0/(1 + \alpha t)$$

where α is a constant. Calculate the components of the velocity and acceleration vectors by applying the Lagrangian and Eulerian approaches.

Solution

Lagrangian components of velocity:

$$u = \frac{dx}{dt} = \alpha x_0 \qquad v = \frac{dy}{dt} = \frac{-\alpha y_0}{(1 + \alpha t)^2}$$

Eulerian components of velocity: These are obtained by the elimination of x_0 and y_0 from the Lagrangian expressions. According to the pathline equations,

$$x_0 = \frac{x}{1 + \alpha t} \qquad y_0 = y(1 + \alpha t)$$

We introduce these equations into the Lagrangian expressions of the velocity components to obtain

$$u = \frac{\alpha x}{1 + \alpha t} \qquad v = \frac{-\alpha y}{1 + \alpha t}$$

Lagrangian components of the acceleration:

$$a_x = \frac{d^2 x}{dt^2} = 0 \qquad a_y = \frac{d^2 y}{dt^2} = \frac{2\alpha^2 y_0}{(1 + \alpha t)^3}$$

Eulerian components of the acceleration: It is possible to introduce x_0 and y_0 into the Lagrangian expressions by x, y, t to obtain the Eulerian expressions of the acceleration components. Alternatively, the accelerations are obtained by direct application of the substantial derivative, as

$$a_x(x, y, t) = \frac{\partial u}{\partial t} + u\frac{\partial u}{\partial x} + v\frac{\partial u}{\partial y} = \frac{-\alpha^2 x}{(1 + \alpha t)^2} + \frac{\alpha^2 x}{(1 + \alpha t)^2} = 0$$

$$a_y(x, y, t) = \frac{\partial v}{\partial t} + u\frac{\partial v}{\partial x} + v\frac{\partial v}{\partial y} = \frac{\alpha^2 y}{(1 + \alpha t)^2} + \frac{\alpha^2 y}{(1 + \alpha t)^2} = \frac{2\alpha^2 y}{(1 + \alpha t)^2}$$

Problem 2.6   Derive the differential form of the continuity equation directly by considering a small fluid element as shown in Fig. 2.23. Density ρ and fluid velocity (u, v, w) are defined at the center of the element. Use a Taylor series expansion to express the densities and velocities on each face in terms of ρ, u, v, and w.

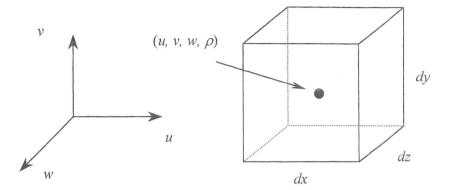

Figure 2.23 Definition sketch, Problem 2.6.

Solution

A general statement of conservation of mass for any control volume is the rate of change of mass in the volume is equal to the rate at which mass is transported into the volume across the control surface, minus the rate at which mass is transported out of the volume, plus or minus the rates at which mass is either created or destroyed in the volume. When applied to the fluid element shown in Fig. 2.23 and noting that water is neither created nor destroyed, this statement is written in mathematical terms as

$$\frac{\partial(\rho \forall)}{\partial t} = \left\{ \left[\rho u - \frac{\partial(\rho u)}{\partial x}\frac{dx}{2} \right] - \left[\rho u + \frac{\partial(\rho u)}{\partial x}\frac{dx}{2} \right] \right\} dy\,dz$$
$$+ \left\{ \left[\rho v - \frac{\partial(\rho v)}{\partial y}\frac{dy}{2} \right] - \left[\rho v + \frac{\partial(\rho v)}{\partial y}\frac{dy}{2} \right] \right\} dx\,dz$$
$$+ \left\{ \left[\rho w - \frac{\partial(\rho w)}{\partial z}\frac{dz}{2} \right] - \left[\rho w + \frac{\partial(\rho w)}{\partial z}\frac{dz}{2} \right] \right\} dx\,dy$$

where $\forall = dx\,dy\,dz$ is the element volume and higher order terms in the Taylor series expansions have been neglected, with the assumption that dx, dy, and dz are all small. Each of the terms on the right-hand side of this equation represents the net transport of fluid mass across the control surface in each of the three coordinate directions. Nothing that the volume is independent of time, then by combining terms and simplifying, we have

$$\frac{\partial \rho}{\partial t}(dx\,dy\,dz) = -\frac{\partial(\rho u)}{\partial x}(dx\,dy\,dz) - \frac{\partial(\rho v)}{\partial y}(dx\,dy\,dz) - \frac{\partial(\rho w)}{\partial z}(dx\,dy\,dz)$$

Dividing by the volume, $dx\,dy\,dz$, and bringing all terms to the left-hand side then leads to Eq. (2.5.6), which is the desired continuity equation.

Problem 2.7   Figure 2.24 shows a reservoir of volume U, which includes for time $t \leq 0$ pure water with density ρ_0. At time $t = 0$, effluent with volumetric discharge $2Q$ and density ρ_1 starts flowing into the reservoir. The reservoir volume is kept constant due to infiltration of the reservoir water into the ground with volumetric discharge Q, and evaporation of pure water (density ρ_0), also with volumetric discharge rate Q. What is the value of the reservoir fluid density as a function of time? What is the value of that density as $t \to \infty$? Assume that the fluid is kept completely mixed in the reservoir.

Solution

The fluid is incompressible, and density is subject to variation due to dissolved solids, which are assumed to not affect the volume of the water. Therefore, we may refer to Eq. (2.5.2) with regard to volumetric quantities, namely, the reservoir volume is kept constant, and volumetric discharge into the reservoir is identical to the total flow out of the reservoir. Using the integral equation of mass conservation (2.5.2), we obtain

$$\frac{d\rho}{dt}U + Q(\rho + \rho_0) - 2Q\rho_1 = 0$$

Using separation of variables, this expression yields

$$\frac{d\rho}{2\rho_1 - \rho_0 - \rho} = \frac{Q}{U}\,dt$$

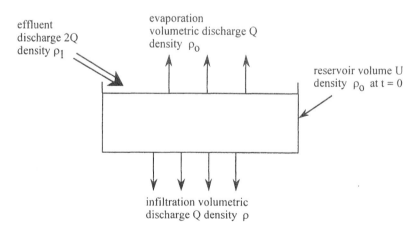

effluent
discharge $2Q$
density ρ_1

evaporation
volumetric discharge Q
density ρ_0

reservoir volume U
density ρ_0 at t = 0

infiltration volumetric
discharge Q density ρ

Figure 2.24   Definition sketch, Problem 2.7.

Direct integration of this expression, while considering that $\rho = \rho_0$ at $t = t_0$, yields

$$\frac{2\rho_1 - \rho_0 - \rho}{2\rho_1 - 2\rho_0} = \exp\left(-\frac{Q}{U}t\right)$$

For $t \to \infty$ the RHS of this expression vanishes. Therefore the asymptotic limit for the fluid density is

$$\rho = 2\rho_1 - \rho_0$$

Problem 2.8 Figure 2.25 shows a system of two stagnant plates and a plate that moves downward with velocity V. Due to the movement of the third plate, the incompressible fluid, which is located between the plates, is subject to flow. The velocity in the x-direction is distributed uniformly between the two horizontal plates. Calculate the velocity distribution in the fluid domain when the gap between the two horizontal plates is h. Find the expression for the stream function. Is the fluid domain subject to steady flow?

Solution

The velocity u in the x-direction is independent of the y-coordinate. The integral equation of continuity (2.5.3), applied to the control volume (C.V.)

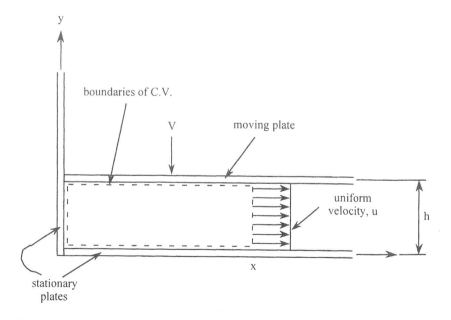

Figure 2.25 Definition sketch, Problem 2.8.

shown in the figure yields

$$- Vx + uh = 0 \Rightarrow u = \frac{V}{h}x$$

Introducing this expression into the differential equation of continuity (2.5.8), we obtain

$$\frac{\partial v}{\partial y} = -\frac{\partial u}{\partial x} = -\frac{V}{h}$$

Direct integration of this expression yields

$$v = -\frac{V}{h}y + f(x, t)$$

Considering that at $y = 0$, the velocity component v vanishes, we obtain

$$v = -\frac{V}{h}y$$

According to Eq. (2.5.9), the following relationship for the stream function is found:

$$\Psi = \int u \, dy = \frac{V}{h}xy + f(x)$$

The derivative of this expression with regard to the y coordinate is

$$\frac{\partial \Psi}{\partial x} = \frac{V}{h}y + f'(x) = -v = \frac{V}{h}y \Rightarrow f'(x) = 0 \Rightarrow f(x) = const$$

By choosing $f(x) = 0$, we obtain

$$\Psi = \frac{V}{h}xy$$

The flow is subject to unsteady state, as the value of h is time dependent.

Problem 2.9 An incompressible fluid flows past a corner making an angle of $(3\pi/4)$ as shown in Fig. 2.26. It is proposed to describe this flow by a stream function,

$$\Psi = 2r^{4/3} \sin\left(\frac{4}{3}\theta\right)$$

(a) What is the magnitude of the velocity at any point in the flow field (as a function of r)?

(b) Show that there is no flow across the solid boundaries shown.

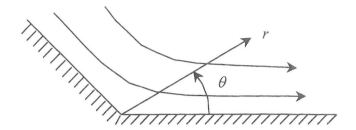

Figure 2.26 Flow of incompressible fluid past a corner, Problem 2.9.

Solution

(a) From Eq. (2.5.13), the velocity components are

$$u_r = \frac{1}{r}\frac{\partial \Psi}{\partial \theta} = \frac{8}{3}r^{1/3}\cos\left(\frac{4}{3}\theta\right)$$

$$v_\theta = -\frac{\partial \Psi}{\partial r} = -\frac{8}{3}r^{1/3}\sin\left(\frac{4}{3}\theta\right)$$

The velocity magnitude is the square root of the sum of the squares of each of the velocity components,

$$V = \frac{8}{3}r^{1/3}$$

(b) Since both boundaries represent radial arms with respect to the origin at the corner, it is sufficient for this problem to show simply that $v_\theta = 0$ when $\theta = 0$ or $\theta = 3\pi/4$. That this is the case is immediately seen when we use the expression for v_θ from part (a). It should also be noted that this result shows that the proposed stream function satisfactorily describes the flow past this corner.

Problem 2.10 Figure 2.27 shows a cylinder, with weight W_c, with a piston standing on a table. Due to a downward movement of the piston, fluid flows out of the cylinder through a nozzle located at the bottom of the cylinder. The cross-sectional area of the cylinder is A_1, the cross-sectional area of the nozzle outlet is A_2, and the fluid density is ρ. Calculate the forces F_H and F_V, which are needed to hold the cylinder, when the depth of the fluid volume is h.

Solution

A Cartesian coordinate system (x, y) is added for reference. We start with the choice of the control volume $(C.V.)$ as shown in Fig. 2.27. It should be noted that other types of control volumes could be used as well.

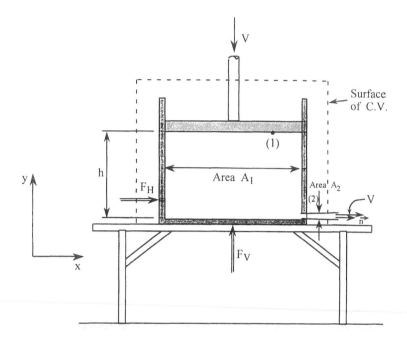

Figure 2.27   Definition sketch, Problem 2.10.

In the x-direction there is no momentum of the control volume. The force F_H is applied by the "solid hand" which is cut by the surface of the control volume. At the exit of the nozzle, the velocity vector and the normal vector have identical directions. Due to continuity, the speed of the jet flowing out of the nozzle is $V(A_1/A_2)$. At the nozzle exit the pressure is equal to atmospheric pressure. Equation (2.6.7) yields for the x-direction:

$$\rho \left[V \frac{A_1}{A_2} \right]^2 = F_H$$

In the y-direction there is a negative momentum. Its values at times t and $t + \Delta t$ are given, respectively, by:

$$(\text{Momentum})_t = -\rho h A_1 V \qquad (\text{Momentum})_{t+\Delta t} = -\rho(h - V\Delta t)A_1 V$$

The difference in momentum between times t and $t + \Delta t$, divided by Δt, provides the first RHS term of Eq. (2.6.7), namely,

$$\frac{\partial}{\partial t} \int_U \rho V_y \, dU = \lim_{\Delta t \to 0} \frac{-\rho(h - V\Delta t)A_1 V + \rho h A_1 V}{\Delta t} = \rho V^2 A_1$$

Two solid surfaces comprise a portion of the surface of the control volume. Through these surfaces two forces are applied. One of them is F_V and the other one is applied through the shaft of the piston. The force applied through the shaft of the piston can be calculated using the Bernoulli equation. We consider that the piston movement is slow, and approximately steady state conditions prevail in the fluid. Then Bernoulli's equation applied between point (1) and point (2) yields

$$\frac{V^2}{2g} + \frac{p_1}{\gamma} + h = \frac{[V(A_1/A_2)]^2}{2g} \Rightarrow p_1 = \rho\frac{V^2}{2}\left[\left(\frac{A_1}{A_2}\right)^2 - 1\right] - \rho gh$$

Considering equilibrium of the piston, we obtain

$$p_1 A_1 = W_p + F_p; \Rightarrow F_p = p_1 A_1 - W_p$$
$$= \rho\frac{V^2 A_1}{2}\left[\left(\frac{A_1}{A_2}\right)^2 - 1\right] - \rho gh A_1 - W_p$$

where W_p is the weight of the piston and F_p is the force applied through the shaft of the piston.

Introducing all the expressions developed in the preceding paragraphs with regard to the y-direction into Eq. (2.6.7), we obtain

$$\rho V^2 A_1 = -\rho gh A_1 - W_c - W_p - F_p + F_V$$

where W_c is the weight of the cylinder. Therefore the force F_V is given by

$$F_V = \rho V^2 A_1 + \rho gh A_1 + W_c + W_p + \rho\frac{V^2 A_1}{2}\left[\left(\frac{A_1}{A_2}\right)^2 - 1\right]$$
$$- \rho gh A_1 - W_p$$
$$= W_c + \rho\frac{V^2 A_1}{2}\left[\left(\frac{A_1}{A_2}\right)^2 + 1\right]$$

Problem 2.11   Figure 2.28 shows a small cart moving with velocity V_v due to the impact of a two-dimensional water jet on a plate oriented at an angle α with respect to the jet direction. The velocity and width of the jet are V_1 and b_1, respectively. The water jet is divided into two smaller jets, whose widths are b_1 and b_2. The force applied by the water jet on the cart is perpendicular to the impacted plate. Assuming that the effect of gravitation is negligible when applying Bernoulli's equation, calculate (a) The widths b_1 and b_2 of the two jets-(b) the vertical and horizontal forces acting on the cart, and (c) the power transferred to the moving cart.

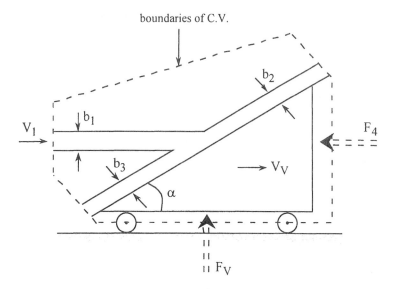

Figure 2.28 Water jet driving cart motion, Problem 2.11.

Solution

We apply a coordinate system that moves with the cart. In such a coordinate system the domain is subject to steady state, and Bernoulli's equation is applicable. The velocity of the jet that hits the cart, in the new coordinate system, is $V_1 - V_v$. As the effect of gravitation is negligible at the jet division, the velocities of the two jets created by the jet division are also $V_1 - V_v$. We apply the control volume with boundaries as shown in the figure. The forces F_H and F_V are needed to keep the control volume in its appropriate position. By applying the equation of momentum conservation (2.6.1) in the horizontal direction, we obtain

$$- \rho(V_1 - V_v)^2 b_1 + \rho(V_1 - V_v)^2 (b_2 - b_3) \cos\alpha = -F_H$$

Applying the conservation of momentum in the y-direction gives

$$\rho(V_1 - V_v)^2 (b_2 - b_3) \sin\alpha = F_V$$

As the resultant force is perpendicular to the oblique plate, we obtain

$$\frac{F_V}{F_H} = \tan\alpha$$

From continuity, we have

$$b_1 = b_2 + b_3$$

The last four equations allow the determination of the four unknown quantities $b_2, b_3, F_H,$ and F_V. The results of the calculation are

$$b_2 = \frac{b_1}{2}\left(1 + \frac{1}{2\cos\alpha}\right) \quad b_3 = \frac{b_1}{2}\left(1 - \frac{1}{2\cos\alpha}\right)$$

$$F_H = \rho(V_1 - V_v)^2 \frac{b_1}{2} \quad F_V = \rho(V_1 - V_v)^2 \frac{b_1}{2}\tan\alpha$$

The force that leads to the cart movement is equal to F_H and acts in the positive x direction. The power transferred from the water jet into the cart is equal to the product of this force with the velocity V_v of the cart in the horizontal direction. Therefore the power N is given by

$$N = \rho(V_1 - V_v)^2 \frac{b_1}{2}V_v$$

Problem 2.12 Figure 2.29 shows a rocket fired from rest in outer space along a horizontal straight line where air friction is negligible. The mass of the body of the rocket is M and it carries an original fuel mass M_f which burns at a mass flow rate α. The exhaust cross-sectional area and velocity relative to the rocket are A_e and V_e, respectively, and the density of the fluid at the exhaust is ρ_e. The velocity of the rocket relative to a fixed observer is V. Our objective is to determine the value of V as a function of time.

Solution

We apply Eq. (2.6.15) to solve this problem. The momentum due to the flow inside the control volume is assumed to be negligible. Therefore the first LHS term of this equation vanishes. Also, all terms of the RHS of Eq. (2.6.16) vanish, except for the volume integral associated with the translational acceleration. Therefore we obtain

$$-\rho_e V_e^2 A_e = -(M + M_f - \alpha t)\frac{dV}{dt}$$

Conservation of mass yields

$$\rho_e V_e A_e = \alpha$$

We introduce this relationship into the equation of momentum conservation. Separation of variables of the resulting expression yields.

$$\frac{dV}{\alpha V_e} = \frac{dt}{M + M_f - \alpha t}$$

Direct integration of this expression and assuming that $V = 0$ at $t = 0$ yields

$$V = V_e \ln\left(\frac{M + M_f}{M + M_f - \alpha t}\right)$$

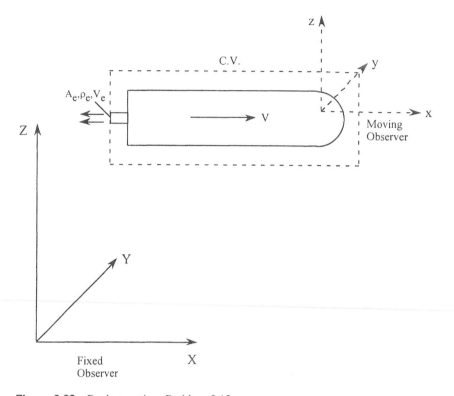

Figure 2.29 Rocket motion, Problem 2.12.

According to this expression, the maximum value of the rocket velocity is obtained when all the fuel is burnt, namely when $t = M_f/\alpha$. At that time the rocket velocity is given by

$$V = V_e \ln\left(\frac{M + M_f}{M}\right)$$

Problem 2.13 Figure 2.30 shows a pump that delivers a water discharge Q from a tank through a pipe of total length L, which is ended with a nozzle. The pipe diameter is D_1, the nozzle diameter D_2. The Darcy–Weissbach friction coefficient for the pipe flow is f. Water level in the tank is h_1 and its value is given. The exit of the nozzle is located at an elevation h_2 above the pump, which is also given. Calculate the power delivered by the pump into the flowing water. The system is at constant temperature.

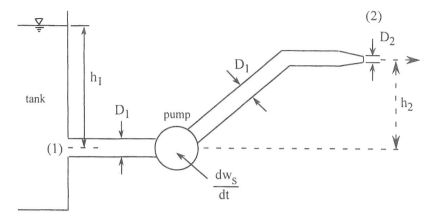

Figure 2.30 Pumped water jet, Problem 2.13.

Solution

The total head at the exit of the nozzle, at cross section (2), is

$$H_2 = \frac{V_2^2}{2g} + h_2 \qquad \text{where} \qquad V_2 = \frac{4Q}{\pi D_2^2}$$

The total head at the entrance cross section (1) is equal to h_1. The power of the pump is needed to increase the water head from its initial value h_1 to its final value at the exit cross section. Part of this power is converted to heat, which is transferred into the surroundings (so that the system remains at constant temperature). The head loss between the entrance and exit multiplied by the weight discharge is equal to the rate of heat transferred into the environment. Therefore the power delivered from the pump into the flowing water is given by

$$N = \rho g Q \left(H_2 - h_1 + f \frac{L}{D_1} \frac{V_1^2}{2g} \right) \qquad \text{where} \qquad V_1 = \frac{4Q}{\pi D_1^2}$$

Problem 2.14   Considering the flow given in Problem 2.9, find the pressure at any point in the flow field, relative to $p = p_0$ at the corner. Neglect gravity.

Solution

It is already known that this flow is steady and incompressible. It can also be shown to be irrotational. In this case, pressures are found using Bernoulli's equation. Since gravity effects are neglected, we have

$$\frac{p_0}{\gamma} + \frac{V_0^2}{2g} = \frac{p}{\gamma} + \frac{V^2}{2g}$$

From the velocity components found in Problem 2.9, it is easily seen that $V_0 = 0$. Then, substituting the general expression for the velocity, we find

$$p = p_0 - \frac{1}{2}\rho V^2 = p_0 - \frac{32}{9}\rho r^{2/3}$$

Unsolved Problems

Problem 2.15   A two-dimensional flow field is given by the following velocity distribution:

$$u = a(y - b) \qquad v = a(x - b)$$

where a and b are constant coefficients.

    (a)   Develop the expression for the pathlines in the domain.
    (b)   Develop the expression for the streamlines. Show that streamlines and pathlines have the same shape. Provide a schematic of the streamlines.

Problem 2.16   Using the velocity distribution of Problem 2.15,

    (a)   Calculate values of components of the rate of strain tensor.
    (b)   Show that the fluid is subject to irrotational flow and develop the expression for the potential function.

Problem 2.17   A two-dimensional flow field is given by

$$u = -a(y - b) \qquad v = a(x - b)$$

where a and b are constant coefficients.

    (a)   Determine values of components of the rate of strain tensor and the vorticity tensor.
    (b)   Calculate the value of the circulation along a circle whose center is at point (b, b), with radius b.

Problem 2.18   The velocity field for a two-dimensional flow is given by

$$u_1 = U \exp\left(-\frac{x_1}{L}\right) \sec h^2\left(\frac{x_2}{L}\right) \qquad \text{and}$$

$$u_2 = Cx_2 + U \exp\left(-\frac{x_1}{L}\right) \tanh\left(\frac{x_2}{L}\right)$$

Where U, C, and L are constants. Find

   (a)   Acceleration of a fluid particle
   (b)   Variation of density of a fluid particle
   (c)   Components of fluid vorticity
   (d)   Components of fluid rate of strain

Problem 2.19   For each of the velocity distributions of Problems 2.15 and 2.17,

   (a)   Calculate the Lagrangian components of the velocity and acceleration.
   (b)   Calculate the Eulerian components of the velocity and acceleration.

Problem 2.20   Velocity components of the flow and density of a fluid are given by

$$u = x(1 + \alpha xy) \qquad v = -y(1 + \alpha xy) \qquad \rho = \frac{\rho_0}{1 + \alpha xy}$$

where α and ρ_0 are constants.

   (a)   Calculate the components of the acceleration.
   (b)   Calculate the rate of change of density of the fluid particles, assuming that α is small and has negligible effect on u and v.

Problem 2.21   Starting with the fluid element shown in Fig. 2.31, demonstrate graphically that the divergence $(\vec{\nabla} \cdot \vec{V})$ must be zero if the fluid is incompressible. Is it necessary that $\dfrac{\partial u}{\partial x} = \dfrac{\partial v}{\partial y} = \dfrac{\partial w}{\partial z} = 0$ in order to make the same conclusion?

Problem 2.22   For each of the velocity distributions of Problems 2.15 and 2.17,

   (a)   Show that continuity is satisfied for incompressible flow.
   (b)   Determine the expression for the stream function.
   (c)   Calculate the flow rate between two streamlines of your choice.

Problem 2.23   For each of the velocity and density distributions of Problem 2.20,

   (a)   Show that the equation of mass conservation is satisfied.
   (b)   Develop the expression for the stream function of the mass flux.

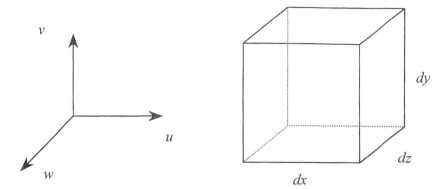

Figure 2.31 Fluid element, Problem 2.21.

Problem 2.24 Derive an integral statement of the equation expressing conservation of dissolved mass (concentration C) following the Reynolds transport theorem approach. Where might such an equation be useful?

Problem 2.25 Figure 2.32 shows a section of a two-dimensional channel with walls described by

$$y = \pm \frac{h}{2(x + 1)}$$

where h is the width of the channel at its entrance where $x = 0$. A fluid of constant density flows through the channel. The velocity component in the x-direction is solely a function of x. At the channel entrance the velocity in the x-direction is given by $u = u_0$.

  (a) Determine the velocity component in the x-direction.
  (b) Determine the velocity component in the y-direction.
  (c) Develop the expression for the stream function in the channel. What are reasonable values of the stream function along the walls of the channel?

Problem 2.26 Fluid is subject to steady-state flow in an infinite domain. In every vertical cross section of the domain, the velocity component in the horizontal x-direction is not a function of y. In every horizontal cross section of the domain, the velocity component in the vertical y-direction is not a function of x. At the point $(x = 8$ m, $y = 0)$ it was found that there is only velocity in the x-direction, whose value is $u = 0.1$ m/s. At the point $(x = -12$ m, $y = 0)$ it was found that there is also only velocity in the x-direction, with a value of $u = -0.1$ m/s.

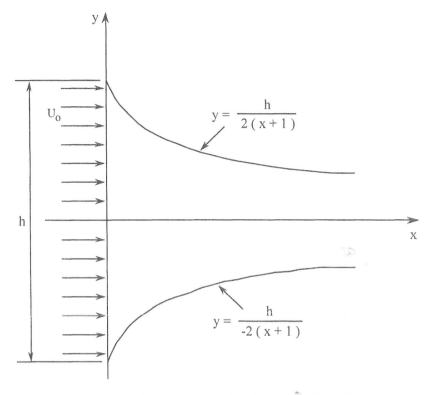

Figure 2.32 Two-dimensional converging flow, Problem 2.25.

(a) Apply the integral continuity equation to determine the distribution of the velocity component in the y-direction.

(b) Apply the differential continuity equation to determine the distribution of the velocity component in the x-direction.

(c) Develop the expression for the stream function. Find stagnation points, and provide a schematic of the streamlines.

(d) Check whether the flow is a potential flow. If it is a potential flow, determine the expression for the potential function.

(e) Determine components of the rate of strain tensor.

(f) Determine components of the rate of strain tensor in the entire domain in the coordinate system $(\bar{x}, \bar{y})$ whose $\bar{x}$ axis bisects the angle between the axes x and y.

Problem 2.27 A water reservoir has a volume $U = 50,000$ m^3. At time $t = 0$ the density of the water is $\rho_0 = 1000$ kg/m^3. At that time two effluent sources start to divert water into the reservoir. Both sources provide an identical volumetric discharge of $Q = 36$ m³/s (for each source). The density of the fluid of the first source is $\rho_1 = 1,020$ kg/m³. The density of the fluid of the second source is $\rho_2 = 1,010$ kg/m³. These sources may be assumed to be rapidly mixed throughout the reservoir. Fluid percolates into the ground through the bottom of the reservoir with flow rate Q and with density equal to that of the reservoir water ρ. At the reservoir surface water evaporates, with discharge Q and density ρ_0.

(a) Prove that the reservoir volume is kept constant.
(b) Develop a general equation for the variation of the density of the reservoir water. What is the value of this density for time $t \to \infty$?
(c) Substitute numerical values of the variables, and find the time at which the water density becomes 99% of its value at $t \to \infty$.

Problem 2.28 A model is needed to predict the transient response of a constant volume mixing tank due to a step change in influent concentration of a conservative substance. The model is to be used to quantify the degree of mixing and short-circuiting in the tank. Assume that a fraction m of the total tank volume V is actually well mixed and that only a fraction n of the inflow Q enters the zone of perfect mixing, while the remaining portion of the inflow short-circuits directly to the outlet (i.e., it is not mixed at all inside the tank). The concentration at any time t in the mixed zone is C'. The material exiting from this zone is mixed with the portion of inflow that is short-circuited and the mixture leaves the tank at flow rate Q and concentration C. The initial concentration in the tank is C_0 (everywhere). At time $t = 0$ the influent concentration C_i is changed suddenly from $C_i = C_0$ to $C_i = 0$.

(a) Sketch the problem, showing how n and m are incorporated.
(b) Show that, in general, the outflow concentration may be calculated as $C = nC' + (1 - n)C_i$
(c) Write the general mass balance equation for C' (in the mixed zone) — include C_i in the formulation.
(d) Substitute the result from part (b) into your result from part (c) and develop a differential equation that describes the rate of change of C with time.
(e) Solve the equation to calculate (C/C_0) as a function of n, m, and (t/t^*), where $t^* = V/Q$ is the overall tank residence time.
(f) Using the experimental data plotted in Fig. 2.33, estimate the values for n and m (note that values for C' are obtained from the middle of the tank, which is expected to be in the fully mixed region).

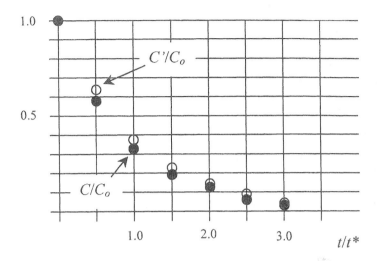

Figure 2.33 Nondimensional concentration data, Problem 2.28.

Problem 2.29 A shallow lake has mean (depth-averaged) horizontal velocity components U and V, in the x and y coordinate directions, respectively, and U and V are in general functions of (x, y, t), where $t = $ time (see Fig. 2.34). Seepage out the bottom of the lake takes place at a rate f, where f is assumed to be directly proportional to the depth, h, so $f = kh$, and h is also a function of (x, y, t). Rain falls at rate i (units of length/time) and $i = i(x, y, t)$. The lake bottom may be assumed to be flat and horizontal. Derive the two-dimensional continuity equation for this problem.

Problem 2.30 Figure 2.35 shows a section of a two-dimensional channel, with walls that are described by

$$y = \pm 0.5(h + x)$$

where $h = 1$ m is the width of the channel at its entrance, where $x = 0$. A fluid of constant density flows through the channel. The velocity component in the x-direction is solely a function of x. At $x = 0$, the velocity in the x-direction is given by $u = u_0 = 1$ m/s.

    (a)   Determine the velocity component in the x-direction.
    (b)   Determine the velocity component in the y-direction.
    (c)   Calculate the discharge per unit width of the channel.

Problem 2.31 Water is subject to unsteady flow in an open channel, as shown in Fig. 2.36. A discharge per unit area, q, flows into the channel through the

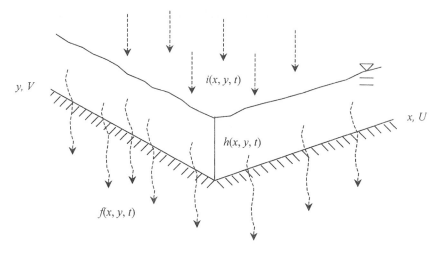

Figure 2.34   Two-dimensional lake schematic, Problem 2.29.

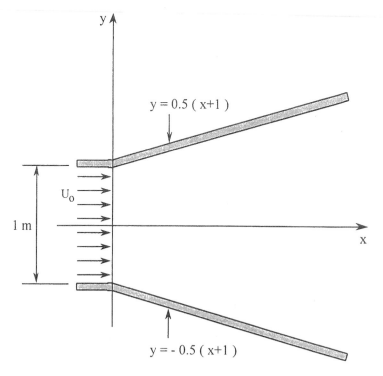

Figure 2.35   Expanding two-dimensional flow, Problem 2.30.

free surface. The water depth in the elementary control volume is h. The width of the channel at the free surface is B.

(a) Refer to the elementary control volume of the open channel shown in part (a) of Fig. 2.36 and develop the differential equation that represents the variation of the water depth along the channel.

(b) Part (b) of Fig. 2.36 indicates that at $x = x_0$, the water depth is h_0. The channel has a rectangular cross section, in which $B = const$. It is found that the water depth downstream of x_0 is represented by $h = h_0 + h_1 \sin(\alpha x + \omega t)$, where, h_0, h_1, α, and ω are constants. It

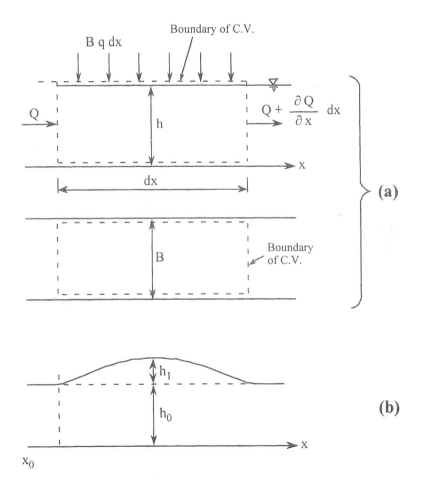

Figure 2.36 Open channel flow, Problem 2.31.

is also given that $q = 0$, and $Q = Q_0$ at x_0. Find the discharge as a function of x and t.

Problem 2.32 Figure 2.37 shows two containers that contain fluids. The volume of container 1 is $U_1 = 30$ m^3 and it contains at time $t < 0$ pure water, with density $\rho_1 = 1,000$ kg/m^3. The volume of container 2 is $U_2 = 20$ m^3 and it includes for time $t < 0$ salt water, with density $\rho_2 = 1,020$ kg/m^3. At time $t = 0$ two pumps start to circulate water between the two containers. Each pump delivers a volumetric discharge of $Q = 10$ m^3/s. A mixer is submerged in each container to insure well-mixed conditions.

  (a) What is the final density of the water in both containers?
  (b) Develop expressions for the variation of the water density in each container as functions of t.
  (c) Show that the expressions that you developed in part (b) converge to the result of part (a) when $t \to \infty$.
  (d) Calculate the value of the time t at which the density of the water in container 1 is equal to 99% of the density when $t \to \infty$. What is the density of the water in container 2? By how many percent is it larger than the density at $t \to \infty$?

Problem 2.33 Figure 2.38 shows a two-dimensional incompressible flow between two long plates. Plate (a) is stagnant. Plate (b) rotates around the origin with constant angular velocity Ω. The radial flow in the domain is not a function of the angular coordinate θ. At time $t = 0$, the angle between the two plates is π.

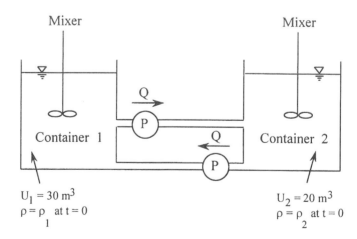

Figure 2.37 Definition sketch, Problem 2.32.

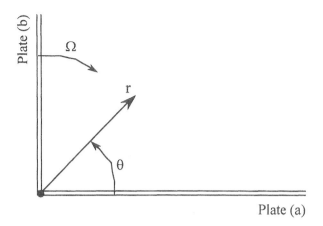

Figure 2.38 Definition sketch, Problem 2.33.

(a) Determine the velocity distribution in the flow domain at time t_1, where $t_1 < \pi/\Omega$.
(b) Determine the expression for the stream function.
(c) Is the value of the stream function at the two plates subject to variation with time? Explain.

Problem 2.34 Figure 2.39 shows viscous incompressible fluid between three plates. Plates (a) and (b) are stagnant, while plate (c) moves downward with constant velocity V. Due to the movement of plate (c), the fluid is subject to flow. There is a parabolic distribution of the velocity component in the x-direction, and it vanishes at plates (a) and (c).

(a) Determine the expressions for the velocity components in the flow domain when the distance between plates (a) and (c) is h.
(b) Determine the expression for the stream function in the domain.
(c) Calculate the variation of the discharge flowing between plates (a) and (c) as a function of time and x-coordinate.

Problem 2.35 A two-dimensional velocity field (u, v) may be defined in terms of a stream function, Ψ, where

$$\vec{V} = \vec{\nabla} \times \Psi(\hat{k})$$

Calculate $\vec{\nabla} \times \vec{V}$, $\nabla^2 \vec{V}$, and $\vec{V} \cdot \vec{\nabla}\vec{V}$ in terms of Ψ.

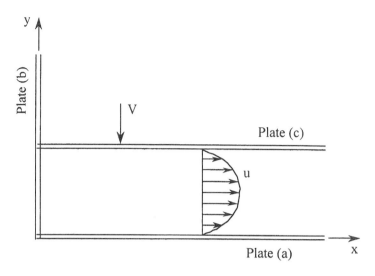

Figure 2.39   Definition sketch, Problem 2.34.

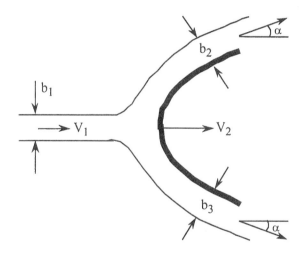

Figure 2.40   Definition sketch, Problem 2.36.

Problem 2.36   A fluid two-dimensional jet of width b_1 and velocity V_1 is directed at a concave plate, which moves with velocity V_2, as shown in Fig. 2.40. Due to the impact with the plate, the fluid jet is divided into two identical jets, which are oriented with angle α to the longitudinal x-direction at the edges of the plate. The fluid density is ρ.

(a) Calculate the thickness of the two jets created by the impact of the jet with the concave plate.
(b) Determine the velocity of the two jets at the edges of the plate.
(c) Determine the power delivered from the jet to the plate.
(d) What should be the relationship between V_1 and V_2 to deliver maximum power?

Problem 2.37 Repeat Problem 2.36 for jet impact with a convex plate, as shown in Fig. 2.41.

Problem 2.38 Consider plane Couette flow, with one wall ($y = 0$) fixed and a second rigid wall ($y = h$) moving at constant speed U in its own plane. Sketch the flow and solve the Navier–Stokes equations for the case of constant density (also no rotation), to show that a possible flow is $\vec{u} = (Uy/h)(\hat{i})$. Also calculate the shear stress on each wall.

Problem 2.39 A *line sink* (large width-to-height ratio) drains a large water reservoir by a rectangular conduit as shown in Fig. 2.42. Assuming the flow is fully developed in the conduit (i.e., at some distance downstream of the reservoir), calculate the following:

(a) Velocity distribution (neglect side wall effects).
(b) Magnitude of shear stress at upper and lower surfaces and at middle of conduit.
(c) Considering the entire length (L) as a control volume, verify that there is zero net force acting on the fluid in the direction along the pipe.

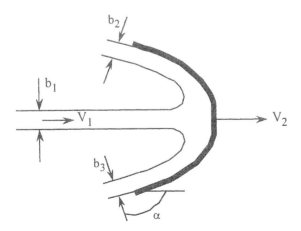

Figure 2.41 Definition sketch, Problem 2.37.

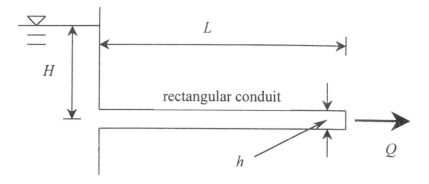

Figure 2.42 Flow through long rectangular conduit, Problem 2.39.

Problem 2.40 Figure 2.43 shows a hose of diameter D_1, which is connected to a nozzle by a flange. The diameter of the nozzle exit is D_2. Water (density ρ) flows through the hose and nozzle with discharge Q. F_H and F_V represent the horizontal and vertical forces applied by the fireman, to keep the hose and nozzle in the appropriate position.

  (a) Determine the force needed to hold the two parts of the flange together.
  (b) Determine the horizontal, vertical, and total forces applied by the fireman.

Problem 2.41 Water flows in an open rectangular channel with a constriction, as shown in Fig. 2.44. The water depth and channel width before the

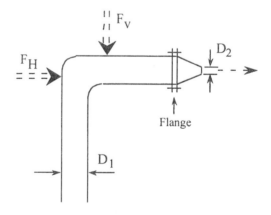

Figure 2.43 Flow around a bend, Problem 2.40.

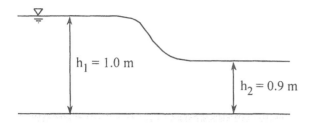

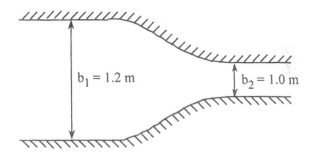

Figure 2.44 Open channel flow constriction, Problem 2.41.

constriction are $h_1 = 1.0$ m and $b_1 = 1.2$ m, respectively. At the constriction, the water depth and channel width are $h_2 = 0.9$ m and $b_2 = 1.0$ m, respectively. The pressure distribution in each vertical cross section is hydrostatic.

(a) Determine the discharge flowing through the channel.
(b) Determine the force applied by the water on the constriction.

Problem 2.42 Water flows through a gate as shown in Fig. 2.45. The channel has a rectangular cross section and its width is 1 m. The water depth upstream of the gate is $h_1 = 0.8$ m. The water depth downstream of the gate is $h_2 = 0.2$ m. At that location, the flow velocity is $V_2 = 3$ m/s.

(a) Determine the discharge in the channel.
(b) Determine the force of the water on the gate.

Problem 2.43 A cart carries a container with water. It moves freely on an inclined area, whose slope is $\alpha = 30°$, as shown in Fig. 2.46. The width of the container is $b = 2$ m, and its length is $L = 2$ m. The top of the container is

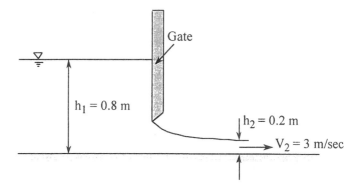

Figure 2.45 Flow under a sluice gate, Problem 2.42.

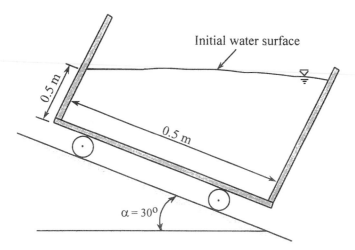

Figure 2.46 Water containing cart on a sloping surface, Problem 2.43.

open, and its side walls are very tall. The initial water depth, measured along the upper wall of the container, is 0.5 m.

(a) Determine the orientation angle between the free surface of the water with respect to horizontal.

(b) Determine the horizontal and vertical components of the pressure gradient in the water.

(c) Determine the total force applied on the front wall, back wall, and bottom of the container.

Problem 2.44 Figure 2.47 shows fluid with density ρ flowing through a two-dimensional conduit, whose width and length are b and h, respectively. At the entrance of the conduit, the velocity is u_0 and is uniformly distributed. The pressure at the entrance is p_A. At the exit of the conduit, the velocity profile is a parabola, given by

$$u = U\left[1 - \left(\frac{2y}{b}\right)^2\right]$$

where U is the maximum value of the velocity at the exit cross section and $y = 0$ represents the centerline. The pressure at the exit cross section is given by

$$p_B = p_A - 2.25\rho\frac{U^2}{2}$$

    (a)   Determine the relationship between u_0 and U.

    (b)   Determine the force applied per unit width of the conduit.

Problem 2.45 A jet aircraft flies at a constant speed V. The jet engine pumps air with volumetric discharge Q_0 and density ρ_0. The mixture of fuel and air has a density almost identical to that of the air. After the burning of the mixture, it flows out with the volumetric discharge $Q_1 = (2/3)Q_0$ and unknown density ρ_1. The inlet cross section area is A_0. The outlet cross section area is $A_1 = 0.1A_0$. The flow velocity through the inlet cross section is identical to that of the outlet cross section. The volumetric discharges Q_0 and Q_1 are independent of the flow velocity V.

    (a)   Determine the fluid density ρ_1 at the outlet cross section.

    (b)   Determine the drag force that is overcome by the jet engine.

    (c)   What is the power of the jet engine?

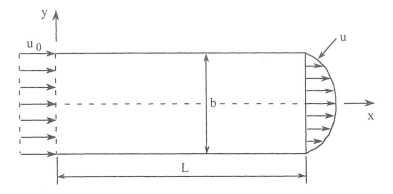

Figure 2.47 Definition sketch, Problem 2.44.

Problem 2.46 Consider that the length of the equator is 40,000 km, and that the earth makes a complete rotation in 24 hours.

  (a) Calculate the value of the effective gravitational acceleration at a point on the earth's surface, whose inclination angle (latitude) is 30°.
  (b) Provide the two equations of motion, based on Eq. (2.7.20), for a two-dimensional horizontal flow at a point on the ocean with an inclination angle of 30°.

Problem 2.47 A mass discharge of dry steam with $Q_m = 1$ kg/s flows through a turbine and delivers a power $N = 1,000$ W through the shaft of the turbine. The entrance and exit flow velocities are $V_1 = 20$ m/s and $V_2 = 10$ m/s, respectively. The entrance and exit specific enthalpy values are $h_1 = 80$ m^2/s^2 and $h_2 = 100$ m^2/s^2, respectively. The entrance elevation is higher than that of the exit of the turbine by 1 m.

  (a) What are the values of ρA (where ρ is the density and A is the cross-sectional area) at the entrance and exit of the turbine?
  (b) Determine the net heat transferred from the turbine into the environment per unit mass of flow.
  (c) Determine the rate of heat transferred from the turbine into the environment.

Problem 2.48 A 3 m diameter tank is filled with water to a depth $h = 10$ m. A value on a 30 cm pipe at the bottom of the tank is opened suddenly and water is allowed to drain as shown in Fig. 2.48. Estimate the time needed for the tank to drain halfway (until $h = 5$ m). State all assumptions.

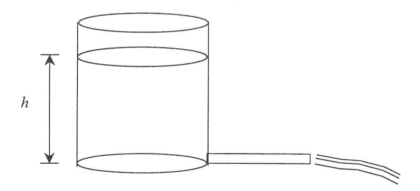

Figure 2.48 Water drainage from tank, Problem 2.48.

Problem 2.49   The pressure at the water surface of a container is 4×10^4 Pa. The water is pumped from the container through a pipe that ends with a nozzle with exit diameter $D = 100$ mm. The water flows as a free jet through the nozzle. As shown in Fig. 2.49, the elevation of the water surface in the container is higher by 1.5 m than the pump. Also, the exit nozzle is elevated by 1.5 m above the pump. The free water jet leaves the nozzle with an angle of $45°$ and it reaches its maximum elevation 3 m above the nozzle exit. Effects of friction between the air and the free jet are negligible.

(a)   Determine the velocity of the water jet at the exit of the nozzle.

(b)   Determine the distance between the exit of the nozzle and the point at the same elevation, through which the water jet passes.

(c)   Assuming that the efficiency of the water pump is 0.8, determine the power needed to operate the pump.

(d)   Draw a schematic of the total and piezometric heads between the container and the exit of the nozzle.

Problem 2.50   Show that for a steady one-directional flow field ($u_1 = u$) of an incompressible fluid with no horizontal variations (i.e., in the x_1 or x_2 directions) of any property, the energy equation can be simplified to

$$\mu \left(\frac{\partial u}{\partial z} \right)^2 = \frac{\partial \varphi_z}{\partial z}$$

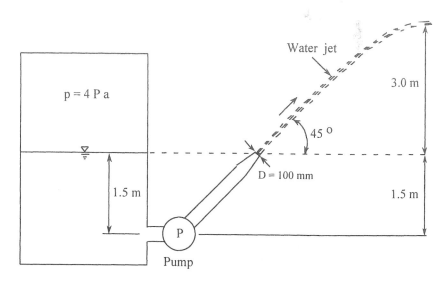

Figure 2.49   Definition sketch, Problem 2.49.

(i.e., viscous dissipation is balanced by radiative heating). Note that the result could be written in terms of ordinary derivatives, since variations occur only in the $x_3 = z$ direction.

Problem 2.51   A horizontal circular pipe 1 m in diameter carries water at a flow rate of 10 m^3/s. Neglecting heat transfer through the walls, find the temperature increase for the water traveling a length of pipe corresponding to a pressure drop of 5 atm. (about 500 kPa). Hint: apply the integral energy equation to a control volume bounded by the pipe walls and sections separated by the distance indicated above. Use $c_v = 4200$ J/kg-$^\circ$C.

Problem 2.52
   (a)   Write the conservation equations in rectangular coordinate form for mass, momentum, and energy for an incompressible fluid with no motion and no horizontal variation of any quantity. Also assume an inertial reference system.
   (b)   Repeat part (a), but for conditions of steady motion in one horizontal direction (x, or x_1) only and, like all other quantities, uniform in horizontal directions.

Problem 2.53   Show that heat energy changes in a fixed volume $d\forall$ are given, for a temperature change of dT, by $(\rho c_v dT d\forall)$. You may assume that (ρc_v) is constant. Following the basic procedures in deriving the basic conservation equations, develop an equation for temperature in a fluid at rest. Although there is no advective flow, assume that there is an average molecular velocity U that must be considered in the balance. Also assume there is a source of heat $Q(x_i, t)$ per unit volume, per unit time at each point of the fluid. Your final result should look like

$$\frac{\partial \theta}{\partial t} = -\nabla(U\theta) + \frac{Q}{\rho c_v}$$

Problem 2.54   A rotating table is built for testing a scale model of a large lake. If the horizontal length scale ratio is $1 : 10^5$, the vertical length scale ration is $1 : 800$ (this is a *distorted* scale model), and the lake is at latitude 44° (N), how fast (in rpm) should the table be rotated in order to simulate Coriolis effects? (Hint: first decide which are the important dimensionless numbers for this problem, arising from scaling of the momentum equations.)

Problem 2.55   Show that in a natural water body with characteristic horizontal dimension L and vertical dimension H, with $H \ll L$, the characteristic vertical velocity W should be much less than the characteristic horizontal velocity U.

SUPPLEMENTAL READING

Aris, R., 1962. *Vectors, Tensors and the Basic Equations of Fluid Mechanics*, Prentice-Hall, Englewood Cliffs, New Jersey. (Provides a comprehensive description of differences between streamlines, pathlines, etc., as well as derivation of the basic equations of motion and the Reynolds transport theorem.)

Batchelor, G. K., 1967. *An Introduction to Fluid Dynamics*, Cambridge University Press, London. (Provides a comprehensive presentation of the basic equations of conservation.)

Pedlosky, J., 1987. *Geophysical Fluid Dynamics*, Springer-Verlag, New York. (A very good intermediate level text for geostrophic flows.)

3

Viscous Flows

3.1 VARIOUS FORMS OF THE EQUATIONS OF MOTION

Viscous flows are mathematically represented by solutions of the equations of motion, based on momentum transfer in an elementary fluid volume. The equations of motion for viscous flows are the Navier–Stokes equations introduced in the previous chapter. For convenience, we repeat these equations here, for cases in which variations in viscosity are negligible:

$$\frac{\partial \vec{V}}{\partial t} + (\vec{V} \cdot \nabla)\vec{V} = -\frac{1}{\rho}\nabla(p + \rho g Z) + v\nabla^2 \vec{V} \qquad (3.1.1a)$$

where V is the velocity, t is time, ∇ is the gradient vector, ρ is the density, p is the pressure, g is gravitational acceleration, Z is the elevation with regard to an arbitrary reference, and v is the kinematic viscosity. In Appendix 1, tables of Navier–Stokes equations for Cartesian, cylindrical, and spherical coordinate systems are listed. In Appendix 2, relationships are given between stress components and velocity components, as implied by the Navier–Stokes equations.

Using Cartesian tensor notation, Eq. (3.1.1a) is represented as

$$\frac{\partial u_i}{\partial t} + u_k \frac{\partial u_i}{\partial x_k} = -\frac{1}{\rho}\frac{\partial p}{\partial x_i} - g\frac{\partial Z}{\partial x_i} + v\frac{\partial^2 u_i}{\partial x_k^2} \qquad (3.1.1b)$$

where u_i represents components of the velocity vector and x_i represents the coordinates. This equation incorporates four unknown quantities: three components of the velocity vector and the pressure. Along with the continuity equation, we thus have a system of four differential equations with four unknowns. The solution of this system subject to appropriate initial and boundary conditions provides the required information about the distribution of the unknown quantities in the domain.

The distributions of velocities and pressure depend on the three space coordinates, x, y and z, and the time coordinate, t. It should be noted that the order of the differential equation (3.1.1) varies with regard to the unknown quantities, as well as with regard to the various coordinates. The velocity components contribute terms of first order with regard to time and of both first and second order with regard to the space coordinates. The pressure contributes terms of first order with regard to the space coordinates. The order of the partial derivatives indicates the number of boundary conditions needed for the solution of this system of partial differential equations. The pressure should be given at a certain point in the domain during all times. The velocity distribution at initial conditions should be given for the whole domain. The velocity at a sufficient number of boundaries should be given for the required time period of the simulation. There are several typical boundary conditions for the velocity vector, or its derivatives. The latter are related to shear stresses. Generally, there are four typical boundary conditions for the velocity and the shear stresses:

> Boundary between the viscous fluid and a solid boundary — fluid velocity is identical to that of the solid boundary, as the viscous fluid adheres to the solid boundary.
>
> Boundary between two viscous immiscible fluids — velocity and shear stress at both sides of the interface are identical.
>
> Boundary between two immiscible fluids with an extremely large difference of viscosity, e.g., liquid and gas — shear stress vanishes at the interface between the two fluids. (An exception to this rule is with wind-driven flows, where boundary shear stress is significant. Momentum transfer at the air/water interface is discussed in Chap. 12, and a particular application, in a geophysical context, is discussed in Chap. 9.)
>
> Finite domain — the velocity has finite value at every point of the domain.

As viscous fluid flow is basically a rotational flow, the equation of motion (3.1.1) can be represented as an equation of vorticity transport. The rotationality of the flow is represented by the distribution and intensity of the vorticity. The vorticity is a kinematic tensorial characteristic of the flow field. The tensor of vorticity is a second-order asymmetric tensor. Such a tensor has three pairs of components. Each pair incorporates two components of identical absolute value and opposite sign. Therefore the vorticity also can be represented by a vector with three components. Each component of this vector represents one pair of components of the vorticity tensor. By the employment of Cartesian tensor notation, the vorticity vector is defined as

$$\omega_j = \frac{\partial u_i}{\partial x_k} - \frac{\partial u_k}{\partial x_i} \tag{3.1.2}$$

where $i, j, k = 1, 2, 3$. One half of the vorticity represents the angular rotation rate of an elementary fluid volume, as previously noted.

By cross differentiation and subtraction of component equations of Eq. (3.1.1b), the pressure is eliminated from the equation of motion. Then the expression of Eq. (3.1.2) can be introduced to obtain a vorticity equation,

$$\frac{D\omega_j}{Dt} - \omega_k \frac{\partial u_j}{\partial x_k} = v \frac{\partial^2 \omega_j}{\partial x_k^2} \qquad (3.1.3)$$

where

$$\frac{D\omega_j}{Dt} = \frac{\partial \omega_j}{\partial t} + u_k \frac{\partial \omega_j}{\partial x_k} \qquad (3.1.4)$$

The first term on the LHS of Eq. (3.1.3) represents the total rate of change of vorticity. The second term represents the deformation of a vortex tube. The term on the RHS of Eq. (3.1.3) represents the diffusion of vorticity due to the viscosity of the fluid.

In cases of two-dimensional flow the vorticity vector has a single component, and the term representing the deformation of the vortex tube vanishes. Then, Eq. (3.1.3) yields

$$\frac{\partial \omega}{\partial t} + u_k \frac{\partial \omega}{\partial x_k} = v \frac{\partial^2 \omega}{\partial x_k^2} \qquad (3.1.5)$$

Also, for two-dimensional flows, it is possible to apply the expression for the stream function, ψ. The stream function is related to components of the velocity vector according to (see Chap. 2)

$$u = u_1 = \frac{\partial \psi}{\partial y} \qquad v = u_2 = -\frac{\partial \psi}{\partial x} \qquad (3.1.6)$$

By using the stream function, the vorticity in a two-dimensional flow field is given by

$$\omega = \frac{\partial v}{\partial x} - \frac{\partial u}{\partial y} = -\left(\frac{\partial^2 \psi}{\partial x^2} + \frac{\partial^2 \psi}{\partial y^2} \right) = -\nabla^2 \psi = -\Delta \psi \qquad (3.1.7)$$

Introducing Eqs. (3.1.6) and (3.1.7) into Eq. (3.1.5), we obtain

$$\frac{\partial \Delta \psi}{\partial t} + \frac{\partial \psi}{\partial y} \frac{\partial \Delta \psi}{\partial x} - \frac{\partial \psi}{\partial x} \frac{\partial \Delta \psi}{\partial y} = v \Delta \Delta \psi \qquad (3.1.8)$$

where $\Delta\Delta = \nabla^4$.

In order to obtain the essential parameters governing the physical phenomena described by the Navier–Stokes equations, we nondimensionalize these equations by the employment of characteristic quantities of the flow field (also see Sec. 2.9). As before, these quantities are

$$L, U, \rho, v \tag{3.1.9}$$

where L is a characteristic length of the domain, U is a characteristic velocity of the flow, ρ is the density, and v is the kinematic viscosity of the fluid. The following dimensionless parameters, symbolized with an asterisk, are then obtained:

$$
\begin{aligned}
&t^* = \frac{tU}{L} \qquad x_i^* = \frac{x_i}{L} \qquad u_i^* = \frac{u_i}{U} \\
&p^* = \frac{p + \rho g Z}{\rho U^2} \qquad \omega^* = \frac{\omega L}{U} \qquad \psi^* = \frac{\psi}{LU}
\end{aligned}
\tag{3.1.10}
$$

By introducing these dimensionless variables into Eqs. (3.1.1), (3.1.5), and (3.1.8), we obtain, respectively,

$$\frac{\partial u_i^*}{\partial t^*} + u_k^* \frac{\partial u_i^*}{\partial x_k^*} = -\frac{\partial p^*}{\partial x_i^*} + \frac{1}{\text{Re}} \frac{\partial^2 u_i^*}{\partial x_k^{*2}} \tag{3.1.11}$$

$$\frac{\partial \omega^*}{\partial t^*} + u_k^* \frac{\partial \omega^*}{\partial x_k^*} = \frac{1}{\text{Re}} \frac{\partial^2 \omega^*}{\partial x_k^{*2}} \tag{3.1.12}$$

$$\frac{\partial \Delta^* \psi^*}{\partial t^*} + \frac{\partial \psi^*}{\partial y^*} \frac{\partial \Delta^* \psi^*}{\partial x^*} - \frac{\partial \psi^*}{\partial x^*} \frac{\partial \Delta^* \psi^*}{\partial y^*} = \frac{1}{\text{Re}} \Delta^* \Delta^* \psi^* \tag{3.1.13}$$

where Re is the Reynolds number and Δ^* represents the dimensionless Laplacian operator:

$$\text{Re} = \frac{UL}{v} \qquad \Delta^* = \frac{\partial^2}{\partial x^{*2}} + \frac{\partial^2}{\partial y^{*2}} \tag{3.1.14}$$

The various forms of the equations of motion represented in the preceding paragraphs are used to classify types of solutions of these equations in the following sections. Generally, the Navier–Stokes equations are nonlinear equations with often quite complicated solutions. It is therefore convenient to make some classifications of families of solutions of these equations, as shown below.

3.2 ONE-DIRECTIONAL FLOWS

One-directional flows are characterized by parallel streamlines. For convenience, consider that the flow is along the x coordinate direction. Flow variables may depend on space and time in cases of unsteady flow conditions. They depend only on the space coordinates for steady state conditions. Cartesian coordinate systems are usually applied to describe domains characterized by one- and two-dimensional flows. By applying cylindrical coordinates, we refer either to domains with one-directional axisymmetric flows or to domains with one-directional circulating flows.

3.2.1 Domains Described by Cartesian Coordinates — Steady-State Conditions

At this stage we refer to a two-dimensional domain in which y is the coordinate perpendicular to the flow direction. The continuity equation is

$$\frac{\partial u}{\partial x} + \frac{\partial v}{\partial y} = 0 \tag{3.2.1}$$

where u is the velocity in the x direction, and v is the velocity in the y direction. According to the definition of one-directional flow, the velocity component, v, vanishes in the entire domain. Therefore, Eq. (3.2.1) reduces to

$$v = 0 \qquad \frac{\partial u}{\partial x} = 0 \qquad u = u(y, t) \tag{3.2.2}$$

We now introduce a quantity called piezometric pressure, defined by

$$p' = p + \rho g Z \tag{3.2.3}$$

Substituting Eqs. (3.2.2) and (3.2.3) into Eq. (3.1.1), we obtain the general differential equations representing one-directional flows in a two-dimensional domain,

$$\frac{\partial u}{\partial t} = -\frac{1}{\rho}\frac{\partial p'}{\partial x} + v\frac{\partial^2 u}{\partial y^2} \tag{3.2.4}$$

$$0 = \frac{\partial p'}{\partial y} \Rightarrow p' = p'(x, t) \tag{3.2.5}$$

For steady-state conditions, the LHS of Eq. (3.2.4) vanishes. Then Eqs. (3.2.4) and (3.2.5) yield

$$\frac{d^2 u}{d y^2} = \frac{1}{\mu}\frac{d p'}{dx} \tag{3.2.6}$$

where μ is the viscosity ($\mu = \rho v$).

Note that in cases of steady state $u = u(y)$ and $p' = p'(x)$ only. Therefore the derivative expressions of Eq. (3.2.6) are not partial derivatives. If a derivative of a function depending on y is identical to the derivative of a function depending on x, then both derivatives must be equal to a constant.

Therefore Eq. (3.2.6) implies that each one of its terms is equal to a constant, and after integrating twice we find

$$\frac{du}{dy} = \frac{y}{\mu}\frac{dp'}{dx} + C_1 \tag{3.2.7}$$

$$u = \frac{y^2}{2\mu}\frac{dp'}{dx} + C_1 y + C_2 \tag{3.2.8}$$

where C_1 and C_2 are integration constants determined by the boundary conditions of the flow domain. Thus two boundary conditions with regard to the velocity field are needed to obtain a complete description of the velocity distribution in the domain. Another set of boundary conditions is needed to obtain the piezometric pressure gradient and the pressure distribution in the domain.

Multiplying Eq. (3.2.7) by the viscosity, we obtain the expression for the shear stress distribution. Integrating Eq. (3.2.8) between y_1 and y_2, which represent locations of two different streamlines, we obtain the expression for the discharge per unit width flowing between these two streamlines. The expressions for the shear stress (τ) and the discharge per unit width (q) are given, respectively, by

$$\tau = y\frac{dp^*}{dx} + \mu C_1 \tag{3.2.9}$$

and

$$q = \frac{1}{6\mu}\frac{dp^*}{dx}(y_2^3 - y_1^3) + \frac{C_1}{2}(y_2^2 - y_1^2) + C_2(y_2 - y_1) \tag{3.2.10}$$

Now, instead of piezometric pressure, we may refer to the following quantities:

$$h = \frac{p'}{\rho g} \qquad J = -\frac{dh}{dx} \tag{3.2.11}$$

where h is the *piezometric head*, and J is the *hydraulic gradient*. With regard to pressure distribution in the domain, Eq. (3.2.5) yields

$$\frac{\partial p}{\partial y} + \rho g\frac{\partial Z}{\partial y} = 0 \tag{3.2.12}$$

Direct integration of this expression and the use of Eq. (3.2.6) gives

$$p = p_0 - \rho g(Z - Z_0) + \mu\frac{d^2u}{dy^2}(x - x_0) \tag{3.2.13}$$

where subscript 0 is associated with a point of reference, representing the boundary condition for pressure.

In summary, the family of steady-state one-directional flows is well represented by simple analytical solutions. Differences between solutions, or members of this family, originate from the different boundary conditions that determine the values of the integration constants C_1 and C_2. The special case of laminar flow between parallel flat plates, called *plane Poiseuille flow*, is often used to approximate flow through porous media. Physical models, called Hele–Shaw models, have been used extensively to simulate flow in aquifers. Such a model consists of parallel vertical plates, separated by a small gap within which a viscous liquid flows. Although this is viscous laminar flow, namely rotational flow, the average velocity in the cross section of the gap is closely represented as if it originated from a potential function given by the piezometric head. Such a presentation is consistent with basic modeling of homogeneous flow through porous media. It also is interesting to note that flows through fractures in geological formations are usually considered in terms of flow between parallel flat plates.

3.2.2 Domains Described by Cylindrical Coordinates — Steady-State Conditions

With regard to cylindrical coordinate systems, two types of flow with parallel streamlines can be identified. One type incorporates axial flows and the other incorporates circulating flows. For axial one-directional flow in the x direction, the Navier–Stokes equations are

$$\frac{\partial u}{\partial t} = -\frac{1}{\rho}\frac{\partial p'}{\partial x} + v\frac{1}{r}\frac{\partial}{\partial r}\left(r\frac{\partial u}{\partial r}\right) \tag{3.2.14}$$

$$0 = -\frac{1}{\rho}\frac{\partial p'}{\partial r} \tag{3.2.15}$$

where x is the axial coordinate, r is the radial coordinate, and u is the axial flow velocity.

In cases of steady-state conditions, Eq. (3.2.14) simplifies to

$$\frac{d}{dr}\left(r\frac{du}{dr}\right) = \frac{r}{\mu}\frac{dp'}{dx} \tag{3.2.16}$$

The LHS of this equation is a function of r, and the RHS is a function of x. Therefore each side of this equation must be a constant, and after integrating twice we find

$$\frac{du}{dr} = \frac{r}{2\mu}\frac{dp'}{dx} + \frac{C_1}{r} \tag{3.2.17}$$

$$u = \frac{r^2}{4\mu}\frac{dp'}{dx} + C_1\ln r + C_2 \tag{3.2.18}$$

where C_1 and C_2 are integration constants determined by the boundary conditions of the problem.

In the case of viscous pipe flow, termed *Poiseuille flow*, C_1 should vanish, to allow finite values of the velocity in the entire cross-sectional area of the pipe (i.e., when r approaches 0), and the value of C_2 is determined by the vanishing value of the velocity at the wall of the pipe. Therefore, for viscous pipe flow, Eq. (3.2.18) yields

$$u = -\frac{R^2}{4\mu}\frac{dp'}{dx}\left[1 - \left(\frac{r}{R}\right)^2\right]$$

(3.2.19)

where R is the pipe radius. Integrating this result over the pipe cross section, we obtain the discharge flowing through the pipe,

$$Q = -\frac{\pi R^4}{8\mu}\frac{dp'}{dx}$$

(3.2.20)

This equation is called the *Poiseuille–Hagen law*. It was derived by Poiseuille from experiments with small glass tubes that were designed to simulate blood flow through blood vessels. Ironically, Poiseuille flow is very different from real blood flow, which is subject to strong pressure variations (pulsating flow) and flows through flexible tubes. Nonetheless, experiments of Reynolds, Stanton, and others have indicated that Eq. (3.2.20) is applicable as long as the Reynolds number (Re $= VD/v$) is smaller than about 2000. In addition, flow through porous media is often simulated as a flow through stochastic bundles of capillaries. Such a simulation has been shown to provide an adequate characterization of flow and transport processes in porous matrices.

By dividing Eq. (3.2.20) by the cross-sectional area and applying Eq. (3.2.11), the average velocity is obtained as

$$V = \frac{D^2 g J}{32v}$$

(3.2.21)

where D is the pipe diameter. This expression can be represented in the form of the *Darcy–Weissbach equation* as

$$J = \frac{64}{Re}\frac{1}{D}\frac{V^2}{2g}$$

(3.2.22)

The term (64/Re) represents the Darcy–Weissbach friction coefficient for laminar pipe flow.

In the case of annular flow, the velocity vanishes at the inner tube (where $r = r_1$), as well as at the outer tube (where $r = r_2$). Introducing these boundary conditions into Eq. (3.2.18), we obtain the following expressions for the constants of Eq. (3.2.18):

$$C_1 = \frac{r_2^2 - r_1^2}{4\mu \ln(r_2/r_1)} \frac{dp^*}{dx} \tag{3.2.23}$$

$$C_2 = \frac{dp^*}{dx} \left[-\frac{r_2^2 + r_1^2}{8\mu} + \frac{r_2^2 - r_1^2}{8\mu} \frac{\ln(r_2 r_1)}{\ln(r_2/r_1)} \right] \tag{3.2.24}$$

For two-dimensional circulating flow, there is only a single component of the velocity in the θ-direction. The Navier–Stokes equations yield, when there is no pressure gradient in the flow direction,

$$-\frac{v^2}{r} = -\frac{1}{\rho} \frac{\partial p'}{\partial r} \tag{3.2.25}$$

$$0 = \mu \left(\frac{d^2 v}{dr^2} + \frac{1}{r} \frac{dv}{dr} - \frac{v}{r^2} \right) = \mu \frac{d}{dr} \left[\frac{1}{r} \frac{d}{dr} (rv) \right] \tag{3.2.26}$$

$$0 = -\frac{\partial p'}{\partial z} \tag{3.2.27}$$

where v is the rotation velocity (velocity in the θ direction), r is the radial coordinate, and z is the vertical coordinate. Equations (3.2.25) and (3.2.27) indicate that p' is a function only of r. Integration of Eq. (3.2.26) provides the velocity distribution,

$$v = Ar + \frac{B}{r} \tag{3.2.28}$$

where A and B are constants that must be determined by the boundary conditions.

If the fluid occupies the space between two coaxial rotating cylinders, whose angular velocities are Ω_1 and Ω_2, respectively, then the values of A and B are given by

$$A = \frac{\Omega_2 r_2^2 - \Omega_1 r_1^2}{r_2^2 - r_1^2} \tag{3.2.29}$$

$$B = \frac{(\Omega_1 - \Omega_2) r_1^2 r_2^2}{(r_2^2 - r_1^2)} \tag{3.2.30}$$

(recall that r_1 and r_2 are the radii of the inner and outer cylinders, respectively).

In the limiting case of $r_2 = \infty$, Eqs. (3.2.28)–(3.2.30) refer to steady flow in an infinite domain around a rotating cylinder whose radius and angular velocity are r_1 and Ω_1, respectively. In such a case, these equations yield

$$v = \frac{\Omega_1 r_1^2}{r} \tag{3.2.31}$$

This expression is identical to the velocity distribution in a potential (irrotational) vortex with circulation Γ, given by

$$\Gamma = 2\pi\Omega_1 r_1^2 \tag{3.2.32}$$

The solution of the Navier–Stokes equations given by Eq. (3.2.31) is an interesting case in which the potential flow solution is identical to that of the viscous flow solution.

In the limiting case of $\Omega_1 = r_1 = 0$, Eqs. (3.2.28)–(3.2.30) represent steady flow inside a cylindrical rotating tank, whose radius and angular velocity are r_2 and Ω_2, respectively. In this case, the result is

$$v = \Omega_2 r \tag{3.2.33}$$

This expression represents a rotational vortex.

3.3 CREEPING FLOWS

For very small Reynolds number, namely with small flow velocities and small size of the body, or with large viscosity of the fluid, the nonlinear inertial terms of the Navier–Stokes equations are much smaller than the viscous friction terms. Such flows are called creeping flows. In these flows, the Navier–Stokes equations can be approximated by the Stokes equations,

$$\rho\frac{\partial u_i}{\partial t} = -\frac{\partial p'}{\partial x_i} + \mu\frac{\partial^2 u_i}{\partial x_k^2} \tag{3.3.1}$$

These equations (for each component), along with the equation of continuity, represent the basic equations for creeping flows. Considering a solid body subject to slow movement in the domain, or slow movement of fluid around a stationary solid body, the fluid velocity at the body surface is equal to that of the solid surface. This provides a convenient boundary condition. Also, by taking the divergence of Eq. (3.3.1), we obtain

$$\frac{\partial^2 p}{\partial x_k^2} = 0 \tag{3.3.2}$$

This indicates that the pressure is a harmonic function in creeping flows.

In two-dimensional, steady creeping flow, Eq. (3.3.1) becomes

$$\nabla^4\psi = 0 \tag{3.3.3}$$

indicating that the stream function is a biharmonic function (for the assumed conditions).

Considering a very slow motion of a sphere of radius r_0, with velocity U in the x direction, the pressure function is given by

$$p = -\frac{3}{2}\frac{\mu U r_0 x}{r^3} \tag{3.3.4}$$

where the center of the sphere represents the origin of the coordinate system and $p \to 0$ for $r \to \infty$ has been assumed. Incorporating both the net pressure force implied by Eq. (3.3.4) and skin friction drag, the *drag coefficient* for the sphere is

$$C_D = \frac{F_D}{(\rho/2)\pi r_0^2 U^2} = \frac{24}{Re} \tag{3.3.5}$$

where F_D is the total drag force applied to the moving sphere. Equation (3.3.5) can be used to measure the viscosity of fluids. It is useful with regard to settling of solid particles in a fluid medium (see Chap. 15).

Experimental results indicate that expression (3.3.5) is accurate for extremely small values of Reynolds number. However, the velocity distribution obtained using the Stokes equation (3.3.1) is not usually very accurate, particularly at larger distances from the sphere. This is because of the formation of a wake region behind the sphere. The solution of the Stokes equation yields a velocity distribution that is symmetrical with regard to a plane perpendicular to the flow direction and passing through the center of the sphere. In other words, it does not incorporate a wake region. This result is also seen by considering the orders of magnitude of the inertial and viscous terms of the Navier–Stokes equations,

$$\rho u_k \frac{\partial u_i}{\partial x_k} = O\left(\rho \frac{U^2}{r}\right) \qquad \mu \frac{\partial^2 u_i}{\partial x_k^2} = O\left(\mu \frac{U}{r^2}\right) \tag{3.3.6}$$

These expressions indicate that the ratio between the inertial and viscous terms is proportional to r. Therefore for distances much greater than r_0 the viscous terms become relatively unimportant, and it may be concluded that the solution of the Stokes equation is not applicable at large distances from the sphere.

An improvement of Stokes' analysis was provided by Oseen, who considered the deviation imposed on the uniform flow U by the presence of the sphere. Therefore he considered a velocity distribution,

$$u = U + u' \qquad v = v' \qquad w = w' \tag{3.3.7}$$

where u', v', and w' are the velocity deviations in the x, y, and z directions, respectively. By introducing Eq. (3.3.7) into the Navier–Stokes equations and neglecting the second-order terms with regard to the velocity deviations, Oseen obtained

$$\frac{\partial u_i'}{\partial t} + U\frac{\partial u_i'}{\partial x} = -\frac{1}{\rho}\frac{\partial p}{\partial x_i} + v\frac{\partial^2 u_i'}{\partial x_k^2} \tag{3.3.8}$$

Here, x represents the direction of the uniform flow U, and x_i represents each of the coordinates. The terms of Eq. (3.3.8) which were added to Eq. (3.3.1) have been shown to improve the calculation of creeping flow at large distances from the center of the sphere.

Applying the divergence operation on Eq. (3.3.7), the continuity equation is written as

$$\frac{\partial u'_k}{\partial x_k} = 0 \tag{3.3.9}$$

(since the uniform flow also must follow continuity). For steady flows, it is possible to consider that each component of the velocity deviation from the uniform flow velocity, U consists of two parts, given by

$$u_i = u'_{1i} + u'_{2i} \tag{3.3.10}$$

where u'_{1i} is a potential flow component, originating from a potential function ϕ. Therefore

$$u_{1i} = -\frac{\partial \phi}{\partial x_i} \qquad \frac{\partial^2 \phi}{\partial x_i^2} = 0 \tag{3.3.11}$$

It is considered that u'_{1i} is associated with the balance of the pressure gradient term of Eq. (3.3.8), whereas u'_{2i} is associated with the frictional force. By applying these assumptions, and introducing Eq. (3.3.11) into Eq. (3.3.8), we obtain

$$p = \rho U \frac{\partial \phi}{\partial x} \tag{3.3.12}$$

The components u'_{2i} are represented by

$$u_{2i} = \frac{\partial W}{\partial x_i} - \delta_i W \frac{U}{v} \tag{3.3.13}$$

where $\delta_1 = 1$, and $\delta_2 = \delta_3 = 0$. The function W must satisfy

$$\frac{\partial W}{\partial x} = \frac{v}{U} \frac{\partial^2 W}{\partial x_k^2} \tag{3.3.14}$$

The appropriate solution of Eqs. (3.3.11) and (3.3.14) represents the essence of Oseen's analysis. Such solutions were obtained for a sphere moving at a uniform speed U. In this case the drag coefficient is

$$C_D = \frac{24}{Re} \left(1 + \frac{3}{16} Re \right) \tag{3.3.15}$$

Generally, the drag coefficient can be expressed in terms of a series expansion of the Reynolds number. Equation (3.3.15) represents the first and second terms of such a series. Additional terms have been developed in more recent studies. Stokes' solution of Eq. (3.3.5) is considered to be applicable in cases of Reynolds numbers smaller than one. Oseen's solution given in Eq. (3.3.15) is applicable up to Reynolds numbers equal to 2. For higher Reynolds numbers more terms should be added to the power series given by Eq. (3.3.15). Flow through porous media can be considered as creeping flow around the solid particles that comprise the porous matrix. When the Reynolds number of the flow, based on a characteristic size of the matrix particle, is smaller than unity, then *Darcy's law* is useful (see Sec. 4.4), and the gradient of the piezometric head is proportional to the average interstitial flow velocity, as well as the specific discharge.

3.4   UNSTEADY FLOWS

There are several exact solutions of the Navier–Stokes equations for unsteady flows. Examples of such flows in the present section also are used to visualize the basic concept of the *boundary layer*.

3.4.1   Quasi-Steady-State Oscillations of a Flat Plate

Consider a flat plate subject to cosinusoidal oscillations. The domain is subject to a uniform pressure distribution. Therefore the Navier–Stokes equations (3.1.1) reduce to

$$\frac{\partial u}{\partial t} = v\frac{\partial^2 u}{\partial y^2} \qquad p' = \text{constant} \tag{3.4.1}$$

It should be noted that Eq. (3.4.1) is identical to the diffusion equation, which is applicable in problems of heat conduction or mass diffusion. The exact solution of Eq. (3.4.1) given in the following paragraphs is similar to some particular solutions of heat conduction in solids. Further discussion of diffusion is presented in Chap. 10.

The differential Eq. (3.4.1) is subject to the boundary conditions,

$$u = U_0 \cos(\omega t) \qquad \text{at} \qquad y = 0$$
$$u = 0 \qquad \text{at} \qquad y \to \infty \tag{3.4.2}$$

Noting that we are looking for a quasi-steady-state solution, only two spatial boundary conditions are required to solve this equation. We assume that the solution is of the form

$$u = \text{Re}[U(y)\exp(i\omega t)] \tag{3.4.3}$$

Here, Re represents the real part of the complex quantity. We introduce Eq. (3.4.3) into Eq. (3.4.1) to obtain

$$\frac{d^2U}{dy^2} - \frac{i\omega}{v}U = 0 \tag{3.4.4}$$

By solving this differential equation and presenting the boundary conditions for U, which are implied by Eq. (3.4.2), we obtain

$$U = U_0 \exp\left[-y\sqrt{\frac{\omega}{2v}}(1+i)\right] \tag{3.4.5}$$

Finally, introducing Eq. (3.4.5) into Eq. (3.4.3), the complete solution is obtained,

$$u = U_0 \exp\left(-y\sqrt{\frac{\omega}{2v}}\right) \cos\left(\omega t - y\sqrt{\frac{\omega}{2v}}\right) \tag{3.4.6}$$

Equation (3.4.6) indicates that the amplitude of the velocity oscillations is subject to exponential decrease with the coordinate y. The practical outcome of this expression may be evaluated by considering the value of $y = \delta$, where the amplitude is 1 percent of its value at the flat plate. From Eq. (3.4.6),

$$\delta = \sqrt{\frac{v}{\pi f}}\ln(100) \tag{3.4.7}$$

where f is the frequency of the plate oscillations ($\omega = 2\pi f$). For water, with kinematic viscosity $v = 10^{-6}$ m^2/s, and assuming a frequency $f = 1$ s^{-1}, we obtain $\delta = 2.6 \times 10^{-3}$ m. This result indicates that only a very thin layer of fluid adjacent to the flat plate is subject to oscillations induced by the flat plate motion. The layer in which the oscillation amplitude is larger than 1 percent of the flat plate amplitude can be termed as a boundary layer. The phenomena of boundary layers is typical of regions close to solid boundaries of flow domains occupied by fluid with low viscosity. Boundary layers are discussed in more detail in Chap. 6.

3.4.2 Unsteady Motion of a Flat Plate

Consider a flat plate at rest at time $t \le 0$ but moving at constant velocity U for $t > 0$. The basic differential Eq. (3.4.1) also is applicable in this case, but the boundary conditions are different. In this case

$$
\begin{aligned}
u = 0 \quad &\text{at} \quad t \le 0 \quad &&\text{for all values of } y \\
u = U \quad &\text{at} \quad t > 0 \quad &&\text{for } y = 0 \\
u = 0 \quad & &&\text{for } y \to \infty
\end{aligned}
\tag{3.4.8}
$$

It is convenient to define a new dimensionless coordinate,

$$\eta = \frac{y}{2\sqrt{vt}} \tag{3.4.9}$$

The modified boundary conditions, in terms of η, are

$$
\begin{aligned}
u &= U &\text{at} &\quad \eta = 0 \\
u &= 0 &\text{at} &\quad \eta \to \infty
\end{aligned}
\tag{3.4.10}
$$

The second boundary condition of Eq. (3.4.10) incorporates both the first and the third boundary conditions of Eq. (3.4.8).

Using the definition (3.4.9), it is easy to find

$$\frac{\partial u}{\partial y} = \frac{du}{d\eta}\frac{\eta}{y} \qquad \frac{\partial^2 u}{\partial y^2} = \frac{d^2 u}{d\eta^2}\left(\frac{\eta}{y}\right)^2 \qquad \frac{\partial u}{\partial t} = \frac{du}{d\eta}\left(-\frac{\eta}{2t}\right) \tag{3.4.11}$$

Introducing Eq. (3.4.11) into Eq. (3.4.1), integrating twice, and introducing the boundary conditions of Eq. (3.4.10), we obtain

$$u = U\left(1 - \frac{2}{\sqrt{\pi}}\int_0^\eta e^{-\xi^2}\,d\xi\right) = U(1 - \text{erf}(\eta)) = U\,\text{erfc}(\eta) \tag{3.4.12}$$

where erf and erfc are the error and complementary error functions, respectively, and ξ is a dummy variable of integration. Again referring to water, as an example, we find that only a thin layer adjacent to the flat plate takes part in the flow, even up to extremely large times.

3.5 NUMERICAL SIMULATION CONSIDERATIONS

Numerical schemes aiming at the solution of the mass conservation and Navier–Stokes equations are usually based on finite difference or finite element methods. By these methods the numerical grid and the basic equations of mass and momentum conservation are used to create a set of approximately linear equations, which incorporate the unknown values of various variables at all grid points. The basic four equations of mass and momentum conservation incorporate four unknown variables for each grid point. These unknown values, for the three-dimensional domain, are the three components of the velocity vector and the pressure. If the domain is two-dimensional, or axisymmetrical, then the two components of the velocity vector can be replaced by the stream function.

As previously noted, the number of boundary conditions needed to solve a differential equation is determined by its order and the dimensions of the domain. With regard to the spatial derivatives of the velocity components, the Navier–Stokes equations are second-order partial differential equations. Therefore two boundary conditions are needed for each velocity component, with regard to each relevant coordinate. Velocity components also are subject to the first derivative in time. Therefore the initial distribution of all velocity components in the entire domain is needed. The pressure is subject to the first spatial derivative. Therefore boundary conditions also are required for the pressure, with regard to each relevant coordinate. If the stream function is applied, in a two-dimensional or axisymmetrical domain, then the basic set of four differential equations can be replaced by the fourth-order differential equation, which is given by Eq. (3.1.8). The solution of this equation requires four boundary conditions for the stream function with regard to each relevant coordinate, and initial distribution of the stream function in the domain.

For numerical simulation of the Navier–Stokes equations, it is common to consider applying the vorticity tensor, as shown in Eq. (3.1.3), or the vorticity vector, as given by Eq. (3.1.5). However, boundary conditions for vorticity are derived from appropriate considerations based on values of the velocity components.

Typical boundary conditions for the solution of the Navier–Stokes equations have been considered in Sec. 3.1. However, at this point it is appropriate to review the various types of boundary conditions, useful for the numerical solution of the various forms of these equations.

3.5.1 Basic Presentation

The solution of Eq. (3.1.1) is based on the following considerations:

At a solid surface, all velocity components are identical to those of the solid surface; if the solid surface is at rest then all velocity components vanish.

At the interface between two immiscible fluids, pressure and components of the velocity and shear stress are identical at both sides of the interface; shear stress components are proportional to the gradients of the velocity components.

At the interface between two immiscible fluids with large differences in viscosity, e.g., liquid and gas, the shear stress vanishes (except for the case of wind-driven flows).

At the entrance of the domain and/or exit cross sections the distribution of the velocity components is prescribed.

At the entrance or exit cross section of the domain the pressure distribution is prescribed.

The initial distribution of velocity components should be given.

3.5.2 Presentation with the Stream Function

For the solution of Eq. (3.1.8), the following considerations hold:

At a solid surface, spatial derivatives of the stream function are identical to velocity components of the solid surface; if the solid surface is at rest, spatial derivatives of the stream function vanish. The solid boundary represents a streamline at which the stream function has a constant value.

At the interface between two immiscible fluids, the first and second gradients of the stream function are identical on both sides of the interface. The interface represents a streamline, at which the stream function has a constant value.

At the interface between two immiscible fluids with large viscosity difference, e.g., liquid and gas (the interface is considered as the free surface of the liquid), the second gradient of the stream function vanishes. The free surface of the fluid is a streamline.

The initial distribution of the stream function in the domain should be given.

It should be noted that interfaces and free surfaces usually represent a sort of nonlinear boundary condition with regard to the velocity components, since the position of the boundary itself (where the boundary condition is to be applied) is part of the solution to the problem. Furthermore, determination of the exact location of free surfaces is very complicated.

Difficulties in solving the Navier–Stokes equations are very often associated with the nonlinear second term of Eq. (3.1.1), or the second and third terms of Eq. (3.1.8). If the flow is dominated by the nonlinear terms, then the numerical simulation is extremely complex, and some methods should be used to obtain a convergent numerical scheme. Furthermore, if boundary conditions are nonlinear, then the numerical solution may require significant approximations to assure convergence of the simulation process. The topic of "computational fluid mechanics" refers to different methods of solving these differential equations. For the present section, we consider only the numerical solution of creeping flows. In such flows the right-hand side terms of Eq. (3.1.8) are very small. Therefore the Navier–Stokes equations are approximated by

$$\Delta\Delta\Psi = 0 \qquad \frac{\partial^4\Psi}{\partial x^4} + 2\frac{\partial^2\Psi}{\partial x^2}\frac{\partial^2\Psi}{\partial y^2} + \frac{\partial^4\Psi}{\partial y^4} = 0 \qquad (3.5.1)$$

This is an *elliptic* differential equation (see Sec. 1.3.3).

As an example, consider a domain bounded on a square, where

$$\Psi = 0 \qquad \text{at} \qquad x = 0, 1 \qquad y = 0, 1$$

$$\frac{\partial \Psi}{\partial n} = 0 \qquad \text{at} \qquad x = 0, 1 \qquad y = 0 \qquad\qquad (3.5.2)$$

$$\frac{\partial \Psi}{\partial n} = 1 \qquad \text{at} \qquad y = 1$$

and a derivative with regard to n is the normal derivative. We introduce a new variable $w(x, y)$, which is defined by

$$\Delta \Psi = \frac{\partial^2 \Psi}{\partial x^2} + \frac{\partial^2 \Psi}{\partial y^2} = w$$

$$\Delta w = \frac{\partial^2 w}{\partial x^2} + \frac{\partial^2 w}{\partial y^2} = 0 \qquad\qquad (3.5.3)$$

The terms of these expressions can be approximated using the following finite difference approximations:

$$\left(\frac{\partial \Omega}{\partial x}\right)_{i,j} \approx \frac{\Omega_{i+1/2,j} - \Omega_{i-1/2,j}}{\Delta x} \qquad\qquad (3.5.4)$$

$$\left(\frac{\partial \Omega}{\partial y}\right)_{i,j} \approx \frac{\Omega_{i,j+1/2} - \Omega_{i,j-1/2}}{\Delta y} \qquad\qquad (3.5.5)$$

$$\left(\frac{\partial^2 \Omega}{\partial x^2}\right)_{i,j} \approx \frac{1}{\Delta x}\left[\left(\frac{\partial \Omega}{\partial x}\right)_{i+1/2,j} - \left(\frac{\partial \Omega}{\partial x}\right)_{i-1/2,j}\right]$$

$$\approx \frac{\Omega_{i+1/2,j} - 2\Omega_{i,j} + \Omega_{i-1/2,j}}{(\Delta x)^2} \qquad\qquad (3.5.6)$$

$$\left(\frac{\partial^2 \Omega}{\partial y^2}\right)_{i,j} \approx \frac{1}{\Delta y}\left[\left(\frac{\partial \Omega}{\partial y}\right)_{,j+1/2} - \left(\frac{\partial \Omega}{\partial y}\right)_{i,j+1/2}\right]$$

$$\approx \frac{\Omega_{i,j+1/2} - 2\Omega_{i,j} + \Omega_{i,j+1/2}}{(\Delta y)^2} \qquad\qquad (3.5.7)$$

where Ω is a dummy variable representing Ψ or w. Subscripts i, j refer to the point i, j of the finite difference grid shown in Fig. 3.1.

Since the numerical grid shown in Fig. 3.1 consists of small squares, for simplicity we assume that $\Delta x = \Delta y = k$. Therefore by introducing these values and Eqs. (3.5.6) and (3.5.7) into Eq. (3.5.3), we obtain

$$\Psi_{i+1,j} + \Psi_{i-1,j} + \Psi_{i,j+1} + \Psi_{i,j-1} - 4\Psi_{i,j} = k^2 w$$

$$w_{i+1,j} + w_{i-1,j} + w_{i,j+1} + w_{i,j-1} - 4w_{i,j} = 0 \qquad\qquad (3.5.8)$$

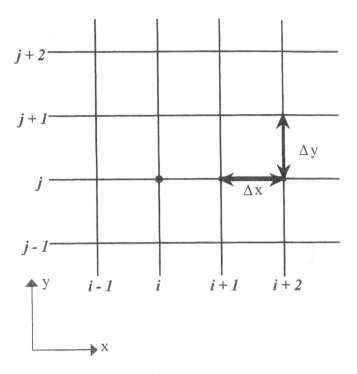

Figure 3.1 The finite difference grid.

Also, the boundary conditions of Eq. (3.5.2) become

$$\Psi = 0 \qquad \text{on all boundaries}$$

$$w = \frac{\partial^2 \Psi}{\partial n^2} \quad \text{on all boundaries} \tag{3.5.9}$$

The set of linear equations obtained by considering all grid points and using Eqs. (3.5.8) and (3.5.9) can be solved by an appropriate iterative procedure. Basically the set of two differential equations given by Eq. (3.5.3) is solved very similarly to the solution of the Laplace equation, which is discussed in greater detail in the following chapter.

PROBLEMS

Solved Problems

Problem 3.1 Introduce the expression for the vorticity vector into Eq. (3.1.5), to obtain an equation of motion based on the velocity components.

Solution

The vorticity vector in a two-dimensional flow field is given by

$$\omega = \frac{\partial v}{\partial x} - \frac{\partial u}{\partial y}$$

Introducing this expression into Eq. (3.1.5), we obtain

$$\frac{\partial^2 v}{\partial x \partial t} - \frac{\partial^2 u}{\partial y \partial t} + u \left(\frac{\partial^2 v}{\partial x^2} - \frac{\partial^2 u}{\partial y \partial x} \right) + v \left(\frac{\partial^2 v}{\partial x \partial y} - \frac{\partial^2 u}{\partial y^2} \right)$$
$$= v \left(\frac{\partial^3 v}{\partial x^3} - \frac{\partial^3 u}{\partial y \partial x^2} + \frac{\partial^3 v}{\partial x \partial y^2} - \frac{\partial^3 u}{\partial y^3} \right)$$

Problem 3.2 Figure 3.2 shows a plate with an orientation angle α, on which a fluid layer with thickness b is subject to flow with a free surface. The viscosity and density of the fluid are μ and ρ, respectively.

(a) Determine the value of the gradient of the piezometric head in the x-direction.
(b) Determine the value of the pressure gradient in the y-direction. What is the value of the pressure at the channel bottom?
(c) Determine the velocity and shear stress distributions.
(d) Determine the discharge per unit width and the average velocity.

Solution

(a) From Fig. 3.2,

$$\frac{\partial Z}{\partial x} = -\sin \alpha : \quad \frac{\partial Z}{\partial y} = \cos \alpha$$

The gradient of the piezometric pressure in the x-direction is given by

$$\frac{d p^*}{d x} = \frac{\partial p}{\partial x} + \frac{\partial Z}{\partial x} \frac{\partial p}{\partial x} - \rho g \sin \alpha = -\rho g J$$

Along the streamline representing the free surface of the fluid, the pressure vanishes. Therefore the pressure gradient in the x-direction is zero along that streamline, as well as along other streamlines, and the piezometric head gradient in the x-direction is given by $J = \sin \alpha$.

(b) According to Eq. (3.2.12) and the value of the partial derivatives of Z, as given in the previous part of this solution, we obtain

$$\frac{\partial p}{\partial y} + \rho g \cos \alpha = 0 \quad \Rightarrow \quad \frac{\partial p}{\partial y} = -\rho g \cos \alpha$$

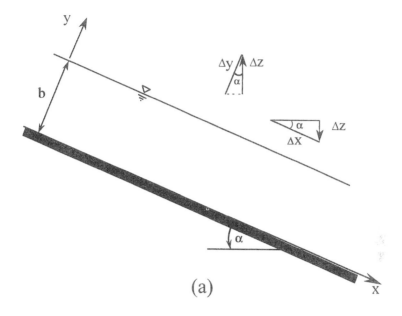

(a)

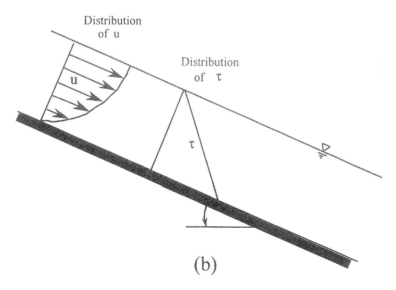

(b)

Figure 3.2 Definition sketch, Problem 3.2.

Direct integration of this expression, while considering that the pressure vanishes at the free surface of the fluid layer (at $y = b$), results in

$$p = \rho g(b - y)\cos \alpha$$

This expression indicates that the pressure at the fluid layer bottom is $(p)_{y=0} = \rho g b \cos \alpha$.

(c) Due to the very low viscosity of air, the shear stress vanishes at the free surface of the fluid layer. Therefore according to Eq. (3.2.9), we obtain

$$0 = -b\rho g \sin \alpha + \mu C_1 \quad \Rightarrow \quad C_1 = \frac{b\rho g}{\mu}\sin \alpha = \frac{bg}{\nu}\sin \alpha$$

At the bottom of the fluid layer ($y = 0$), the velocity vanishes. Therefore Eq. (3.2.8) yields $C_2 = 0$. By introducing values of the piezometric head gradient and those of C_1 and C_2 into Eqs. (3.2.8) and (3.2.9), we obtain the following expressions for the velocity and shear stress distributions, respectively:

$$u = \frac{g \sin \alpha}{\nu}\left(by - \frac{y^2}{2}\right) \qquad \tau = (b - y)\rho g \sin \alpha$$

(d) While referring to Eq. (3.2.10), we may consider that $y_1 = 0$, and $y_2 = b$. By introducing values of the piezometric head gradient and those of C_1 and C_2 into Eqs. (3.2.10), we obtain the following expression for the discharge per unit width and the average flow velocity, respectively:

$$q = \frac{gb^3 \sin \alpha}{3\nu} \qquad V = \frac{q}{b} = \frac{gb^2 \sin \alpha}{3\nu}$$

Problem 3.3 A fluid layer flows between two plates, with orientation angle α with respect to horizontal. The thickness of the fluid layer is b. The lower plate is stationary. The upper plate moves upward with velocity U. The pressure at the bottom of the fluid layer is given at two points: at $x = 0$ the pressure is p_0, and at $x = L$ the pressure is p_L. The viscosity and density of the fluid are μ and ρ, respectively.

(a) Determine the value of the gradient of the piezometric head in the x-direction.
(b) Determine the pressure distribution in the entire domain.
(c) Determine the velocity and shear stress distributions.
(d) Determine the discharge per unit width and the average velocity.
(e) Determine the power per unit area that is needed to move the upper plate.

Solution

(a) From geometrical considerations,

$$\frac{\partial Z}{\partial x} = -\sin\alpha : \frac{\partial Z}{\partial y} = \cos\alpha$$

The gradient of the piezometric pressure in the x-direction is then given by

$$\frac{dp^*}{dx} = \frac{\partial p}{\partial x} + \frac{\partial Z}{\partial x} = \frac{p_L - p_o}{L} - \rho g \sin\alpha = -\rho g J$$

$$\Rightarrow \quad J = \frac{p_o - p_L}{\rho g L} + \sin\alpha$$

(b) From part (a),

$$\frac{\partial p}{\partial x} = \frac{p_L - p_0}{L} : \quad \Rightarrow \quad p = p_0 + \frac{p_L - p_0}{L} x + f(y)$$

where $f(y)$ is a function of y that vanishes at $y = 0$. Differentiation of the last expression yields

$$\frac{\partial p}{\partial y} = f'(y)$$

According to Eq. (3.2.12) and the value of the partial derivatives of Z, as given in part (a) of this solution, we obtain

$$\frac{\partial p}{\partial y} + \rho g \cos\alpha = 0 \quad \Rightarrow \quad \frac{\partial p}{\partial y} = -\rho g \cos\alpha = f'(y)$$

Direct integration of this expression yields

$$f(y) = -\rho g y \cos\alpha \quad \Rightarrow \quad p = p_0 + \frac{p_L - p_0}{L} x - \rho g y \cos\alpha$$

This expression indicates that the pressure at $x = 0$ at the top of the fluid layer is

$$(p)_{y=b} = p_0 - \rho g b \cos\alpha$$

(c) At the fluid layer bottom ($y = 0$), the velocity vanishes. Therefore by using Eq. (3.2.8), we find $C_2 = 0$. At the upper plate the fluid velocity is identical to that of the moving plate. Therefore Eq. (3.2.8) yields for $y = b$,

$$-U = \frac{b^2}{2\mu}\left(\frac{p_L - p_0}{L} - \rho g \sin\alpha\right) + C_1 b$$

$$\Rightarrow \quad C_1 = \frac{b}{2\mu}\left(\frac{p_0 - p_L}{L} + \rho g \sin\alpha\right) - \frac{U}{b}$$

By introducing values of the piezometric pressure gradient and those of C_1 and C_2 into Eqs. (3.2.8) and (3.2.9), we obtain the following expressions for the velocity and shear stress distributions, respectively:

$$u = \frac{b}{2\mu}\left(\frac{p_0 - p_L}{L} + \rho g \sin\alpha\right)(by - y^2) - \frac{U}{b}y$$

$$\tau = \frac{b}{2}\left(\frac{p_0 - p_L}{L} + \rho g \sin\alpha\right)(b - 2y) - \mu\frac{U}{b}$$

(d) While referring to Eq. (3.2.10) we consider that $y_1 = 0$ and $y_2 = b$. By introducing values of the piezometric pressure gradient and those of C_1 and C_2 into Eqs. (3.2.10), we obtain the following expressions for the discharge per unit width and the average flow velocity, respectively:

$$q = \frac{b^3}{12\mu}\left(\frac{p_0 - p_L}{L} + \rho g \sin\alpha\right) - \frac{Ub}{2}$$

$$\Rightarrow \quad V = \frac{b^2}{12\mu}\left(\frac{p_0 - p_L}{L} + \rho g \sin\alpha\right) - \frac{U}{2}$$

(e) The power per unit width that is needed to move the upper plate is given by

$$N = (\tau u)_{y=b} = \frac{b^2 U}{2}\left(\frac{p_0 - p_L}{L} + \rho g \sin\alpha\right) + \frac{U^2 b}{2}$$

Problem 3.4   Determine the settling velocity of a sand particle in water. The particle may be assumed to be approximately spherical, with a diameter $d = 0.2$ mm. Its density is $\rho_s = 2,400$ kg/m^3. The density and kinematic viscosity of the water are $\rho_w = 1,000$ kg/m^3 and $\nu = 10^{-6}$ m^2/s, respectively.

Solution

The settling velocity is found by setting up an equilibrium force balance. First, the submerged weight of the sand particle is

$$W = \frac{4}{3}\pi r_0^3(\rho_s - \rho_w)g = \frac{4}{3}\pi(0.1 \times 10^{-3})^3(2,400 - 1,000)$$
$$= 5.86 \times 10^{-9} \text{ N}$$

where $r_0 = d/2$ is the radius of the particle. This expression is equal to the drag force during steady-state settling of the sand particle. According to Eq. (3.3.5),

$$W = \frac{24\nu}{Ud}\frac{\rho_w}{2}\pi r_0^2 U^2$$

$$\Rightarrow \quad U = \frac{W}{6\pi\rho_w \nu r_0}$$

$$= \frac{5.86 \times 10^{-9}}{6\pi \times 1,000 \times 10^{-6} \times 0.1 \times 10^{-3}} = 3.1 \times 10^{-3} \text{ m/s}$$

However, in order to use this equation, the Reynolds number must be checked. The value of the Reynolds number is

$$Re = \frac{Ud}{\nu} = \frac{3.1 \times 10^{-3} \times 0.2 \times 10^{-3}}{10^{-6}} = 0.62$$

which is less than 1. Therefore, use of the Stokes approximation was appropriate.

Problem 3.5 A flat plate is subject to oscillatory motions, with velocity given by

$$U_0 \sin(\omega t)$$

On top of the plate there is a semi-infinite fluid domain with uniform pressure distribution. The density and kinematic viscosity of the fluid are ρ and ν, respectively.

(a) Determine the velocity distribution in the domain.
(b) Determine the shear stress distribution. What is the phase lag between the maximum values of the shear stress and that of the velocity?
(c) What are the force and power per unit area needed to move the plate? What are the maximum values of these parameters?

Solution

(a) This problem is represented by the differential Eq. (3.5.1), subject to the following boundary conditions:

$$u = U_0 \sin(\omega t) \qquad \text{at} \qquad y = 0$$

$$u = 0 \qquad \text{at} \qquad y \to \infty$$

These boundary conditions suggest consideration of the following expression for the velocity:

$$u = \text{Im}\left[U(y)\exp(i\omega t)\right]$$

Similarly as in Eqs. (3.5.4)–(3.5.6), the velocity distribution is found as

$$u = U_0 \exp\left(-y\sqrt{\frac{\omega}{2\nu}}\right) \sin\left(\omega t - y\sqrt{\frac{\omega}{2\nu}}\right)$$

(b) The shear stress is given by

$$\tau = \rho v \frac{\partial u}{\partial y} = -\rho v \sqrt{\frac{\omega}{2v}} U_0 \left[\exp\left(-y\sqrt{\frac{\omega}{2v}}\right)\right]$$

$$\left[\sin\left(\omega t - y\sqrt{\frac{\omega}{2v}}\right) + \cos\left(\omega t - y\sqrt{\frac{\omega}{2v}}\right)\right]$$

The maximum value of μ is obtained when

$$\sin\left(\omega t - y\sqrt{\frac{\omega}{2v}}\right) = 1 \quad \Rightarrow \quad \omega t - y\sqrt{\frac{\omega}{2v}} = \frac{\pi}{2}$$

The maximum value of τ is obtained when

$$\sin\left(\omega t - y\sqrt{\frac{\omega}{2v}}\right) + \cos\left(\omega t - y\sqrt{\frac{\omega}{2v}}\right) \rightarrow \text{max}$$

$$\Rightarrow \quad \omega t - y\sqrt{\frac{\omega}{2v}} = \frac{\pi}{4}$$

Therefore the phase difference between u_{max} and τ_{max} is $\pi/4$.

(c) The force per unit area needed to move the plate is equal to the negative value of the shear stress at $y = 0$, namely

$$\frac{F}{A} = -(\tau)_{y=0} = \rho\sqrt{\frac{\omega v}{2}} U_0[\sin(\omega t) + \cos(\omega t)]$$

where F is the force and A is the area of the plate. The power needed to move the plate is equal to the product of that force with the velocity of the plate, or

$$\frac{N}{A} = \frac{F}{A}(u)_{y=0} = \rho\sqrt{\frac{\omega v}{2}} U_0^2[\sin(\omega t)][\sin(\omega t) + \cos(\omega t)]$$

The maximum value of this parameter is obtained when $\{\sin(\omega t)\,[\sin(\omega t) + \cos(\omega t)] \rightarrow \text{max}\}$. Differentiation of this expression indicates that the maximum value of the power is obtained when

$$\omega t = \frac{\pi}{2}\left(n - \frac{1}{4}\right) \quad \text{where} \quad n = 1, 2 \ldots$$

Unsolved Problems

Problem 3.6 The velocity distribution for flow between two plates is given by

$$u = U\left[1 - \left(\frac{2y}{b}\right)^2\right]$$

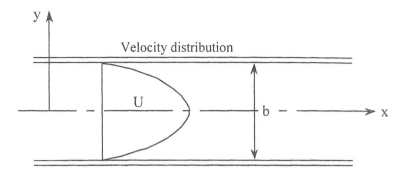

Figure 3.3 Flow between two plates, Problem 3.6.

where b is the gap between the two plates, U is the velocity at the centerline of the fluid layer, and y is the distance from the centerline (see Fig. 3.3).

(a) Show that the flow is a rotational flow. What is the vorticity distribution in the fluid layer?
(b) What boundary conditions are satisfied by the velocity distribution?
(c) Considering that the characteristic length and velocity are the gap between the plate and the average flow velocity, respectively, what is the expression for the dimensionless velocity distribution?
(d) What is the expression for the Reynolds number?

Problem 3.7   Water flows on an oblique plate forming a roof, as shown in Fig. 3.4. The water flows as a fluid layer with thickness $b = 10^{-3}$ m. The water density is $\rho = 1,000$ kg/m^3. Its kinematic viscosity is $\nu = 10^{-6}$ m^2/s. The slope of the roof is $\alpha = 30°$. Due to wind gusts, the surface of the flowing

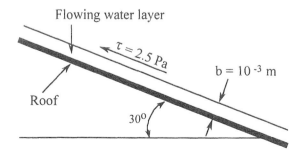

Figure 3.4 Flow along a sloping surface, Problem 3.7.

water layer is subject to a shear stress $\tau = 2.5$ Pa, in the upward direction of the roof.

(a) What is the water discharge per unit width of the roof?
(b) Prove that the flow is laminar.
(c) What is the shear stress applied on the roof?

Problem 3.8 A gap of thickness $b = 5 \times 10^{-4}$ m separates two vertical belts and is occupied by viscous oil, whose density is $\rho = 800$ kg/m^3, as shown in Fig. 3.5. The viscosity of the oil is $\mu = 8 \times 10^{-2}$ Pa s. One belt moves upward with a velocity of $V_1 = 2$ m/s. The other belt moves downward. The gravitational forces and the movement of the belts only affect the flow of the oil layer. The net discharge of the oil is zero.

(a) What is the velocity of the second belt?
(b) Draw a schematic of the velocity and shear stress distributions in the oil layer.

Problem 3.9 Figure 3.6 shows a "belt pump", which diverts oil from a lower tank to an upper one. The density of the oil is $\rho = 800$ kg/m^3 and its viscosity is $\mu = 8 \times 10^{-2}$ Pa s. The belt moves with velocity $U = 0.2$ m/s. The thickness of the oil layer is $b = 2 \times 10^{-3}$ m. The orientation angle of the belt is $\theta = 45°$. The horizontal distance between the two tanks is $L = 5$ m.

(a) What is the discharge delivered by the pump?
(b) What is the power needed to operate the pump?

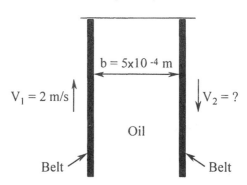

Atmospheric pressure

$b = 5 \times 10^{-4}$ m

$V_1 = 2$ m/s

$V_2 = ?$

Oil

Belt   Belt

Atmospheric pressure

Figure 3.5   Flow of oil between two belts, Problem 3.8.

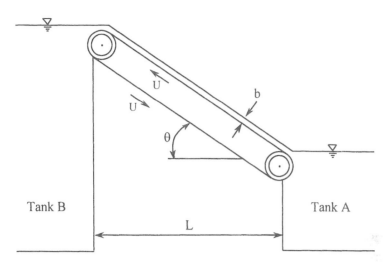

Figure 3.6 Belt pump, Problem 3.9.

(c) What is the efficiency of the pump?
(d) What should be the thickness of the fluid layer which maximizes the discharge?

Problem 3.10 The domain for a flow of oil is defined by the following stream function:

$$\Psi = U \left(y - \frac{y^3}{3b^2} \right)$$

where $U = 0.1$ m/s and $b = 0.05$ m. The density of the oil is $\rho = 800$ kg/m³, and its kinematic viscosity is $\nu = 8 \times 10^{-5}$ m²/s.

(a) Prove that the flow domain is the gap between two parallel plates, where the size of that gap is $2b$.
(b) What are the velocity and shear stress distributions in the flow domain?
(c) What is the gradient of the piezometric head?
(d) What is the power loss along a unit length of the flow domain?

Problem 3.11 Figure 3.7 shows oil flowing steadily along a vertical wall in a thin layer of thickness $b = 3 \times 10^{-3}$ m, with a discharge per unit width $q = 3 \times 10^{-3}$ m²/s. The density of the oil is $\rho = 800$ kg/m³.

(a) What are the viscosity and kinematic viscosity of the oil?
(b) What is the shear stress applied on the wall by the flowing oil?

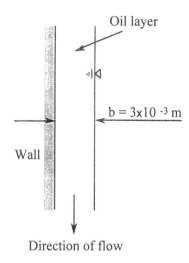

Oil layer

b = 3x10 $^{-3}$ m

Wall

Direction of flow

Figure 3.7   Flow of oil along a vertical wall, Problem 3.11.

Problem 3.12   Oil flows due to gravity on an oblique plate in a layer of thickness $b = 3 \times 10^{-3}$ m, as shown in Fig. 3.8. The angle of orientation of the plate is $\theta = 30°$ with respect to horizontal. The plate moves upward with velocity $V = 0.1$ m/s. The kinematic viscosity of the oil is $v = 8 \times 10^{-5}$ m^2/s, and its density is $\rho = 800$ kg/m^3.

(a)   Calculate and draw a schematic of the velocity and shear stress distributions.

(b)   What is the direction and value of the discharge per unit width?

Problem 3.13   Oil is located between two flat plates, as shown in Fig. 3.9. The kinematic viscosity of the oil is $v = 8 \times 10^{-5}$ m^2/s, and its density is $\rho =$

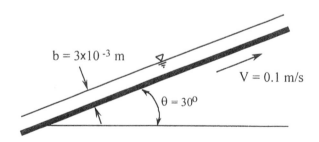

b = 3x10 $^{-3}$ m

V = 0.1 m/s

$\theta = 30°$

Figure 3.8   Flow of oil on a sloping surface, Problem 3.12.

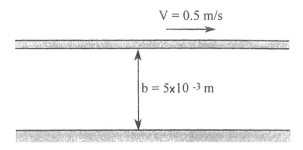

Figure 3.9   Viscous flow between two plates, Problem 3.13.

800 kg/m^3. The upper plate moves to the right with a velocity $V = 0.5$ m/s. The lower plate is stationary. The gap between the plates has thickness $b = 5 \times 10^{-3}$ m. The net discharge of the oil is zero.

(a)   What is the pressure gradient between the plates?
(b)   What is the shear stress at each one of the plates?
(c)   Where does the shear stress obtain its maximum and minimum absolute values?
(d)   Draw a schematic of the velocity and shear stress distributions.

Problem 3.14   Oil flows out of a tank, as shown in Fig. 3.10. The oil density is $\rho = 800$ kg/m^3 and its viscosity is $\mu = 8 \times 10^{-2}$ Pa s. The difference in elevation between the oil-free surface in the tank and the outlet is $\Delta h = 12.2$ m. The oil flows out through a pipe whose diameter and length are

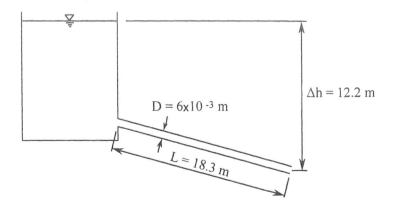

Figure 3.10   Definition sketch, Problem 3.14.

$D = 6 \times 10^{-3}$ m and $L = 18.3$ m. Piezometric head loss at the pipe entrance is negligible.

    (a)   What is the gradient of the piezometric head along the pipe?
    (b)   What is the oil discharge?
    (c)   Is the flow laminar? Why or why not?
    (d)   What is the power loss due to the flow through the pipe?

Problem 3.15   A motor shaft, with diameter $D_1 = 5 \times 10^{-2}$ m, rotates at a rate of $n = 1,200$ rpm, inside a bearing, as shown in Fig. 3.11. The internal diameter of the bearing is $D_2 = 5.02 \times 10^{-2}$ m. Its length is $L = 0.1$ m. The viscosity of the oil is $\mu = 10^{-2}$ Pa s. It occupies the gap between the bearing and the shaft. The shaft and the bearing form a system of coaxial cylinders.

    (a)   What is the shear stress applied on the oil?
    (b)   What is the power loss in the bearing?

Problem 3.16   Helium flows through a pipe of diameter D. The flow of helium is different from that of other fluids in that the usual no-slip condition at a solid boundary does not apply. There is some sliding at the pipe wall, and the helium has some velocity at that location. The boundary condition at the pipe wall is

$$(u^3)_{r=R} = \left(K \frac{du}{dr} \right)_{r=R}$$

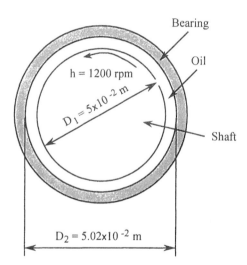

Figure 3.11 Definition sketch, Problem 3.15.

where r is the radial coordinate, R is the pipe radius, and K is a constant.

(a) Determine the velocity profile of the helium pipe flow.
(b) Determine the shear stress distribution.
(c) Determine the relationship between the discharge and the gradient of the piezometric pressure.

Problem 3.17 Oil flows from container A to container B through a pipe of length $L = 10$ m and diameter $d = 10^{-2}$ m, as shown in Fig. 3.12. The kinematic viscosity of the oil is $v = 2 \times 10^{-4}$ m^2/s. The diameter of each container is $D = 1$ m. When the flow starts, at $t = 0$, the oil level in container A is $H_1 = 1$ m, and in container B it is $H_2 = 0.1$ m. Steady flow conditions may be assumed.

(a) Calculate the initial discharge and average flow velocity. Prove that the flow is laminar.
(b) Develop the expression for the variation of oil levels in the containers. Find at what time the oil level in container B is equal to 0.5 m.

Problem 3.18 Figure 3.13 indicates two containers holding oil, with kinematic viscosity $v = 8 \times 10^{-5}$ m^2/s and density $\rho = 800$ kg/m^3. The containers are connected by a pipe with length $L = 500$ m and diameter $d = 5 \times 10^{-2}$ m. The oil level in container A is $H_1 = 55$ m, and in container B the oil level is $H_2 = 50$ m. Assume that oil levels in both containers are kept constant. Local head losses and the velocity head loss at the pipe exit may be neglected.

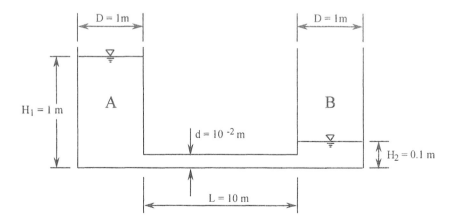

Figure 3.12 Definition sketch, Problem 3.17.

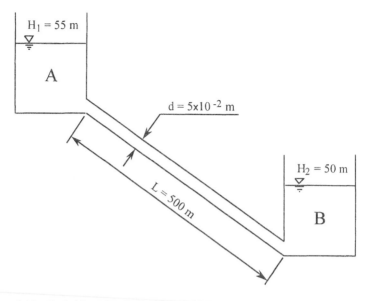

Figure 3.13 Definition sketch, Problem 3.18.

    (a)   What are the oil discharge and the average flow velocity?
    (b)   What is the Reynolds number of the flow?
    (c)   What is the shear stress at the pipe wall?
    (d)   What is the power loss due to the oil flow?

Problem 3.19   Figure 3.14 shows a laboratory system similar to an infusion system. At time $t = 0$, the fluid level is at point A. The initial fluid volume in the container is $U = 10^{-3}$ m^3. The container is a top open cylinder, with initial fluid depth $h_0 = 0.1$ m. The kinematic viscosity of the fluid is $\nu = 10^{-5}$ m^2/s, and its density is $\rho = 1{,}020$ kg/m^3. The fluid flows out of the container through a tube whose length is $L = 2$ m and whose diameter is $d = 10^{-3}$ m. At the exit of the pipe the pressure is kept constant, at $p = 10^4$ Pa. The bottom of the container is elevated to a level of $H = 1.5$ m above the pipe exit. It may be assumed that the flow is steady, with variable head loss.

    (a)   What is the initial, maximum fluid discharge (at time $t = 0$, when the fluid level is at point A)?
    (b)   What is the final, minimum discharge (when the fluid level is at point B)?
    (c)   How much time is required to empty the container?

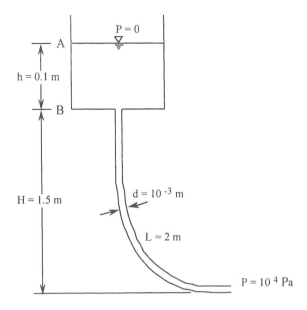

Figure 3.14 Fluid drainage by gravity, Problem 3.19.

Problem 3.20 Two types of fluids occupy the gap between two parallel horizontal flat plates, as shown in Fig. 3.15. There is no pressure gradient along the flow direction. The width of the gap between the plates is $b = 10^{-2}$ m. The lower half is occupied by a fluid whose density and viscosity are $\rho = 900$ kg/m^3 and $\mu = 0.1$ Pa s, respectively. The upper half of the gap is occupied by a second fluid, whose density and viscosity are $\rho = 700$ kg/m^3

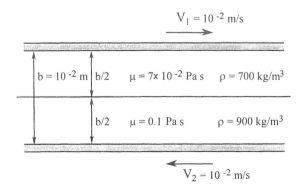

Figure 3.15 Flow of two fluids between plates, Problem 3.20.

and $\mu = 7 \times 10^{-2}$ Pa s, respectively. The upper plate moves to the right with a velocity of $V_1 = 10^{-2}$ m/s. The lower plate moves to the left with a velocity $V_2 = 10^{-2}$ m/s.

(a) Calculate and draw a schematic of the velocity and shear stress distributions in the fluid layers.

(b) Calculate the net discharge of each fluid.

Problem 3.21   Figure 3.16 shows oil flowing from container (1) to container (2) through a tube whose length and diameter are $L = 1.2$ m and $D = 4 \times 10^{-3}$ m, respectively. The oil flow is driven by a constant pressure $p = 10^4$ Pa, maintained in the free space of container (1), as well as the difference between the elevations of the oil free surfaces in both containers. Initially, that difference of elevation is $H_0 = 1$ m. Container (2) is open to the atmosphere.

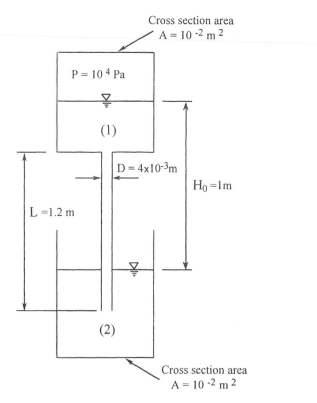

Figure 3.16   Definition sketch, Problem 3.21.

The cross-sectional area of each container is $A = 10^{-2}$ m^2. The density and viscosity of the oil are $\rho = 900$ kg/m^3 and $\mu = 0.1$ Pa s, respectively.

(a) Determine the general expression representing the relationship between the discharge Q, which flows from container (1) to container (2), and the parameters H, p, D, L, ρ, and μ.

(b) Determine the maximum, initial value of the discharge.

(c) Determine the time T, during which H will reduce from $H_0 = 1$ m to $H = 0.5$ m.

Problem 3.22 A viscous fluid flows from container A to container B through an annulus, as shown in Fig. 3.17. The annulus consists of a steel member, whose diameter is $D_1 = 0.02$ m, and a pipe, whose internal diameter is $D_2 = 0.022$ m. The difference between fluid levels in containers A and B is kept constant, at $H = 5$ m. The length of the annulus is $L = 100$ m. The fluid density and viscosity are $\rho = 900$ kg/m^3 and $\mu = 5 \times 10^{-3}$ Pa s, respectively.

(a) Determine the distributions of the velocity and shear stress in the annulus cross section.

(b) Determine the discharge, which flows through the annulus.

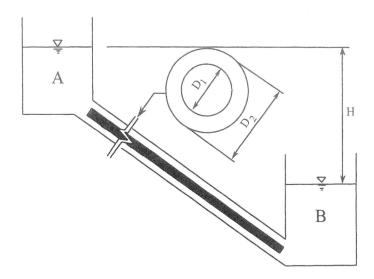

Figure 3.17 Definition sketch, Problem 3.22.

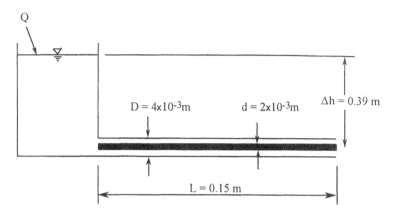

Figure 3.18   Definition sketch, Problem 3.23.

Problem 3.23   In Fig. 3.18, fluid flows out of the container through a horizontal pipe, whose length is $L = 0.15$ m and inner diameter is $D = 4 \times 10^{-3}$ m. Inside the pipe a metal member is inserted. The diameter of the metal member is $d = 2 \times 10^{-3}$ m, and it is coaxial with the pipe. The density and viscosity of the fluid are $\rho = 800$ kg/m^3 and $\mu = 5 \times 10^{-3}$ Pa s, respectively. The difference between the elevation of the free surface of the container fluid and the pipe exit is $\Delta h = 0.39$ m. A discharge Q flows into the container, to maintain the free surface of the container fluid at a constant level.

(a)   What is the value of the discharge Q, flowing into the container?
(b)   By how much should Q be increased, if the metal member is taken out of the pipe?
(c)   What is the total shear force applied on the metal member?

SUPPLEMENTAL READING

Batchelor, G. K., 1967. *An Introduction to Fluid Dynamics*, Cambridge University Press, London.

Streeter, V. L., 1961. *Handbook of Fluid Dynamics*, McGraw-Hill, New York. (Summarizes the different presentations and uses of the viscous flow equations.)

Wendt, J. F., ed., 1996. *Computational Fluid Dynamics*, Springer Verlag, Berlin. (Includes an introduction and survey of different approaches aimed at appropriate use of numerical approaches for the solution of the equations of motion.)

4

Inviscid Flows and Potential Flow Theory

4.1 INTRODUCTION

The vorticity form of the Navier–Stokes Eq. (3.1.3) implies that if the flow of a fluid with constant density initially has zero vorticity, and the fluid viscosity is zero, then the flow is always irrotational. Such a flow is called an ideal, irrotational, or inviscid flow, and it has a nonzero velocity tangential to any solid surface. A real fluid, with nonzero viscosity, is subject to a no-slip boundary condition, and its velocity at a solid surface is identical to that of the solid surface.

As indicated in Sec. 3.4, in fluids with small kinematic viscosity, viscous effects are confined to thin layers close to solid surfaces. In Chap. 6, concerning boundary layers in hydrodynamics, viscous layers are shown to be thin when the Reynolds number of the viscous layer is small. This Reynolds number is defined using the characteristic velocity, U, of the free flow outside the viscous layer, and a characteristic length, L, associated with the variation of the velocity profile in the viscous layer. Therefore the domain can be divided into two regions: (a) the inner region of viscous rotational flow in which diffusion of vorticity is important, and (b) the outer region of irrotational flow. The outer region can be approximately simulated by a modeling approach ignoring the existence of the thin boundary layer and applying methods of solution relevant to nonviscous fluids and irrotational flows. Following the calculation of the outer region of irrotational flow, viscous flow calculations are used to represent the inner region, with solutions matching the solution of the outer region. However, in cases of phenomena associated with boundary layer separation, matching between the inner and outer regions cannot be done without the aid of experimental data.

The present chapter concerns the motion of inviscid, incompressible, and irrotational flows. In cases of such flows the velocity vector is derived from a potential function. The vorticity of a vector derived from a potential function is zero, or

$$\vec{V} = \nabla\Phi \qquad \nabla \times \nabla\Phi = 0 \tag{4.1.1}$$

This expression indicates that every potential flow is also an irrotational flow.

In the following sections, special attention will be given to two-dimensional flows, which are the most common situation for analysis using potential flow theory. There also is some discussion of axisymmetric flows, and numerical solutions of two- and three-dimensional flows.

4.2 TWO-DIMENSIONAL FLOWS AND THE COMPLEX POTENTIAL

4.2.1 General Considerations

In cases of potential, incompressible, two-dimensional flows, velocity components are derived from the potential function, due to lack of vorticity, as well as from the stream function, due to the incompressibility of the fluid. Therefore the velocity components can be represented by

$$u = \frac{\partial\Phi}{\partial x} = \frac{\partial\Psi}{\partial y} \qquad v = \frac{\partial\Phi}{\partial y} = -\frac{\partial\Psi}{\partial x} \tag{4.2.1}$$

These relationships between the partial derivatives of the potential and stream functions are called the *Cauchy–Riemann equations*.

According to Eq. (4.2.1), the potential function can be determined by direct integration of the expressions for the velocity components,

$$\Phi = \int u\,dx + f(y) \qquad \text{or} \qquad \Phi = \int v\,dy + g(x) \tag{4.2.2}$$

The expression for $f(y)$ can be determined by

$$v = \frac{\partial}{\partial y}\left[\int u\,dx + f(y)\right] \Rightarrow \qquad f'(y) = v - \frac{\partial}{\partial y}\left[\int u\,dx\right] \Rightarrow$$

$$f(y) = \int \left\{v - \frac{\partial}{\partial y}\left[\int u\,dx\right]\right\} dy \tag{4.2.3}$$

By the same approach, the expression for $g(x)$ can be determined by

$$g(x) = \int \left\{u - \frac{\partial}{\partial x}\left[\int v\,dy\right]\right\} dx \tag{4.2.4}$$

If the expression for the potential function is given, then the expression for the stream function can be obtained by applying Eq. (4.2.1). The stream function expression also can be obtained by direct integration of the expressions of the velocity components, using

$$\Psi = -\int v\, dx + h(y) \quad \text{or} \quad \Psi = \int u\, dy + k(x) \tag{4.2.5}$$

where

$$
\begin{aligned}
h(y) &= \int \left\{ u + \frac{\partial}{\partial x} \left[\int v\, dy \right] \right\} dy \\
k(x) &= \int \left\{ v - \frac{\partial}{\partial x} \left[\int u\, dy \right] \right\} dx
\end{aligned}
\tag{4.2.6}
$$

According to Eq. (4.1.1) the velocity vector is defined as the gradient of the function Φ. Therefore the velocity vector is perpendicular to the equipotential contour lines. According to Eq. (2.5.10), contour lines with a constant value of Ψ are streamlines, namely, lines that are tangential to the velocity vector. Therefore equipotential lines are perpendicular to the streamlines. A schematic of several streamlines and equipotential lines, called a *flow-net*, is presented in Fig. 4.1. The differences in value between each pair of adjacent streamlines is $\Delta\Psi$. The difference in value between each pair of adjacent

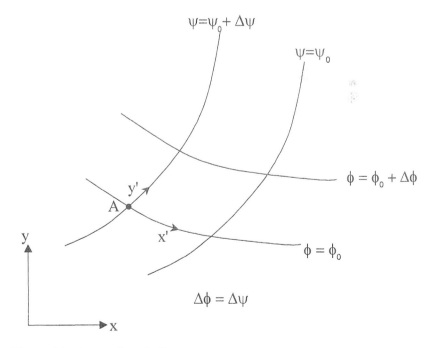

Figure 4.1 Schematics of a flow-net.

equipotential lines is $\Delta\Phi$. Usually, flow-nets are drawn so that $\Delta\Psi = \Delta\Phi$. Therefore, if at the point A of an intersection between a streamline and an equipotential line we adopt a Cartesian coordinate system, in which y' is tangential to the streamline and x' is tangential to the equipotential line, then according to Eq. (4.2.1), the small rectangle of the flow-net is a square.

By considering the incompressibility of the flow, as given by Eq. (2.5.7) or Eq. (2.5.8), and applying Eq. (4.1.1) or Eq. (4.2.1) with regard to the potential function, we obtain

$$\nabla \cdot (\nabla\Phi) = 0 \Rightarrow \qquad \nabla^2\Phi = 0 \Rightarrow \qquad \frac{\partial^2\Phi}{\partial x^2} + \frac{\partial^2\Phi}{\partial y^2} = 0 \qquad (4.2.7)$$

This expression indicates that the potential function must satisfy the Laplace equation.

Consider now the irrotational flow condition, which is given by vanishing values of all components of vorticity, $\vec{\omega}$ in Eqs. (2.3.11) and (2.3.12), and apply Eq. (4.2.1) with regard to the stream function, so

$$\frac{\partial^2\Psi}{\partial x^2} + \frac{\partial^2\Psi}{\partial y^2} = 0 \qquad (4.2.8)$$

indicating that the stream function also satisfies the Laplace equation. Therefore either the stream function or the potential function can be used for the presentation of the streamlines or equipotential lines.

If polar coordinates are used for the calculation of two-dimensional potential flow, then we may apply the following form of the Cauchy–Riemann equations,

$$u_r = \frac{\partial\Phi}{\partial r} = \frac{1}{r}\frac{\partial\Psi}{\partial\theta} \qquad v_\theta = \frac{1}{r}\frac{\partial\Phi}{\partial\theta} = -\frac{\partial\Psi}{\partial r} \qquad (4.2.9)$$

where u_r and v_θ are components of the velocity vector in the r and θ directions, respectively. The potential and stream functions can be determined if expressions for the velocity components are given, according to the method represented by Eqs. (4.2.2)–(4.2.6).

The discussion in the previous paragraphs has indicated that equipotential lines (lines of constant value of Φ) are orthogonal to streamlines (lines of constant value of Ψ). Therefore it is possible to consider the complex function w, as given by Eq. (1.3.91), which incorporates both functions in the complex domain. We may consider the plane w, which is depicted by the coordinates Φ and Ψ, as shown in Fig. 4.2. Equipotential lines and streamlines in the w plane of that figure represent the schematic of the flow-net. The plane of the complex variable z is depicted by applying the coordinates x and y. Streamlines and equipotential lines depicted in the z plane represent the common flow-net. The transformation of Φ–Ψ mapping in the w plane to x–y mapping in the

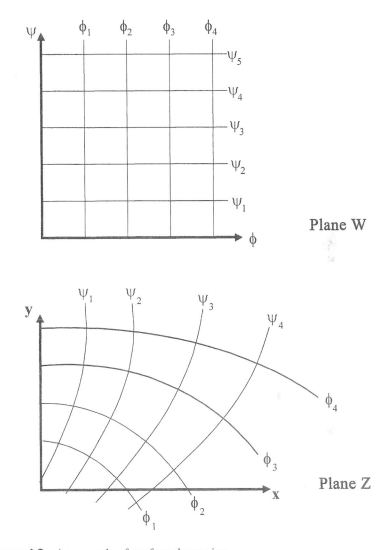

Figure 4.2 An example of conformal mapping.

z plane is called *conformal mapping*. An example of conformal mapping is represented in Fig. 4.2. Small squares in the w plane are transformed into small squares in the z plane by this procedure. The function w is called *the complex potential* and is represented by

$$w = \Phi + i\Psi \qquad (4.2.10)$$

The major properties of the complex potential and its implications with regard to Ψ and Φ are presented in Eqs. (1.3.90)–(1.3.99). The complex potential function is an analytical function, namely, a function of z. Various functions of z can be useful for the description and depiction of different flow domains, in terms of equipotential lines and streamlines.

As shown by Eqs. (4.2.7) and (4.2.8), the potential function and stream function satisfy the Laplace equation. Therefore the complex potential function also satisfies the Laplace equation, as it represents a linear combination of Φ and Ψ. Also, the Laplace equation is a linear differential equation. Therefore, if the complex potential w_1 represents a potential flow domain, and w_2 represents another potential flow domain, then any linear combination such as $\alpha w_1 + \beta w_2$ also represents a potential flow domain.

As shown by Eqs. (1.3.92)–(1.3.97),

$$\frac{dw}{dz} = \frac{\partial w}{\partial x} = -i\frac{\partial w}{\partial y} = \frac{\partial \Phi}{\partial x} + i\frac{\partial \Psi}{\partial x} = u - iv = \tilde{V} \tag{4.2.11}$$

This expression indicates that the derivative of w is equal to the conjugate of the velocity.

One further point to note is that, in a potential flow domain, the Bernoulli equation is satisfied between any two points of reference, as shown by Eqs. (2.6.10)–(2.6.12). This provides an important tool for analyzing pressure distributions in potential flows, as will be seen in the following subsections, where we review several special cases of two-dimensional potential flows.

4.2.2 Uniform Flow

Consider a flow with constant speed U, parallel to the x coordinate. This might represent, for example, the flow of air above the earth. Components of the velocity vector are then given by

$$u = U \qquad v = 0 \tag{4.2.12}$$

By applying Eqs. (4.2.2)–(4.2.6), we obtain

$$\Phi = Ux \qquad \Psi = Uy \qquad w = U(x + iy) = Uz \tag{4.2.13}$$

These expressions indicate that streamlines are parallel horizontal lines. For each streamline, the value of the y coordinate is constant. Equipotential lines are vertical lines. For each equipotential line the value of the x coordinate is kept constant. Also, according to the Bernoulli equation (2.6.12), the pressure is constant along horizontal streamlines and varies as hydrostatic pressure in the vertical direction.

If the parallel flow streamlines make an angle α with respect to the x coordinate, then the complex potential is given by

$$w = Uz e^{-i\alpha} \tag{4.2.14}$$

4.2.3 Flow at a Corner

Consider the flow domain represented by the complex potential function

$$w = Az^2 = A\left[(x^2 - y^2) + i2xy\right] \tag{4.2.15}$$

where A is a constant positive coefficient. The conjugate velocity is given by

$$\tilde{V} = u - iv = \frac{dw}{dz} = 2Az = 2A(x + iy) \tag{4.2.16}$$

Equations (4.2.15) and (4.2.16) imply

$$\Phi = A(x^2 - y^2) \qquad \Psi = 2Axy \qquad u = 2Ax \qquad v = -2Ay \tag{4.2.17}$$

Therefore equipotential lines and streamlines are hyperbolas, as shown in Fig. 4.3. On the streamlines, small arrows show the flow direction. They are depicted according to signs of the velocity components implied by Eq. (4.2.17). This equation indicates that the velocity vanishes at the coordinate origin. Therefore this point is a singular stagnation point. At a singular point, the velocity vanishes or becomes infinite. If the velocity vanishes, the point is a stagnation point. If the velocity has infinite value, it is a cavitation point. Streamlines or equipotential lines may intersect only at singular points. Eq. (4.2.17) also indicates that the velocity increases with distance from the origin. However, there is no particular singular point of infinite velocity.

By employing the Bernoulli equation, the distribution of pressure along the x coordinate is

$$p = p_0 - 2\rho A^2 x^2 \tag{4.2.18}$$

where p_0 is the pressure at the origin. In Fig. 4.3, a parabolic curve shows the pressure distribution along the x-direction. It indicates that the flow at the corner cannot persist for large distances from the origin, since according to Eq. (4.2.18), at some distance from the origin the pressure is too low to afford the streamline pattern of Eq. (4.2.17).

If the flow takes place at a corner of angle $\alpha = \pi/n$, then the complex potential is given by

$$w = Az^n \tag{4.2.19}$$

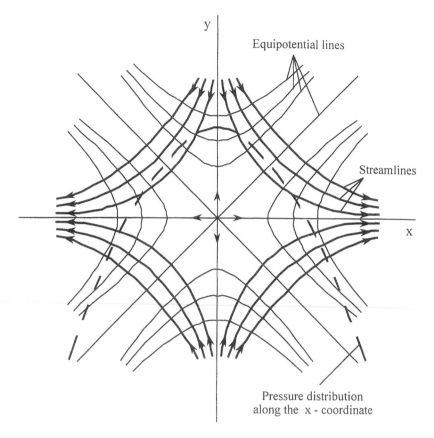

y

Equipotential lines

Streamlines

x

Pressure distribution
along the x - coordinate

Figure 4.3 Flow at a 90° corner.

4.2.4 Source Flow

The complex potential function for a source flow is

$$w = \frac{q}{2\pi} \ln z = \frac{q}{2\pi} \ln(r\, e^{i\theta}) = \frac{q}{2\pi}(\ln r + i\theta) \qquad (4.2.20)$$

Therefore the potential and stream functions are given, respectively, by

$$\Phi = \frac{q}{2\pi} \ln r \qquad \Psi = \frac{q}{2\pi}\theta \qquad (4.2.21)$$

These expressions indicate that streamlines are straight lines radiating outward from the origin. For each streamline, the value of Ψ is kept constant. Equipotential lines are concentric circles surrounding the coordinate origin. For each equipotential line, the value of r is kept constant.

It is possible to use the expressions for the potential function, the stream function, or the complex potential function for the calculation of the velocity components. We exemplify here application of the complex potential function:

$$\tilde{V} = u - iv = \frac{dw}{dz} = \frac{q}{2\pi z} = \frac{q\tilde{z}}{2\pi z \tilde{z}} = \frac{q}{2\pi} \left(\frac{x - iy}{x^2 + y^2} \right) \tag{4.2.22}$$

Therefore the complex velocity is given by

$$V = \frac{q}{2\pi} \left(\frac{x + iy}{x^2 + y^2} \right) = \frac{q}{2\pi} \left(\frac{\cos\theta + i\sin\theta}{r} \right) = \frac{q}{2\pi r} e^{i\theta} \tag{4.2.23}$$

This result indicates that the absolute velocity is kept constant in a circle surrounding the origin, i.e., the fluid flows in the radial direction. The velocity is infinite at the origin and vanishes at a large distance from the origin.

If a circle of radius r is drawn around the coordinate origin, then the radial flow velocity of the fluid that penetrates the circle is given by

$$V = u_r = \frac{q}{2\pi r} \tag{4.2.24}$$

It should be noted that the complex velocity of Eq. (4.2.23) is different from the absolute velocity of Eq. (4.2.24). Equation (4.2.24) indicates that the source strength q represents the total flow rate penetrating the circle surrounding the origin.

If the flow domain is horizontal, then Bernoulli's equation yields

$$p = p_\infty - \rho \frac{V^2}{2} = p_\infty - \frac{\rho}{2} \left(\frac{q}{2\pi} \right)^2 \frac{1}{r^2} \tag{4.2.25}$$

where p_∞ is the pressure at an infinite distance from the source point. At the origin the pressure is infinitely negative. Therefore the origin is a singular cavitation point.

Figure 4.4 shows the flow-net and pressure distribution along a radial coordinate of a source flow.

4.2.5 Simple Vortex

We consider the flow domain represented by the complex potential,

$$w = -\frac{i\kappa}{2\pi} \ln z = \frac{\kappa}{2\pi} (\theta - i \ln r) \tag{4.2.26}$$

According to this expression,

$$\Phi = \frac{\kappa}{2\pi} \theta \qquad \Psi = -\frac{\kappa}{2\pi} \ln r \tag{4.2.27}$$

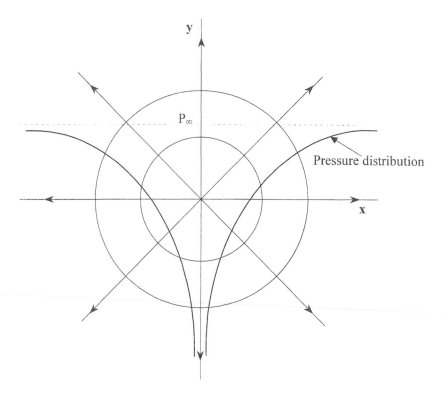

Figure 4.4   Source of flow.

These relations indicate that equipotential lines are straight radial lines emanating from the coordinate origin, while streamlines are circles surrounding the origin.

By appropriate differentiation of either of the expressions given by Eq. (4.2.27), expressions for the velocity components may be obtained as

$$u_r = 0 \qquad v_\theta = \frac{\kappa}{2\pi r} \qquad\qquad (4.2.28)$$

These expressions indicate that the velocity is proportional to the inverse of the distance from the coordinate origin, its value is constant along circles surrounding the origin, and its direction is counterclockwise. At the origin, the velocity is infinite. Therefore this point is a singular cavitation point. The pressure distribution along a radial coordinate is identical to that given by Eq. (4.2.25) for the source flow, where κ replaces q. Figure 4.5 shows the flow-net and pressure distribution along a radial coordinate of a simple vortex flow.

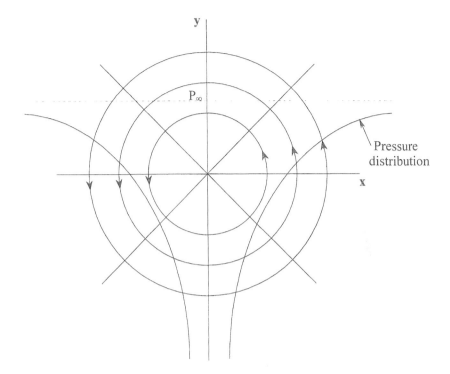

Figure 4.5 Simple vortex.

If we depict a circle of radius r about the origin and calculate the circulation by the integral of Eq. (2.3.14), we obtain

$$\Gamma = \oint \vec{V} \cdot d\vec{s} = \int_0^{2\pi} v_\theta r \, d\theta = \int_0^{2\pi} \frac{\kappa}{2\pi r} r \, d\theta = \kappa \qquad (4.2.29)$$

This expression indicates that κ represents the circulation of the vortex, namely, the *vortex strength*. According to Eq. (2.3.15), the circulation is zero for a potential flow domain. However, if in a potential flow domain the closed curve of the integral of Eq. (2.3.14) surrounds singular points of circulating flows, then the circulation does not vanish. It represents the strength of the circulating flow, in the domain surrounding that singular point.

4.2.6 Doublet

Doublet flow is obtained due to the superposition of a positive and a negative source of equal strength. The distance between the sources is a, the strength of each source is q, and the following conditions take place in the flow domain:

$$a \to 0$$
$$q \to \infty \qquad (4.2.29)$$
$$\frac{aq}{\pi} \to \lambda$$

The complex potential function of the doublet is developed as follows:

$$
\begin{aligned}
w &= \frac{q}{2\pi} \ln \left[\frac{z+a}{z-a} \right] = \frac{q}{2\pi} \ln \left[\frac{z^2 + 2az + a^2}{z^2 - a^2} \right] \\
&= \frac{q}{2\pi} \ln \left[\frac{1 + (2a/z)}{1 - (a/z)^2} + \frac{a^2}{z^2 - a^2} \right] \qquad (4.2.30) \\
w &\to \frac{q}{2\pi} \ln \left[\left(1 + \frac{2a}{z} \right) \left(1 + \frac{a^2}{z^2} \cdots \right) \right] \to \frac{q}{2\pi} \frac{2a}{z} = \frac{\lambda}{z}
\end{aligned}
$$

The doublet of Eq. (4.2.30) incorporates a positive source, located to the left of the origin (at $x = -a$), and a negative source, located to the right of the origin (at $x = a$).

According to Eq. (4.2.30), we can find the potential and stream functions as follows:

$$w = \Phi + i\Psi = \frac{\lambda}{z} = \frac{\lambda}{r} e^{-i\theta} = \frac{\lambda}{r}(\cos\theta - i\sin\theta) \qquad (4.2.31)$$

Therefore

$$\Phi = \frac{\lambda \cos\theta}{r} = \frac{\lambda x}{x^2 + y^2} \qquad \Psi = -\frac{\lambda y}{x^2 + y^2} \qquad (4.2.32)$$

The equation of equipotential lines is

$$\left(x - \frac{\lambda}{2\Phi} \right)^2 + y^2 = \left(\frac{\lambda}{2\Phi} \right)^2 \qquad (4.2.33)$$

This expression indicates that equipotential lines are circles, which pass through the origin, and have their centers located on the x axis. By applying the expression for Ψ in Eq. (4.2.32), the equation for the streamlines is

$$x^2 + \left(y + \frac{\lambda}{2\Psi} \right)^2 = \left(\frac{\lambda}{2\Psi} \right)^2 \qquad (4.2.34)$$

This expression indicates that streamlines are circles, passing through the origin, whose centers are located on the y axis.

The conjugate velocity is obtained by differentiating Eq. (4.2.31) to obtain

$$\tilde{V} = -\frac{\lambda}{z^2} = -\frac{\lambda}{r^2} e^{-2i\theta} = \frac{\lambda}{r^2}[-\cos(2\theta) + i\sin(2\theta)] \qquad (4.2.35)$$

Therefore components of the velocity are given by

$$u = -\frac{\lambda}{r^2}\cos(2\theta) = -\frac{\lambda}{x^2+y^2}\left(\frac{x^2-y^2}{x^2+y^2}\right) = \frac{\lambda(y^2-x^2)}{(x^2+y^2)^2}$$

$$v = -\frac{\lambda}{r^2}\sin(2\theta) = -\frac{\lambda}{x^2+y^2}\left(\frac{2xy}{x^2+y^2}\right) = -\frac{2\lambda xy}{(x^2+y^2)^2}$$

(4.2.36)

The flow net for a doublet is sketched in Fig. 4.6.

4.2.7 The Image Method

The flow domain given by the potential, stream, and complex potential functions is basically infinite. Considerations of solid boundaries in such a domain are usually made by assuming that solid boundaries are represented by particular streamlines (note that there is no flow *across* a streamline). Representation of solid boundaries by particular streamlines often requires the superposition of several simple potential flows. The presentation of flow around a cylinder,

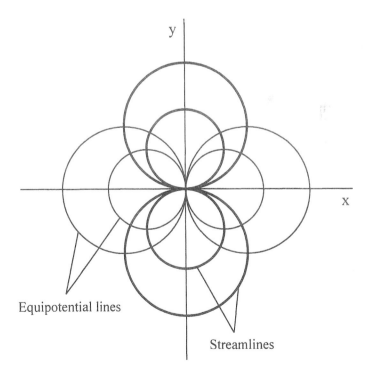

Equipotential lines

Streamlines

Figure 4.6   Flow associated with a doublet.

as shown in Sec. 4.5, is obtained by the superposition of a uniform flow and a doublet flow. Very often, adequate superposition is obtained by trial-and-error experiments, but in some particular cases the appropriate superposition is obtained by straightforward calculations.

Figure 4.7 shows a source located at a distance $x = a$ from a solid wall. There is no flow perpendicular to the wall. Therefore to obtain a velocity tangential to the wall at point A, a second source must be added, of identical strength, at $x = -a$. The complex potential describing the flow created by a source of strength q, located at a distance a from a wall, is given by

$$w = \frac{q}{2\pi} \ln[(z - a)(z + a)] \tag{4.2.36}$$

Figure 4.8 shows a source located at a corner between two solid walls. The distance of the source from one wall is $x = a$. The distance from the other wall is $y = b$. In this case, to represent the two walls as streamlines, the superposition should incorporate four sources, as indicated by Fig. 4.8.

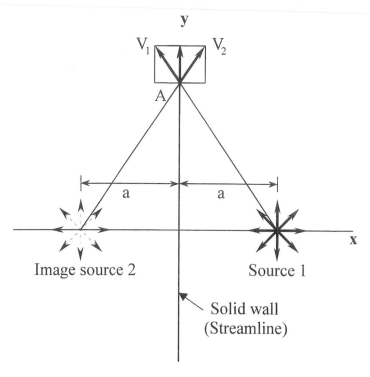

Figure 4.7   Source located at a wall.

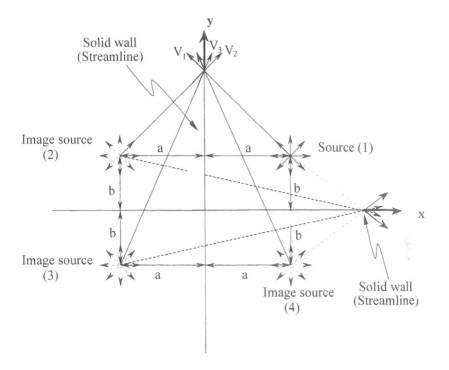

Figure 4.8 Source at the corner between two walls.

Therefore the complex potential function is given by

$$w = \frac{q}{2\pi} \ln[(z - a - ib)(z + a - ib)(z + a + ib)(z - a + ib)] \quad (4.2.37)$$

Figure 4.9 shows a source of strength q located at a distance $x = a$ from an equipotential straight line given by $x = 0$. Practically, such a case can be useful for the calculation of groundwater flow at an injection well, which is located close to a river. Section 4.3 provides details concerning the application of the potential flow theory to calculations of flow through porous media. To keep the line $x = 0$ as an equipotential line, another negative source of equal strength should be added at $x = -a$, as shown in Fig. 4.9. Therefore the complex potential function is given by

$$w = \frac{q}{2\pi} \ln \left(\frac{z - a}{z + a} \right) \quad (4.2.38)$$

Figure 4.10 shows a vortex of circulation κ, located at the corner between two solid walls, given by $x = 0$ and $y = 0$. Its distance from one wall is $x = a$,

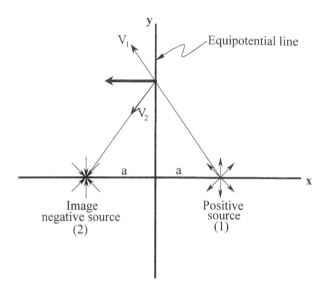

Figure 4.9   Source at an equipotential line.

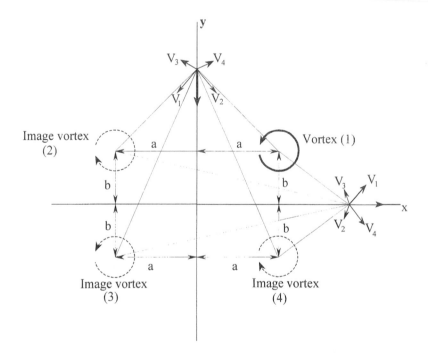

Figure 4.10   Vortex at the corner between two solid walls.

and its distance from the other wall is $y = b$. To represent the lines $x = 0$ and $y = 0$ as streamlines, three vortices of equal circulation should be added, as shown in Fig. 4.10. Therefore the complex potential function is given by

$$w = -\frac{i\kappa}{2\pi} \ln \left[\frac{(z - a - ib)(z + a + ib)}{(z + a - ib)(z - a + ib)} \right] \qquad (4.2.40)$$

It should be noted that for relevance to real-world problems, positive and negative sources are kept in a stable position, whereas the vortex of Fig. 4.10 is subject to movement in the domain. The position change of the vortex of this figure is caused by the flow velocity components of the image vortices.

4.3 FLOW THROUGH POROUS MEDIA

Flow through porous media such as aquifers, alluvial material, sand, small gravel, etc. is usually laminar flow, associated with very small Reynolds numbers. The definition of the Reynolds number for flow through porous media is

$$\text{Re} = \frac{q d_p}{\nu} \qquad (4.3.1)$$

where q is the specific discharge (with dimensions of LT^{-1}); d_p is a characteristic pore size, usually considered as a representative average diameter of the particles comprising the matrix, or derived from the permeability (another concept that will be defined later) of the porous matrix, and ν is the kinematic viscosity of the fluid. The specific discharge, called also filtration velocity, is related to the average interstitial flow velocity by

$$q = V\phi \qquad (4.3.2)$$

where ϕ is the porosity of the matrix. In an isotropic material the volumetric and surface porosity are identical. It should be noted that V represents the velocity of advection of contaminants migrating with the flowing fluid through the porous matrix. The quantity q represents the flow rate per unit surface of the porous matrix.

In most cases of environmental flow through porous media, the value of the Reynolds number, defined in eq. (4.3.1), is smaller than unity. Therefore flow through porous media in most cases may be considered as laminar creeping flow (Section 3.3). However, there are also examples in which the Reynolds number is higher, as with flows through coarse gravel, flows through rock fill, wave breakers, etc. The present section refers only to creeping flow through porous media; other topics in porous media flow are discussed in Chap. 11.

In creeping flows, the equations of motion (Navier–Stokes) reduce to

$$\nabla p' = \mu \nabla^2 \vec{V} \tag{4.3.3}$$

where $p = \rho g$ is the piezometric pressure, V is the flow velocity, and μ is the viscosity of the fluid.

4.3.1 Darcy's Law

The laminar flow through a porous matrix can be visualized as a flow through many parallel flat plates, or through a bundle of capillaries. With regard to a single capillary of diameter d and length L, we may apply the solution of Poiseuille–Hagen to Eq. (4.3.3) to obtain

$$J = \frac{\Delta h}{L} = \frac{1}{\rho g}\frac{\Delta p^*}{L} = \frac{32 v}{g d^2} V \tag{4.3.4}$$

where h is the piezometric head and J is the hydraulic gradient. The capillary diameter, d, may be considered as a characteristic pore size of the porous matrix.

Considering that the porosity, ϕ, represents the ratio between the total area of cross sections of the bundle of capillaries and the cross section of the porous matrix, Eq. (4.3.4) implies

$$q = KJ \tag{4.3.5}$$

where K is the hydraulic conductivity of the porous matrix, given by

$$K = \frac{g d^2 \phi}{32 v} \tag{4.3.6}$$

This result shows that the hydraulic conductivity depends on properties of the porous matrix, namely the porosity, the characteristic pore size, and also the kinematic viscosity of the fluid. The permeability is a parameter associated with the flow through the porous matrix and depends solely on the matrix properties. Its definition and relation to the hydraulic conductivity are given as

$$k = \frac{\phi d^2}{32} \qquad K = \frac{g k}{v} \tag{4.3.7}$$

For three-dimensional domains, Eq. (4.3.5) can be generalized as

$$\vec{q} = -K \nabla h \tag{4.3.8}$$

This proportionality between the specific discharge and the gradient of the piezometric head is called *Darcy's law*.

4.3.2 Relevance of Potential Flow Theory

Equation (4.3.8) implies that, in cases of constant hydraulic conductivity, the specific discharge vector originates from a gradient of a potential function Φ, which is equal to Kh. In cases of two-dimensional flow, with negligible compression of the fluid and the solid matrix, it is possible to define a stream function, Ψ, that satisfies continuity and has constant values along the streamlines. The relationships between the components of the specific discharge and the functions Φ and Ψ are

$$q_x = -\frac{\partial \Phi}{\partial x} = -\frac{\partial \Psi}{\partial y}$$

$$q_x = -\frac{\partial \Phi}{\partial y} = \frac{\partial \Psi}{\partial x} \qquad (4.3.9)$$

The negative sign for the derivatives in Φ shows that the flow is in the direction of decreasing values of Φ. These relations are basically Cauchy–Riemann equations, as introduced earlier in Sec. 4.2.1. The continuity, represented by Ψ, and the potential function Φ, both satisfy the Laplace equation,

$$\nabla^2 \Phi = 0 \qquad \nabla^2 \Psi = 0 \qquad \nabla^2 h = 0 \qquad (4.3.10)$$

Therefore all techniques applicable to the solution of the Laplace equation can be used for the calculation of incompressible flow through porous media. The function theory with the employment of complex variables is useful for the evaluation of practical issues associated with flow through porous media. In potential fluid flows, the potential function has no physical meaning. In flow through porous media, the potential function, Φ, is derived from the piezometric head.

On the basis of Eq. (4.3.9), flow-nets can often be defined to obtain quick estimates of the intensity of the flow through a limited-size porous medium. They also can easily provide estimates of uplift forces exerted on structures. The flow-net incorporates a grid of small squares whose boundaries are equipotential lines and streamlines, as noted previously. Calculations of uplift forces and total flow through the domain are based on the number of small squares in the grid and the hydraulic conductivity of the domain. Flow-nets can easily be used for the evaluation of seepage underneath a dam, uplift forces on the dam, the effect of cut-off walls, etc.

4.3.3 Anisotropic Porous Medium

Expressions referring to flow through porous media in the preceding paragraphs consider the hydraulic conductivity as a scalar parameter and property. In cases of anisotropy of the domain, the hydraulic conductivity can be represented as a second-order tensor. As an example, in natural sandy soils, the average hydraulic conductivity in a horizontal direction can be from two to ten times the value for the vertical direction. In cases of anisotropy of the porous medium, the last part of Eq. (4.3.10) is written as

$$K_H \frac{\partial^2 h}{\partial x^2} + K_V \frac{\partial^2 h}{\partial y^2} = 0 \qquad (4.3.11)$$

where K_H and K_V are the horizontal and vertical hydraulic conductivity, respectively.

It is convenient to define a new coordinate x_1 by

$$x_1 = x \sqrt{\frac{K_V}{K_H}} \qquad (4.3.12)$$

Introducing Eq. (4.3.12) into Eq. (4.3.11), the piezometric head again satisfies the Laplace equation. Therefore, a modification in the construction of the flow-net is necessary to allow consideration of domains with different horizontal and vertical hydraulic conductivity. This involves drawing the domain of reference and its boundary conditions with the horizontal dimensions reduced by the factor $\sqrt{K_V/K_H}$. Then the flow-net is drawn for the distorted boundaries and the discharge is computed using the average harmonic hydraulic conductivity,

$$K = \sqrt{K_H K_V} \qquad (4.3.13)$$

4.3.4 Flow-Nets

Nowadays, quick solutions of the Laplace equation can be obtained by numerical approaches, which will be reviewed in subsequent chapters. However, it is appropriate to consider at this stage some particular examples of possible uses of flow-net construction. By these examples, some characteristics of flow through porous media can be visualized. For example, in the case of percolation under a dam through the porous layer of alluvial material which overlies an impervious layer, the flow pattern is independent of the upstream and downstream water levels. The difference, H, in these levels only determines the scale of the flow, as shown in Fig. 4.11. Since $\Delta\Phi$ is constant between adjacent equipotential lines, the total drop in piezometric head (equal to H) is divided along any flow line into increments, ΔH. Thus with n unit squares in each channel of the flow-net, the decrease in piezometric head, or uplift pressure head along the base of the dam, follows from the values of the piezometric head at the points of intersection of the equipotential lines with the base.

The effectiveness of cutoff walls and sheet piling in various locations and of upstream and downstream aprons in reducing uplift pressures can be evaluated by means of the flow-net. Each of these devices lengthens the seepage

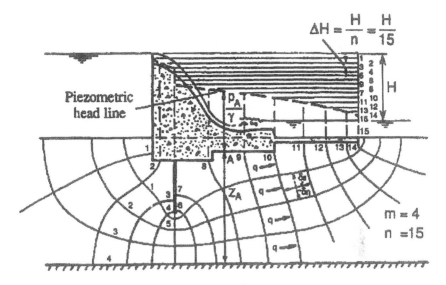

Figure 4.11   Flow-net under a dam.

paths, with cutoff walls producing a vertical drop in the piezometric head and aprons decreasing its gradient. Points of high velocity at the downstream end of the net, where "piping" may occur, can be identified and remedial measures can be evaluated.

The rate of flow through a unit square of one channel per meter width of the dam shown in Fig. 4.11 is

$$Q_s = -KA\frac{dh}{ds} = K\,\Delta n\frac{H/n}{\Delta s} = K\frac{H}{n} \tag{4.3.14}$$

where A is the cross-sectional area of a single channel, which is also the height of the small square of the flow-net, whose value is Δn. The length of the small square is Δs. The value of H is equally divided along the n lengths of the small squares. For m channels, each carrying an equal flow rate Q_s, the total flow-rate Q is mQ_s, or

$$Q = K\frac{m}{n}H \tag{4.3.15}$$

With regard to the total flow rate, the flow-net determines the ratio m/n. In its construction, the number of channels m is arbitrarily selected. The number of squares per channel varies with the number of channels, but the total flow-rate determinations for different values of m should agree with each other. The construction of the flow-net proceeds upstream and downstream

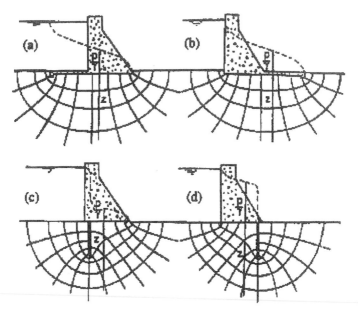

Figure 4.12 Possible effect of apron and cutoff wall on piezometric head distribution: (a) horizontal apron at head of dam; (b) apron at toe of dam; (c) vertical cut off wall near head of dam; and (d) vertical wall near toe of dam.

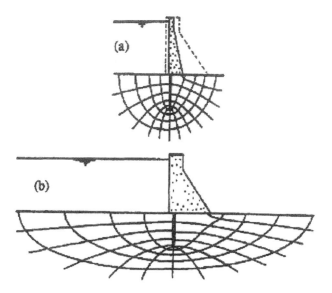

Figure 4.13 Flow net for anisotropic porous media.

from trial locations of the portions of the streamlines in the narrowest region of the flow path.

Figure 4.12 provides several examples concerning the possible effect of apron and cutoff wall on the distribution of the piezometric head in the alluvial layer. Figure 4.13 exemplifies use of the flow-net for anisotropic porous material.

4.4   CALCULATION OF FORCES

4.4.1   Force on a Cylinder

Figure 4.14 shows a cylinder of arbitrary cross section in a two-dimensional flow field. The fluid is assumed to be inviscid. The pressure force acting on an element of the surface is $p \, ds$ and it is normal to the surface element ds. The cylinder width, perpendicular to the paper plane of Fig. 4.14, is unity. The components of the pressure force in the x and y-directions are

$$F_x = -p \, ds \cos \left(\theta - \frac{\pi}{2} \right) = -p \, ds \sin \theta = -p \, dy$$

$$F_y = -p \, ds \sin \left(\theta - \frac{\pi}{2} \right) = p \, ds \cos \theta = p \, dx$$

(4.4.1)

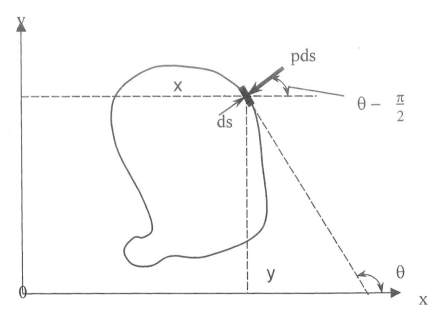

Figure 4.14   Pressure force acting on an elementary surface.

where θ is the angle made by the surface element with the x axis. The total pressure force components in the x- and y-directions are obtained by integrating over the cylinder surface,

$$F_x = \oint_c -p\,dy \qquad F_y = \oint_c p\,dx \qquad\qquad (4.4.2)$$

4.4.2 Steady Flow Around a Circular Cylinder Without Circulation

Steady flow around a circular cylinder without circulation can be expressed as a superposition of uniform flow and a doublet, with velocity potential given by

$$w = U\left(z + \frac{a^2}{z}\right) = U\left[r\exp(i\theta) + \frac{a^2}{r}\exp(-i\theta)\right] \qquad (4.4.3)$$

Following the procedures of Sec. 4.2, this complex potential is separated into the potential and stream functions,

$$\Phi = U\left(r + \frac{a^2}{r}\right)\cos\theta \qquad \Psi = U\left(r - \frac{a^2}{r}\right)\sin\theta \qquad (4.4.4)$$

Figure 4.15 represents a schematic description of several streamlines of the flow around a cylinder without circulation.

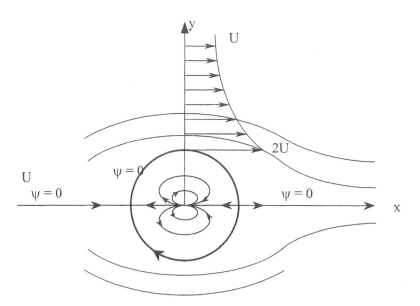

Figure 4.15 Steady flow around a cylinder without circulation.

The complex velocity for this flow field is

$$\frac{dw}{dz} = U\left(1 - \frac{a^2}{z^2}\right) = U\left[1 - \frac{a^2}{r^2}\exp(-i2\theta)\right] \tag{4.4.5}$$

This expression indicates that there are two stagnation points in the domain, defined by $r = a$ and $\theta = 0, \pi$. Then, according to Eq. (4.4.4), the stagnation points are located on the streamline defined by $\Psi = 0$. This line separates the fluid associated with the uniform flow from the fluid associated with the doublet and causes the flow field to behave as if there were a solid surface coincident with this streamline. According to Eq. (4.4.5), the absolute velocity along the separating streamline is

$$V^2 = \left|\frac{dw}{dz}\right|^2 = U^2\left[(1 - \cos 2\theta)^2 + (\sin 2\theta)^2\right] = 4U^2\sin^2\theta \tag{4.4.6}$$

Figure 4.15 shows the velocity distribution along the y axis above the cylinder. According to Bernoulli's equation,

$$\frac{p_s}{\rho g} = \frac{p}{\rho g} + \frac{V^2}{2g} = \frac{p_\infty}{\rho g} + \frac{U^2}{2g} \tag{4.4.7}$$

where p_s is the pressure at the stagnation point and p_∞ is the pressure far from the cylinder. We now refer to the surface of the circular cylinder, $r = a$. The surface element for this cylinder is $ds = a\,d\theta$. Introducing this quantity, along with Eqs. (4.4.6) and (4.4.7) into Eq. (4.4.1), the pressure distribution is obtained along the surface as shown in Fig. 4.16. By integrating the pressure

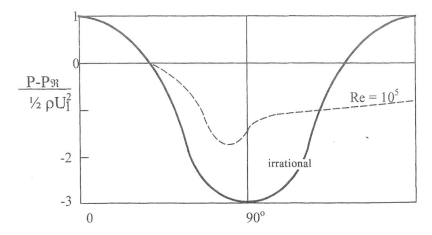

Figure 4.16   Pressure distribution around a cylinder without circulation.

distribution over the cylinder surface, it is easy to show that there is zero resultant force acting on the cylinder. This result indicates that according to potential flow theory, no drag or lift forces act on a body moving in a domain of inviscid fluid. This result is called *d'Alembert's paradox*. In real fluids there is always drag force acting on the body. The drag force originates from friction and separation of the flow from the sides of the body. The flow separation results in a wake and eddies migrating downstream of points of separation. The pressure at the wake is approximately equal to the pressure at the separation point, which is smaller than that predicted by potential flow theory. Therefore real fluid flow around the cylinder is always associated with drag force. A schematic description of the pressure distribution around the circular cylinder with a real fluid is shown in Fig. 4.16.

4.4.3 Steady Flow Around a Circular Cylinder with Circulation

In this case, a clockwise potential vortex with circulation Γ is added to the previous situation of a doublet in a uniform flow. The complex potential is now given by

$$w = U\left(z + \frac{a^2}{z}\right) + \frac{i\Gamma}{2\pi}\ln z \qquad (4.4.8)$$

This expression can be separated to provide expressions for the potential and stream functions, and differentiated to yield an expression for the complex velocity, as before,

$$\Phi = U\left(r + \frac{a^2}{r}\right)\cos\theta - \frac{\Gamma}{2\pi}\theta;$$

$$\Psi = U\left(r - \frac{a^2}{r}\right)\sin\theta + \frac{\Gamma}{2\pi}\ln r \qquad (4.4.9)$$

$$\frac{dw}{dz} = U\left(1 - \frac{a^2}{z^2}\right) + \frac{i\Gamma}{2\pi z} \qquad (4.4.10)$$

The streamline $\Psi = \Gamma/(2\pi)\ln a$ represents the circular cylinder $r = a$. Therefore the complex potential of Eq. (4.4.8) refers to uniform flow around a circular cylinder. At a large value of z, Eq. (4.4.10) indicates that the velocity is U. Referring to the surface of the circular cylinder, Eq. (4.4.10) yields the value of the absolute velocity,

$$
\begin{aligned}
V = \left| \frac{dw}{dz} \right|_{r=a} &= \left| \left[2U \sin\theta + \frac{\Gamma}{2\pi} \right] (\sin\theta + i\cos\theta) \right| \\
&= \left| 2U\sin\theta + \frac{\Gamma}{2\pi} \right|
\end{aligned}
\tag{4.4.11}
$$

This expression indicates that the velocity vanishes if

$$
\sin\theta = -\frac{\Gamma}{4\pi a U}
\tag{4.4.12}
$$

Results of this expression are schematically represented by Fig. 4.17. There are two stagnation points if $\Gamma < 4\pi a U$. If $\Gamma = 4\pi a U$ then there is a single stagnation point at the cylinder surface. If $\Gamma > 4\pi a U$ then the stagnation point moves downward into the flow.

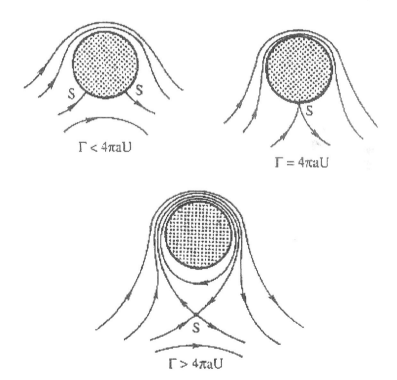

Figure 4.17 Flow around a cylinder with circulation.

The pressure distribution along the cylinder surface is again obtained by using the Bernoulli equation,

$$p + \frac{1}{2}\rho V^2 = p_\infty + \frac{1}{2}\rho U^2 \tag{4.4.13}$$

where p_∞ is the pressure far from the cylinder. Introducing Eq. (4.5.11) into Eq. (4.5.13), the pressure distribution on the cylinder surface is found to be

$$p_{r=a} = p_\infty + \frac{1}{2}\rho\left[U^2 - \left(2U\sin\theta + \frac{\Gamma}{2\pi a}\right)^2\right] \tag{4.4.14}$$

The symmetry of the flow about the y axis indicates that there is no drag (net pressure force in the x-direction) for this flow field. On the other hand, the circulation leads to lift force (net pressure force applied on the cylinder in the y-direction). The calculation of the lift force is obtained by integration of Eq. (4.4.14),

$$F_y = -\int_0^{2\pi} p_{r=a}\sin\theta a\, d\theta = \rho U\Gamma \tag{4.4.15}$$

This expression is valid for potential flow around any two-dimensional body and is known as the *Kutta–Zhukhovski lift theorem*. This theorem is discussed further in Sec. 4.4.5.

In flow of real fluids around bodies, circulation is created due to the viscosity of the fluid. However, the magnitude of the circulation does not depend on the viscosity. Rather, it depends on the free flow velocity U and the shape of the body. In terms of potential flow theory, circulation around a circular cylinder can be created only by rotating the cylinder, around which fluid flows with a uniform flow velocity U.

4.4.4 The Theorem of Blasius

Equation (4.4.1) can be represented as a complex quantity,

$$dF_x - i\, dF_y = d\tilde{F} = -p\, dy - i p\, dx = -i p\, d\tilde{z} \tag{4.4.16}$$

where the wavy overbar denotes the complex conjugate. By integrating Eq. (4.4.16) over the entire surface of the cylinder, we obtain

$$F_x - iF_y = \tilde{F} = -i\oint_c p\, d\tilde{z} \tag{4.4.17}$$

where c denotes integration over the entire surface of the cylinder in the counter-clockwise direction.

By applying the Bernoulli equation between a reference point far from the cylinder and any other point in the flow domain,

$$p_\infty + \frac{1}{2}\rho U^2 = p + \frac{1}{2}\rho(u^2 + v^2) = p + \frac{1}{2}\rho(u + iv)(u - iv) \qquad (4.4.18)$$

Introducing this expression for p into Eq. (4.4.16), we obtain

$$\tilde{F} = -i \oint \left[p_\infty + \frac{1}{2}\rho U^2 - \frac{1}{2}\rho(u + iv)(u - iv) \right] d\tilde{z} \qquad (4.4.19)$$

The integral of $(p_\infty + \rho U^2/2)$ vanishes, as this term has a constant value. With regard to other terms of the integral in Eq. (4.4.19), first note that

$$u + iv = V \exp(i\theta)$$
$$(u - iv)(u + iv) d\tilde{z} = (u^2 + v^2) d\tilde{z} \qquad (4.4.20)$$

Introducing these expressions into Eq. (4.4.19), we obtain

$$\tilde{F} = i\frac{\rho}{2} \oint_c \left| \frac{dw}{dz} \right|^2 d\tilde{z} \qquad (4.4.21)$$

This equation is called the *Blasius theorem*. It expresses the total pressure force applied on a cylinder of any shape that is submerged in a fluid subject to potential flow.

The Cauchy integral theorem states that the line integral of a complex function around any closed curve is zero, provided that no singular point is present in the region enclosed by the curve. If one or more singular points are present in that region, then the closed line integral does not vanish. The value of that integral does not depend on the closed curve chosen for the calculation of the integral, provided that the number of singular points in the enclosed region is kept constant. According to the theory of complex variables, the closed line integral around the singular points is equal to $2\pi i$ multiplied by the sum of the residues of all singular points in the enclosed region.

4.4.5 The Lift Theorem of Kutta–Zhukovski

According to the Blasius and Chauchy integral theorems, we may perform the integral of Eq. (4.4.21) at a large distance from the center of the cylinder, which can be represented by a superposition of uniform flow with sources, sinks, and doublets. From a large distance, all singular points are considered to be close to the origin. Therefore the complex potential function is given by

$$w = Uz + \frac{q}{2\pi} \ln z + \frac{i\Gamma}{2\pi} \ln z + \frac{\chi}{z} + \cdots \qquad (4.4.22)$$

As the superposition refers to a closed curve representing the cylinder, the net flux of sources and sinks should be zero. Therefore by introducing Eq. (4.4.22) into Eq. (4.4.21) we obtain

$$\tilde{F} = i\frac{\rho}{2} \oint \left| U + \frac{i\Gamma}{2\pi z} - \frac{\chi}{z^2} + \cdots \right|^2 dz \tag{4.4.23}$$

The residue of the complex function subject to integration in Eq. (4.4.23) is the coefficient of the term incorporating $1/z$ in the power series expansion of that function. This coefficient is equal to $i\Gamma U/\pi$. Therefore Eq. (4.4.23) yields

$$\tilde{F} = i\frac{\rho}{2} \left[2\pi i \left(i\frac{\Gamma U}{\pi} \right) \right] = -i\rho U \Gamma \tag{4.4.24}$$

This expression indicates that the potential flow theory predicts that no drag force is applied on the cylinder, and the lift force is proportional to ρ, U, and Γ, as

$$F_x = 0 \qquad F_y = \rho U \Gamma \tag{4.4.25}$$

This result is called the Kutta–Zhukhovski lift theorem, as previously noted.

4.5 NUMERICAL SIMULATION CONSIDERATIONS

Numerical simulations of potential incompressible flows are based on the solution of the Laplace equation, in terms of the potential or the stream function,

$$\nabla^2 \Phi = 0 \qquad \nabla^2 \Psi = 0 \tag{4.5.1}$$

In a two-dimensional Cartesian coordinate system this equation for Φ is

$$\frac{\partial^2 \Phi}{\partial x^2} + \frac{\partial^2 \Phi}{\partial y^2} = 0 \tag{4.5.2}$$

This expression is a second-order partial differential equation (PDE). As discussed in Sec. 1.3.3, the order of a PDE is determined by the highest order derivative in the equation. Furthermore, Eq. (4.5.2) is a linear PDE. In a linear PDE, the coefficients of the highest order derivatives are constants or functions of the independent variables x and y.

The general format of a second-order linear PDE in a two-dimensional domain can be written as

$$a\frac{\partial^2 \varphi}{\partial x^2} + b\frac{\partial^2 \varphi}{\partial x \partial y} + c\frac{\partial^2 \varphi}{\partial y^2} = f \tag{4.5.3}$$

where f represents a linear combination of coefficients multiplied by lower order derivatives of the dependent variable φ. The method and form of the solution of a PDE subject to initial and boundary conditions depend on the type of the PDE. As discussed in Chap. 1, it is common to classify a PDE according to the relationships between the coefficients of Eq. (4.5.3), as follows:

If $b^2 - 4ac > 0$ then the PDE is hyperbolic $\hspace{2cm}$ (4.5.4a)

If $b^2 - 4ac = 0$ then the PDE is parabolic $\hspace{2cm}$ (4.5.4b)

If $b^2 - 4ac < 0$ then the PDE is elliptic $\hspace{2cm}$ (4.5.4c)

According to Eq. (4.5.4), the Laplace Eq. (4.5.2) is an elliptic PDE.

For elliptic PDEs there are no initial conditions (note that time does not appear as an independent variable in Eq. 4.5.2), but only boundary conditions, which must be expressed in terms of some property of the dependent variable. For the present case, four boundary conditions are needed, two in each coordinate direction. In Eq. (4.5.2), Φ is the dependent variable. In the two-dimensional $x-y$ domain, any time-dependent phenomenon associated with the value of Φ is introduced, under unsteady-state conditions, through the boundary conditions. However, at this point we consider steady-state flows only. Because Eq. (4.5.2) is a second-order PDE with regard to x as well as with regard to y, there are three types of linear boundary conditions that can be applied to its solution (see also Sec. 1.3.3):

(1) All values of Φ are specified on the boundaries of the flow domain, or

$$\Phi = f(x, y) \quad \text{where} \quad (x, y) \in G \hspace{2cm} (4.5.5)$$

and G is the surface of the domain. With regard to the surface shown in Fig. 4.18 we may write

$$\begin{aligned} \Phi = f_1(x, y_1) \quad & \Phi = f_2(x_2, y); \\ \Phi = f_3(x, y_2) \quad & \Phi = f_4(x_1, y) \end{aligned} \hspace{2cm} (4.5.6)$$

so that the required four boundary conditions are provided. Boundary conditions of the type represented by Eqs. (4.5.5) and (4.5.6) are referred to as *Dirichlet boundary conditions*.

(2) All values of the gradient of Φ, i.e., the velocity components, are specified on the boundaries of the domain, so

$$\frac{\partial \Phi}{\partial n} = f(x, y) \quad \text{where} \quad (x, y) \in G \hspace{2cm} (4.5.7)$$

and n represents a coordinate normal to the boundary G, and pointing away from it. Boundary conditions of this type are called *Neumann boundary conditions*.

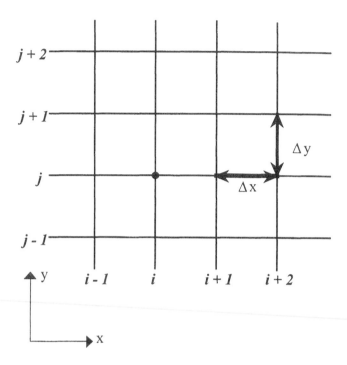

Figure 4.18   Example of a domain for an elliptic PDE.

(3) A general linear combination of Dirichlet and Neumann boundary conditions is written as

$$a\Phi + b\frac{\partial \Phi}{\partial n} = c \tag{4.5.8}$$

where a, b, and c are functions of position (x, y). Again, four such boundary conditions must be written.

Common linear boundary conditions are represented by solid boundaries. As previously noted, a solid boundary may be considered as a streamline. Therefore the velocity component perpendicular to the solid boundary vanishes. Complete analysis and simulation of a flow domain concerns the determination of the distribution of the velocity components and the pressure. Determination of the potential function Φ in the entire domain basically yields the velocity distribution. Then by using the Bernoulli equation, we obtain the pressure distribution. However, very often the pressure is the variable specified on some portions of the domain. As an example, consider the case of free surface flow. For this type of flow the free surface is a streamline, on which the pressure vanishes. The given pressure provides a nonlinear specification of the velocity at the surface, through the Bernoulli equation. Furthermore, the location of the free surface may be one of the unknown variables, which must be determined as part of the overall solution to the problem. If there is a simultaneous flow of immiscible fluids, then the employment of potential flow theory can sometimes be considered. In such cases, the interface between adjacent fluid domains represents a boundary, on both sides of which the pressure and normal flow velocity are identical. Again, this represents a sort of nonlinear boundary condition, since the position of the interface may not be known. Topics of nonlinear boundary conditions are beyond the scope of the present text.

Singular points in a flow domain can sometimes be introduced by simple means, based on measurable parameters. Typical examples are sources and sinks. Sometimes combinations of source sheets are used to represent solid bodies immersed in the flow domain. By such a presentation, the streamline shape of the immersed body can be simulated with small amounts of computer resources, and limited requirements for boundary conditions. Vortices cannot be created in a numerical simulation unless they are artificially introduced. The common boundary conditions of solid boundaries do not produce singular points typical of vortices. Therefore numerical simulation with simple Dirichlet or Neumann boundary conditions cannot simulate lift forces. The introduction of artificial vortices or vortex sheets is commonly used for the simulation of lift forces.

In the framework of the present section, we provide a basic presentation of finite difference solutions of the Laplace equation or the *Poisson equation*, which is the nonhomogeneous form of the Laplace equation. Figure 4.19 represents a portion of the domain covered by a finite difference grid. The grid is made of small squares, with equal spacing Δx and Δy in the x- and y-directions, respectively. For each nodal point, subscript i refers to the number of the x interval and subscript j refers to the number of the y interval. The finite *central difference* (i.e., nodal values are used from both sides of the node at which the derivative is to be evaluated) approximations of the first- and second-order derivatives of Φ for the nodal point (i, j) are given as

$$\left(\frac{\partial \Phi}{\partial x} \right)_{i,j} \approx \frac{\Phi_{i+1/2,j} - \Phi_{i-1/2,j}}{\Delta x} \tag{4.5.9a}$$

$$\left(\frac{\partial \Phi}{\partial y} \right)_{i,j} \approx \frac{\Phi_{i,j+1/2} - \Phi_{i,j-1/2}}{\Delta y} \tag{4.5.9b}$$

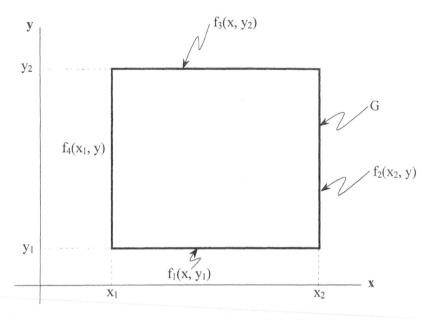

Figure 4.19 Portion of the domain covered by the finite difference grid.

$$\left(\frac{\partial^2 \Phi}{\partial x^2}\right)_{i,j} \approx \frac{1}{\Delta x}\left[\left(\frac{\partial \Phi}{\partial x}\right)_{i+1/2,j} - \left(\frac{\partial \Phi}{\partial x}\right)_{i-1/2,j}\right]$$

$$\approx \frac{\Phi_{i+1,j} - 2\Phi_{i,j} + \Phi_{i-1,j}}{(\Delta x)^2} \tag{4.5.10a}$$

$$\left(\frac{\partial^2 \Phi}{\partial y^2}\right)_{i,j} \approx \frac{1}{\Delta y}\left[\left(\frac{\partial \Phi}{\partial y}\right)_{i,j+1/2} - \left(\frac{\partial \Phi}{\partial y}\right)_{i,j-1/2}\right]$$

$$\approx \frac{\Phi_{i,j+1} - 2\Phi_{i,j} + \Phi_{i,j-1}}{(\Delta y)^2} \tag{4.5.10b}$$

As Eqs. (4.5.9) and (4.5.10) are obtained by a central difference approximation, their truncation error is of second order with respect to the grid interval. These representations also are valid when sources of strength q are located at some of the nodal points, in which case the Laplace equation is modified as the *Poisson equation*,

$$\frac{\Phi_{i+1,j} - 2\Phi_{i,j} + \Phi_{i-1,j}}{(\Delta x)^2} + \frac{\Phi_{i,j+1} - 2\Phi_{i,j} + \Phi_{i,j-1}}{(\Delta y)^2} = q_{i,j} \tag{4.5.11}$$

For convenience, since the numerical grid consists of small squares, it is assumed that $\Delta x = \Delta y = k$. Equation (4.5.11) then yields

$$\Phi_{i-1,j} + \Phi_{i+1,j} + \Phi_{i,j-1} + \Phi_{i,j+1} - 4\Phi_{i,j} = k^2 q_{i,j} \qquad (4.5.12\text{a})$$

$$\Phi_{i,j} = \frac{1}{4}(\Phi_{i+1,j} + \Phi_{i-1,j} + \Phi_{i,j+1} + \Phi_{i,j-1} - k^2 q_{i,j}) \qquad (4.5.12\text{b})$$

For simplicity, we assume there are no sources present in the domain, so that $q_{ij} = 0$ (i.e., solutions to the Laplace equation will be determined). Then, Eq. (4.5.12) indicates that the value of Φ at the i, j nodal point is the average of the four nodal points around that point. Each internal nodal point associated with subscripts $i < i_{max} - 1$, and j not too close to the bottom and top boundaries of the domain shown in Fig. 4.5.3, leads to an equation with five unknown values of Φ, associated with the i, j point and the four nodal points around that point. Figure 4.20a illustrates a case in which Dirichlet boundary conditions are used. Figure 4.20b shows a case where Neumann boundary conditions are used.

Considering the case of Dirichlet boundary conditions, shown in Fig. 4.20a, grid points with subscript i_{max} are associated with a prescribed value of $\Phi = \Phi_1$. Therefore there is no need to use Eq. (4.5.12) for the determination of Φ at these boundary nodal points. For nodal points with subscript $i_{max} - 1$, Eq. (4.5.12) incorporates the known value of $\Phi_{i_{max},j}$. Therefore for the Laplace equation, only nodal points located in the proximity of the boundary have RHS values different from zero. A similar arrangement should be considered with regard to all other boundaries at which the value of the potential function is specified.

Considering the case of Neumann boundary conditions, shown in Fig. 4.20b, at grid points with subscript i_{max} the value of the derivative of Φ is given. The finite (central) difference approximation for that derivative can be represented by

$$\left(\frac{\partial \Phi}{\partial x}\right)_{i_{max},j} = u_1 \approx \frac{\Phi_{i_{max}+1,j} - \Phi_{i_{max}-1,j}}{2\Delta x} \qquad (4.5.13)$$

where the subscript $i_{max} + 1$ represents an artifical extension of the numerical grid beyond the simulated flow domain. This is rearranged to solve for Φ at position $i_{max} + 1$,

$$\Phi_{i_{max}+1,j} = \Phi_{i_{max}-1,j} + 2u_1 \Delta x \qquad (4.5.14)$$

The linear equation set represented by Eq. (4.5.12) incorporates nodal points with subscript i_{max}. The RHS of equations associated with these nodal

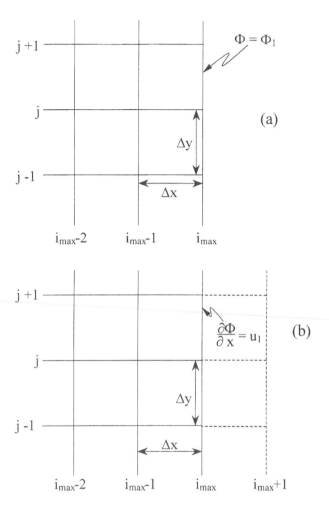

Figure 4.20 Numerical representation of boundary conditions: (a) Dirichlet boundary condition; and (b) Neumann boundary condition.

points is represented by a nonvanishing coefficient, provided that u_1 is different from zero. If $u_1 = 0$, then the line of i_{max} probably represents a solid boundary. In this case, nonvanishing RHS coefficients are provided at other boundaries of the domain. Expression similar to Eqs. (4.5.13) and (4.5.14) can be applied to all other boundaries of the flow domain at which the gradient of the potential function is specified.

The maximum number of unknown values of Φ incorporated in each nodal point in Eq. (4.5.12) is five. Therefore each row of the matrix of coefficients of these unknowns includes a maximum number of five nonzero coefficients. The solution of the set of linear equations represented by Eq. (4.5.12) can be obtained using a noniterative method such as Gauss elimination. However, due to the large number of zero-valued coefficients in each row of the coefficient matrix, an iterative solution is more efficient than the noniterative approach in this case. The diagonal term in the coefficient matrix is the dominant term in each row of that matrix. Therefore convergence of the iterative procedure is guaranteed. In the following paragraphs, we present several common iterative procedures that can be applied to solve the set of linear equations represented by Eq. (4.5.12) when $q_{i,j} = 0$.

According to the iterative method of *Jacobi*, each new value of $\Phi_{i,j}$, at iteration $n + 1$, is obtained by using values of Φ, at nodal points around point (i, j), which were obtained at iteration n. Therefore, Jacobi's method implies that Eq. (4.5.12) should be modified as

$$\Phi_{i,j}^{(n+1)} = \frac{1}{4}\left[\Phi_{i-1,j}^{(n)} + \Phi_{i+1,j}^{(n)} + \Phi_{i,j-1}^{(n)} + \Phi_{i,j+1}^{(n)}\right] \qquad (4.5.15)$$

where the superscript (in parentheses) indicates the iteration number at which the variable is calculated.

The method of *Gauss–Seidel* uses the latest computed values of Φ, as they become available during the iteration process. This method is slightly more efficient than that of Jacobi. The iteration algorithm for the solution of the Laplace equation by the Gauss–Seidel method is

$$\Phi_{i,j}^{(n+1)} = \frac{1}{4}\left[\Phi_{i-1,j}^{(n+1)} + \Phi_{i+1,j}^{(n)} + \Phi_{i,j-1}^{(n+1)} + \Phi_{i,j+1}^{(n)}\right] \qquad (4.5.16)$$

The rate of convergence of the Gauss–Seidel method can be improved by using *successive over-relaxation* (SOR). According to this method, the provisional value Φ_p of the function Φ at the nodal point (i, j) and at iteration $(n + 1)$ is calculated by the Gauss–Seidel algorithm, but this value is modified at the $(n + 1)$ iteration by means of a *relaxation parameter* ω,

$$\Phi_{i,j}^{(n+1)} = \Phi_{i,j}^{(n)} + \omega\left[\Phi_p - \Phi_{i,j}^{(n)}\right] \qquad (4.5.17)$$

where Φ_p is given by Eq. (4.5.16). If $\omega = 1$, then the SOR method is identical to the method of Gauss–Seidel. Equation (4.5.17) also can be written as

$$\Phi_{i,j}^{(n+1)} = \frac{\omega}{4}\left[\Phi_{i-1,j}^{(n+1)} + \Phi_{i+1,j}^{(n)} + \Phi_{i,j-1}^{(n+1)} + \Phi_{i,j+1}^{(n)}\right]$$
$$+ (1 - \omega)\Phi_{i,j}^{(n)} \qquad (4.5.18a)$$

or

$$\Phi_{i,j}^{(n+1)} = \Phi_{i,j}^{(n)} + \frac{\omega}{4}\left[\Phi_{i-1,j}^{(n+1)} + \Phi_{i+1,j}^{(n)} + \Phi_{i,j-1}^{(n+1)}\right]$$
$$+ \Phi_{i,j+1}^{(n)} - 4\Phi_{i,j}^{(n)}\right] \qquad (4.5.18b)$$

Finally, it should be noted that the solution of the Laplace or Poisson equation is based on an iterative solution of a set of linear equations, generally presented by

$$[A]\{\Phi\} = \{b\} \qquad\qquad (4.5.19)$$

where $[A]$ is the matrix of coefficients of Φ values, $\{\Phi\}$ is the vector of Φ values, and $\{b\}$ is the vector of RHS coefficients of all the equations. There are many different ways to iterate and solve Eq. (4.5.19). The choice of the most appropriate method of solution depends mainly on convergence properties for a particular set of conditions.

PROBLEMS

Solved Problems

Problem 4.1   Confirm for each of the following flow fields that incompressible flow is indicated. Which of these represent potential flow? Why?

(a) $u = \alpha x$ $v = -\alpha y$
(b) $u = \alpha y$ $v = -\alpha x$
(c) $u = \alpha y$ $v = \alpha x$
(d) $u_r = \dfrac{\alpha}{r}$ $v = 0$

Solution

In a potential (and incompressible) flow, the following relationships should be satisfied:

$$\nabla \times \vec{V} = 0 \qquad \nabla \cdot \vec{V} = 0$$

In a Cartesian coordinate system, these expressions imply

$$\frac{\partial v}{\partial x} - \frac{\partial u}{\partial y} = 0 \qquad \frac{\partial u}{\partial x} + \frac{\partial v}{\partial y} = 0$$

In a plane polar coordinate system, these expressions yield

$$\frac{1}{r}\frac{\partial(rv_\theta)}{\partial r} - \frac{1}{r}\frac{\partial u_r}{\partial \theta} = 0 \qquad \frac{1}{r}\frac{\partial(ru_r)}{\partial r} + \frac{1}{r}\frac{\partial v_\theta}{\partial \theta} = 0$$

Upon considering each of the given velocity fields, by substituting into the above differential equations, we find

(a) Potential incompressible flow
(b) Rotational incompressible flow

(c)   Potential incompressible flow
(d)   Potential incompressible flow

Problem 4.2   Find the potential and stream functions, if possible, for each of the flows given in solved problem (4.1.1).

Solution

We apply Eqs. (4.2.5) and (4.2.6) to determine the expressions for the required functions:

(a) $\Phi = \int u\, dx = \int \alpha x\, dx = \dfrac{\alpha x^2}{2} + f(y)$

$\dfrac{\partial \Phi}{\partial y} = f'(y) = v = -\alpha y; \quad \Rightarrow f(y) = -\dfrac{\alpha y^2}{2} + C;$ assume $C = 0$

$\Phi = \dfrac{\alpha}{2}(x^2 - y^2)$

$\Psi = \int u\, dy = \int \alpha x\, dy = \alpha xy + f(x)$

$\dfrac{\partial \Psi}{\partial x} = \alpha y + f'(x) = -v = \alpha y \Rightarrow f'(x) = 0 \Rightarrow f(x) = C = 0$

$\Psi = \alpha xy$

(b)   Using the same approach as above, we obtain

$\Psi = \dfrac{\alpha}{2}(x^2 + y^2)$

Φ does not exist, since the flow is rotational

(c) $\Phi = \alpha xy \qquad \Psi = \dfrac{\alpha}{2}(y^2 - x^2)$

(d) $\Phi = \alpha \ln r = \dfrac{\alpha}{2}\ln(x^2 + y^2) \qquad \psi = \alpha\theta = \arctan\left(\dfrac{y}{x}\right)$

Problem 4.3   Determine the flow-net for the flow field created by the superposition of a uniform flow with speed U and a source of strength q, as shown in Fig. 4.21.

Solution

The complex potential is given by the sum of the potential functions for uniform flow and a source,

$$w = Uz + \frac{q}{2\pi}\ln z = Ur\cos\theta + \frac{q}{2\pi}\ln r + i\left(Ur\sin\theta + \frac{q}{2\pi}\theta\right)$$

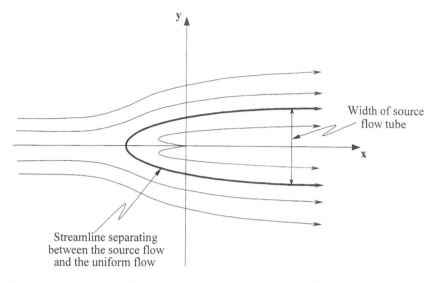

Figure 4.21 Superposition of source and uniform flow, Problem 4.3.

Therefore the potential and stream functions are

$$\Phi = Ur\cos\theta + \frac{q}{2\pi}\ln r \qquad \Psi = Ur\sin\theta + \frac{q}{2\pi}\theta$$

By differentiation of the complex potential, or either of the stream or potential functions, we obtain the following components of the velocity vector:

$$u_r = U\cos\theta + \frac{q}{2\pi r} \qquad v_\theta = -U\sin\theta$$

These expressions indicate that besides the cavitation singular point at $r = 0$, there is another singular stagnation point at $\theta = \pi, r = q/(2\pi U)$. We introduce the values of θ and r for the stagnation point into the expression of the stream function, to obtain the value of this function along the streamline, which separates the region between the uniform flow and the source flow. The result is

$$\Psi = \frac{q}{2}$$

Therefore, the streamline that separates the uniform flow from the source flow satisfies the following relationships:

$$r\sin\theta = \frac{q}{2U}\left(1 - \frac{\theta}{\pi}\right) \qquad \text{or} \qquad y = \frac{q}{2U}\left[1 - \frac{1}{\pi}\arctan\left(\frac{y}{x}\right)\right]$$

According to this expression, the width of the stream tube occupied by the source flow approaches the value $q/(2U)$ at a large distance from the origin.

Problem 4.4 If in Fig. 4.3.1, $H = 10$ m, and $K = 1$ m/day, provide an estimate of the seepage flow rate per one meter width of dam.

Solution

The solution to this problem is obtained by direct use of Eq. (4.3.15),

$$Q = K\frac{m}{n}H = 1 \times \frac{4}{15} \times 10 = 2.67 \text{ m}^3/(\text{day} \cdot \text{m}) \qquad (4.3.16)$$

Problem 4.5 Water, with kinematic viscosity $v = 10^{-6}$ m²/s, flows through sandy soil. The characteristic pore size of the soil is $d = 0.1 \times 10^{-3}$ m, and its porosity is $\phi = 0.3$.

(a) What are the permeability and hydraulic conductivity of the soil?
(b) Darcy's law is applicable up to a Reynolds number of one, where the Reynolds number is based on the characteristic pore size and the specific discharge. Determine the maximum value of the hydraulic gradient, for which Darcy's law can be applied.

Solution

(a) $K = \dfrac{gd^2\phi}{32v} = \dfrac{9.81 \times (0.1 \times 10^{-3})^2 \times 0.3}{32 \times 10^{-6}}$

$= 9.2 \times 10^{-4}$ m/s $= 79.5$ m/day

(b) $\text{Re} = \dfrac{qd}{v} = 1 \Rightarrow q = \text{Re}\dfrac{v}{d} = 1 \times \dfrac{10^{-6}}{0.1 \times 10^{-3}} = 10^{-2}$ m/s

$q = KJ \Rightarrow J = \dfrac{q}{K} = \dfrac{10^{-2}}{9.2 \times 10^{-4}} = 10.9$

Unsolved Problems

Problem 4.6 Consider the Navier–Stokes equation for the x-direction (horizontal) velocity component (for simplicity, neglect rotation effects):

$$\frac{Du}{Dt} = -\frac{1}{\rho}\frac{\partial p}{\partial x} + v\nabla^2 u$$

where u is the velocity, p is pressure, ρ is density, and v is kinematic viscosity. For inviscid flow the viscosity is 0 and pressure is the only force to consider. Show that the viscous term also drops out, i.e., $\nabla^2 u = 0$, for the case of irrotational flow.

Problem 4.7   Consider the flow field created by a superposition of a source with strength q and a simple vortex, whose circulation Γ has the same magnitude as q.

    (a)   Find the complex potential, potential, and stream functions.
    (b)   Plot the flow-net.
    (c)   Find the pressure distribution along a radial coordinate.

Problem 4.8   Consider the flow field created by a superposition of a uniform flow with speed U, in the positive x-direction, a positive source, of strength q, located at $x = -a$, and a negative source (sink), located at $x = a$.

    (a)   Find the complex potential, potential, and stream functions.
    (b)   Find the location and types of singular points.
    (c)   Plot the flow-net.
    (d)   Find the pressure distribution along the x axis.

Problem 4.9   A simple vortex, with circulation Γ, is located at $x = a$, $y = b$, which is in a $90°$ corner between two solid walls. Determine the pathline of the vortex movement.

Problem 4.10   A flow field is created by two sources at a solid wall, which is represented by the y axis. One source, of strength q, is located at $x = a$. The second source, of strength $q/2$, is located at $x = 2a$.

    (a)   Find the complex potential, potential, and stream functions.
    (b)   Determine the locations and types of singular points.
    (c)   Plot the flow-net.
    (d)   Find the pressure distribution along the x axis.

Problem 4.11   Consider the flow domain represented by

$$z = a\cos(w)$$

where

$$z = x + iy \qquad w = \Phi + i\Psi$$

    (a)   Find the complex potential, potential, and stream functions.
    (b)   Determine the locations and types of singular points.
    (c)   Plot the flow-net.
    (d)   Find the pressure distribution along the x axis.

Problem 4.12   Wind blowing over a bluff is to be simulated using potential flow theory. A uniform wind, $U = 20$ m/s, is combined with a source, having a

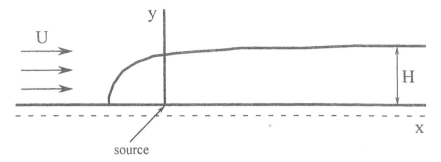

Figure 4.22 Simulation of flow over a buffer, Problem 4.12.

flow rate $q = 6,000$ m^2/s. The resulting flow is sketched in Fig. 4.22 (only the top half of the bounding streamline is shown). Note that the stream function is

$$\Psi = Ur \sin \theta + m\theta$$

where m is $q/(2\pi)$.

    (a)   What is the equation for the bounding streamline?

    (b)   How high is the bluff (H)?

    (c)   What is the velocity along the surface of the bluff, directly above the source?

Problem 4.13 Flow over a hump is to be analyzed using potential flow theory. The mathematical expression for the flow field is developed by considering one-half of the field created by superimposing a doublet in a uniform flow. As shown in Fig. 4.23, the flow far from the hump has velocity U and

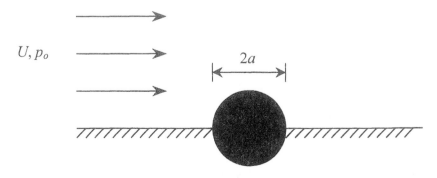

Figure 4.23 Definition sketch, Problem 4.13.

pressure p_o. The fluid density is ρ and it is incompressible. The streamline corresponding with the solid surface has $\Psi = 0$, and gravity effects may be neglected.

(a) What are the maximum and minimum pressures for this flow, in terms of ρ, U, and p_o?

(b) Develop the equation for the streamline passing through the point $r = 2a, \theta = \pi/2$.

Problem 4.14   Consider some possible values of the Rossby number in practical cases that you may encounter with regard to environmental flows in lakes and reservoirs. Choose real values for at least three examples.

Problem 4.15   The engineer of a municipality has suggested that a treated effluent can be disposed of by pumping it into an injection well. The well should be able to accept a flow of 500 m³/h. It is drilled in an aquifer whose thickness is 40 m and hydraulic conductivity is 100 m/day. Upstream of the planned injection well, the municipality pumps its water supply needs from a pumping well with a capacity of 600 m³/h. The natural flow in the aquifer is achieved with a gradient of 0.1%.

(a) Consider several hypothetical cases in which the effluent could possibly arrive at the pumping well. In other words, under what circumstances could this occur?

(b) Consider several possible values of the distance between the pumping and injection wells, and provide suggestions to the municipality on how migration of the effluent into the pumping well could be avoided.

Problem 4.16   Water flows through a confined aquifer of thickness 20 m. The hydraulic gradient is 0.1%, the hydraulic conductivity is 50 m/day, and the porosity is 0.3. The characteristic particle size of the aquifer sediment is 0.1 mm.

(a) Determine the specific discharge.

(b) Find the flow velocity.

(c) What is the Reynolds number of the flow?

Problem 4.17   Consider the schematics of flow under a concrete dam, as shown in Fig. 4.24. Depict the flow-net and provide an estimate of the total flow underneath the dam.

Problem 4.18   A function, which in the neighborhood of $z = a$ has an expansion that contains negative powers of $(z - a)$, is singular at $z = a$. In this case

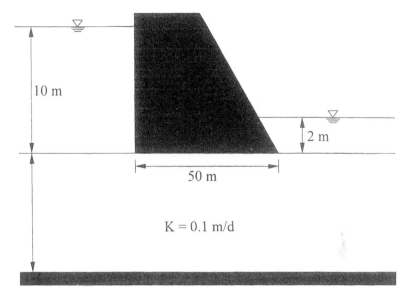

Figure 4.24 Definition sketch, Problem 4.17.

the coefficient of $(z - a)^{-1}$ is called the *residue* of the function at $z = a$.
Determine the residue of the following functions:

(a) $(z - a)^n$ where $n = -1, 1, 2 \ldots$

(b) $A_2(z - a)^2 + A_1(z - a) + A_0 + \dfrac{B_1}{(z - a)} + \dfrac{B_2}{(z - a)^2}$

Problem 4.19 According to *Cauchy's residue theorem*, the integral along a
closed line C of a holomorphic function (a function of z) is given by

$$\oint_c f(z)\, dz = 2\pi i (a_1 + a_2 + a_3 + \cdots)$$

where $a_1, a_2, \ldots$ are the residues at the singular points of the area enclosed
by the line C. Find the poles and corresponding residues of the following
functions, as well as the integral of Chauchy's residue theorem:

(a) $\dfrac{z}{z + 1}$ (b) $\dfrac{z + 2}{z^2 - 1}$ (c) $\dfrac{z + z^2}{z^2 + 1}$ (d) $\dfrac{4z^2}{z^4 - 1}$

Problem 4.20 A source of strength q is placed at a point $(a, 0)$ outside the
circle $|z| = b$, as shown in Fig. 4.25.

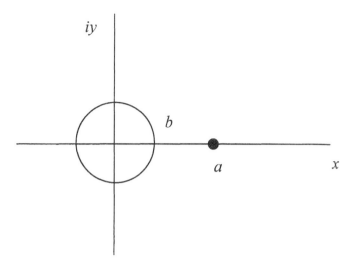

Figure 4.25 Definition sketch, Problem 4.20.

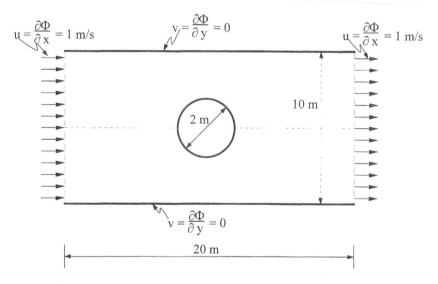

Figure 4.26 Bounded flow past a cylinder, Problem 4.21.

(a) Show that the complex potential describing the flow field is given by

$$w = \frac{q}{2\pi}\left[\ln(z-a) + \ln\left(z - \frac{b^2}{a}\right) - \ln z\right]$$

(b) Use the theorem of Blasius to prove that there is no moment about the center of the circle, and that the circle is urged towards the source by a force equal to

$$\frac{2\rho q^2 b^2}{\pi a(a^2 - b^2)}$$

(c) Find the corresponding result when the source is replaced by a vortex.

Problem 4.21 Consider the bounded potential flow around a cylinder as shown in Fig. 4.26. Provide a numerical solution and sketch the flow-net. Compare your results with the analytical solution for an infinite flow field.

SUPPLEMENTAL READING

Milne-Thomson, L. M., 1966. *Theoretical Aerodynamics*, Macmillan, London. (Contains a thorough coverage of all aspects of inviscid flow theory and applications.)

Wendt, J. F., ed. 1996. *Computational Fluid Dynamics*, Springer-Verlag, Berlin, Germany. (Includes a review of various approaches of numerical simulation of potential flows.)

5

Introduction to Turbulence

5.1 INTRODUCTION

Turbulence results from a breakdown in stability of a fluid flow and is normally associated with large Reynolds numbers. The instability may be generated by an infinitesimal disturbance in an otherwise laminar flow, and the stability of the flow depends on whether the disturbance grows or dies out, as discussed in Sec. 5.4. Almost all natural flows are turbulent to some extent, and it is important to understand and be able to represent turbulence effects when modeling a given system. Turbulence itself is difficult to define, though certain of its characteristics may be summarized as follows:

Turbulence is generally three-dimensional and is thought of as consisting of eddies superimposed on the mean flow; these eddies are represented as fluctuations in the flow field properties.

The eddy motions are irregular and *vortical* (they have vorticity).

The motions are mostly random in nature and are usually described in statistical terms.

In *fully developed turbulence* there is a continuous spectrum of eddy sizes, with a cascade of energy from larger eddies to successively smaller ones, until the kinetic energy of the eddies is dissipated by viscosity into heat by the smallest eddies.

From an environmental point of view, one of the most important effects of turbulence is to enhance greatly the mixing of fluid properties. This is accomplished through the action of the eddies, which are much more effective than molecular motions in redistributing fluid particles within a given flow field (see Chap. 10 for further discussion of molecular and turbulent diffusivities). A full analysis of turbulence is made difficult by a number of factors, primarily because of its random nature and wide range of scales of motion, from the largest eddies, which scale with the mean flow geometry, to the smallest eddies, at which viscosity dissipates the kinetic energy of the turbulent motions. This means, for instance, that a complete description of a turbulent flow field at any instant in time requires specification of flow values at a very fine spatial resolution, to capture the smallest eddies. Thus specification of initial conditions for application of the governing equations describing a velocity field is tedious at best, for any problem of practical interest. Turbulent flows are governed by the same equations of motion as were presented in Chap. 2, but the usual difficulties in solving the Navier–Stokes equations are compounded by the introduction of the velocity fluctuation terms, which add three additional unknowns to be solved for. Thus the basic conservation equations are no longer sufficient to generate a solution for all the variables involved, and the problem of *closure* of the system arises. There has been considerable effort to investigate ways of developing additional equations for the fluctuating quantities. A few examples of these methods are described in Sec. 5.5.

The present chapter is meant to provide an introduction to the analysis of turbulent motions, to define some of the most common terms involved in the study of turbulence, and to describe its effect on transport and mixing in a fluid system. The intent here is to develop a basis for understanding how to incorporate turbulence in practical modeling applications. The reader is directed to references listed in the back of the chapter for more in-depth discussions.

5.2 DEFINITIONS

In the analysis of turbulent flows, it is helpful to think of a time record for any fluid property of interest (velocity, temperature, pressure, salinity, etc.) as consisting of a mean, time-averaged component and a fluctuating component that is a function of time. It is normally assumed that the statistical properties of the fluctuating component remain constant, or *stationary* over the averaging period used to define the mean. Turbulent velocity fluctuations are illustrated in Fig. 5.1. In this figure is shown a time record of one component of velocity measured at a point in a flow field. Although the mean flow (U) is constant, or steady over the time period of measurement shown, there are random fluctuations superimposed on the mean. These fluctuating values are denoted by primes (i.e., u'), so that the total velocity at any point in time is

$$u = U + u' \qquad (5.2.1)$$

and U is defined as the time average of u,

$$U = \frac{1}{T} \int_0^T u \, dt \qquad (5.2.2)$$

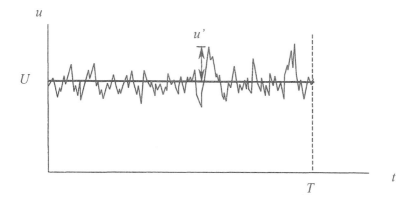

Figure 5.1 Illustration of turbulent velocity record.

where T is the averaging period. By definition, the time average of $u' = 0$, so

$$\int_0^T u' \, dt = 0 \qquad (5.2.3)$$

Similarly, the average of u' multiplied by a constant is also zero.

Turbulent fluctuations are treated as realizations of a *random process*. A random process in general is associated with certain temporal and spatial scales. Consider first the time variability of a random process of interest, relative to a range of time scales, as sketched in Fig. 5.2. This variability can be illustrated with a simple example of sunlight intensity. Characteristic time scales for sunlight intensity consist of one day (diurnal variability) and several months (seasonal variability). Relative to the daily time scale, solar intensity variations over periods of a few seconds or a few minutes

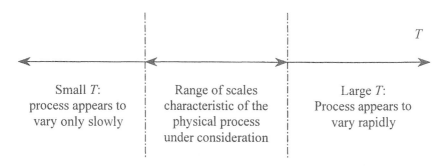

Figure 5.2 Illustration of time scale variability of a random process.

would appear to be approximately constant. In other words, the variability
of the process is very small over this much smaller time period, relative to
a day, and the process could be considered as being approximately steady.
On the other hand, over a period of perhaps several weeks or months, the
solar intensity variation will have completed a number of cycles and would
appear to vary rapidly. In that case, the daily variability would appear as
fluctuations superimposed on the mean value over the period of interest.
Depending on the particular analysis involved, it might be useful to consider
averaged values over this period. As an alternative to the time scales shown
in Fig. 5.2, process variations over a range of frequencies can be evaluated,
where frequency, ω, is related to the inverse of T. Then large T implies small
ω and vice-versa.

An equivalent description can be made in spatial terms. For example,
consider a wavy water surface where the wavelength is λ (Fig. 5.3). In other
words, the water surface variations are characterized by a process with length
scale λ. Then, over length scales (distances) L much smaller than λ, the water
surface would appear to be approximately constant, while for L much greater
than λ the water surface appears to be rapidly varying. In this case the waves
represent fluctuations on the mean water surface, and it may make sense to
consider only the average water surface, or perhaps the average wave height.
Just as with temporal variations, we may also consider frequency variations in
space. In this case a spatial frequency or *wave number* k is defined, proportional to the inverse of L.

Due to the random nature of turbulent motions, it is usually more
convenient to deal with statistical or averaged properties of the flow field.
An *ensemble average* is defined as the arithmetic average of a number of
measurements of a random process. For example, the ensemble average of a
set of velocity measurements made at a number of locations in a flow field is

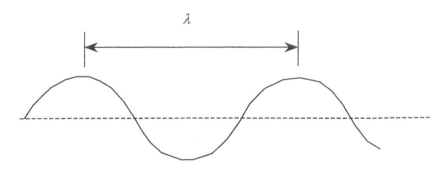

Figure 5.3   Characteristic length scale for wavy water surface.

defined as

$$\overline{\overline{u_i(\vec{x}, t)}} = \lim_{N \to \infty} \left\{ \frac{1}{N} \sum_{j=1}^{N} u_i(\vec{x}, t; j) \right\} \tag{5.2.4}$$

where the double overbar indicates a spatial average, the "j" notation indicates the realization or measurement identification number, and N is the total number of measurements or tests made.

A process is *stationary* if the statistics describing the process (i.e., mean, standard deviation, correlations, etc.) are independent of time. It is common to assume stationarity for the analysis of turbulent fluctuations. For stationary processes, the ensemble average is the same as a time average such as is defined in Eq. (5.2.2). In more general terms, a *time average* is defined by

$$\overline{u_i(\vec{x}, t_0)} = \lim_{T \to \infty} \left\{ \frac{1}{T} \int_{t_0 - T/2}^{t_0 + T/2} u_i(\vec{x}, t) \, dt \right\} \tag{5.2.5}$$

where T is the period of measurement (for averaging) and t_0 is the point in time for which the average is calculated. A single overbar in general is used to denote an averaged quantity, usually a time average. The value of the average is independent of t_0 when the process is stationary.

A process is *homogeneous* when the statistics are independent of position, and *isotropic* when independent of orientation (i.e. invariant to rotation about any coordinate axis). For example, if the average square of the fluctuating velocities is a homogeneous process, then

$$\overline{u_i'(\vec{x}_0, t)^2} = \overline{u_i'(\vec{x}_0 + \vec{x}', t)^2} \tag{5.2.6}$$

where $\vec{x}'$ represents an arbitrary displacement vector. In other words, the value of the average does not depend on position.

Just as the fluid is considered as a continuum, the turbulent eddies that are manifested by the velocity fluctuations are thought to occupy the fluid fully, with smaller eddies embedded in larger ones. The turbulence consists of a continuous spectrum of eddy sizes and can be represented by Fourier integrals (Sec. 5.3). The eddies fully interact with each other, exchanging energy and momentum, and the movement of any one eddy affects the fluid surrounding it. This implies that the fluctuating velocity at a given point in the fluid is statistically *correlated* with that at neighboring points. This correlation decreases with separation, eventually reaching zero at a sufficiently large distance. When this happens, the corresponding distance may be considered as an estimate for the size of the largest eddy. A similar argument can be made with respect to temporal fluctuations, in which case the largest correrlation time, at which the correlation approaches zero, is considered as an estimate of the longest time scale associated with an eddy.

A general *covariance function* (for velocity fluctuations) is defined by $\overline{u_i'(\vec{x}_1, t_1)u_j'(\vec{x}_2, t_2)}$, which is simply the average values of the product of two values calculated at two locations $\vec{x}_1$ and $\vec{x}_2$ and two times t_1 and t_2. The *correlation coefficient* is

$$R_{ij} = \frac{\overline{u_i'(\vec{x}_1, t_1)u_j'(\vec{x}_2, t_2)}}{\left[\overline{u_i'(\vec{x}_1, t_1)^2} \cdot \overline{u_j'(\vec{x}_2, t_2)^2}\right]^{1/2}} \qquad (5.2.7)$$

and $R_{ij} \leq 1$. A closely related parameter is the *autocorrelation coefficient*, which may be defined over time or space. The autocorrelation is calculated on the basis of the covariance between a value and itself, but displaced in either time or space. For a series of data measured by a velocity probe at a fixed location in space, the autocorrelation coefficient is

$$R_{ii}(\Delta t) = \frac{\overline{u_j'(\vec{x}, t)u_i'(\vec{x}, t + \Delta t)}}{\overline{u_i'(\vec{x})^2}} \qquad (5.2.8)$$

where Δt is the time interval or *time lag* over which the autocorrelation is calculated. In other words, the *autocovariance function* (numerator of Eq. 5.2.8) is calculated as the average of all values of u_i' multiplied by itself, but lagged in time by Δt. This value is then normalized by the mean square value of u_i' (denominator of Eq. 5.2.8), which is the autocovariance function for $\Delta t = 0$. Since the correlation of any value with itself is highest for zero lag, then $R_{ii}(\Delta t) \leq 1$. In performing the calculation in Eq. (5.2.8), it is assumed that u_i' is a stationary process. It also should be noted that $R_{ii} = R_{ij}(\Delta t) = R_{ii}(-\Delta t)$, and $R_{ii}(0) = 1$. A typical curve for R_{ii} as a function of Δt is illustrated in Fig. 5.4.

Until recently, making temporal measurements at a relatively small number of locations (maybe only one) has been by far the most common procedure in fluids experiments, due to the available instrumentation. This limits statistical calculations to the temporal domain. The point at which R_{ii} reaches zero may be interpreted in a similar manner as a length autocorrelation, and the time ($\tau_{\max}$ in Fig. 5.2.4) corresponds to the longest eddies. Assuming stationarity, temporal scales are converted to length scales by multiplying with an appropriate velocity scale, often the root-mean-square fluctuating velocity. In the past decade or so, however, the advent of particle tracking velocimetry (PTV) and particle image velocimetry (PIV) has provided sufficiently detailed spatial resolution for velocity data that calculations such as Eq. (5.2.8) can be performed directly for spatial lags rather than temporal lags (i.e., substituting $\Delta \vec{x}$ for Δt). For now, we continue the discussion for temporal analysis.

Once R_{ii} is known, the *integral time scale* is

$$\tau = \int_0^\infty R_{ii}(\Delta t)\, d(\Delta t) \qquad (5.2.9)$$

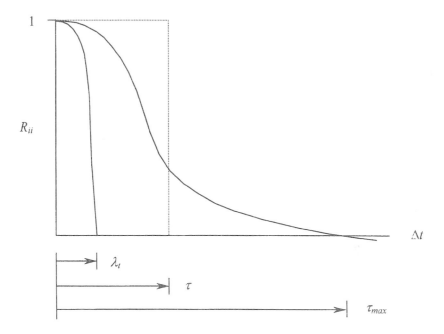

Figure 5.4 Variation of R_{ii} with time lag, illustrating calculations for temporal integral and micro scales.

This calculation gives another measure of the time over which statistics of the process are correlated, and for turbulent velocities, τ is considered to be a representative time scale for the main energy-containing eddies in a turbulent velocity field. These are known as the *integral-scale eddies*. For a real data set, this calculation cannot be computed to the limit of infinity. Instead, the upper limit is set by the number of measurements made (N). If the time interval between each measurement is Δt, the calculation is

$$\tau = \Delta t \sum_{n=0}^{N-1} R_{ii}(n\, \Delta t) \tag{5.2.10}$$

Better estimates for τ are obtained with longer data records, which involve larger numbers of data for the averaging process. In practice, values for R_{ii} tend to oscillate around 0 after some time, and the calculation for τ normally includes values for R_{ii} only up to the first zero crossing. It is easily seen that τ is equivalent to the area under the autocorrelation coefficient curve.

An estimate for the *turbulence temporal microscale* (λ_t, corresponding to the smallest eddies expected in the flow field) is found by fitting a parabola to the autocorrelation coefficient curve at $\Delta t = 0$ (Fig. 5.4),

$$\lambda_t = \left[-\frac{2}{\left. \dfrac{\overline{d^2 R_{ii}}}{dt^2} \right|_{\Delta t=0}} \right]^{1/2} \tag{5.2.11}$$

This value also is shown in Fig. 5.4.

The equivalent calculations for spatial scales, based on measurements taken simultaneously at a set of different locations, leads to a corresponding spatial microscale called the *Taylor microscale*. In other words, the autocorrelation coefficient in space is calculated as

$$R_{ii} = \frac{\overline{u_i'(\vec{x}, t) u_i'(\vec{x} + \Delta \vec{x}, t)}}{\overline{u_i'(t)^2}} \tag{5.2.12}$$

where it is assumed that the process is homogeneous, at least over the area of measurement. Then, equations equivalent to Eqs. (5.2.9) or (5.2.10), and

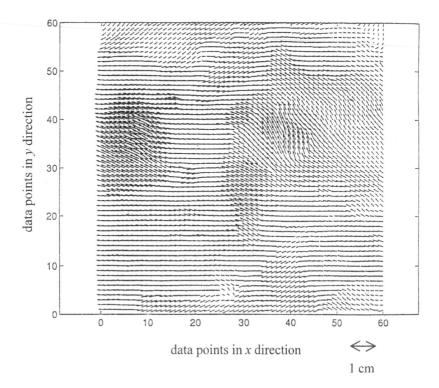

Figure 5.5 Example of velocity field measurement in a 2-liter mixing jar, obtained by PIV. (From Cheng et al., 1997.)

(5.2.11) are used to obtain integral and microlength scales, respectively. If only temporal data are available, the usual procedure is to apply Taylor's *"frozen turbulence"* hypothesis, which states that temporally varying measurements obtained at a single location in space may be related to spatial variations through the mean velocity. In other words, this procedure assumes that the turbulence properties are changing only very slowly relative to the mean flow position. In some cases, the velocity probe itself is moved through the fluid, providing its own mean velocity. The relation between a time step Δt and a spatial step Δx is then $\Delta x = U \Delta t$. This procedure is not always valid, particularly in situations when U is small.

As already noted, PTV and PIV enable simultaneous measurements of turbulent velocities at many different points in a flow field. For example, Fig. 5.5 shows a PIV image result from a flow field inside a 2-liter mixing jar, similar to those used in studying flocculation processes for water treatment studies. The fine spatial detail is immediately evident, and these data enable spatial statistics to be calculated directly. In fact, PIV systems now provide direct digital images of turbulent eddy motions, as shown in Fig. 5.6. This

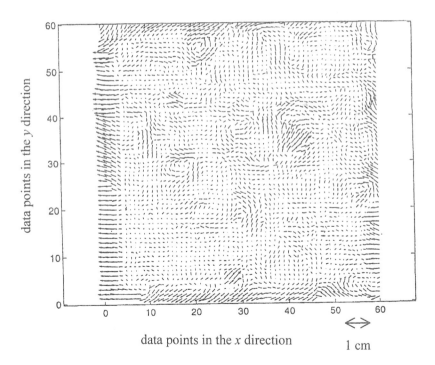

Figure 5.6 High-pass velocity field obtained from the data of Figure 5.5, showing turbulent eddy structures. (From Cheng et al., 1997.)

latter figure was obtained from the data of Fig. 5.5 through a high-pass filtering process, which removes the low-frequency (mean) flow and reveals only the high-(spatial)-frequency turbulence motions. Unfortunately, while these data have provided interesting results in the lab, with few exceptions PIV has not yet been widely adapted for field studies.

5.3 FREQUENCY ANALYSIS

As noted previously, the continuous spectrum of eddy size and time scales in a fully developed turbulence field allows analyis by Fourier integrals. This leads to calculation of the power spectrum, as described below. The end result is a determination of the main frequencies of importance in a set of measured flow data and an indication of the manner in which turbulent energy is transported from larger scales (lower frequencies) to smaller scales (higher frequencies). The procedures are based on the concept that a continuous data series may be represented by a *Fourier series*, which in general is an infinite sum of sine and cosine terms. As an example, consider a simple periodic function, as shown in Fig. 5.7. This function may be represented by

$$f(t) = a_0 + a\,\sin(\omega t) \tag{5.3.1}$$

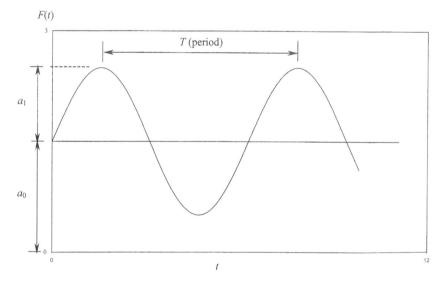

Figure 5.7   Illustration of simple periodic function of time (sine wave).

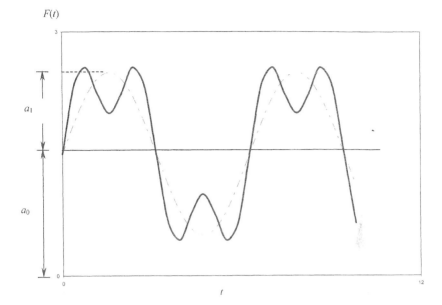

Figure 5.8 Dual periodic function, consisting of two sine waves; a second sine wave, with half the amplitude and one-third the period of the wave in Fig. 5.7, has been added to generate the bold-faced curve.

where a_0 is a constant, a is the amplitude of the sine term, and $\omega = 2\pi/T$ is the frequency. Adding a second sine term Fig. 5.8,

$$f(t) = a_0 + a_1 \sin(\omega_1 t) + a_2 \sin(\omega_2 t) \tag{5.3.2}$$

where a_1 and a_2 are the amplitudes and $\omega_1 = 2\pi/T_1$ and $\omega_2 = 2\pi/T_2$ are the frequencies, respectively, of the two sine curves. Equivalent expressions in terms of spatial frequencies or *wave numbers* are possible for spatially varying data.

This idea may be extended to include as many terms as necessary to represent a given function. In general, both sine and cosine terms are included in the summation. If a function is measured over either a time period T or spatial length L (as would be done in a realistic measurement), it is assumed for Fourier series representation that the length of record is one cycle of a cyclic process. For example, a function $f(x)$ is shown in Fig. 5.9, where observations are made only over the interval between x_0 and $x_0 + L$. Figure 5.10 shows the Fourier series representation for $f(x)$, written as

$$f(x) = a_0 + \sum_{n=1}^{\infty} \left(a_n \cos \frac{2\pi n x}{L} + b_n \sin \frac{2\pi n x}{L} \right) \tag{5.3.3}$$

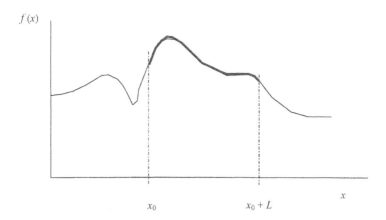

Figure 5.9   A function $f(x)$, measured between x_0 and $x_0 + L$.

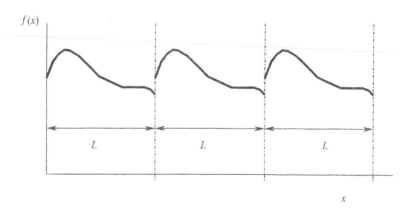

Figure 5.10   Periodic representation of the function $f(x)$ measured as in Fig. 5.9.

where a_0 is a constant magnitude function, equal to the long-term mean of $f(x)$, and the a_n and b_n are the amplitudes associated with each frequency. These magnitudes are found from the *Fourier transform functions*, written as

$$a_0 = \frac{1}{L} \int_{x_0}^{x_0+L} f(x)\,dx \tag{5.3.4}$$

$$a_n = \frac{2}{L} \int_{x_0}^{x_0+L} f(x)\cos\frac{2\pi n x}{L}\,dx \tag{5.3.5}$$

$$b_n = \frac{2}{L} \int_{x_0}^{x_0+L} f(x)\sin\frac{2\pi n x}{L}\,dx \tag{5.3.6}$$

It is usual to define a *wave number*, $k_n = 2\pi n / L$, as a spatial frequency, so the arguments for the sine and cosine terms may be written simply as $(k_n x)$. It also should be noted that a finite sum in Eq. (5.3.3) is an exact representation of $f(x)$ when $f(x)$ is a true periodic function (of period L).

The statistical representation of a random process $\tilde{x}$ (in other words, $\tilde{x}$ represents a set of values measured for a given parameter of interest, such as velocity, pressure, depth, concentration, etc., and x represents one of the values in the set) depends largely on the distribution of values as given by the *probability density function*, $p(x)$, where $0 \le p(x) \le 1$, and

$$\int_{-\infty}^{\infty} p(x)\,dx = 1 \tag{5.3.7}$$

The *cumulative probability distribution function*, $F(x)$, for a given x, is just the area under the curve of $p(x)$,

$$F(x) = \int_{-\infty}^{\infty} p(x)\,dx \tag{5.3.8}$$

This value indicates the percentage of all values in $\tilde{x}$ that are less than x.

The *expected value* or *mean* of $\tilde{x}$ is

$$E(\tilde{x}) = \bar{x} = \int_{-\infty}^{\infty} x p(x)\,dx \tag{5.3.9}$$

and the *variance* is

$$\sigma^2(\tilde{x}) = \int_{-\infty}^{\infty} (x - \bar{x})^2 p(x)\,dx = E(\tilde{x}^2) - [E(\tilde{x})]^2 \tag{5.3.10}$$

As previously noted, it is usual in turbulence analysis to assume that the processes being measured are *stationary*. With this assumption, the mean of a process can be defined in terms of a finite version of Eq. (5.3.9),

$$E(\tilde{x}) = \frac{1}{T} \int_0^T x(t)\,dt \tag{5.3.11}$$

where T is the time of measurement. For digital data the mean is simply

$$E(\tilde{x}) = \frac{1}{N} \sum_{n=1}^{N} x_n \tag{5.3.12}$$

where N is the total number of observations and the x_n are individual measurements of the process $\tilde{x}$. Similarly, the variance for a stationary process can be calculated as

$$\sigma^2(\tilde{x}) = \frac{1}{T} \int_0^T [x(t) - E(\tilde{x})]^2 dt \tag{5.3.13}$$

or, for digital data,

$$\sigma^2(\tilde{x}) = \frac{1}{N-1} \sum_{n=1}^{N} [x_n - E(\tilde{x})]^2 \tag{5.3.14}$$

Note that the division in Eq. (5.3.14) is carried out by $(N-1)$, rather than by N, to obtain an *unbiased* estimate of the variance. The *autocorrelation* is

$$R(\Delta t) = \frac{1}{T} \int_0^T x(t)x(t+\Delta t)\,dt = \frac{1}{N'} \sum_{n=1}^{N'} (x_n)(x_{n+m}) \tag{5.3.15}$$

where m is the number of steps (between data points) corresponding to Δt, and N' is the number of terms that can be included in the averaging procedure. For a fixed record length T, N' is smaller for larger Δt. For example, a record of 10 values ($N = 10$) can have nine terms in the sum of Eq. (5.3.16) when $m = 1$, but only one term when $m = 9$.

Finally, the *power spectrum function*, $S(\omega)$, is related to the autocorrelation as

$$S(\omega) = \int_0^{\infty} R(\Delta t) \cos(\omega\,\Delta t)\,d(\Delta t) \tag{5.3.16}$$

and

$$R(\Delta t) = \frac{2}{\pi} \int_0^{\infty} S(\omega) \cos(\omega\,\Delta t)\,d\omega \tag{5.3.17}$$

These last two relationships form the *Fourier transform pair*. Equation (5.3.16), in particular, provides a direct means of identifying specific frequencies of interest in the signal for a given record of observation, since an *amplitude function* can be determined by

$$P(\omega) = [\omega S(\omega)]^{1/2} \tag{5.3.18}$$

which gives the amplitude of the function at frequency ω.

Figure 5.11 shows a power spectrum calculated for the longitudinal (i.e., in the direction of mean flow) turbulent fluctuating velocity measured along the centerline of a surface jet. Figure 5.12 shows the corresponding amplitude function. It is easy to see certain peaks in these figures, which indicate the frequencies of the most energetic motions. Another example of this approach is with respect to water surface elevations measured at the mouth of an estuary. Application of frequency analysis to a set of such data should be able to provide an indication of the normal tidal period (or frequency), as well as the dominant wave period.

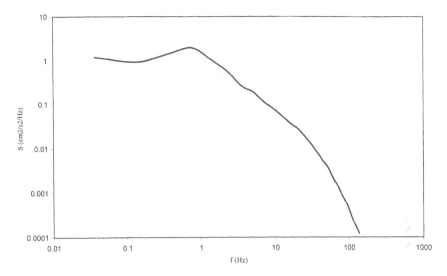

Figure 5.11 Power spectrum calculated for longitudinal turbulent velocity fluctuations measured along the centerline of a surface jet in the laboratory.

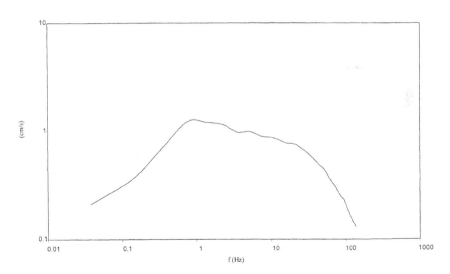

Figure 5.12 Amplitude function corresponding to the power spectrum from Fig. 5.11.

5.4   STABILITY ANALYSIS

The essence of stability theory is to define whether a small disturbance imposed
on a flow will grow or not. In this section we consider *linear stability*, in which
a stable, steady flow represented by velocities and pressure $(\overline{u}_i, \overline{p})$ is subjected
to small disturbances u_i' and p'. Thus

$$u_i = \overline{u}_i + u_i' \qquad p = \overline{p} + p' \tag{5.4.1}$$

where $u_i (i = 1, 2, 3)$ represents the three velocity components in a Cartesian
coordinate system. Note that this definition for u_i appears similar to Eq. (5.2.1),
except here our interest is the initial growth of the disturbances, rather than
the properties of the disturbances once they are fully developed. In general,
expressions similar to Eq. (5.4.1) can be written for other fluid properties such
as density, temperature, or mass concentration (dissolved mass), but for now
we are primarily interested in the velocities and pressure; density variations
are considered only in connection to buoyancy terms, consistent with the
Boussinesq approximation.

   We first consider the Navier–Stokes equations for incompressible flow,
assuming a Cartesian coordinate system (refer to Chap. 2),

$$\frac{\partial u_i}{\partial x_i} = 0 \tag{5.4.2}$$

$$\frac{\partial u_i}{\partial t} + u_j \frac{\partial u_i}{\partial x_j} = -\frac{1}{\rho} \frac{\partial p}{\partial x_i} - g_i + \nu \nabla^2 u_i \tag{5.4.3}$$

where Eq. (5.4.2) expresses continuity and Eq. (5.4.3) expresses momentum
conservation, with $i, j = 1, 2, 3$, $g = $ gravity, and $\nu = $ kinematic viscosity.
Following normal convention, the Coriolis term is not included here, as it
is not important for a first-order linear stability analysis.

   It is assumed that the steady, stable flow $(\overline{u}_i, \overline{p})$ satisfies

$$\frac{\partial \overline{u}_i}{\partial x_i} = 0 \tag{5.4.4}$$

and

$$\overline{u}_j \frac{\partial \overline{u}_i}{\partial x_j} = -\frac{1}{\rho} \frac{\partial \overline{p}}{\partial x_i} - g_i + \nu \nabla^2 \overline{u}_i \tag{5.4.5}$$

This solution also is assumed to satisfy all relevant boundary conditions
of the problem. For the continuity equation, substitution of Eq. (5.4.1) into
Eq. (5.4.2) gives

$$\frac{\partial u_i}{\partial x_i} = \frac{\partial (\overline{u}_i + u_i')}{\partial x_i} = \frac{\partial \overline{u}_i}{\partial x_i} + \frac{\partial u_i'}{\partial x_i} = 0 \tag{5.4.6}$$

The time average of this expression is

$$\overline{\frac{\partial \overline{u}_i}{\partial x_i}} + \overline{\frac{\partial u'_i}{\partial x_i}} = \frac{\partial \overline{u}_i}{\partial x_i} + \overline{\frac{\partial u'_i}{\partial x_i}} = 0 \tag{5.4.7}$$

However, because of Eq. (5.4.4) and also because, by definition, the time average of a fluctuating quantity is 0 (since u'_i is a fluctuating quantity, so is $\partial u'_i / \partial x_i$), then

$$\overline{\frac{\partial u'_i}{\partial x_i}} = 0 \tag{5.4.8}$$

Also, from Eq. (5.4.6), it is seen that the fluctuating part of the velocity field has zero divergence, i.e.,

$$\frac{\partial u'_i}{\partial x_i} = 0 \tag{5.4.9}$$

For the momentum equations, substitution of Eq. (5.4.1) into Eq. (5.4.3) results in

$$\frac{\partial}{\partial t}(\overline{u}_i + u'_i) + (\overline{u}_j + u'_j)\frac{\partial}{\partial x_j}(\overline{u}_i + u'_i)$$

$$= -\frac{1}{\rho_0}\frac{\partial}{\partial x_i}(\overline{p} + p') + g_i + \nu\nabla^2(\overline{u}_i + u'_i) \tag{5.4.10}$$

Then, by subtracting Eq. (5.4.5) and noting that the assumed stable solution is steady, an equation for the disturbances is obtained as

$$\frac{\partial u'_i}{\partial t} + \overline{u}_j\frac{\partial u'_i}{\partial x_j} + u'_j\frac{\partial \overline{u}_i}{\partial x_j} = -\frac{1}{\rho}\frac{\partial p'}{\partial x_i} + \nu\nabla^2 u'_i \tag{5.4.11}$$

Equations (5.4.9) and (5.4.11) are the governing equations for the disturbances.

The simplest case to consider for the disturbances is that of natural sinusoidal oscillations of small amplitude. Since Eqs. (5.4.9) and (5.4.11) are linear in the disturbances and the flow given by $(\overline{u}_i, \overline{p})$ is independent of time, formulations for the disturbances can be written as

$$u'_i = f_i(x_1, x_2, x_3)e^{-i\sigma t} \qquad p' = f'(x_1, x_2, x_3)e^{-i\sigma t} \tag{5.4.12}$$

where f_i and f' are functions only of position and σ is a complex function having units of time^{-1}. From Eq. (5.4.12), the disturbances are seen to grow or decay with time, depending on whether the imaginary part of $\sigma(\sigma_i)$ is positive or negative, respectively. Thus the flow is unstable when $\sigma_i > 0$ and stable when $\sigma_i < 0$. When $\sigma_i = 0$, a state of *neutral stability* results, in which the disturbances neither grow nor decay with time. Substituting Eq. (5.4.12) into Eqs. (5.4.9) and (5.4.11) gives, respectively,

$$\frac{\partial f_i}{\partial x_i} = 0 \tag{5.4.13}$$

and

$$-i\sigma f_i + \bar{u}_j \frac{\partial f_i}{\partial x_j} + f_j \frac{\partial \bar{u}_i}{\partial x_j} = -\frac{1}{\rho}\frac{\partial f'}{\partial x_i} + \nu \nabla^2 f_i \tag{5.4.14}$$

Since the flow field $(\bar{u}_i, \bar{p})$ satisfies the boundary conditions of the originally posed problem, the boundary conditions for Eqs. (5.4.13) and (5.4.14) must be homogeneous, which also means the boundary conditions for the f_i are homogeneous. A nontrivial solution for these equations then will be possible for certain values of σ, i.e., an eigenvalue problem results for σ and f_i.

The spatial stability problem can be addressed by considering the disturbances as an oblique traveling wave. For convenience, the wave is assumed to travel in the x_1–x_3 plane, and functional forms for f_i and f' are written as

$$\begin{aligned} f_i(x_1, x_2, x_3) &= \varphi_i(x_2)\, e^{i(k_1 x_1 + k_3 x_3)} \\ f'(x_1, x_2, x_3) &= \zeta(x_2)\, e^{i(k_1 x_1 + k_3 x_3)} \end{aligned} \tag{5.4.15}$$

where k_1 and k_3 are complex wave numbers in the x_1 and x_3 directions, respectively, with units of length^{-1}. After substituting Eq. (5.4.15) into Eqs. (5.4.13) and (5.4.14), rearranging and simplifying, we have

$$\frac{\partial \varphi_2}{\partial x_2} + i(k_1 \varphi_1 + k_3 \varphi_3) = 0 \tag{5.4.16}$$

$$i(-\sigma + k\bar{u} + k_3 \bar{u}_3)f_i + \bar{u}_2 \frac{\partial \varphi_i}{\partial x_2} + \varphi_j \frac{\partial \bar{u}_i}{\partial x_j}$$

$$= -\frac{1}{\rho}\frac{\partial \zeta}{\partial x_i}\delta_{2i} - i\frac{\zeta}{\rho}(k_1 \delta_{1i} + k_3 \delta_{3i}) + \nu\left[\frac{\partial^2 \varphi}{\partial x_2^2} - (k_1^2 + k_2^2)\varphi_i\right] \tag{5.4.17}$$

where δ_{ij} is the Kronecker delta. Now, for fixed σ, Eqs. (5.4.16) and (5.4.17) form an eigenvalue problem for k_1 and k_3. Similar to the situation with σ, if either k_1 or k_3 has a negative imaginary component, the flow will be spatially unstable.

It is helpful to consider the equations in nondimensional form. Following the development in Sec. 2.9, this is done by designating characteristic velocity (U') and length (L) scales, to define nondimensional variables as

$$x = \frac{x_1}{L} \qquad y = \frac{x}{L} \qquad z = \frac{x_3}{L} \qquad U = \frac{\bar{u}_1}{U'} \qquad V = \frac{\bar{u}_2}{U'}$$

$$W = \frac{\bar{u}_3}{U'} \qquad \alpha = Lk_1 \qquad \beta = Lk_3 \qquad \tau = \frac{U'}{L}t \qquad (5.4.18)$$

$$K = \frac{\zeta}{\rho U'^2} \qquad \Omega = \frac{\sigma L}{U'} \qquad \psi_i = \frac{\varphi_i}{U'} \qquad \text{Re} = \frac{U'L}{\nu}$$

Introducing these into Eqs. (5.4.16) and (5.4.17) then gives

$$\frac{d\psi_2}{dy} + i(\alpha\psi_1 + \beta\psi_3) = 0 \qquad (5.4.19)$$

$$i\psi_1(-\Omega + \alpha U + \beta W) + V\frac{d\psi_1}{dy} + \psi_1\frac{\partial U}{\partial x} + \psi_2\frac{\partial U}{\partial y} + \psi_{31}\frac{\partial U}{\partial z}$$

$$= -i\alpha K + \frac{1}{\text{Re}}\left[\frac{d^2\psi_1}{dy^2} - (\alpha^2 + \beta^2)\psi_1\right] \qquad (5.4.20)$$

$$i\psi_2(-\Omega + \alpha U + \beta W) + V\frac{d\psi_2}{dy} + \psi_1\frac{\partial V}{\partial x} + \psi_2\frac{\partial V}{\partial y} + \psi_{31}\frac{\partial V}{\partial z}$$

$$= -\frac{dK}{dy} + \frac{1}{\text{Re}}\left[\frac{d^2\psi_2}{dy^2} - (\alpha^2 + \beta^2)\psi_2\right] \qquad (5.4.21)$$

$$i\psi_3(-\Omega + \alpha U + \beta W) + V\frac{d\psi_3}{dy} + \psi_1\frac{\partial W}{\partial x} + \psi_2\frac{\partial W}{\partial y} + \psi_3\frac{\partial W}{\partial z}$$

$$= -i\beta K + \frac{1}{\text{Re}}\left[\frac{d^2\psi_3}{dy^2} - (\alpha^2 + \beta^2)\psi_3\right] \qquad (5.4.22)$$

5.4.1 Stability of Plane Laminar Flows

In general, the system of equations (5.4.19)–(5.4.22) is very difficult to solve. However, there are certain simplified cases for which further development is possible. Consider a flow that consists of parallel streamlines in the x–z plane, so that y indicates the direction normal to the mean flow direction. This type of flow might exist between two infinite parallel planes, for example. The variations of flow properties in the x and z directions are assumed to be very small relative to the y direction, so that the flow field may be considered as a function of y only [i.e., $U = U(y)$, $V = V(y)$, $W = W(y)$]. Then, from the continuity equation (5.4.4), $V = 0$. For simplicity, it also is assumed that $W = 0$ in the following. In other words, a straight flow is considered, where the coordinate system is aligned so that the x axis points along the direction of the mean flow. With these assumptions, Eq. (5.4.19) is unchanged, and

Eqs. (5.4.20)–(5.4.22) become, respectively,

$$i\psi_1(-\Omega + \alpha U) + \psi_2 \frac{\partial U}{\partial y} = -i\alpha K + \frac{1}{Re}\left[\frac{d^2\psi_1}{dy^2} - (\alpha^2 + \beta^2)\psi_1\right] \quad (5.4.23)$$

$$i\psi_2(-\Omega + \alpha U) = -\frac{dK}{dy} + \frac{1}{Re}\left[\frac{d^2\psi_2}{dy^2} - (\alpha^2 + \beta^2)\psi_2\right] \quad (5.4.24)$$

$$i\psi_3(-\Omega + \alpha U) = -i\beta K + \frac{1}{Re}\left[\frac{d^2\psi_3}{dy^2} - (\alpha^2 + \beta^2)\psi_3\right] \quad (5.4.25)$$

These equations are combined by first eliminating K between Eqs. (5.4.23) and (5.4.24) and substituting for ψ_1 from Eq. (5.4.19) to obtain an equation with ψ_2 and ψ_3. A second equation in ψ_2 and ψ_3 is developed by eliminating K between Eqs. (5.4.24) and (5.4.25), again using Eq. (5.4.19) to substitute for ψ_1. The two equations in ψ_2 and ψ_3 are then combined to give

$$(U - c)\left[\frac{d^2}{dy^2} - (\alpha^2 + \beta^2)\right]\psi_2 - \psi_2\frac{d^2U}{dy^2}$$

$$= -\frac{i}{\alpha Re}\left[\frac{d^4}{dy^4} - 2(\alpha^2 + \beta^2)\frac{d^2}{dy^2} + (\alpha^2 + \beta^2)^2\right]\psi_2 \quad (5.4.26)$$

where

$$c = \frac{\Omega}{\alpha} \quad (5.4.27)$$

is a complex velocity (nondimensional), called the *phase speed* of the disturbance. The corresponding dimensional complex velocity is the ratio of the time parameter σ to the wave number.

We now consider three-dimensional disturbances in the form of an oblique wave in the x–z plane, with amplitude as a function of y. This is called a *Tollmien–Schlichting wave*, and we further assume that the coordinate system is oriented so that the direction of travel of the wave is along the x-axis. This implies $\beta = 0$. Under this condition, the continuity equation (5.4.19) simplifies to

$$\frac{d\psi_2}{dy} + i\alpha\psi_1 = 0 \quad (5.4.28)$$

A general solution to this equation may be expressed in terms of a function $\psi(y)$, such that

$$\psi_1 = \frac{d\psi}{dy} \qquad \psi_2 = -i\alpha\psi \quad (5.4.29)$$

Substituting Eq. (5.4.29) into Eq. (5.4.26) then gives

$$(U - c)\left(\frac{d^2}{dy^2} - \alpha^2\right)\psi - \psi\frac{d^2 U}{dy^2}$$
$$= -\frac{i}{\alpha \text{Re}}\left[\frac{d^4}{dy^4} - 2\alpha^2\frac{d^2}{dy^2} + \alpha^4\right]\psi \qquad (5.4.30)$$

This is known as the *Orr–Sommerfeld equation*. This equation also can be derived directly from the two-dimensional Navier–Stokes equations, assuming disturbances of the form $\psi(y)\exp[i\alpha(x - ct)]$, from which the velocity c may be interpreted as the *velocity of wave propagation*.

Solution of the Orr–Sommerfeld equation depends on boundary conditions for a specific problem and is usually accomplished by numerical integration. Once the solution for ψ is obtained, the velocity perturbations are found from

$$u' = \frac{u_1'}{U'} = \frac{d\psi}{dy}e^{i(\alpha x - \Omega \tau)} \qquad v' = \frac{u_2'}{U'} = -i\alpha\psi\, e^{i(\alpha x - \Omega \tau)} \qquad (5.4.31)$$

which can be seen from the definition of ψ in Eqs. (5.4.28) and (5.4.29). These perturbations can grow in either time or space, if the imaginary parts of α or Ω are negative. In dimensional terms, the perturbations are

$$u_1' = U'\frac{d\psi}{dy}e^{i(kx_1 - \sigma t)} \qquad u_2' = -ikLU'\psi\, e^{i(kx_1 - \sigma t)} \qquad (5.4.32)$$

(recall that this solution is for waves traveling in the x-direction, and k is the corresponding wave number).

As previously mentioned, the solution for α and Ω (or, in dimensional terms, k and σ) forms an eigenvalue problem. Consider, for example, the case of *temporal instability*. In this case, α is assumed to be real and α_r (= real part of α) is specified, along with Re and $U(y)$, which are the main parameters of Eq. (5.4.30). Solution of the differential equation then produces one eigenfunction ψ and one complex eigenvalue c for each pair of values (α, Re). The condition of *neutral stability* is then of interest, since curves of neutral stability stable from unstable regions in the parameter space (α, Re). A representative curve of neutral stability is shown in Fig. 5.13. The point on the curve that corresponds to the lowest value for Re gives the critical value, Re_c = critical Reynolds number. For $\text{Re} < \text{Re}_c$, the disturbances are stable for all values of α_r. For larger Re, unstable solutions appear, although there are still wave numbers for which the solutions are stable, even for high Re.

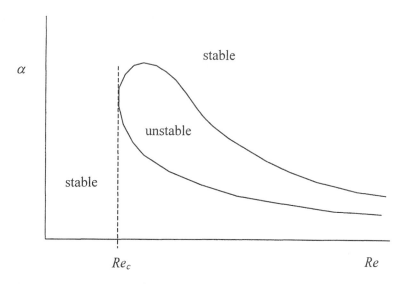

Figure 5.13   Neutral stability curve for wavelike disturbances superimposed on plane laminar flow.

5.5   TURBULENCE MODELING

As previously mentioned, nearly all natural flows are turbulent. In the previous few sections we described some of the basic techniques used in describing turbulent quantities, from a statistical point of view, as well as analysis that leads to a prediction of conditions under which turbulence will occur. However, empirical evidence is still commonly used to describe turbulent flow, either the initial transition to turbulence or properties of the fully turbulent flow field.

Perhaps the most common example of this is the early observation of Reynolds that pipe flow becomes turbulent for a Reynolds number of about 2000, where Reynolds number is defined with the pipe diameter and mean flow velocity. Empirical data also play a role in the analyses described in the present section, where we present several methods used to obtain closure for the set of equations describing a turbulent flow. These approaches provide a basis for modeling such flows.

5.5.1   Reynolds Averaging

The basic governing equations for turbulent flow are the same as those developed in Chap. 2, except here the fluctuating nature of the various properties of the system is included explicitly. One of the most important consequences of these turbulent fluctuations is their effect on transport, and one of the most common methods of incorporating turbulent transport for each of the fluid properties is through definition of a *turbulent diffusivity* or, as in the case of the momentum equation, a *turbulent eddy viscosity*. This is done through the process of averaging the effects (over time) of the fluctuating components of the fluid properties. That is, we consider various properties of the fluid system as consisting of mean and fluctuating parts, as in Eq. (5.2.1),

$$u_i = \bar{u}_i + u_i' \qquad \rho = \bar{\rho} + \rho' \qquad p = \bar{p} + p' \qquad T = \bar{T} + T' \quad (5.5.1)$$

where the u_i are the velocity components, ρ is density, p is pressure and T is temperature. An overbar is used to denote the mean, and a primed quantity is a fluctuating part. These have basically the same meaning as in the previous section (5.4), although here we take it for granted that the fluctuations are present. Following the approach in previous sections, statistical properties of the turbulence are assumed to be stationary.

First consider the continuity equation. Using index notation and assuming incompressible flow, this equation comes from Eq. (2.5.7) and is written as

$$\frac{\partial u_i}{\partial x_i} = \frac{\partial(\bar{u}_i + u_i')}{\partial x_i} = \frac{\partial \bar{u}_i}{\partial x_i} + \frac{\partial u_i'}{\partial x_i} = 0 \quad (5.5.2)$$

Taking a time average of this expression gives

$$\overline{\frac{\partial \bar{u}_i}{\partial x_i}} + \overline{\frac{\partial u_i'}{\partial x_i}} = \frac{\partial \bar{u}_i}{\partial x_i} + \overline{\frac{\partial u_i'}{\partial x_i}} = 0 \quad (5.5.3)$$

and the time average of a fluctuating quantity is 0, so

$$\overline{\frac{\partial u_i'}{\partial x_i}} = 0 \quad (5.5.4)$$

Also, the time average of a mean quantity is just the mean itself. Combining Eqs. (5.5.3) and (5.5.4) shows that the mean flow field must satisfy the continuity relation (in fact, this is the steady, stable flow U considered in Sec. 5.4). Then, from Eq. (5.5.2), we know that

$$\frac{\partial u_i'}{\partial x_i} = 0 \quad (5.5.5)$$

For the general form of the momentum equation, consider Eq. (2.9.17), rewritten here for convenience as

$$\frac{\partial u_i}{\partial t} + u_j \frac{\partial u_i}{\partial x_j} + 2\varepsilon_{ijk}\Omega_j u_k = \frac{\rho}{\rho_0}g_i - \frac{1}{\rho_0}\frac{\partial p}{\partial x_i} + \nu\nabla^2 u_i \quad (5.5.6)$$

where gradients in the reference level (h_r) have been neglected. Substituting fluctuating variables for velocity, density, and pressure (Eq. 5.5.1), we obtain

$$\frac{\partial}{\partial t}(\bar{u}_i + u_i') + (\bar{u}_j + u_j')\frac{\partial}{\partial x_j}(\bar{u}_i + u_i') + 2\varepsilon_{ijk}\Omega_j(\bar{u}_k + u_k')$$

$$= \frac{\bar{p} + \rho'}{\rho_0}g_i - \frac{1}{\rho_0}\frac{\partial}{\partial x_i}(\bar{p} + p') + \nu\frac{\partial}{\partial x_j}\left[\frac{\partial}{\partial x_j}(\bar{u}_i + u_i')\right] \quad (5.5.7)$$

where the Boussinesq approximation has been used to neglect density variations except in the buoyancy term. Multiplying the terms and time-averaging then gives

$$\frac{\partial \bar{u}_i}{\partial t} + \bar{u}_j\frac{\partial \bar{u}_i}{\partial x_j} + \overline{u_j'\frac{\partial u_i'}{\partial x_j}} + 2\varepsilon_{ijk}\Omega_j\bar{u}_k$$

$$= \frac{\bar{\rho}}{\rho_0}g_i - \frac{1}{\rho_0}\frac{\partial \bar{p}}{\partial x_i} + \nu\frac{\partial}{\partial x_j}\left(\frac{\partial \bar{u}_i}{\partial x_j}\right) \quad (5.5.8)$$

Note that although the mean of a fluctuating quantity is zero, the mean of the product of two fluctuating quantities is not usually zero. Also, the mean fluctuating term (third term on the left-hand side) can be rewritten using

$$\overline{u_j'\frac{\partial u_i'}{\partial x_j}} = \frac{\partial}{\partial x_j}\overline{(u_i'u_j')} - \overline{u_i'\frac{\partial u_j'}{\partial x_j}} \quad (5.5.9)$$

where, from Eq. (5.5.5), the last term on the right-hand side of this result is zero. Thus after substituting Eq. (5.5.9) back into Eq. (5.5.8) and rearranging, we obtain

$$\frac{\partial \bar{u}_i}{\partial t} + \bar{u}_j\frac{\partial \bar{u}_i}{\partial x_j} + 2\varepsilon_{ijk}\Omega_j\bar{u}_k$$

$$= \frac{\bar{\rho}}{\rho_0}g_i - \frac{1}{\rho_0}\frac{\partial \bar{p}}{\partial x_i} + \nu\frac{\partial^2 \bar{u}_i}{\partial x_j\partial x_j} - \frac{\partial}{\partial x_j}\overline{(u_i{}^\prime u_j')} \quad (5.5.10)$$

This is the *Reynolds averaged equation* for mean momentum transport.

The last term on the right-hand side of Eq. (5.5.10), when multiplied by ρ_0, represents the *Reynolds stresses*. This term produces an effect similar to that of viscous stresses, though it should be kept in mind that the physical basis for viscous stress is fluid viscosity, while turbulent shear stress (Reynolds stresses) results from the fluctuating nature of the velocity field. In other words, the turbulent eddies transport various fluid properties by their random three-dimensional motions, superimposed on top of the mean flow (advective) transport. This process is illustrated in Fig. 5.14.

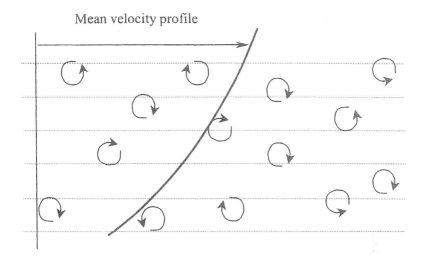

Mean velocity profile

Figure 5.14 Illustration of turbulent transport by small-scale eddy motions superimposed on top of the mean flow field.

The problem now is to find ways to evaluate the Reynolds stresses, since the equation set at this point only relates these stresses to the mean flow. One of the simplest approaches is to express the magnitude of the Reynolds stresses in terms of a gradient formulation, similar to molecular diffusion, that is,

$$-\overline{u'_i u'_j} = \nu_{t(j)} \frac{\partial \overline{u}_i}{\partial x_j} \qquad (5.5.11)$$

where $\nu_{t(j)}$ is defined as a *turbulent kinematic eddy viscosity* in the j-direction (note that summation comvention is not used here). This formulation allows the molecular and Reynolds stresses in Eq. (5.5.10) to be combined, since they both depend on the gradient of mean velocity. By substituting Eq. (5.5.11) into Eq. (5.5.10) we obtain

$$\frac{\partial \overline{u}_i}{\partial t} + \overline{u}_j \frac{\partial \overline{u}_i}{\partial x_j} + 2\varepsilon_{ijk}\Omega_j \overline{u}_k$$

$$= \frac{\overline{\rho}}{\rho_0} g_i - \frac{1}{\rho_0} \frac{\partial \overline{p}}{\partial x_i} + \frac{\partial}{\partial x_j}\left[(\nu + \nu_{t(j)})\frac{\partial \overline{u}_i}{\partial x_j}\right] \qquad (5.5.12)$$

Then, since turbulent transport is generally much stronger than molecular transport, $\nu \ll \nu_{t(j)}$, and the molecular term is usually neglected in writing the mean momentum equation,

$$\frac{\partial \overline{u}_i}{\partial t} + \overline{u}_j \frac{\partial \overline{u}_i}{\partial x_j} + 2\varepsilon_{ijk}\Omega_j \overline{u}_k = \frac{\overline{\rho}}{\rho_0} g_i - \frac{1}{\rho_0} \frac{\partial \overline{p}}{\partial x_i} + \frac{\partial}{\partial x_j}\left[\nu_{t(j)}\frac{\partial \overline{u}_i}{\partial x_j}\right] \qquad (5.5.13)$$

Estimates for $\nu_{t(j)}$ may be obtained by direct measurements of the Reynolds stresses and mean velocity gradients, using Eq. (5.5.11). However, this is normally a difficult and time-consuming procedure, requiring a large number of measurements in order to obtain reasonable averages. Other estimates of turbulent viscosities may be made by conducting experiments where the diffusion of a conservative tracer is observed over time and values for $\nu_{t(j)}$ are chosen so that the equation fits the observations (assuming that at least the mean flow is well known). This procedure requires using a solution to the turbulent advection–diffusion equation for dissolved mass, which is described in Chap. 10, along with the application of the *Reynolds analogy*, which is an assumption that the turbulent diffusivities for momentum, mass, and heat are the same, since they all depend on the same eddies for transport. Turbulent viscosities also may be estimated as the product of a typical turbulent velocity scale, such as the root-mean-square velocity fluctuation, $\overline{u'^2}^{1/2}$, and an appropriate length scale, usually the integral length scale (analogous to the integral time scale defined in Eq. 5.2.9). This approach is similar to the mixing length approach used in defining molecular diffusivities, described in Sec. 10.3.1.

In general, eddy viscosity is dependent on direction, as indicated by the directional subscript j in Eq. (5.5.11). This is because the eddy viscosity is directly related to the turbulence structure, as expressed through the fluctuating velocity components, which in general may have different characteristics (length, frequency, magnitude) in different coordinate directions. For the case of *isotropic turbulence*, the eddy viscosity values are the same for the three coordinate directions, and in the case of *homogeneous turbulence*, ν_t is independent of location but may still have directional differences. For this case, ν_t can be brought outside of the gradient operator in Eq. (5.5.13).

A more formal estimate for the eddy viscosity can be obtained using a two-equation model, one for transport of turbulent kinetic energy (K) and one for the dissipation rate (ε). Both k and ε are per unit mass and are defined formally in the following section. Based on physical considerations, Prandtl and others have argued that the eddy viscosity ν_t should depend on K. From dimensional considerations, a length scale for the turbulence (l) also must be introduced, so that

$$\nu_t = c_1 l K^{1/2} \tag{5.5.14}$$

where c_1 is a constant. In other words, $K^{1/2}$ provides the velocity scale referred to above. The characteristic length l also is related to dissipation, since it may be argued that ε does not explicitly depend on molecular viscosity, because the energy of the turbulence is mostly associated with larger eddies. In other words, ε can be expressed as (see also Sec. 5.6)

$$\varepsilon = \frac{c_2 K^{3/2}}{l} \tag{5.5.15}$$

where c_2 is another constant. Using this result to substitute for l in Eq. (5.5.14) then gives

$$v_t = \frac{c_3 K^2}{\varepsilon} \tag{5.5.16}$$

where $c_3 = c_1 c_2$ has been found to have a value of approximately 0.09. This result applies mainly for isotropic turbulence, since there is no directional association with K, and l has been assumed as an isotropic quantity. Formulation of the equations for K and ε are described below.

Before continuing with the discussion for turbulence, it is of interest to examine the effect of turbulent transport for other properties of the system. Here, we show the temperature equation, though a similar analysis holds for transport of other properties, such as dissolved mass. Following the same procedures as above, fluctuating variables are introduced into Eq. (2.9.33). The result, written in index notation, is

$$\frac{DT}{Dt} = -\frac{1}{\rho_0 c}\frac{\partial \varphi_{ri}}{\partial x_i} + \frac{\partial}{\partial x_j}\left[k_T \frac{\partial \overline{T}}{\partial x_j} - \overline{T' u_{j'}} \right] \tag{5.5.17}$$

where T is temperature and φ_r is radiation flux. A turbulent thermal diffusivity can be defined for the turbulent transport term on the right-hand side, similar to the eddy viscosity defined in Eq. (5.5.11), i.e.,

$$k_{T(j)} = -\frac{\overline{u_{j'} T'}}{\partial \overline{T}/\partial x_j} \tag{5.5.18}$$

Turbulent diffusivities may have different magnitudes in different directions, since they also depend directly on the turbulence structure. Using the *Reynolds analogy*, it often is assumed that $k_{T(j)} \cong v_{t(j)}$, i.e., the *turbulent Prandlt number* (ratio of momentum diffusivity to thermal diffusivity) is approximately equal to 1.

5.5.2 Turbulent Kinetic Energy Equation

The Reynolds stresses are directly related to the kinetic energy of the fluctuating components of velocity. When there is a higher kinetic energy level for the fluctuating velocity components, they are more active in transporting fluid properties in the flow field. We first develop an equation for the mean kinetic energy, by multiplying (dot product) the mean momentum equation (5.5.10) by $\bar{u}_i$. This gives

$$\frac{\partial}{\partial t}\left(\frac{1}{2}\bar{u}_i^2\right) + \bar{u}_j\frac{\partial}{\partial x_j}\left(\frac{1}{2}\bar{u}_i^2\right) = \frac{\bar{\rho}}{\rho_0}g_i\bar{u}_i - \frac{1}{\rho_0}\frac{\partial}{\partial x_i}(\bar{u}_i\bar{p})$$

$$+ v\left[\frac{\partial}{\partial x_j}\left(\bar{u}_i\frac{\partial\bar{u}_i}{\partial x_j}\right) - \left(\frac{\partial\bar{u}_i}{\partial x_j}\right)^2\right] - \frac{\partial}{\partial x_j}(\bar{u}_i\overline{u_i'u_j'}) + \overline{u_i'u_j'}\frac{\partial\bar{u}_i}{\partial x_j} \qquad (5.5.19)$$

After rearranging, this result may be written as

$$\frac{D}{Dt}\left(\frac{\bar{u}_i^2}{2}\right) = \frac{\bar{\rho}}{\rho_0}g_i\bar{u}_i - \frac{\partial}{\partial x_j}\left(\bar{u}_i\frac{\bar{p}}{\rho_0} - v\frac{\partial}{\partial x_j}\left(\frac{\bar{u}_i^2}{2}\right) + \overline{u_i u_i' u_j'}\right)$$

$$+ \overline{u_i'u_j'}\frac{\partial\bar{u}_i}{\partial x_j} - v\left(\frac{\partial\bar{u}_i}{\partial x_j}\right)^2 \qquad (5.5.20)$$

The left-hand side of this equation is the time rate of change of mean kinetic energy (per unit mass, following a fluid element). The first term on the right-hand side is gravity, or buoyancy work, the second term is referred to as the *flux divergence* term and refers to redistribution and transport of mean kinetic energy by pressure and shear stresses (both viscous and turbulent), the third term is the rate of work of the Reynolds stresses to convert mean kinetic energy to turbulent kinetic energy (TKE), called *shear production*, and the last term is *dissipation* of mean kinetic energy directly into heat. It is interesting to note that the Coriolis term drops out of the mean kinetic energy equation. This is because the Coriolis term does no work, since the force is acting at a right angle to the velocity vector. However, the Coriolis term does affect the distribution of energy among the different velocity components.

The TKE conservation equation is derived by multiplying the original (nonaveraged) momentum equation (5.5.7) by u_i, time averaging, and subtracting the mean kinetic energy equation (5.5.20). After some rearranging, the result is

$$\frac{DK}{Dt} = \frac{\overline{\rho'u_i'}}{\rho_0}g_i - \frac{\partial}{\partial x_j}\left(\frac{\overline{u_i'p'}}{\rho_0} - v\frac{\partial K}{\partial x_j} + \frac{1}{2}\overline{u_i'u_i'u_j'}\right)$$

$$- \overline{u_i'u_j'}\frac{\partial\bar{u}_i}{\partial x_j} - v\overline{\left(\frac{\partial u_i'}{\partial x_j}\right)^2} \qquad (5.5.21)$$

where $K = 1/2\overline{u_i'^2}$ is the turbulent kinetic energy per unit mass. The terms in Eq. (5.5.21) have analogous interpretations as in Eq. (5.5.20), though the shear production term has opposite sign. Normally, this is a sink of mean energy and a source of TKE. Only under certain conditions does this term change sign (there is some evidence, for instance, that the flow of energy is in the opposite direction in rotating flows). Once again, note that the Coriolis term drops out of the total TKE equation, for the same reason as it did for the mean kinetic energy equation. Additional averaged fluctuating terms do not appear because fluctuations in Ω are not considered. However, as with the mean flow terms, the Coriolis acceleration does affect the redistribution of TKE among the various velocity components.

For high Re flows the viscous terms are all usually neglected, except the dissipation term for TKE, which is the ultimate sink for mechanical energy. The mean kinetic energy equation is then

$$\frac{D}{Dt}\left(\frac{\overline{u_i^2}}{2}\right) = \frac{\overline{\rho}}{\rho_0}g_i\overline{u_i} - \frac{\partial}{\partial x_j}\left(\overline{u_i}\frac{\overline{p}}{\rho_0} + \overline{u_i}\overline{u_i'u_j'}\right) + \overline{u_i'u_j'}\frac{\partial \overline{u_i}}{\partial x_j} \tag{5.5.22}$$

and the TKE equation is

$$\frac{DK}{Dt} = \frac{\overline{\rho'u_i'}}{\rho_0}g_i - \frac{\partial}{\partial x_j}\left(\frac{\overline{u_i'p'}}{\rho_0} + \frac{1}{2}\overline{u_i'u_i'u_j'}\right) - \overline{u_i'u_j'}\frac{\partial \overline{u_i}}{\partial x_j} - \varepsilon \tag{5.5.23}$$

where ε is the dissipation of TKE per unit mass and has been substituted for the last term on the right-hand side of Eq. (5.5.21), i.e.,

$$\varepsilon = \nu\overline{\left(\frac{\partial u_i'}{\partial x_j}\right)^2} \tag{5.5.24}$$

5.5.3 Reynolds Stress Equations

Equations for the Reynolds stresses are developed using procedures similar to the above for TKE. First, subtract the average momentum equation (5.5.10) from the full momentum equation (5.5.7), neglecting the Coriolis and gravity terms, to obtain

$$\frac{\partial u_i'}{\partial t} + \overline{u_k}\frac{\partial u_i'}{\partial x_k} + u_k'\frac{\partial \overline{u_i}}{\partial x_k} + u_k'\frac{\partial u_i'}{\partial x_j}$$

$$= -\frac{1}{\rho_0}\frac{\partial u_i'}{\partial x_k} + \nu\frac{\partial^2 u_i'}{\partial x_k \partial x_k} + \frac{\partial}{\partial x_k}\overline{u_i'u_k'} \tag{5.5.25}$$

where subscript k has been used in place of j for convenience in the following. Multiplying Eq. (5.5.25) by u_j' gives

$$\frac{\partial}{\partial t}(u_i'u_j') - u_i'\frac{\partial u_j'}{\partial t} + \overline{u_k}\left[\frac{\partial}{\partial x_k}(u_i'u_j') - u_i'\frac{\partial u_j'}{\partial x_k}\right] + u_j'u_k'\frac{\partial \overline{u_i}}{\partial x_k} + u_j'u_k'\frac{\partial u_i'}{\partial x_k}$$

$$= -\frac{1}{\rho_0}u_j'\frac{\partial p'}{\partial x_i} + u_j'\nu\frac{\partial^2 u_i'}{\partial x_k \partial x_k} + u_j'\frac{\partial}{\partial x_k}(\overline{u_i'u_k'}) \tag{5.5.26}$$

Now let us write an equation similar to Eq. (5.5.25) but using subscript j instead of i and multiplying by u'_i. Adding this equation to Eq. (5.5.26) and rearranging gives

$$\frac{\partial}{\partial t}(u'_i u'_j) + \bar{u}_k \frac{\partial}{\partial x_k}(u'_i u'_j) + \frac{\partial}{\partial x_k}(u'_i u'_j u'_k) + u'_i u'_k \frac{\partial \bar{u}_j}{\partial x_k} + u'_j u'_k \frac{\partial \bar{u}_i}{\partial x_k}$$

$$= -\frac{1}{\rho_0}\left[-p'\left(\frac{\partial u'_i}{\partial x_j} + \frac{\partial u'_j}{\partial x_i}\right) + \frac{\partial}{\partial x_i}(u'_j p') + \frac{\partial}{\partial x_j}(u'_i p')\right]$$

$$+ v\frac{\partial^2(u'_i u'_j)}{\partial x_k \partial x_k} - 2v\frac{\partial u'_i}{\partial x_k}\frac{\partial u'_j}{\partial x_k} + u'_j \frac{\partial}{\partial x_k}\overline{(u'_i u'_k)} + u'_i \frac{\partial}{\partial x_k}\overline{(u'_j u'_k)} \quad (5.5.27)$$

Taking the time average and letting $R_{ij} = \overline{u'_i u'_j}$, this equation becomes

$$\frac{\partial R_{ij}}{\partial t} + \bar{u}_k \frac{\partial R_{ij}}{\partial x_k} = \left\{\frac{\overline{p'}}{\rho_0}\left(\frac{\partial u'_i}{\partial x_j} + \frac{\partial u'_j}{\partial x_i}\right)\right\} + \left\{-R_{ik}\frac{\partial \bar{u}_j}{\partial x_k} - R_{jk}\frac{\partial \bar{u}_i}{\partial x_k}\right\}$$

$$+ \left\{-\frac{1}{\rho_0}\left(\frac{\partial \overline{u'_j p'}}{\partial x_i} + \frac{\partial \overline{u'_i p'}}{\partial x_j}\right) - \frac{\partial}{\partial x_k}\overline{(u'_i u'_j u'_k)}\right\}$$

$$+ \left\{v\frac{\partial^2 R_{ij}}{\partial x_k \partial x_k}\right\} + \left\{-2v\overline{\frac{\partial u'_i}{\partial x_k}\frac{\partial u'_j}{\partial x_k}}\right\} \quad (5.5.28)$$

This is the transport equation for Reynolds stress. The left-hand side is the total rate of change of R_{ij}. The terms on the right-hand side are grouped according to their physical interpretation. The first term is the *pressure–strain correlation*, which plays an important role in the distribution of R_{ij}. The second term is the *production* of R_{ij}, similar to the production term in the TKE equation. The third term is turbulent diffusion and redistribution of R_{ij} by pressure. The fourth term is molecular diffusion, and the fifth term is *dissipation*, again similar to the TKE equation. Solution of Eq. (5.5.28) provides values for R_{ij} that can be input directly into Eq. (5.5.10), thus closing the system of equations.

An alternative to solving the differential equation for R_{ij} is the *algebraic stress model*. This is obtained directly from Eq. (5.5.28). To simplify the notation, let

$$P_{ij} = \frac{\overline{p'}}{\rho_0}\left(\frac{\partial u'_i}{\partial x_j} + \frac{\partial u'_j}{\partial x_i}\right) \quad (5.5.29a)$$

$$Q_{ij} = -\left(R_{ik}\frac{\partial \bar{u}_i}{\partial x_k} + R_{jk}\frac{\partial \bar{u}_j}{\partial x_k}\right) \quad (5.5.29b)$$

$$F_{ij} = -\frac{1}{\rho_0}\left(\frac{\partial \overline{u'_j p'}\partial x_i}{+} \frac{\partial \overline{u'_i p'}}{\partial x_j}\right) - \frac{\partial}{\partial x_k}\overline{(u'_i u'_j u'_k)} \quad (5.5.29c)$$

$$d_{ij} = F_{ij} + \nu \frac{\partial^2 R_{ij}}{\partial x_k \partial x_k} \tag{5.5.29d}$$

$$\varepsilon_{ij} = 2\nu \overline{\frac{\partial u'_i}{\partial x_k} \frac{\partial u'_j}{\partial x_k}} \tag{5.5.29e}$$

With these definitions, Eq. (5.5.28) can be rewritten as

$$\frac{DR_{ij}}{Dt} = P_{ij} + Q_{ij} - \varepsilon_{ij} + d_{ij} \tag{5.5.30}$$

Using these definitions also allows the TKE equation (5.5.21) to be rewritten as

$$\frac{DK}{Dt} = p - \varepsilon + d \tag{5.5.31}$$

where $P = P_{ii}/2$, $\varepsilon = \varepsilon_{ii}/2$, and $d = d_{ii}/2$. Note that, upon contraction of the indices, $Q_{ii} = 0$. For simplicity, the buoyancy work term also has been omitted in Eq. (5.5.31).

We now define $T_{ij} = R_{ij}/K$ and substitute into Eq. (5.5.30) to obtain

$$T_{ij}(P - \varepsilon + d) + K\frac{DT_{ij}}{Dt} = P_{ij} + Q_{ij} - \varepsilon_{ij} + d_{ij} \tag{5.5.32}$$

The derivatives of T_{ij} are normally small, relative to the other terms in the equation, and may be neglected. Furthermore, it is assumed that $d_{ij} = T_{ij}d$; the viscous transport term also is neglected, so that Eq. (5.5.32) becomes

$$T_{ij}(P - \varepsilon) = P_{ij} + Q_{ij} - \varepsilon_{ij} \tag{5.5.33}$$

This provides a direct nondifferential equation to evaluate the Reynolds stresses. Of course, the other terms in the equation, notably Q_{ij} and ε_{ij}, must first be obtained. However, algebraic expressions also have been developed for these terms. The derivations are not presented here, but the final result, as shown by Warsi (1993), is

$$T_{ij} = \frac{2}{3}\delta_{ij} + \frac{\gamma_0}{P + a_0 \varepsilon}\left(P_{ij} - \frac{2}{3}P\delta_{ij}\right) \tag{5.5.34}$$

where $\gamma_0 = 1 - 10c_2$, $a_0 - 3c_1 - 1$, $c_1 = 0.5$, and $c_2 = 0.06$.

5.5.4 Dissipation Equation

A formal equation for the rate of change of ε is obtained by differentiating Eq. (5.5.25) with respect to x_j, multiplying by $\partial u'_i/\partial x_j$ and taking the time average. The result is

$$
\begin{aligned}
\frac{D\varepsilon}{Dt} = -2\nu &\left\{ \frac{1}{\rho_0} \frac{\partial}{\partial x_k} \left(\overline{\frac{\partial u_k'}{\partial x_l} \frac{\partial p'}{\partial x_l}} \right) + \frac{1}{2} \frac{\partial}{\partial x_k} \left[\overline{u_k' \left(\frac{\partial u_j'}{\partial x_k} \right)^2} \right] + \overline{\frac{\partial u_j'}{\partial x_k} \frac{\partial u_j'}{\partial x_l} \frac{\partial u_k'}{\partial x_l}} \right\} \\
&- 2\nu \left\{ \frac{\partial^2 \overline{u}_j}{\partial x_l \partial x_k} \left(\overline{u_k' \frac{\partial u_j'}{\partial x_l}} \right) + \frac{\partial \overline{u}_j}{\partial x_k} \left[\overline{\frac{\partial u_i'}{\partial x_l} \frac{\partial u_k'}{\partial x_l}} + \overline{\frac{\partial u_i'}{\partial x_j} \frac{\partial u_i'}{\partial x_k}} \right] \right. \\
&+ \nu \left. \overline{\left(\frac{\partial^2 u_j'}{\partial x_k \partial x_l} \right)^2} \right\} + \nu \frac{\partial^2 \varepsilon}{\partial x_k \partial x_k}
\end{aligned} \tag{5.5.35}
$$

This can be solved in connection with either the Reynolds stress model or the $K-\varepsilon$ approach.

5.5.5 $K-\varepsilon$ Model

The basis of the two-equation $K-\varepsilon$ model is solving Eqs. (5.5.23) and (5.5.35) for K and ε, respectively. Once these are known, Eq. (5.5.16) can be used to obtain ν_t, which is then used in Eqs. (5.5.11) and (5.5.10) to solve for the momentum transport.

5.6 SCALES OF TURBULENT MOTION

One of the main characteristics of turbulence is the presence of a full spectrum of scales in both length and time in a fully developed flow. As noted in the previous section (see Eqs. 5.5.21 and 5.5.23), the flow turbulence gains kinetic energy by the Reynolds stresses acting on the mean velocity gradient, or by gravity work, in the case of convection-driven flow in which there is an unstable density gradient. This energy feeds into relatively large eddies, with a size or length scale that depends on the size of the system in which the flow occurs. For example, cooling at the top of a column of water that is initially well mixed will cause water near the surface to become heavier than the underlying water. This heavier water then drops through the depth and is replaced by cooler water swept upwards, thus generating a circulation throughout the depth of the column. This is what happens, for example, during the fall and spring "overturns" in temperate lakes. Large-scale eddies also are generated from instabilities in mean flow (mean shear), with a typical size that depends on the scale of the mean flow.

These large eddies interact with each other and with the boundaries of the flow, breaking down to produce smaller eddies. These smaller eddies interact with each other, with the larger eddies, and with the system boundaries to

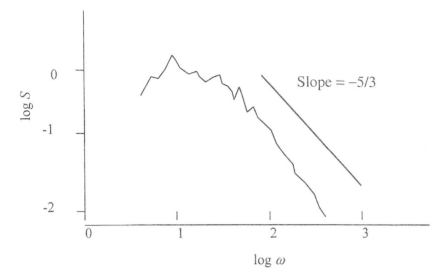

Figure 5.15 Example of power spectrum plot for fully developed turbulent flow.

produce even smaller eddies and so on, until the smallest eddies in the system cannot support their own motions and are dissipated by viscosity. This process is usually described as a *turbulence cascade*, where energy continually flows from larger to smaller eddies and, at the smallest eddy scale, there is an ultimate sink of energy by viscous dissipation. It can be represented by a distribution of energy scales on a plot of the power spectrum, as indicated in Fig. 5.15. The slope of -5/3 for this plot is derived from consideration of the energy transfer process in the range of the integral length scales.

The range of possible scales of motion in a flow is illustrated in Fig. 5.16. The mean flow velocity and length scales are denoted by U and L, respectively. These imply a characteristic time scale, $T \approx L/U$. The largest eddies in the flow have about the same characteristic time scale, so $\delta/\hat{u} \approx T$, where δ and $\hat{u}$ are the length and velocity scales, respectively, for the largest eddies. Typical estimates for the magnitudes of $\hat{u}$ and δ are $\hat{u}/U \cong \delta/L \cong 0.1$–$0.5$. The scales of the main energy-containing eddies may be estimated from the integral calculations described in Sec. 5.2 (i.e., leading to the integral length scale l) and a characteristic velocity, often taken as the root-mean-square value of the fluctuating velocity,

$$u_{\text{rms}} \approx (\overline{u'^2})^{1/2} \qquad (5.6.1)$$

where the average is taken over either time or space.

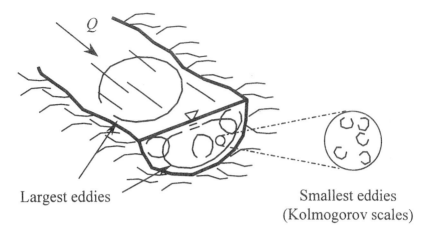

Figure 5.16   Illustration of range of turbulent eddy scales in an open channel flow.

From observations, the magnitude of transport by the Reynolds stresses is of the same order as the mean flow terms in the mean momentum equation, i.e.,

$$\frac{\partial}{\partial x_j}(\overline{u_i' u_j'}) \approx \frac{u_{rms}^2}{l} \approx \frac{U^2}{L} \tag{5.6.2}$$

Also, considering that an eddy with energy proportional to u_{rms}^2 will transfer its energy to smaller eddies at a rate proportional to $1/T$, an estimate for the dissipation rate is obtained,

$$\varepsilon \approx \frac{u_{rms}^2}{T} = \frac{u_{rms}^2}{1/u_{rms}} = \frac{u_{rms}^3}{l} \tag{5.6.3}$$

This result is similar to the dimensional argument leading to Eq. (5.5.15).

The process of energy transfer from larger to successively smaller eddies continues until the eddies become so small that viscous effects become important and the energy is dissipated. This stage is characterized by an eddy Reynolds number approximately equal to one, where the eddy Reynolds number is defined using the characteristic length and velocity of the smallest eddies. This reflects the idea that at these smallest scales of motion, the inertial strength of the eddy is approximately equal to its viscous transport strength, or the eddy "viscosity," is approximately equal to the kinematic viscosity. Thus

$$\frac{v\eta}{v} \approx 1 \tag{5.6.4}$$

where v and η are the velocity and length scales, respectively, for the smallest eddies. We also know (see Eq. 5.5.24)

$$\varepsilon \approx v \left(\frac{v}{\eta} \right)^2 \tag{5.6.5}$$

By combining Eqs. (5.6.4) and (5.6.5), we obtain estimates for the smallest eddies, in terms of the dissipation rate. These are called the *Kolmogorov microscales*,

$$\eta = \frac{v^{3/4}}{\varepsilon^{1/4}} \qquad v = (v\varepsilon)^{1/4} \tag{5.6.6}$$

A micro-time scale, t', also can be defined as the ratio of η to v,

$$t' = \frac{\eta}{v} \tag{5.6.7}$$

As will be seen in later chapters, the scales of turbulent motion have strong influences on transport and mixing properties for water quality modeling. They also control a number of processes of direct interest in environmental flow modeling, such as particle–particle interactions and contaminant desorption phenomena. Some of these applications are described further in Part 2 of this text.

PROBLEMS

Solved Problems

Problem 5.1   Consider a turbulent flow of water with a measured power spectral density curve as shown and listed in Fig. 5.17.

(a) A common estimate for the turbulent velocity scale (denoted by u for this problem) is the root-mean-square value of the fluctuations, $u = u'_{rms} = (u'^2)^{1/2}$, where the u' are the fluctuating velocities. Calculate u for this flow, using the fact that the average value for u'^2 is the autocorrelation for a time lag of 0.

(b) One specification of the capabilities of a flow-measuring instrument is its time factor, or frequency response. This value tells how fast the instrument is able to respond to fluctuations in the signal being measured. Suppose an anemometer were used to measure turbulence in the flow considered here, but that it could resolve signal frequencies only up to 20 Hz (1 Hz = 1 cps). What value of u would be calculated using data from this instrument?

Solution

(a) The autocorrelation for zero time lag is simply the area under the power spectrum curve. For simplicity, the area is estimated here from the tabulated

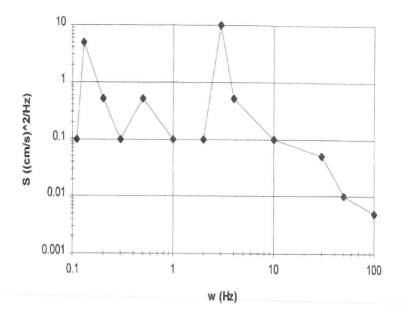

w (Hz)	S		Δw (Hz)	S_{avg}	area
0.11	0.1		-	-	-
0.13	5		0.03	2.55	0.0765
0.2	0.5		0.07	2.75	0.1925
0.3	0.1		0.1	0.3	0.03
0.5	0.5		0.2	0.3	0.06
1	0.1		0.5	0.3	0.15
2	0.1		1.0	0.1	0.1
3	10		1.0	5.05	5.05
4	0.5		1.0	5.25	5.25
10	0.1		6.0	0.3	1.8
30	0.05		20.0	0.075	1.5
50	0.01		20.0	0.03	0.6
100	0.005		50.0	0.0075	0.375

Figure 5.17   Data for Problem 5.1.

data using the average value for S within each frequency range (set of data points). The results of these calculations are shown in the right-hand side of the table in Fig. 5.17. The sum of the last column on the right-hand side of the table is 15.184 (cm^2/s^2) and the square root of this is $u = 3.9$ cm/s. (Note, however, that better averaging schemes could be used to obtain improved estimates.)

(b) Assuming that the rest of the spectrum remains the same, we estimate the value for S at 20 Hz as 0.075 (cm/s)^2 (again, better averaging and extrapolation procedures could be used here, but we use a simple approach to illustrate the procedures). Then, using the same procedure as in part (a), but including the sum only up to $w = 20$ Hz, we obtain $u = 3.8$ cm/s. The difference between this result and that of part (a) is not very large, since most of the signal is contained in frequencies less than 20 Hz.

Problem 5.2 Using the data from solved problem 5.1, and assuming that the characteristic length scale for the eddies with velocity u is $l = 5$ cm, calculate the microturbulence length and velocity scales (Kolmogorov scales) for the flow.

Solution

First we need an estimate for the dissipation rate. This is

$$\varepsilon = \frac{u^3}{l} = \frac{3.9^3}{5} = 11.86 \text{ cm}^2/\text{s}^3$$

Then, using 10^{-2}cm^2/s as the kinematic viscosity for water,

$$\eta = \frac{v^{3/4}}{\varepsilon^{1/4}} = 0.047 \text{ cm} \quad \text{and} \quad v = (v\varepsilon)^{1/4} = 0.59 \text{ cm/s}$$

Unsolved Problems

Problem 5.3 Show that the derivative of a time-averaged quantity is the time average of the derivative of that quantity, i.e.,

$$\frac{\overline{\partial f}}{\partial x} = \overline{\frac{\partial f}{\partial x}}$$

Problem 5.4 Consider a turbulent flow, with velocity field and concentration given by

$$u_i = U_i + u'_i \quad \text{and} \quad c = C + c',$$

respectively, where capital letters indicate mean (time-averaged) quantities and primes indicate turbulent fluctuations. Apply the Reynolds averaging procedure to derive the advection–diffusion equation for mean transport of the concentration c, including explicit terms for turbulent diffusive transport. Describe how you might conduct an experiment to measure the turbulent diffusivities directly.

Problem 5.5 Isotropic turbulence has been produced in laboratory experiments by placing a grid or screen in a wind or water tunnel. It can be shown that the initial decay of the turbulence can be described by

$$\frac{d}{dt}(\overline{u'^2})^{1/2} = -C_1 \frac{\overline{u'^2}}{L} \qquad \frac{dL}{dt} = C_2 (\overline{u'^2})^{1/2}$$

where u' is the turbulent fluctuating velocity, L is a longitudinal length scale, and C_1 and C_2 are constants.

(a) Use the substitution $t = x/U$, where U is the mean velocity in the tunnel, to rewrite these equations in terms of derivatives in x.

(b) At the location of the screen, assume $x = x_0$, $(\overline{u'^2})^{1/2} = [(\overline{u'^2})^{1/2}]_0$, and $L = C_3 d$, where d is the diameter of the screen wires and C_3 is another constant. Solve the equations developed in part (a), along with these boundary conditions, to show that

$$\frac{\left[(\overline{u'^2})^{1/2}\right]_0}{(\overline{u'^2})^{1/2}} = \left[1 + (C_1 + C_2)\frac{[(\overline{u'^2})^{1/2}]_0}{C_3 U}\frac{(x - x_0)}{d}\right]^{C_1/(C_1+C_2)}$$

$$\frac{L}{C_3 d} = \left[1 + (C_1 + C_2)\frac{[(\overline{u'^2})^{1/2}]_0}{C_3 U}\frac{(x - x_0)}{d}\right]^{C^2/(C_1+C_2)}$$

Problem 5.6 Develop the Reynolds-averaged equations in a cylindrical coordinate system.

Problem 5.7 At a solid wall, the velocity fluctuations vanish just as the mean velocity components do. In addition, the gradients of the fluctuations tangent to the wall vanish. Using the coordinate system of Fig. 5.18, then,

$$u' = v' = w' = \frac{\partial \lambda}{\partial x} = \frac{\partial \lambda}{\partial z} = 0 \qquad \text{at} \qquad y = 0$$

Use the continuity equation for the fluctuations to write an expression for the second derivative of v' with respect to y. Then show that the first and second derivatives of the Reynolds stress $(\overline{u'v'})$ with respect to y are both zero.

Problem 5.8 Write the TKE equation for the condition of zero horizontal gradients. All viscous terms except dissipation may be neglected. Further

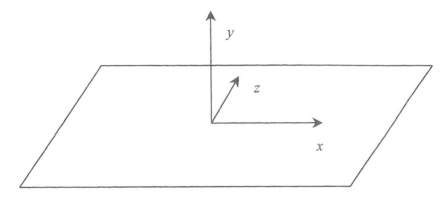

Figure 5.18 Definition sketch, Problem 5.7.

simplify the equation for a steady-state constant-density flow with negligible flux divergence. What is the physical interpretation of the resulting equation?

Problem 5.9 Estimate the characteristic scales (length, velocity, and time) for the largest and smallest turbulent eddies created by wind swirling around the corners of buildings, past doorways, etc. State any assumptions you make. Assume the kinematic viscosity of air is 1.5×10^{-5} m^2/s.

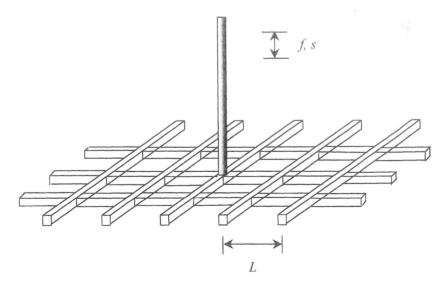

Figure 5.19 Oscillating grid for generating turbulence, Problem 5.10.

Problem 5.10 A well-known laboratory experimental setup for studying turbulent mixing involves using an oscillating grid to generate the turbulence, as sketched in Fig. 5.19. The grid oscillates with frequency f and stroke s. The mesh spacing between the grid elements is L and the grid elements have diameter d. There is no mean flow.

(a) Assuming that a steady state can be reached, what is the basic balance in the TKE equation? In other words, what are the important physical processes to consider?

(b) Estimate the length and velocity scales of the eddies generated at the grid.

(c) Estimate the rate of transport of TKE away from the grid.

SUPPLEMENTAL READING

Batchelor, G. K., 1953. *The Theory of Homogeneous Turbulence*. Cambridge University Press, London.

Hinze, H. O., 1959. *Turbulence*. McGraw-Hill, New York.

Joseph, D. D., 1976. *Stability of Fluid Motion*. Springer-Verlag, New York.

Lin, C. C., 1955. *The Theory of Hydrodynamic Stability*. Cambridge University Press, London.

Monin, A. S. and Yaglom, A. N. M., 1971. *Statistical Fluid Mechanics: Mechanics of Turbulence*. M.I.T. Press, Cambridge, MA.

Tennekes, H. and Lumley, J. L., 1972. *A First Course in Turbulence*. M.I.T. Press, Cambridge, MA.

Townsend, A. A., 1956. *The Structure of Turbulent Shear Flow*. Cambridge University Press, London.

Von Karman, T., 1938. On the theory of turbulence. *Proc. Nat. Acad. Sci.* **23**:98.

Warsi, Z. U. A., 1993. *Fluid Dynamics, Theoretical and Computational Approaches*. CRC Press, Boca Raton, FL.

6
Boundary Layers

6.1 INTRODUCTION

The concept of the boundary layer was introduced in the beginning of the 20th century by Prandtl and Pohlhausen to simplify the analysis and calculation of flow phenomena and interaction with solid boundaries. This concept has since been extended for calculations of heat and mass transfer in various types of domains. Several examples in the previous chapters have already been introduced that involve the concept of boundary layers, where it was shown that often, and especially in environmental flows, only a minor portion of the flow domain is subject to viscous laminar or turbulent flow with significant velocity gradients and shear stresses. Other portions of the domain are practically at rest or subject to potential flow. For example, in Chaps. 3 and 4 we considered the flow domain on top of an oscillating plate and the case of flow around a cylinder, as examples in which the concept of the boundary layer was applicable. For the present chapter, we cover only several cases in which the boundary layer concept is applicable for flow calculations in environmental fluid mechanics, though some consideration of other issues of transport phenomena also are included.

If fluid with small viscosity, like water or air, flows with high Reynolds number around a solid body, then even at a very small distance from the solid body, the effect of the inertial forces is much more significant than the effect of the viscous forces. Therefore, it is possible to consider that in major portions of the domain the flow is virtually unaffected by friction. Frictionless flow is described by potential flow theory (Chap. 4). However, close to the solid boundary, the effect of the viscous shear stresses is significant and cannot be ignored.

In order to exemplify the application of the boundary layer concept to flow phenomena, we first consider the buildup of a boundary layer and the effect of shear stresses associated with the development of the flow field over an infinite horizontal plate. Figure 6.1 shows the various regions of development of the boundary layer over a horizontal flat plate.

At the leading edge of the flat plate (at $x = 0$), the fluid is subject to uniform flow U. Development of the boundary layer begins at this location, and its thickness increases with x. Inside the boundary layer region, there is a significant velocity gradient, and effects of shear stresses should be taken into account. Close to the leading edge of the plate, the flow is laminar. For larger distance downstream of the leading edge, there is a transition zone, which is followed by a boundary layer region, which consists of a turbulent region and a very thin laminar sublayer, as is discussed further below. As the domain is semi-infinite ($0 \leq y \leq \infty$), development of the boundary layer has a negligible effect on the value of the potential flow velocity outside the boundary layer, which maintains the value U.

As the boundary layer grows, it becomes less stable to small perturbations. Consider that at some distance L downstream of the leading edge of the flat plate, the boundary layer flow becomes unstable, and there is a transition from laminar to turbulent flow. The value of L is defined by the value of a Reynolds number, Re_x,

$$\text{Re}_x = \text{Re}_L = \frac{\rho U L}{\mu} = 5 \times 10^5 \tag{6.1.1}$$

where x is the distance from the leading edge. Downstream of $x = L$, where the laminar boundary layer flow starts to become unstable, there is a *transition zone*, as the structure of the boundary layer adjusts to the turbulent condition. In the turbulent region, close to the solid flat plate, velocity fluctuations are small, as the viscous fluid adheres to the solid wall. Furthermore, the presence of the solid wall limits the size of the turbulent vortices. Therefore the flow close to the solid wall is considered as laminar flow. This region of laminar flow comprises the *laminar sublayer*.

6.2 THE EQUATIONS OF MOTION FOR BOUNDARY LAYERS

The analysis of boundary layers starts with a scaling analysis of the governing equations. The steady-state equations of motion in a two-dimensional domain are (refer to Chap. 2)

$$u\frac{\partial u}{\partial x} + v\frac{\partial u}{\partial y} = -\frac{1}{\rho}\frac{\partial}{\partial x}(p + \rho g Z) + v\left(\frac{\partial^2 u}{\partial x^2} + \frac{\partial^2 u}{\partial y^2}\right) \tag{6.2.1a}$$

$$u\frac{\partial v}{\partial x} + v\frac{\partial v}{\partial y} = -\frac{1}{\rho}\frac{\partial}{\partial y}(p + \rho g Z) + v\left(\frac{\partial^2 v}{\partial x^2} + \frac{\partial^2 v}{\partial y^2}\right) \tag{6.2.1b}$$

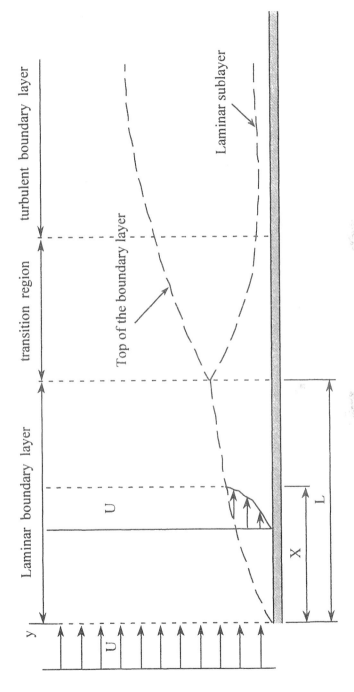

Figure 6.1 Boundary layer development over a horizontal flat plate.

These equations can be simplified for boundary layer flow by consideration of the relative magnitudes of the various terms. We let u denote the characteristic quantity of velocity in the boundary layer and U be the potential flow velocity that exists outside the boundary layer region. We also define a characteristic longitudinal length L, over which u is subject to an appreciable change. Therefore the order of magnitude of the first term of Eq. (6.2.1a), which represents the inertial terms of the equation of motion, is

$$u\frac{\partial u}{\partial x} \approx \frac{U^2}{L} \tag{6.2.2}$$

A characteristic length in the y-direction, over which changes in u are appreciable, is of the order of magnitude of the average thickness of the boundary layer δ. Therefore the last term of Eq. (6.2.1a), which represents the effect of shear stress between laminae of the boundary layer, is of order of magnitude

$$\nu\frac{\partial^2 u}{\partial y^2} \approx \frac{\nu U}{\delta^2} \tag{6.2.3}$$

In the boundary layer, effects of inertia and viscous shear are of the same order of magnitude, so that the terms in Eqs. (6.2.2) and (6.2.3) are approximately the same and

$$\delta \approx \sqrt{\frac{\nu L}{U}} \tag{6.2.4}$$

This result indicates that the boundary layer thickness is of order of magnitude $\sqrt{\nu t}$, where $t = L/U$ is a characteristic time scale for motions in the boundary layer.

From the continuity equation for the boundary layer flow,

$$\frac{\partial u}{\partial x} + \frac{\partial v}{\partial y} = 0 \tag{6.2.5}$$

Applying appropriate scaling quantities, this expression implies

$$\frac{U}{L} \approx \frac{v}{\delta} \quad \Rightarrow \quad v \approx \frac{\delta U}{L} \tag{6.2.6}$$

With L generally much larger than δ, this last result suggests that the velocity normal to the boundary is much smaller than the velocity along the boundary. A further approximation used in boundary layer analysis is to assume that gradients in the normal direction are generally much larger than gradients in the flow direction, so that

$$\frac{\partial}{\partial y} \gg \frac{\partial}{\partial x} \tag{6.2.7}$$

The pressure gradient is of the same order of magnitude as the inertial terms in Eq. (6.2.1). Therefore

$$\frac{\partial}{\partial x}(p + \rho g Z) \approx \rho u \frac{\partial u}{\partial x} \qquad \Rightarrow \qquad p + \rho g Z \approx \rho U^2 \tag{6.2.8}$$

From an examination of equations (6.2.2)–(6.2.8), the following dimensionless variables are suggested:

$$x^* = \frac{x}{L} \qquad y^* = \frac{y}{\delta}$$

$$u^* = \frac{u}{U} \qquad v^* = \frac{vL}{\delta U} \qquad p^* = \frac{p + \rho g Z}{\rho U^2} \tag{6.2.9}$$

where an asterisk indicates a dimensionless quantity. By introducing Eqs. (6.2.9) and (6.2.4) into Eqs. (6.2.1) and (6.2.5), we obtain nondimensional forms of the governing equations (in the following, all variables are dimensionless — the asterisks are omitted for simplicity),

$$u \frac{\partial u}{\partial x} + v \frac{\partial u}{\partial y} = -\frac{\partial p}{\partial x} + \frac{1}{\mathrm{Re}} \frac{\partial^2 u}{\partial x^2} + \frac{\partial^2 u}{\partial y^2} \tag{6.2.10a}$$

$$\frac{1}{\mathrm{Re}}\left(u \frac{\partial v}{\partial x} + v \frac{\partial v}{\partial y}\right) = -\frac{\partial p}{\partial y} + \frac{1}{\mathrm{Re}^2} \frac{\partial^2 v}{\partial x^2} + \frac{1}{\mathrm{Re}} \frac{\partial^2 v}{\partial y^2} \tag{6.2.10b}$$

$$\frac{\partial u}{\partial x} + \frac{\partial v}{\partial y} = 0 \tag{6.2.10c}$$

where Re is the overall Reynolds number,

$$\mathrm{Re} = \frac{UL}{\nu} \tag{6.2.11}$$

In typical boundary layer flow Re is large. Under this condition the terms in Eq. (6.2.10) that are divided by Re can be neglected, leaving

$$u \frac{\partial u}{\partial x} + v \frac{\partial u}{\partial y} = -\frac{\partial p}{\partial x} + \frac{\partial^2 u}{\partial y^2} \tag{6.2.12a}$$

$$0 = -\frac{\partial p}{\partial y} \tag{6.2.12b}$$

$$\frac{\partial u}{\partial x} + \frac{\partial v}{\partial y} = 0 \tag{6.2.12c}$$

An interesting result is immediately obvious from Eq. (6.2.12b), which indicates that the pressure (or *piezometric pressure*) within the boundary layer is (approximately) equal to its value at the top of the boundary layer. Therefore the value of the pressure in the boundary layer can be obtained from the calculation of the pressure in the potential flow region, outside the boundary layer.

There are several measures and definitions of the effect of the boundary layer region on the total flow field. According to the simplest definition, the top of the boundary layer is located where the boundary layer flow obtains a value equal to 99% of the potential mean stream flow, or $u = 0.99U$. Another measure of the boundary layer, termed the *displacement thickness* δ_d, is defined as the thickness of a layer conveying fluid with velocity U, having volumetric flow rate identical to the difference between the flow rate of a potential flow and that of the boundary layer flow. According to this definition,

$$\delta_d = \int_0^\infty \left(1 - \frac{u}{U} \right) dy \qquad (6.2.13)$$

The displacement thickness represents the outward displacement of the potential flow streamlines that results from the presence of the viscous boundary layer. It is useful in defining the thickness by which the actual solid wall or body should be increased before the potential flow theory may be applied.

A third measure, termed *momentum thickness* δ_m, represents the thickness of a layer conveying fluid with velocity U, whose momentum flux is identical to the difference between the momentum of the potential flow and that of the boundary layer flow. This definition is expressed by

$$\delta_m = \int_0^\infty \frac{u}{U} \left(1 - \frac{u}{U} \right) dy \qquad (6.2.14)$$

This boundary layer definition is sometimes used when determining drag on an object.

For convenience, we now return to the analysis of boundary layer development over a flat plate, as in Sec. 6.1. The velocity profile in the boundary layer has been shown to be well approximated by a nondimensional *similarity profile*,

$$\frac{u}{U} = \frac{u}{U}(\eta) \qquad \text{where} \qquad \eta = \frac{y}{\delta} \qquad (6.2.15)$$

Outside the boundary layer region, the velocity U of the potential flow is constant. A dimensionless stream function $f(\eta)$ can be defined by

$$f(\eta) = \frac{\Psi}{U\delta} \qquad (6.2.16)$$

where the stream function Ψ is related to the velocity components by (also see Sec. 2.5.3)

$$u = \frac{\partial \Psi}{\partial y} \qquad v = -\frac{\partial \Psi}{\partial x} \qquad (6.2.17)$$

As previously noted (Eq. 6.2.7), in most cases of boundary layer flow the longitudinal pressure gradient is negligible. Making this assumption and introducing Eq. (6.2.17) into Eq. (6.2.12a) results in

$$\frac{\partial \Psi}{\partial y} \frac{\partial^2 \Psi}{\partial x \partial y} - \frac{\partial \Psi}{\partial x} \frac{\partial^2 \Psi}{\partial y^2} = v \frac{\partial^3 \Psi}{\partial y^3} \tag{6.2.18}$$

This equation is subject to the following boundary conditions of the boundary layer flow over a flat plate:

$$\frac{\partial \Psi}{\partial y} = U \qquad \text{at} \qquad x = 0 \tag{6.2.19a}$$

$$\frac{\partial \Psi}{\partial y} = \Psi = 0 \qquad \text{at} \qquad y = 0 \tag{6.2.19b}$$

$$\frac{\partial \Psi}{\partial y} \to U \qquad \text{at} \qquad y \to \infty \tag{6.2.19c}$$

We now introduce Eq. (6.2.16) to evaluate the various terms of Eq. (6.2.18),

$$\frac{\partial \Psi}{\partial x} = U \frac{d\delta}{dx} (f - f'\eta) \qquad \frac{\partial^2 \Psi}{\partial x \partial y} = -\frac{U\eta f''}{\delta} \frac{d\delta}{dx}$$

$$\frac{\partial \Psi}{\partial y} = U f' \qquad \frac{\partial^2 \Psi}{\partial y^2} = \frac{U f''}{\delta} \qquad \frac{\partial^3 \Psi}{\partial y^3} = \frac{U f'''}{\delta^2} \tag{6.2.20}$$

where $f' = df/d\eta$ and $f'' = d^2 f/d\eta^2$. Thus, Eq. (6.2.18) becomes

$$-\left(\frac{U\delta}{v} \frac{d\delta}{dx} \right) f f' = f''' \tag{6.2.21}$$

Since $f = f(\eta)$ only, this last expression can be true only if

$$\left(\frac{U\delta}{v} \frac{d\delta}{dx} \right) = const \tag{6.2.22}$$

When this constant is chosen to be $\frac{1}{2}$, a single integration with respect to x gives

$$\delta = \sqrt{\frac{vx}{U}} \tag{6.2.23}$$

which is consistent with previous results (Eq. 6.2.4). Introducing this value into Eq. (6.2.21) results in

$$\frac{1}{2} f f'' + f''' = 0 \tag{6.2.24}$$

According to Eq. (6.2.19), the differential equation (6.2.24) is subject to the following boundary conditions:

$$f'(\infty) = 1 \qquad f(0) = f'(0) = 0 \qquad (6.2.25)$$

Blasius developed a power series solution for Eq. (6.2.24), subject to the boundary conditions of Eq. (6.2.25). Numerical solution of Eq. (6.2.24) also can be obtained quite easily using an adequate numerical code. In either case, the solution shows that

$$\frac{u}{U} = 0.99 \qquad \text{at} \qquad y\sqrt{\frac{U}{vx}} = 4.9 \qquad (6.2.26)$$

This gives an indication of the boundary layer thickness according to the "simplest" definition introduced earlier in this section. By substituting δ for y,

$$\delta = 4.9\sqrt{\frac{vx}{U}} \qquad \text{or} \qquad \frac{\delta}{x} = \frac{4.9}{\sqrt{\text{Re}_x}} \qquad (6.2.27)$$

where $\text{Re}_x = Ux/v$. This expression provides an improved estimate for the boundary layer thickness, relative to Eq. (6.2.23)

6.3 THE INTEGRAL APPROACH OF VON KARMAN

In principle, Eq. (6.2.24) may be solved to find $f(\eta)$ and its derivatives, from which values of the shear stress on the flat plate, friction force, and drag coefficient can be obtained. However, we prefer to present the calculation of these quantities by the *integral method of Von Karman*. Results of this approximate method are very similar to those obtained by solving Eq. (6.2.24) directly.

The integral approach of Von Karman is applicable to calculations of laminar as well as turbulent boundary layers. It incorporates several approximations, including steady state, but in many cases its accuracy is sufficient for engineering purposes. According to this approach, at the bottom of the boundary layer (at $y = 0$) the velocity vanishes, and at the top of the boundary layer (at $y = \delta$) the velocity is U (rather than some percentage of U). Figure 6.2 shows the boundary layer conditions considered for this analysis.

Under steady-state flow, the integral basic conservation theorems are applied to the control volume of Fig. 6.2, with unit width, length Δx, and thickness varying between δ and $\delta + \Delta\delta$, where

$$\Delta\delta = \frac{d\delta}{dx} dx \qquad (6.3.1)$$

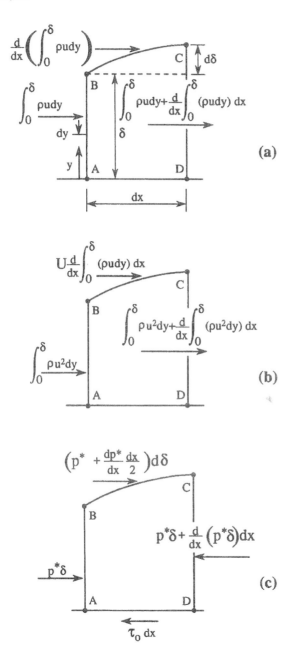

Figure 6.2 Application of basic conservation theorems to control volume of the boundary layer: (a) mass flux; (b) momentum flux; and (c) surface forces.

First consider mass conservation (Fig. 6.2a). The mass flux through section AB is

$$\left(\int_0^\delta \rho u \, dy \right)\Big|_x = \int_0^\delta \rho u \, dy \tag{6.3.2a}$$

Similarly, the mass flux through section CD is

$$\left(\int_0^{\delta+d\delta} \rho u \, dy \right)\Big|_{x+dx} = \int_0^\delta \rho u \, dy + \frac{d}{dx} \left(\int_0^\delta \rho u \, dy \right) dx \tag{6.3.2b}$$

Therefore the mass flux through BC is the difference between the fluxes passing through sections AB and CD, given by

$$\frac{d}{dx} \left(\int_0^\delta \rho u \, dy \right) dx \tag{6.3.2c}$$

Figure 6.2b shows the various momentum fluxes that penetrate into the control volume of the boundary layer. The momentum flux entering the control volume through AB is

$$\left(\int_0^\delta \rho u^2 dy \right)\Big|_x = \int_0^\delta \rho u^2 dy \tag{6.3.3a}$$

and the momentum flux leaving the control volume through CD is

$$\left(\int_0^{\delta+d\delta} \rho u^2 dy \right)\Big|_{x+dx} = \int_0^\delta \rho u^2 dy + \frac{d}{dx} \left(\int_0^\delta \rho u^2 dy \right) dx \tag{6.3.3b}$$

At the top of the boundary layer the flow velocity is U. Therefore the momentum penetrating into the control volume of the boundary layer through BC is given by the product of this velocity and the mass flux crossing BC. Thus the momentum flux through BC is

$$U \frac{d}{dx} \left(\int_0^\delta \rho u \, dy \right) dx \tag{6.3.3c}$$

Figure 6.2c shows the various surface forces acting on the control volume of the boundary layer. In order to generalize the calculation beyond the case of a horizontal flat plate, we refer to the piezometric pressure p^* rather than the pressure p. The force acting on AB is in the positive x-direction, and its value is given by

$$\left(\int_0^\delta p^* dy \right)\Big|_x \approx p^* \delta \tag{6.3.4a}$$

The force acting on CD is in the negative x-direction, with magnitude

$$\left(\int_0^{\delta+d\delta} p^* dy\right)\bigg|_{x+dx} \approx p^*\delta + \frac{d}{dx}(p^*\delta)\,dx$$

$$= p^*\delta + \delta\frac{dp^*}{dx}dx + p^*d\delta \qquad (6.3.4\text{b})$$

The force acting on BC is in the positive x-direction and its value is approximately given by

$$\left(p^* + \frac{dp^*}{dx}\frac{dx}{2}\right)d\delta \approx p^*d\delta \qquad (6.3.4\text{c})$$

The force acting on AD results from wall shear stress and is in the negative x-direction, given by

$$\tau_0\,dx \qquad (6.3.4\text{d})$$

where τ_0 is the shear stress exerted by the flat plate on the control volume.

Since the plate is horizontal, there are no body forces to consider in the x-direction, and the total surface forces acting are given by the sum of the parts of Eq. (6.3.4),

$$dF = -\left(\delta\frac{dp^*}{dx} + \tau_0\right)dx \qquad (6.3.5)$$

According to the Reynolds transport theorem (Sec. 2.4.3), under steady state the net force acting on the control volume in the x-direction is equal to the difference between momentum fluxes leaving the control volume and those entering the control volume. Therefore

$$\delta\frac{dp^*}{dx} + \tau_0 = U\frac{d}{dx}\left(\int_0^\delta \rho u\,dy\right) - \frac{d}{dx}\left(\int_0^\delta \rho u^2 dy\right) \qquad (6.3.6)$$

This is the basic expression obtained by Von Karman for analysis of boundary layer flow.

If the boundary layer is not too thick, then the first term on the left-hand side (LHS) of Eq. (6.3.6) is much smaller than the second term, so that

$$\tau_0 = \frac{d}{dx}\left[\rho U^2\delta \int_0^1 \left(1 - \frac{u}{U}\right)\frac{u}{U}d\eta\right] \qquad (6.3.7)$$

where $\eta = y/\delta$ as before (Eq. 6.2.15). Equation (6.3.7) provides a quantitative connection between the shear stress on the plate and the velocity profile in the boundary layer. In other words, the shear stress applied by the flat plate on the boundary layer can be calculated directly as long as the velocity profile

is known. This equation also shows that there is a direct connection between
the wall shear stress and the rate of development of the boundary layer along
the plate, which can be seen by rewriting Eq. (6.3.7), noting that x and η are
independent of each other:

$$\tau_0 = \left[\rho U^2 \int_0^1 \left(1 - \frac{u}{U} \right) \frac{u}{U} d\eta \right] \frac{d\delta}{dx} \tag{6.3.8}$$

Thus τ_0 depends on $d\delta/dx$.

It is common to assume a similarity profile for the velocity distribution
(as with Eq. 6.2.15), i.e.,

$$\frac{u}{U} = f\left(\frac{y}{\delta} \right) \qquad \text{or} \qquad \frac{u}{U} = f(\eta) \tag{6.3.9}$$

This applies to both laminar and turbulent boundary layers. In fact, it should be
noted that in developing Eq. (6.3.7), no assumption was made about whether
the flow in the boundary layer is laminar or turbulent, so that any of the
results obtained so far apply equally well for either condition. The differences
between laminar and turbulent boundary layers are related to different velocity
profiles and differences between boundary conditions typical of those velocity
profiles, as discussed in the following sections.

6.4 LAMINAR BOUNDARY LAYERS

The boundary layer equations can be solved as long as boundary conditions
are specified. At the plate, the velocity vanishes. Thus

$$\frac{u}{U}(0) = 0 \tag{6.4.1}$$

and at the top of the boundary layer the velocity is assumed to be equal to
that of the potential flow existing outside of the boundary layer, thus

$$\frac{u}{U}(1) = 1 \tag{6.4.2}$$

For laminar flow, shear stress in general is equal to the velocity gradient
multiplied by the fluid viscosity. Therefore, at the flat plate, the shear stress
τ_0 is equal to the viscosity multiplied by the velocity gradient at $\eta = 0$,

$$\tau_0 = \mu \left(\frac{du}{dy} \right) \bigg|_{y=0} = \mu U \left[\frac{d}{d\eta} \left(\frac{u}{U} \right) \left(\frac{\partial \eta}{\partial y} \right) \right] \bigg|_{\eta=0} \tag{6.4.3}$$

At the top of the laminar boundary layer, the velocity profile is tangential to the
uniform potential flow profile located outside the boundary layer. Therefore

shear stress vanishes and

$$\left[\frac{d}{d\eta}\left(\frac{u}{U}\right)\right]_{\eta=1} = 0 \tag{6.4.4}$$

Many different velocity profiles may satisfy the boundary conditions of Eqs. (6.4.1)–(6.4.4). A common approach to deriving the velocity profile is to consider polynomials or power series of sinusoidal functions. As an example, consider the second-order polynomial,

$$\frac{u}{U} = a + b\eta + c\eta^2 \qquad \eta = \frac{y}{\delta} \tag{6.4.5}$$

Due to the boundary conditions of Eqs. (6.4.1)–(6.4.4), $a = 0$, $b = 2$, and $c = -1$. Therefore

$$\frac{u}{U} = 2\eta - \eta^2 \tag{6.4.6}$$

Using this expression to evaluate the integral of Eq. (6.3.8) results in

$$\int_0^1 \left(1 - \frac{u}{U}\right)\frac{u}{U}\,d\eta = \int_0^1 \left[(1 - 2\eta + \eta^2)(2\eta - \eta^2)\right]d\eta = \frac{2}{15} \tag{6.4.7}$$

Also, introducing Eq. (6.4.6) into Eq. (6.4.3) yields

$$\tau_0 = 2\mu\frac{U}{\delta} \tag{6.4.8}$$

Substituting Eqs. (6.4.7) and (6.4.8) into Eq. (6.3.8) and rearranging gives

$$\frac{2}{15}\rho U^2\frac{d\delta}{dx} = 2\mu\frac{U}{\delta} \qquad \Rightarrow \qquad \frac{d}{dx}(\delta^2) = \frac{30\mu}{\rho U} \tag{6.4.9}$$

We integrate this expression, assuming that $\delta = 0$ at $x = 0$, to obtain

$$\delta = 5.48\sqrt{\frac{\mu x}{\rho U}} = 5.48\frac{x}{\sqrt{\text{Re}_x}} \tag{6.4.10}$$

where Re_x is the same as defined earlier in Eq. (6.2.27). According to Eq. (6.4.10), the thickness of the laminar boundary layer is proportional to the square root of the distance from the leading edge of the flat plate.

By introducing Eq. (6.4.10) into Eq. (6.4.8), the wall shear stress can be expressed as

$$\tau_0 = \frac{0.365}{\sqrt{\text{Re}_x}}\rho U^2 \tag{6.4.11}$$

The *friction drag force* applied on the flat plate is obtained by integrating Eq. (6.4.11) over the entire length L of the plate:

$$F_{\mathrm{Df}} = \int_0^L \tau_0 \, dx = \frac{1.46}{\sqrt{\mathrm{Re}_L}} \rho \frac{U^2}{2} L \tag{6.4.12}$$

where Re_L is defined similar to Re_x, but with L in place of x. In addition, the *coefficient of friction drag* for the plate is defined by

$$C_{\mathrm{Df}} = \frac{F_{\mathrm{Df}}}{\rho \dfrac{U^2}{2} L} = \frac{1.46}{\sqrt{\mathrm{Re}_L}} \tag{6.4.13}$$

In general, however, this is not the total drag acting on an object in a flow. For example, recall from Chap. 4 that the total drag force applied on a cylinder is due to both *friction drag* and *form drag*. The latter originates from pressure forces associated with separation of the boundary layer from the cylinder. This phenomenon occurs in cases of flow around solid bodies, expansion of conduits, and other cases in which the flow is associated with positive pressure gradients. Then it is common to express the total drag coefficient as a sum of the friction drag coefficient and the form drag coefficient,

$$C_{\mathrm{D}} = C_{\mathrm{Df}} + C_{\mathrm{Ds}} \tag{6.4.14}$$

where C_{Ds} is the form drag coefficient. However, in the case of flow over a flat plate, there is no form drag, as there is no positive pressure gradient, and thereby the boundary layer is not subject to separation and $C_{\mathrm{Ds}} = 0$.

6.5 TURBULENT BOUNDARY LAYERS

At sufficient distance (L) and potential flow velocity U, the boundary layer becomes turbulent (see Eq. 6.1.1). As shown in Fig. 6.1, the turbulent boundary layer includes a laminar sublayer. In general, the structure of the turbulent boundary layer is more complicated than that of the laminar boundary layer. In the turbulent region, the velocity profile is approximately proportional to the logarithm of the distance from the solid wall, provided that y is not very large. However, the logarithmic velocity profile is never exactly tangential to the uniform profile of the potential flow outside the boundary layer (at $y = \delta$), though this problem is relatively minor and is often neglected in practical applications. In practice, other velocity profiles have been used to approximate the velocity distribution in the turbulent boundary layer, such as the 1/7th *law* discussed below.

Again as in Sec. 6.4, we consider here development of a turbulent boundary layer over a flat plate. As with the laminar boundary layer, a similarity velocity profile is assumed, satisfying the boundary conditions,

$$\frac{u}{U}(0) = 0 \qquad \frac{u}{U}(1) = 1 \tag{6.5.1}$$

Observations have shown that the velocity profile in a turbulent boundary layer can be closely approximated by

$$\frac{u}{u_*} = 8.74 \left(\frac{yu_*}{\nu}\right)^{1/7} \tag{6.5.2}$$

where u_* is the shear velocity

$$u_* = \sqrt{\frac{\tau_0}{\rho}} \tag{6.5.3}$$

At the top of the boundary layer, Eq. (6.5.2) yields

$$\frac{U}{u_*} = 8.74 \left(\frac{\delta u_*}{\nu}\right)^{1/7} \tag{6.5.4}$$

and by dividing Eq. (6.5.2) by Eq. (6.5.4), we obtain

$$\frac{u}{U} = \left(\frac{y}{\delta}\right)^{1/7} = \eta^{1/7} \tag{6.5.5}$$

which is the 1/7th law referred to earlier.

Performing the integral of Eq. (6.3.8) with the velocity profile of Eq. (6.5.5) results in

$$\int_0^1 \left(1 - \frac{u}{U}\right) \frac{u}{U} \, d\eta = \int_0^1 \left(1 - \eta^{1/7}\right) \eta^{1/7} d\eta = \frac{7}{72} \tag{6.5.6}$$

so that

$$\tau_0 = \frac{7}{72} \rho U^2 \frac{d\delta}{dx} \tag{6.5.7}$$

This also can be written, using Eq. (6.5.4), as

$$\tau_0 = \rho u_*^2 = 0.0225 \rho \left(\frac{\nu}{\delta}\right)^{1/4} U^{7/4} \tag{6.5.8}$$

By comparing Eq. (6.5.7) with Eq. (6.5.8), it is seen that

$$\frac{7}{72} \rho U^2 \frac{d\delta}{dx} = 0.0225 \rho U^2 \left(\frac{\nu}{U\delta}\right)^{1/4} \tag{6.5.9}$$

The solution of this equation requires a boundary condition to integrate $d\delta/dx$. Although we know that a laminar boundary develops initially at the leading edge of the flat plate, this region is usually ignored, since the laminar boundary layer comprises a relatively minor portion of the entire boundary layer. Then, assuming that the entire boundary layer is turbulent (so that $\delta = 0$ when $x = 0$), and integrating Eq. (6.5.9), we obtain

$$\delta = 0.37x \left(\frac{\nu}{Ux}\right)^{1/5} = 0.37x\mathrm{Re}_x^{-1/5} \tag{6.5.10}$$

According to this expression, the thickness of the turbulent boundary layer is proportional to $x^{4/5}$. By comparison, the thickness of the laminar boundary layer is proportional to $x^{1/2}$ (Eq. 6.2.27). Thus the turbulent boundary layer growth is much more significant than that of the laminar boundary layer. This is due to the action of turbulent vortices spreading momentum away from the plate more effectively than by viscous stresses alone.

An expression for the wall shear stress is obtained by substituting Eq. (6.5.10) into Eq. (6.5.8):

$$\tau_0 = 0.058\rho\frac{U^2}{2} \left(\frac{\nu}{Ux}\right)^{1/5} \tag{6.5.11}$$

By integrating Eq. (6.5.11) over the entire length L of the plate, the total friction drag is found as

$$F_{\mathrm{Df}} = \int_0^L \tau_0 \, dx = 0.072\rho\frac{U^2}{2}L \left(\frac{\nu}{UL}\right)^{1/5} \tag{6.5.12}$$

Then, according to this expression, the coefficient of friction drag is given by

$$C_{\mathrm{Df}} = \frac{F_{\mathrm{Df}}}{L(\rho U^2/2)}) = 0.072\mathrm{Re}_L^{-1/5} \tag{6.5.13}$$

As with the laminar boundary layer, there is no form drag on a flat plate.

It should be noted that the expressions developed in this section are applicable for values of Re_L up to about 10^7. For higher values of Re_L, more complicated velocity profiles should be considered.

6.6 APPLICATION OF THE BOUNDARY LAYER CONCEPT TO HEAT AND MASS TRANSFER

The use of the boundary layer approximation as an integral method for the solution of partial differential equations dates back to Von Karman and Pohlhausen, who applied this method to phenomena of fluid flow. Since then the boundary layer approximation has been useful in a variety of topics associated with fluid flows, heat, and mass transfer. For the present discussion, we introduce several examples in which the method is applied to solve problems of mass transfer, though the basic approach is the same for cases of heat transfer.

The basic differential equation for mass diffusion in a one-dimensional domain is (see Sec. 10.3)

$$\frac{\partial C}{\partial t} = k_m \frac{\partial^2 C}{\partial y^2} \tag{6.6.1}$$

where C is the *mass concentration*, k_m is the *mass diffusivity*, t is time, and y is the space coordinate. For a one-dimensional heat conduction problem, C would represent the temperature and k_m the heat diffusivity.

Consider the case of diffusion from a flat plate into a semi-infinite domain. Then the appropriate initial and boundary condition are

$$
\begin{array}{llll}
C = 0 & \text{for} & 0 \leq y \leq \infty, & t < 0 \\
C = C_0 & \text{for} & y = 0, & t \geq 0 \\
C = 0 & \text{for} & y \to \infty, & t \geq 0
\end{array}
\tag{6.6.2}
$$

Following the ideas presented in previous sections of this chapter, it is assumed that solute concentration above the flat plate is significant in a range of values for y between the flat plate, where $y = 0$, and some point $y = \delta$, which is considered to represent a boundary layer thickness for mass diffusion (Fig. 6.3). We also assume similarity concentration profiles, so that

$$\frac{C}{C_0} = f(\eta) \qquad \eta = \frac{y}{\delta} \tag{6.6.3}$$

Integrating Eq. (6.6.1) with respect to y gives

$$\int_0^\delta \frac{\partial C}{\partial t} \, dy = -\left[k_m \frac{\partial C}{\partial y} \right]_{y=0} \tag{6.6.4}$$

where it has been assumed that the concentration gradient vanishes at $y = \delta$ and therefore there is zero flux at that location. The right-hand side of Eq. (6.6.4) represents the flux of contaminant at the wall that diffuses into the semi-infinite domain. The LHS represents the rate of change of mass within the boundary layer control volume. This integral can be evaluated by application of *Leibniz's rule* to obtain

$$\frac{\partial}{\partial t} \int_0^\delta C \, dy = \int_0^\delta \frac{\partial C}{\partial t} \, dy + (C)_{y=\delta} \frac{\partial \delta}{\partial t} = \int_0^\delta \frac{\partial C}{\partial t} \, dy \tag{6.6.5}$$

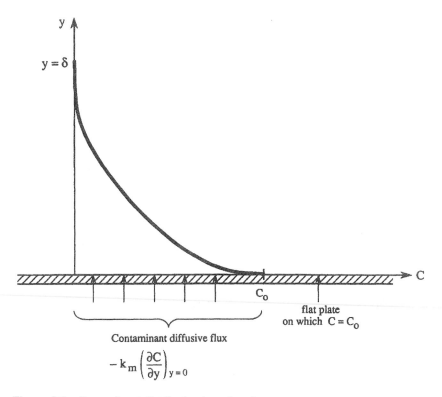

Figure 6.3 Contaminant distribution boundary layer.

since $C = 0$ at $y = \delta$. By introducing Eq. (6.6.3) into Eqs. (6.6.4) and (6.6.5),

$$\frac{d\delta}{dt} \int_0^1 f \, d\eta = -k_m \frac{1}{\delta} f'(0) \tag{6.6.6}$$

Direct integration of this expression then yields

$$\delta^2 = -k_m \frac{2 f'(0)}{\int_0^1 f \, d\eta} t \tag{6.6.7}$$

which shows that the thickness of the boundary layer is proportional to $t^{1/2}$.
The function $f(\eta)$ should satisfy the following boundary conditions:

$$
\begin{aligned}
f(\eta) &= 1 & \text{at} \quad \eta &= 0 \\
f(\eta) &= 0 & \text{at} \quad \eta &= 1 \\
f'(\eta) &= 0 & \text{at} \quad \eta &= 1
\end{aligned}
\tag{6.6.8}
$$

A possible form for $f(\eta)$ that satisfies these conditions is

$$f(\eta) = (1 - \eta)^n \tag{6.6.9}$$

where n is a power coefficient, which should be chosen to provide a best fit of the concentration profile to measured values. By introducing Eq. (6.6.9) into Eq. (6.6.7), we obtain

$$\delta^2 = 2n(n + 1)k_m t \tag{6.6.10}$$

The boundary layer method also can be applied in cases of prescribed contaminant flux at $y = 0$ (rather than specifying concentration). The extension of the boundary layer approximation to such cases is done by the *top specified boundary layer* approach. According to this method, a "region of interest" is identified in which the similarity concentration profile (Eq. 6.6.3) is valid. In this case, however, the value for C_0 is obtained as part of the solution and is generally time dependent.

PROBLEMS

Solved Problems

Problem 6.1   According to Eq. (6.2.9), the equations of motion for boundary layer flow are given by

$$u\frac{\partial u}{\partial x} + v\frac{\partial u}{\partial y} = -\frac{1}{\rho}\frac{\partial p}{\partial x} + v\frac{\partial^2 u}{\partial y^2} \tag{1}$$

$$0 = -\frac{1}{\rho}\frac{\partial p}{\partial y} \tag{2}$$

$$\frac{\partial u}{\partial x} + \frac{\partial v}{\partial y} = 0 \tag{3}$$

Using the equation of continuity (3) and direct integration of the equations of continuity (3) and motion (1), while using *Leibniz's rule*, derive Eq. (6.3.7).

Solution

Equation (2) indicates that the pressure depends only on the x coordinate. However, the effect of the pressure gradient on the structure of the boundary layer flow is usually negligible. We multiply Eq. (3) by u and add it to Eq. (1), to obtain

$$\frac{\partial u^2}{\partial x} + \frac{\partial(uv)}{\partial y} = v\frac{\partial^2 u}{\partial y^2} \tag{4}$$

According to Leibniz's rule:

$$\frac{\partial}{\partial \alpha} \int_{f(\alpha)}^{g(\alpha)} w(\alpha, x)\, dx = \int_{f(\alpha)}^{g(\alpha)} \frac{\partial w}{\partial \alpha}\, dx + (w)_{g(\alpha)} \frac{dg}{d\alpha} - (w)_{f(\alpha)} \frac{df}{d\alpha}$$

We first use Leibniz's rule for the integration of the continuity equation. Integrating the first term of Eq. (3) over the thickness of the boundary layer gives

$$\int_0^\delta \frac{\partial u}{\partial x}\, dy = \frac{\partial}{\partial x} \int_0^\delta u\, dy - U \frac{d\delta}{dx} \tag{5}$$

where U is the longitudinal velocity at the top of the boundary layer. Integrating the second term of Eq. (3) over the boundary layer thickness gives

$$\int_0^\delta \frac{\partial v}{\partial y}\, dy = (v)_{y=\delta}$$

In summary, by integration of the equation of continuity (3) over the boundary layer thickness, we have

$$(v)_{y=\delta} = U \frac{d\delta}{dx} - \frac{\partial}{\partial x} \int_0^\delta u\, dy \tag{6}$$

We now multiply Eq. (4) by the density and integrate all terms over the boundary layer thickness. The first term gives

$$\int_0^\delta \frac{\partial u^2}{\partial x}\, dy = \frac{\partial}{\partial x} \int_0^\delta u^2 dy - U^2 \frac{d\delta}{dx} \tag{7}$$

Integration of the second term of Eq. (4) yields

$$\int_0^\delta \frac{\partial (uv)}{\partial x}\, dy = (uv)_{y=\delta}$$

By introducing Eq. (6) into this expression, we obtain

$$(uv)_{y=\delta} = U \left(U \frac{d\delta}{dx} - \frac{\partial}{\partial x} \int_0^\delta u\, dy \right) \tag{8}$$

Then, by combining this expression (8) with Eq. (7), we find

$$\int_0^\delta \left[\frac{\partial u^2}{\partial x} + \frac{\partial (uv)}{\partial y} \right] dy = \frac{\partial}{\partial x} \int_0^\delta [u(u - U)]\, dy \tag{9}$$

By integration of the right-hand-side term of Eq. (4) over the thickness of the boundary layer,

$$\int_0^\delta \left(v \frac{\partial^2 u}{\partial y^2} \right) dy = \left(v \frac{\partial u}{\partial y} \right) \bigg|_{y=\delta} - \left(v \frac{\partial u}{\partial y} \right) \bigg|_{y=0} = -\frac{1}{\rho} \tau_0$$

This expression is equal to the expression of Eq. (9). Therefore

$$\tau_0 = \frac{\partial}{\partial x}\left\{\rho U^2 \int_0^\delta \left[\frac{u}{U}\left(1 - \frac{u}{U}\right)\right]dy\right\} \tag{10}$$

We consider similar velocity profiles of the boundary layer and introduce a new coordinate of the boundary layer, defined by

$$\eta = \frac{y}{\delta}$$

By introducing this coordinate into Eq. (10), we obtain

$$\tau_0 = \frac{d}{dx}\left\{\rho U^2 \delta \int_0^1 \left[\frac{u}{U}\left(1 - \frac{u}{U}\right)\right]d\eta\right\} = \rho V^2 \int_0^1 \left[\frac{u}{V}\left(1 - \frac{u}{V}\right)\right]d\eta\,\frac{d\delta}{dx} \tag{11}$$

which is the desired expression for the bottom shear stress.

Problem 6.2   Assume that the velocity distribution in the laminar boundary layer over a flat plate is given by

$$\frac{u}{U} = \sin\left(\frac{\pi y}{2\delta}\right)$$

(a) Find the dependence of the boundary layer thickness δ on the velocity U, the kinematic viscosity of the fluid, and the distance from the leading edge of the BL.

(b) What is the shear stress along the plate?

(c) What is the coefficient of friction drag?

Solution

We introduce the velocity profile into Eq. (6.4.7) to obtain

$$\int_0^1 \left(1 - \frac{u}{U}\right)\frac{u}{U}\,d\eta = \int_0^1 \left[1 - \sin\left(\frac{\pi\eta}{2}\right)\right]\sin\left(\frac{\pi\eta}{2}\right)d\eta$$

$$= \int_0^1 \left[\sin\left(\frac{\pi\eta}{2}\right) - \sin^2\left(\frac{\pi\eta}{2}\right)\right]d\eta$$

$$= \int_0^1 \left[\sin\left(\frac{\pi\eta}{2}\right) + \frac{1}{2}\cos(\pi\eta) - \frac{1}{2}\right]d\eta$$

$$= \left[-\frac{2}{\pi}\cos\left(\frac{\pi\eta}{2}\right) + \frac{1}{2\pi}\sin(\pi\eta) - \frac{\eta}{2}\right]_0^1$$

$$= 0.137$$

Introducing this result into Eqs. (6.3.10) and (6.4.8) gives

$$\tau_0 = \mu\frac{\pi U}{2\delta} \qquad \frac{d}{dx}\left(\delta^2\right) = \frac{23\nu}{U}$$

We integrate this expression, and assume that $\delta = 0$ at $x = 0$, to obtain

$$\delta = 4.8\sqrt{\frac{vx}{U}} \qquad \tau_0 = 0.33\rho U^2 \frac{1}{\sqrt{Re_x}}$$

By integrating the expression for τ_0 over the length L of the plate, we obtain

$$F_{Df} = \frac{1.32}{\sqrt{Re_L}}\rho\frac{U^2}{2}L \qquad \text{and} \qquad C_{Df} = \frac{1.32}{\sqrt{Re_L}}$$

Problem 6.3   Consider the development of the velocity profile along a flat plate (this was originally discussed in Chap. 3). Develop an expression for the velocity profile in terms of U = velocity of the plate, kinematic viscosity of the fluid, and δ = boundary layer thickness, using the boundary layer method. Assume that the velocity is described by a similarity profile,

$$\frac{u}{U} = (1 - \eta)^n \qquad \text{where} \qquad \eta = \frac{y}{\delta}$$

Solution

The differential equation of the problem is given by Eq. (3.4.1), which is basically the same as the equation of diffusion, Eq. (6.6.1). This differential equation is subject to the following initial and boundary conditions:

$$u = 0 \qquad \text{at} \qquad t \le 0 \qquad \text{for all values of } y$$
$$u = U \qquad \text{at} \qquad t > 0 \qquad \text{for } y = 0$$
$$u = 0 \qquad\qquad\qquad\qquad \text{for } y \to \infty$$

(also see Eq. 3.4.8). We apply a modified set of the following boundary conditions by incorporating the boundary layer thickness directly:

$$\delta = 0 \qquad \text{at} \qquad t = 0$$
$$u = U \qquad \text{at} \qquad y = 0$$
$$\frac{\partial u}{\partial y} = 0 \qquad \text{at} \qquad y = \delta$$

The velocity profile, suggested by the problem, complies with these boundary and initial conditions. Introducing the velocity profile into the differential equation and applying Leibnitz's rule, we obtain

$$\frac{d}{dt}\left[U\delta \int (1 - \eta)^n d\eta = v\frac{Un}{\delta} \right] \qquad \Rightarrow \qquad \frac{d\delta^2}{dt} = 2vn(n + 1)$$

Direct integration of this expression yields

$$\delta^2 = 2vtn(n + 1) \qquad \Rightarrow \qquad \delta = \sqrt{2vtn(n + 1)}$$

By introducing this result into the expression of the similar velocity profile, we find that

$$\frac{u}{U} = \left(1 - \frac{y}{\sqrt{2vtn(n+1)}}\right)^n$$

Unsolved Problems

Problem 6.4   Assume that the velocity profile for the laminar boundary layer over a flat plate is given by

$$u = \alpha y + \beta y^3$$

(a)   Find appropriate values for α and β.
(b)   Determine the rate of growth of δ versus x.
(c)   Find the variation of τ_0 versus x.
(d)   Find the friction drag and friction drag coefficient for a plate of unit width and length L.

Problem 6.5   Consider the following velocity profile for the laminar boundary layer over a flat plate:

$$\frac{u}{U} = \left(1 - \frac{y}{\delta}\right)^n$$

Determine the difference between values of the coefficients of friction drag by assuming $n = 2$ and $n = 3$.

Problem 6.6   Assume that the velocity distribution in the turbulent boundary layer over a flat plate is given by

$$\frac{u}{u_*} = 2.5 \ln\left(\frac{yu_*}{v}\right) + 5.5$$

(a)   Find the rate of growth of δ versus x.
(b)   Find the variation of τ_0 versus x.
(c)   Find the friction drag and friction drag coefficient for a plate of unit width and length L.

Problem 6.7   Consider a boundary through which a constant mass flux of solute penetrates into the fluid domain. The mass flux per unit width and length of the boundary is q_m. Assume that the penetrating solute flux is small, so that the velocity profile is unaffected by the solute, which spreads into the domain by diffusion only. Also assume that solute concentration in the fluid domain may be described using a similarity profile,

$$\frac{C}{C_b} = F(\eta) \qquad \text{where} \qquad \eta = \frac{y}{\delta}$$

where

$$F = (1 - \eta)^n$$

where n is equal to 2 or 3, and C_b is the solute concentration at the boundary. Note that C_b varies with time.

(a) What is the rate of growth of δ versus t?
(b) Find the difference in values of δ between the assumptions of $n = 2$ and $n = 3$.
(c) Determine the expression for the variation of C_b versus t.

Problem 6.8   A solute diffuses from a contaminated bank of a river into the river, which is considered to be wide. The river bank represents a boundary of constant concentration C_0. Assume that the river depth is uniform, and that the flow velocity V is uniformly distributed. For this case, under steady-state conditions, the equation of contaminant diffusion–advection is given by (see Chap. 10 for development of this equation)

$$V\frac{\partial C}{\partial x} = D\frac{\partial^2 C}{\partial y^2}$$

where C is the solute concentration.

(a) Determine the solute concentration profiles, by applying the method presented in Sec. 3.4, for unsteady motion of a flat plate.
(b) Determine the solute concentration profiles by applying the similarity profiles of Problem 6.7.
(c) Evaluate the accuracy of the similarity solutions, relative to the full boundary layer solution obtained in part (a), while considering $n = 2$ or $n = 3$.

SUPPLEMENTAL READING

Schlichting, H., 1968. *Boundary Layer Theory*. McGraw-Hill, New York. (Contains a thorough coverage of all topics connected with boundary layer flows.)
Ozisik, M. N., 1993. *Heat Conduction*. 2d ed. Wiley Interscience, New York. (Contains a comprehensive and textbook type presentation of heat conduction topics, including the boundary layer application for the solution of the diffusion and heat conduction equations.)

Rubin, H. and Buddemeier, R. W., 1996. A top specified boundary layer (TSBL) approximation approach for the simulation of groundwater contamination processes. *Journal of Contaminant Hydrology*, **22**; 123–144. (Presents a discussion of common uses of the boundary layer approximation to solve the equation of solute advection–diffusion problem, subject to various types of boundary conditions.)

7

Surface Water Flows

7.1 INTRODUCTION

Environmental surface water flows develop as direct runoff, flow in streams and rivers, transfer of lake water to outlet rivers, circulation in lakes and ponds, flow in estuaries, and groundwater supply into streams and rivers. Artificial conduits constructed to transport surface water are common in water and wastewater management, including canals, irrigation systems, and sewer systems. The focus of the present chapter is on a review of elements of free surface flow in open channels, under steady-state and quasi-steady-state conditions. Circulation in lakes and reservoirs is introduced briefly, in Sec. 7.7, though these flows are generally much more complicated and not amenable to the types of analyses typical of open channel flow.

7.2 HYDRAULIC CHARACTERISTICS OF OPEN CHANNEL FLOW

7.2.1 Basic Geometric Parameters of the Channel Cross Section

A defining characteristic of open channel flow is the presence of a free surface, open to the atmosphere and at which the pressure is zero (or, equal to atmospheric pressure in absolute pressure terms). As an example, Fig. 7.1 illustrates two types of common channels: (a) a channel with trapezoidal cross section; and (b) a channel with circular cross section. Much of the discussion that follows concerns basic parameters of channel flow in a trapezoidal channel, since this is the most common shape used for constructed channels and also provides a reasonable approximation for natural streams.

The cross-sectional area, A, of the channel is the area of the channel cross section, measured perpendicular to the flow velocity vector. In the case

(a)

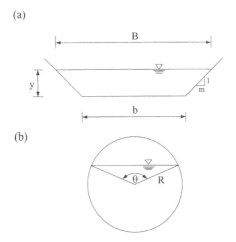

(b)

Figure 7.1   Parameters of the channel cross section: (a) trapezoidal cross section; and (b) circular cross section.

of the trapezoidal cross section, shown in Fig. 7.1,

$$A = (b + my)y = by + my^2 \tag{7.2.1}$$

where y is the depth, b is the base width of the trapezoid, and m is the side slope. Triangular and rectangular cross sections are particular cases of the trapezoidal cross section, with $b = 0$ for a triangular cross section and $m = 0$ for a rectangular cross section.

The *wetted perimeter*, P, is the length of the line of contact between the fluid and the solid wall of the channel. In the case of a trapezoidal cross section,

$$P = b + 2y\sqrt{m^2 + 1} \tag{7.2.2}$$

The *hydraulic radius*, R_h, is defined as the ratio between the cross-sectional area and the wetted perimeter. For the trapezoidal channel,

$$R_h = \frac{A}{P} = \frac{by + my^2}{b + 2y\sqrt{m^2 + 1}} \tag{7.2.3}$$

In the case of a *wide* rectangular channel, $m = 0$, and $y \ll b$. Therefore, a common approximation is

$$R_h \cong y \tag{7.2.4}$$

For a channel of circular cross section of diameter D, which is completely full (Fig. 7.1b), the ratio of area to wetted perimeter gives

$$R_h = \frac{\pi D^2/4}{\pi D} = \frac{D}{4} \qquad (7.2.5)$$

The surface width, B, is written in terms of the trapezoidal channel section shown in Fig. 7.1a as

$$B = b + 2my \qquad (7.2.6)$$

The *hydraulic depth*, y_h, is defined as the cross-sectional area divided by surface width,

$$y_h = \frac{A}{B} \qquad (7.2.7)$$

7.2.2 Parameters of Open Channel Flow

The discharge flowing through the channel is Q and the average flow velocity is V, where

$$V = \frac{Q}{A} \qquad (7.2.8)$$

Basic phenomena of open channel flow are characterized by two dimensionless parameters, the Reynolds number, Re, and the Froude number, Fr. The Reynolds number represents the ratio between inertial and viscous forces (see Sec. 1.4.2). The Reynolds number was originally defined for pipe flow, where the characteristic length scale is the diameter. For open channel flow, the characteristic length is R_h, since it is associated with the friction between flowing water and the solid walls of the channel. In order to be consistent with the definition for pipe flow, Eq. (7.2.5) suggests Re should be defined for open channel flow as

$$Re = \frac{4VR_h}{\nu} \qquad (7.2.9)$$

where ν is the fluid kinematic viscosity.

As with pipes, the value of Re indicates whether the flow is subject to laminar or turbulent conditions. Environmental flows of water in open channels are usually fully turbulent, in which case the effect of variability of Re is of minor importance.

The Froude number represents the ratio between the inertial and gravitational forces (Sec. 1.4.2). Its value is useful as an index for the development of free surface waves. Therefore the characteristic length associated with the definition of the Froude number should represent the freedom of the surface to oscillate. Such a length is the hydraulic depth of the channel, so Fr in open channel flow is defined as

$$Fr = \frac{V}{\sqrt{g y_h}} \qquad (7.2.10)$$

where g is gravitational acceleration.

For a rectangular channel, particularly one that is very wide compared to its width, the flow is often analyzed as a two-dimensional phenomenon. In this case, the discharge per unit width of the channel is defined as

$$q = \frac{Q}{b} = V y \qquad (7.2.11)$$

By introducing Eq. (7.2.11) into Eq. (7.2.10), Fr for wide rectangular channel flow is

$$Fr = \frac{q}{\sqrt{g y^3}} \qquad (7.2.12)$$

7.2.3 The Longitudinal Cross Section of Open Channel Flow

Figure 7.2 shows a longitudinal cross section of an open channel flow. Such a cross section is usually drawn with a distorted scale, with the horizontal scale much larger than the vertical scale. Longitudinal distances are measured in the horizontal direction, though the x coordinate extends along the channel bed. The channel bed has a slope angle α, which is normally assumed to be relatively small. Thus the elevation, water depth, and velocity head are written with respect to the vertical direction, which is almost perpendicular to the channel bed. We consider variations of the channel bed elevation, the water free surface, and the total head along the channel, as shown in Fig. 7.2.

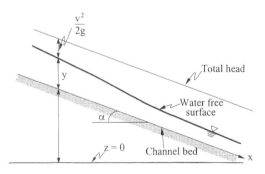

Figure 7.2 Longitudinal cross section of the channel.

The *channel bed slope* is defined by

$$S_0 = -\frac{dz}{dx} = \sin\alpha \tag{7.2.13}$$

where z is the bed elevation, relative to datum.

The slope of the water free surface is

$$i = -\frac{\partial}{\partial x}(z + y) = S_0 - \frac{\partial y}{\partial x} \tag{7.2.14}$$

The partial derivative of y is used since the water depth may be a space- and time-dependent variable.

The slope of the total head line (*energy line*) is called the *energy slope* or *friction slope* and is defined by

$$S_f = -\frac{\partial H}{\partial x} = -\frac{\partial}{\partial x}\left(z + y + \frac{V^2}{2g}\right) = S_0 - \frac{\partial E}{\partial x} \tag{7.2.15}$$

where E is called the *specific energy*, which is the total head measured with respect to the channel bed as a datum. As implied in Eq. (7.2.15), it is defined by

$$E = y + \frac{V^2}{2g} \tag{7.2.16}$$

There are several special cases that simplify the analysis of open channel flows. First, in the case of steady-state conditions, the local water depth and velocity are constant with time, namely,

$$\frac{\partial y}{\partial t} = \frac{\partial V}{\partial t} = 0 \tag{7.2.17}$$

In the case of *uniform flow*, the water depth and velocity are constant along the channel,

$$\frac{\partial y}{\partial x} = \frac{\partial V}{\partial x} = 0 \tag{7.2.18}$$

If the flow is both steady and uniform, then the channel bed, free water surface, and energy line are all parallel.

7.2.4 The Equation of Open Channel Flow

In Fig. 7.3 a control volume is defined that incorporates a portion of the channel fluid, subject to steady, uniform flow. The net forces acting on this control volume are due to gravity and shear stress along the solid boundary.

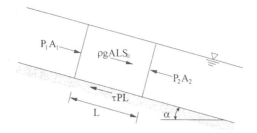

Figure 7.3   Forces acting on a control volume of the channel fluid.

There is no net pressure force since the depth is constant. Thus, applying the concept of momentum conservation,

$$\tau PL = \rho gALS_0 \tag{7.2.19}$$

where τ is the shear stress applied along the wetted perimeter and ρ is the fluid density. With minor modification, this relationship also is applicable in nonuniform flow, substituting S_f for S_0,

$$\tau PL = \rho gALS_f \tag{7.2.20}$$

Solving for τ, we obtain

$$\tau = \rho gR_hS_f \tag{7.2.21}$$

Upon dividing both sides of Eq. (7.2.21) by $\rho V^2/8$, an expression for the dimensionless *Darcy–Weissbach friction coefficient* is obtained,

$$f = \frac{8\tau}{\rho V^2} = \frac{8gR_hS_f}{V^2} \tag{7.2.22}$$

This expression leads to the *Darcy–Weissbach equation of motion*,

$$S_f = \frac{f}{4R_h}\frac{V^2}{2g} \quad \text{or} \quad V = \sqrt{\frac{8g}{f}R_hS_f} \tag{7.2.23}$$

This gives the energy slope in terms of the friction factor, f, and velocity in terms of friction slope.

An alternative representation is by the *Chezy equation*,

$$V = C\sqrt{R_hS_f} \tag{7.2.24}$$

where C is the *Chezy friction coefficient*, which by comparison of Eqs. (7.2.23) and (7.2.24) is related to the Darcy–Weissbach friction coefficient by

$$C = \sqrt{\frac{8g}{f}} \tag{7.2.25}$$

The various expressions for the equation of motion for open channel flow can be derived from the relevant expressions developed and used for pipe flow, with slightly modified coefficients. For commercial pipes, for example, it is common to apply the *Colebrook equation*, which has a modified form for open channels given by

$$\frac{1}{\sqrt{f}} = \frac{C}{\sqrt{8g}} = -2\log_{10}\left(\frac{\varepsilon}{12R_h} + \frac{2.5}{\mathrm{Re}\sqrt{f}}\right) \tag{7.2.26}$$

where ε is the relative roughness of the channel bed. It is defined by $\varepsilon = k/l$, where k is the roughness length, characterizing the small projections of material from the surface of the channel boundary into the flow, and l is the characteristic length of the flow, i.e., the hydraulic radius in open channel flow or the diameter in pipe flow. The *dimensionless roughness parameter* ε^+ is usually used to determine the type of turbulent flow (smooth or rough), and whether some terms of Eq. (7.2.26) can be ignored under certain circumstances. This parameter is defined by

$$\varepsilon^+ = \frac{ku_*}{\nu} \tag{7.2.27}$$

where $u*$ is the shear velocity,

$$u_* = \sqrt{\frac{\tau}{\rho}} \tag{7.2.28}$$

The following ranges of ε^+ determine the type of turbulent flow in the channel:

    (a) Smooth turbulent flow : $\varepsilon^+ < 5$ (7.2.29a)

    (b) Transition between smooth and rough turbulent flow :

       $5 < \varepsilon^+ < 80$ (7.2.29b)

    (c) Rough turbulent flow : $\varepsilon^+ > 80$ (7.2.29c)

Smooth turbulent flow exists when the roughness projections are all submerged within the laminar sublayer (refer to Chap. 6), and rough flow exists when the roughness protrudes entirely through the laminar sublayer. In most cases, environmental open channel flow is characterized as rough turbulent flow. For this condition, experiments with pipes have led to

$$f = 0.113\left(\frac{\varepsilon}{R_h}\right)^{1/3} \tag{7.2.30}$$

By introducing this result into Eq. (7.2.25), the Chezy coefficient becomes

$$C = 26.32\left(\frac{R_h}{\varepsilon}\right)^{1/6} \tag{7.2.31}$$

This indicates that the Chezy coefficient is directly related to ε. It should also be kept in mind that the gravitational acceleration value in SI units was applied to calculate the constant coefficient of the right-hand side of Eq. (7.2.31).

By introducing Eq. (7.2.31) into Eq. (7.2.24), we obtain

$$V = \frac{1}{n} R_{\mathrm{h}}^{2/3} S_{\mathrm{f}}^{1/2} \tag{7.2.32}$$

which applies for rough turbulent flow in open channels. This is called *the Manning equation*, and *Manning roughness coefficient*, n, is given by

$$n = 0.038 \varepsilon^{1/6} \tag{7.2.33}$$

It should be noted that Eq. (7.2.32) is not dimensionally consistent (there are no units of time on the right-hand side, for instance). Since SI units were used (Eq. 7.2.31), V is in m/s when R_{h} is in m. When English units are used (V in ft/s and R_{h} in ft), Eq. (7.2.32) has an added factor of 1.49 in the numerator.

7.2.5 Uniform Flow in Open Channels

Uniform flow through an open channel or natural stream is most often calculated using the Manning equation, where V from Eq. (7.2.32) is multiplied by the cross-sectional area, A,

$$Q = \frac{A R_{\mathrm{h}}^{2/3}}{n} S_{\mathrm{f}}^{1/2} \tag{7.2.34}$$

The Manning equation also is applicable to nonuniform open channel flow. However, in cases of steady nonuniform flow, only Q and n of Eq. (7.2.34) are kept constant along the channel. Therefore numerical procedures are usually needed for the calculation of nonuniform flow. Some further discussion of this approach is presented in Sec. 7.6.

The channel bed may be subject to erosion if the shear stress applied on it by the flowing water exceeds a defined critical value. Using the definition for shear stress, Eq. (7.2.21), with Eq. (7.2.34), then in order to prevent erosion, the following relationship must hold:

$$\frac{Q n \sqrt{g}}{u_{*_\mathrm{c}}} \le A R_{\mathrm{h}}^{1/6} \tag{7.2.35}$$

where u_{*_c} is the critical shear velocity defined as in Eq. (7.2.28), using τ_c in place of τ, and τ_c is the critical shear stress for erosion to occur. This value depends on characteristics of the bed and the material of the bed. Sedimentation is discussed further in Chap. 15.

7.2.6 The Concept of Specific Energy and Critical Depth

From Eq. (7.2.16), the specific energy in open channel flow is defined as

$$E = y + \frac{V^2}{2g} = y + \frac{Q^2}{2gA^2} \tag{7.2.36}$$

For a rectangular channel, this can be written as

$$E = y + \frac{q^2}{2gy^2} \tag{7.2.37}$$

where, as before, q is the discharge per unit width. As discussed further in Sec. 7.3, the concept of specific energy is useful in calculations concerning transitions in open channels, where energy losses are negligible.

Equations (7.2.36) and (7.2.37) indicate that, for a constant value of Q, the specific energy is expressed as a sum of two terms, the water depth and the *velocity head*. The velocity head decreases with water depth from an infinite value at $y \rightarrow 0$, to a negligible value as $y \rightarrow \infty$. Therefore the sum of two such terms should have a minimum value for a certain value of y.

Differentiating Eq. (7.2.36) with respect to y, we obtain

$$\frac{dE}{dy} = 1 - \frac{Q^2}{gA^3}\frac{dA}{dy} = 1 - \frac{V^2}{g}\left(\frac{1}{A}\frac{dA}{dy}\right) \tag{7.2.38}$$

To evaluate the derivative on the right-hand side, note from Fig. 7.4 that

$$\frac{dA}{dy} = B \tag{7.2.39}$$

Introducing this result into Eq. (7.2.38) then gives

$$\frac{dE}{dy} = 1 - \frac{V^2}{g(A/B)} = 1 - Fr^2 \tag{7.2.40}$$

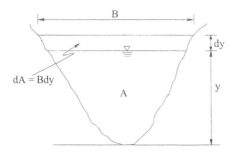

Figure 7.4 Schematic variation of area with depth.

For a constant value of Q, Fr decreases when y increases. In the case of a rectangular channel, the Froude number is inversely proportional to $y^{3/2}$ (Eq. 7.3.12). Therefore in cases of small values of y, Fr is larger than 1, and for large values of y, Fr is smaller than 1. Hence the following features of the curve of E versus y can be deduced:

$$\frac{dE}{dy} < 0 \qquad \text{if } Fr > 1 \qquad \text{and } y \text{ is small} \tag{7.2.41a}$$

$$\frac{dE}{dy} = 0 \qquad \text{if } Fr = 1 \qquad \text{and } y \text{ has a critical value} \tag{7.2.41b}$$

$$\frac{dE}{dy} > 0 \qquad \text{if } Fr < 1 \qquad \text{and } y \text{ is large} \tag{7.2.41c}$$

In addition, from Eq. (7.2.37), there are in general two roots (values of y) possible for a given specific energy level, E. Figure 7.5 provides a schematic of the variation of E and its derivative versus y, for a constant value of q. From this figure, it can be seen that E has two asymptotic limits: (a) the E axis as $y \to 0$; and (b) the straight line that passes through the origin with a slope of $45°$ (unity slope). When Fr $= 1$, the specific energy obtains its minimum value, and the water depth is called *critical depth*.

For critical conditions in a rectangular channel, Eq. (7.2.12) yields

$$y_c = \left(\frac{q^2}{g} \right)^{1/3} \tag{7.2.42}$$

By introducing Eq. (7.2.42) into Eq. (7.2.37), an expression for the minimum value of E in a rectangular channel is found as (recall that Fr $= 1$),

$$E_{min} = \tfrac{3}{2} y_c \tag{7.2.43}$$

This result indicates that all points of minimum value of E in rectangular channels comprise a straight line that passes through the origin, having slope equal to 1.5. For a channel of nonrectangular cross section, under critical flow conditions,

$$E_{min} = y_c + \tfrac{1}{2} y_{h_c} \tag{7.2.44}$$

where y_{hc} is the hydraulic depth under critical flow conditions.

By rearranging the definition for specific energy, Eq. (7.2.36), the flow rate is

$$Q = A\sqrt{2g(E - y)} \tag{7.2.45}$$

The maximum discharge that may flow through the channel is obtained by differentiating Eq. (7.2.45) with respect to y. Setting the derivative equal to 0

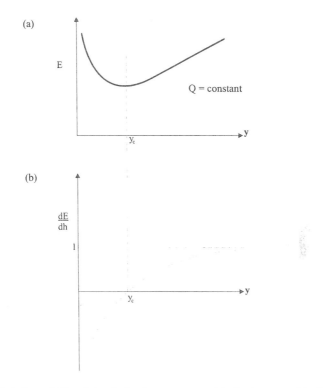

Figure 7.5 Variation of E and its derivative versus y: (a) E versus y; (b) dE/dh versus y.

then gives

$$E = y + \tfrac{1}{2} y_h \qquad (7.2.46)$$

By comparison with Eq. (7.2.44), it is seen that this condition is satisfied only under critical conditions. Thus for a given value of the specific energy, the maximum discharge that can flow through the channel cross section is obtained under critical flow conditions.

7.2.7 The Concept of the Momentum Function

The concept of the *momentum function* is associated with the employment of the conservation of momentum principle for open channel hydraulics. Figure 7.6 shows a *prismatic channel*, namely a channel with constant cross-sectional shape. In the following, we assume steady-state conditions and calculate the forces acting on a fluid control volume of length L.

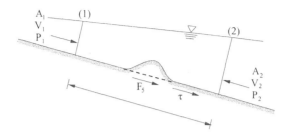

Figure 7.6   Schematics of a fluid control volume.

The equation of momentum conservation with regard to the fluid control volume is

$$\frac{\partial}{\partial t}\int_U \rho\vec{V}\,dU + \int_S \rho\vec{V}(\vec{V}\cdot\vec{n})\,dS = -\int_S p\vec{n}\,dS - \int_U \rho g\vec{k}\,dU$$

$$+ \int_U \vec{\tau}\cdot\vec{n}\,dS + \vec{F}_S \qquad (7.2.47)$$

where $\vec{V}$ is the velocity vector, ρ is the fluid density, $\vec{n}$ is a unit normal vector, $\vec{k}$ is a unit vertical vector, g is gravitational acceleration, $\vec{\tau}$ is the shear stress tensor, F_S is the sum of forces acting on solid surfaces of the control volume, U is the volume of the control volume, and S is the surface of the control volume. Applying the flow and force parameters shown in Fig. 7.6 with regard to the control volume results in

$$-\rho QV_1 + \rho QV_2 = p_1A_1 - p_2A_2 + \int_0^L \rho gAS_0\,dx$$

$$+ \int_0^L \tau P\,dx + F_S \qquad (7.2.48)$$

where p_1 and p_2 are the pressures at the center of gravity of cross sections 1 and 2, respectively. It should be noted that although F_S and τ in Fig. 7.6 are drawn in the positive x-direction, their real directions should be found by solution of the appropriate basic equations.

Rearrangement of Eq. (7.2.48) yields

$$\Omega_2 - \Omega_1 = \int_0^L \rho gAS_0\,dx + \int_0^L \tau P\,dx + F_s \qquad (7.2.49)$$

where

$$\Omega_1 = \rho QV_1 + p_1A_1 \qquad \Omega_2 = \rho QV_2 + p_2A_2 \qquad (7.2.50)$$

and Ω is called the *momentum function*. If the control volume does not have solid surfaces, the length of the control volume is small, and the slope of the

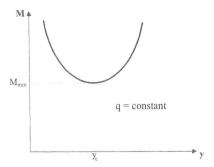

Figure 7.7 Example of momentum function.

channel is also small, then the momentum function is constant between the two cross sections of the channel shown. If the channel has a rectangular cross section, we may define *kinematic momentum per unit width, M*, by

$$M = \frac{\Omega}{\rho g b} = \frac{q^2}{gy} + \frac{y^2}{2} \tag{7.2.51}$$

In Fig. 7.7 is shown a schematic description of the momentum function versus y, for a constant discharge. As indicated by Eq. (7.2.51), the momentum function comprises two terms. One term is proportional to y^2. The other term is inversely proportional to y. Therefore the sum of these two terms should have a minimum. The minimum value of M is obtained by differentiating Eq. (7.2.51) with respect to y:

$$\frac{dM}{dh} = y - \frac{q^2}{gh^2} = y(1 - \mathrm{Fr}^2) \tag{7.2.52}$$

This result indicates that the minimum value of M occurs where $\mathrm{Fr} = 1$ (and $y = y_c$).

As indicated in Fig. 7.7 and Eq. (7.2.49), variation of the momentum function between successive cross sections of the channel can be useful for the calculation of forces acting on objects present in the channel, as well as hydraulic structures. However, in various cases where energy losses are significant, the momentum conservation principle can still be applied.

7.3 APPLICATION OF THE ENERGY CONSERVATION PRINCIPLE

The energy conservation principle is useful for calculations of changes in the velocity and water depth of the channel flow, provided that such changes are associated with negligible energy loss. Examples of such changes of the flow velocity and water depth can be considered in cases of smooth channel transitions, lateral outflow, and others.

7.3.1 Transitions in Open Channel Flow

Contraction — Expansion of the Channel Cross Section

Figure 7.8 describes a rectangular channel with a contraction in width. The transition from cross section 1 to cross section 2 takes place along a short length of the channel and is made in a manner minimizing local head losses. Also, it is assumed that the slope of the channel is very small. Therefore, energy losses between cross sections 1 and 2 are neglected. As the bottom of the channel is almost horizontal, it may be considered as a datum, so that

$$\frac{dE}{dx} = 0 \tag{7.3.1}$$

Using the definition of specific energy from Eq. (7.2.37), Eq. (7.3.1) implies

$$\frac{dy}{dx} - \frac{Q^2}{g(by)^3}\left(b\frac{dy}{dx} + y\frac{db}{dx}\right) = \frac{dy}{dx}(1 - \mathrm{Fr}^2) - \mathrm{Fr}^2\frac{y}{b}\frac{db}{dx} = 0 \tag{7.3.2}$$

The transition of the channel shown in Fig. 7.8 is associated with a negative value of the derivative of b with respect to x. Therefore the derivative of y

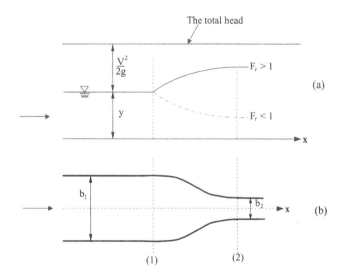

Figure 7.8 Transition in the open channel flow due to a variable width: (a) side view; and (b) top view.

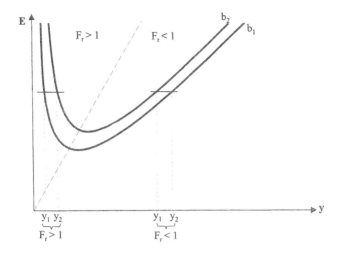

Figure 7.9 Relationships between y_1 and y_2 shown on the diagram of (E versus y) for the transition from wide into a narrow cross section.

with respect to x is negative if Fr is smaller than 1, and the derivative is positive if Fr is greater than 1. Figure 7.9 shows the relationships between y_1 and y_2 on the diagram of E versus y.

As a special case, consider the situation where, following the local constriction of the channel downstream of cross section 2, the channel expands back to its original width, b_1. Equation (7.3.2) indicates that in such a case, if Fr is smaller than 1, then the minimum value of y occurs when b is at its minimum value. If Fr is larger than 1, then the maximum value of y occurs when b is at its minimum value.

If the water depth is known at either cross section 1 or 2, such as shown in Fig. 7.8, then the other unknown water depth can be calculated by solving a third-order algebraic equation, based on the constant value of E for the two cross sections. In the case of rectangular cross sections, the equation is given by

$$f(y_2) = y_2^3 - E y_2^2 + \frac{Q^2}{2gb_2^2} = 0 \tag{7.3.3}$$

If the contraction of the channel width is very significant, then the horizontal line that crosses the two curves of E in Fig. 7.9 does not cross the curve of E for cross section 2. Then the channel constriction acts as a *choke*. Under such conditions, Fr should be equal to 1 where b obtains its minimum value, namely, flow at cross section 2 is kept under critical conditions. Critical flow at cross section 2 and no loss of head between cross sections 1 and 2 dictate a rise of the free surface at cross section 1. In such a case, at cross section 1, the flow is always subcritical (Fr < 1), and the water level is often higher than its original value. The gain of head in cross section 1 is obtained by reducing the head loss upstream of the choke, due to the lower value of the flow velocity. Thus we may conclude that the flow downstream of the choke should be supercritical (with Fr > 1).

Elevation of the Channel Bottom

Figure 7.10 shows a rectangular channel with an elevated bottom (or hump). We again assume that there is no head loss between cross sections 1 and 2, as well as between cross sections 2 and 3, and neglect bottom slope. Referring to a datum at $z = 0$, which is located at the original elevation of the channel bottom, the total head at the elevated portion of the channel is

$$H = z + y + \frac{q^2}{2gy^2} \tag{7.3.4}$$

The negligible head loss along the elevated portion of the channel implies

$$\frac{dH}{dx} = 0 = \frac{dz}{dx} + \frac{dy}{dx}\left(1 - \frac{q^2}{gy^3}\right) = \frac{dz}{dx} + \frac{dy}{dx}(1 - \mathrm{Fr}^2) \tag{7.3.5}$$

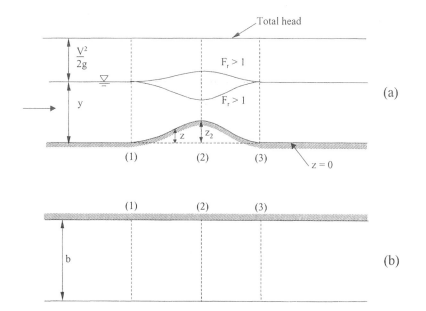

Figure 7.10 A channel with an elevated bottom: (a) side view; and (b) top view.

This result indicates that y increases with z if Fr is larger than 1, and it decreases when z increases if Fr is smaller than 1. Considering Fr > 1, the free surface of the water is elevated by an amount larger than the elevation of the channel bottom, since both z and y have positive derivatives with respect to x. If Fr is smaller than 1, then the derivatives of y and z with regard to x have opposite signs. Under such conditions, the coefficient of the derivative of y is smaller than 1, so the decrease of y is larger than the increase of z. Therefore the elevated bottom causes a depression of the water surface, provided that Fr < 1.

Equation (7.3.5) indicates that in cross section 3, if Fr is smaller than 1, the minimum value of y occurs where z obtains its maximum value. If Fr is larger than 1, then the maximum value of y occurs where z obtains its maximum value.

Figure 7.11 shows the relationships between y_1, y_2, and z_2 due to an elevated channel bottom. If the water depth is given at either cross section 1 or 2, it is possible to calculate the other water depth by solving a third-order algebraic equation, based on the constant total head for cross sections 1 and 2. In the case of rectangular cross sections, the equation is

$$f(y_2) = y_2^3 - (E_1 - z_2)y_2^2 + \frac{q^2}{2g} = 0 \tag{7.3.6}$$

If the elevation of the channel bottom is very significant, then the horizontal line that represents the value of E at cross section 2 in Fig. 7.11 does

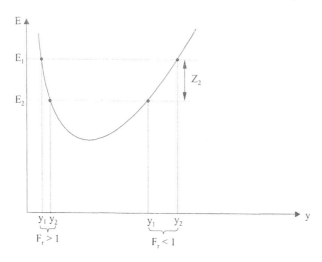

Figure 7.11 Relationships between y_1 and y_2 shown on the diagram of (E versus y) for the transition due to an elevated channel bottom.

not cross the E curve. Then the elevation of the channel bottom acts as a choke. Under such conditions, where z obtains its maximum value, Fr should be equal to 1, namely, critical flow conditions should prevail in cross section 2. Critical flow conditions at cross section 2 and no loss of head between cross sections 1 and 2 dictate a rise of the free water surface at cross section 1. In such a case, at cross section 1 the flow is always subcritical (Fr < 1) and the water head is often higher than its original value. The gain of head in cross section 1 is obtained by reducing the head loss upstream of the choke, due to the lower value of the flow velocity. Thus we may conclude that the flow downstream of the choke should be supercritical (with Fr > 1), similar to the channel contraction discussed above.

7.3.2 Lateral Outflow from the Channel

Figure 7.12 shows a channel with a side weir. The discharge over an interval dx of the weir is given by

$$dQ = C_1\sqrt{2g(y - W)^3}dx \qquad (7.3.7)$$

where C_1 is a coefficient of the weir and W is the weir crest elevation.

If the length L of the weir is small, then it may be assumed that the specific energy is constant along the weir. Applying Eq. (7.2.36), we obtain

$$\frac{dE}{dx} = 0 = \frac{dy}{dx} + \frac{Q}{gA}\frac{dQ}{dx} - \frac{Q^2B}{gA^3}\frac{dy}{dx} \qquad (7.3.8)$$

It should be noted that dQ in Eq. (7.3.8) is the negative value of the weir discharge given by Eq. (7.3.7). Substituting this value from Eq. (7.3.7) into Eq. (7.3.8) then gives

$$\frac{dy}{dx} = \frac{(Q/gA^2)C_1\sqrt{2g(y - W)^3}}{1 - \text{Fr}^2} \qquad (7.3.9)$$

Equation (7.3.9) indicates that if Fr < 1, then the water depth increases along the weir, as shown in Fig. 7.12a. If Fr > 1, then the water depth decreases along the weir, as shown in Fig. 7.12b. In the general case, Eq. (7.3.9) should be solved numerically, using an approach such as is discussed later in Sec. 7.6. In that section the problem of gradually varied flow in nonprismatic channels is addressed. The same basic approach applicable to nonprismatic channels can be used for cases of discharge varying along the channel as represented by Eq. (7.3.9).

If the channel has a rectangular cross section, then the channel flow rate at any cross section is given by

$$Q = by\sqrt{2g(E - y)} \qquad (7.3.10)$$

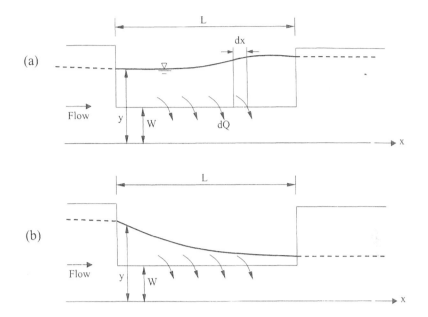

Figure 7.12 Lateral flow out of an open channel by a side weir: (a) case of subcritical flow; and (b) case of supercritical flow.

Using this expression to represent the Froude number, we have

$$\text{Fr}^2 = \frac{Q^2}{gb^2 y^3} = 2\left(\frac{E}{y} - 1\right) \tag{7.3.11}$$

Then, introducing Eqs. (7.3.10) and (7.3.11) into Eq. (7.3.9), we find

$$\frac{dx}{dy} = \frac{b}{2C_1} \frac{\sqrt{(E-y)(y-W)^3}}{(3y - 2E)} \tag{7.3.12}$$

Direct integration of Eq. (7.3.12) yields

$$x = \frac{C_1}{b}\left[\left(\frac{2E - 3W}{E - W}\right)\sqrt{\frac{E-y}{y-W}} - 3\sin^{-1}\sqrt{\frac{E-y}{y-W}}\right] + const \tag{7.3.13}$$

Employment of this solution may be difficult in cases of subcritical flow, where calculations of the water depth, as shown in Sec. 7.6, are performed in the upstream direction, starting from a control section located at the down-stream end of the channel reach. However, since the channel discharge downstream of the weir is not known prior to the calculation of the lateral flow out of the weir, trial-and-error calculations may be needed when using Eq. (7.3.13).

7.4 APPLICATION OF THE MOMENTUM CONSERVATION PRINCIPLE

As shown in Sec. 7.2, the principle of momentum conservation and the concept of the momentum function are related to the calculation of forces that act on a control volume of fluid in open channel flow. In cases of the calculation of transitions in open channels, where energy conservation is applied to calculate the water depth at the transition, the momentum conservation principle can be applied to calculate the forces acting on the constricted portion of the channel. However, the momentum conservation principle and the momentum function also are useful for the calculation of phenomena associated with significant energy losses. In such cases the energy conservation principle should follow the employment of the momentum conservation principle. Then, the momentum principle is used for the calculation of flow conditions at the entrance and exit of the particular control volume. Following this calculation, the energy conservation principle is applied to evaluate possible energy losses.

7.4.1 Forces Acting on a Channel Constriction

Fig. 7.13 shows two types of channel constrictions. Figure 7.13a refers to a rectangular channel, whose width varies from b_1 to b_2, and Fig. 7.13b refers to a rectangular channel, whose bottom is elevated by an amount Δz. In both cases, a force F is applied by the solid portion of the control volume through the solid surfaces that bound the control volume. The direction of F is shown in the (assumed) positive x-direction.

Using the momentum conservation principle, ignoring the effect of τ and assuming zero bottom slope, Eq. (7.2.49) gives

$$F = \Omega_2 - \Omega_1 \tag{7.4.1}$$

The principle of energy conservation provides the value of y_2 for a given value of y_1. Then, Eq. (7.4.1) yields

$$F = \rho g(b_2 M_2 - b_1 M_1) \tag{7.4.2}$$

where M is defined in Eq. (7.2.51).

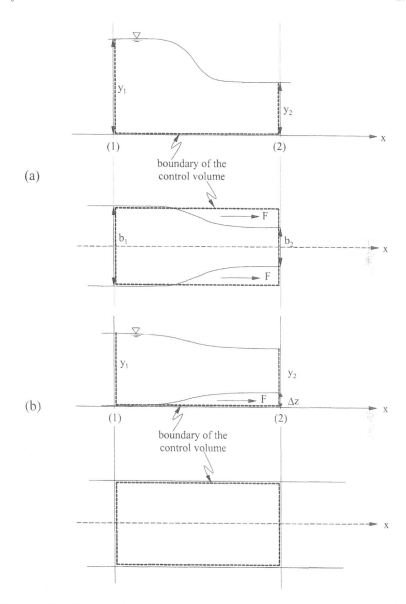

Figure 7.13 Forces acting on a channel constriction: (a) change in channel width; and (b) elevation of channel bottom.

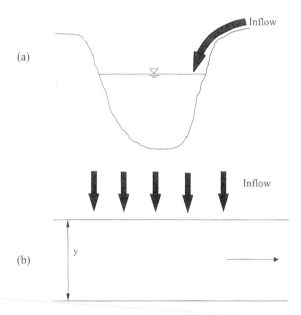

Figure 7.14 Lateral flow into an open channel: (a) channel cross section; and (b) longitudinal cross section of the channel.

7.4.2 Lateral Flow into an Open Channel

Figure 7.14 shows a schematic of lateral flow into an open channel. Such an inflow is usually associated with significant energy losses. Therefore even for a minor portion of the channel, we cannot consider negligible energy loss. On the other hand, if the lateral inflow takes place along a short section of the channel, we may ignore the effect of the bottom shear stress in that portion of the channel. Under this assumption, the momentum function is constant along that portion of the channel subject to lateral inflow, so that

$$\frac{d\Omega}{dx} = 0 \tag{7.4.3}$$

Using the definition for the *momentum function*, given by Eq. (7.2.50), this becomes

$$\frac{dy}{dx} = -\frac{(2Q/gA^2)(dQ/dx)}{1 - \text{Fr}^2} \tag{7.4.4}$$

Equation (7.4.4) is a nonlinear ordinary differential equation, which can be solved using an appropriate numerical procedure. Several possible numerical procedures for this purpose are presented in Sec. 7.6.

7.4.3 The Formation of Surges and the Hydraulic Jump

This topic is presented in Sec. 8.7, as a special case of the formation of positive waves in an open channel. Both surges and hydraulic jumps are characterized by an abrupt change in water depth, which is associated with a transition from supercritical flow to subcritical flow. The surge is a wave subject to movement in the channel, whereas the hydraulic jump is a stationary wave. However, in both cases energy losses are significant. The relationship between the water depths upstream and downstream of the wave can be calculated using the principle of momentum conservation, as explained in Sec. 8.7.

7.5 VELOCITY DISTRIBUTION IN OPEN CHANNEL FLOW

7.5.1 Use of Navier–Stokes Equations for Unidirectional Turbulent Flow

The velocity distribution in open channel flow is affected by the shape of the channel bed. Often, secondary currents develop due to effects originating from the specific geometry of the channel cross section, side walls, curvature of the channel course, etc. These situations are complicated and site specific, and in the present section we consider uniform flow in a wide, straight rectangular channel.

Calculation of the velocity profile starts with the two-dimensional Navier–Stokes equations of motion and the equation of continuity (refer to Chap. 2), with x and z representing the longitudinal and upward vertical directions, respectively. These equations are given by

$$\frac{\partial u}{\partial t} + u\frac{\partial u}{\partial x} + w\frac{\partial u}{\partial z} = -\frac{1}{\rho}\frac{\partial p}{\partial x} + v\left(\frac{\partial^2 u}{\partial x^2} + \frac{\partial^2 u}{\partial z^2}\right) \qquad (7.5.1a)$$

$$\frac{\partial w}{\partial t} + u\frac{\partial w}{\partial x} + w\frac{\partial w}{\partial z} = -\frac{1}{\rho}\frac{\partial p}{\partial z} + v\left(\frac{\partial^2 w}{\partial x^2} + \frac{\partial^2 w}{\partial z^2}\right) - g \qquad (7.5.1b)$$

$$\frac{\partial u}{\partial x} + \frac{\partial w}{\partial z} = 0 \qquad (7.5.2)$$

where u and w are the horizontal and vertical components of the velocity vector, respectively, t is time, p is piezometric pressure, which is equal to the piezometric head multiplied by the specific weight of the fluid, ρ is the density of the fluid, and v is the kinematic viscosity of the fluid.

By multiplying Eq. (7.5.2) by u and combining the resulting expression with Eq. (7.5.1a), we find

$$\frac{\partial u}{\partial t} + \frac{\partial u^2}{\partial x} + \frac{\partial(uw)}{\partial z} = -\frac{1}{\rho}\frac{\partial p}{\partial x} + v\left(\frac{\partial^2 u}{\partial x^2} + \frac{\partial^2 u}{\partial z^2}\right) \qquad (7.5.3a)$$

Similarly, multiplying Eq. (7.5.2) by v and combining with Eq. (7.5.1b) gives

$$\frac{\partial v}{\partial t} + \frac{\partial(wu)}{\partial x} + \frac{\partial w^2}{\partial z} = -\frac{1}{\rho}\frac{\partial p}{\partial z} + v\left(\frac{\partial^2 w}{\partial x^2} + \frac{\partial^2 w}{\partial z^2}\right) - g \qquad (7.5.3b)$$

Since the flow is assumed to be turbulent, the velocity components and pressure in the flow domain can be written as (see Chap. 5)

$$u = U + u' \qquad w = w' \qquad p = P + p' \qquad (7.5.4)$$

where U is the average longitudinal flow velocity, u' and w' are the velocity fluctuations in the longitudinal and vertical direction, respectively, P is the average pressure, and p' is the pressure fluctuation.

Following similar procedures as outlined in Chap. 5, introducing Eq. (7.5.4) into Eq. (7.5.2) and averaging the resulting equation shows that

$$\frac{\partial U}{\partial x} = 0 \qquad (7.5.5)$$

Therefore variations of mean flow velocity in the x-direction are negligible. Also, introducing Eqs. (7.5.4) and (7.5.5) into Eq. (7.5.3) and averaging the resulting expressions gives, respectively,

$$\frac{\partial U}{\partial t} + \frac{\partial(u'w')_{\text{av}}}{\partial z} = -\frac{1}{\rho}\frac{\partial P}{\partial x} + v\left(\frac{\partial^2 U}{\partial z^2}\right) \qquad (7.5.6a)$$

$$0 = -\frac{1}{\rho}\frac{\partial P}{\partial z} - g \qquad (7.5.6b)$$

where the subscript "av" refers to the average value.

Mostly, interest is for steady flow conditions. Therefore the first term of Eq. (7.5.6a) vanishes and

$$\frac{dP}{dx} = \frac{d}{dz}\left[\mu\frac{dU}{dz} - \rho(u'w')_{\text{av}}\right] \qquad (7.5.7)$$

Direct integration of Eq. (7.5.7) yields

$$y\frac{dP}{dx} + C = \mu\frac{dU}{dz} - \rho(u'w')_{\text{av}} \qquad (7.5.8)$$

where C is an integration constant.

The first term of Eq. (7.5.8) is negligible in comparison to the right-hand side terms. The first and second terms on the right-hand side of Eq. (7.5.8) represent the viscous and turbulent shear stresses, respectively. At the bottom of the channel, the turbulent fluctuations become extremely small and the last

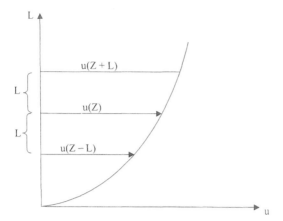

Figure 7.15 Schematic description of the velocity profile and the mixing length.

term on the right-hand side of Eq. (7.5.8) becomes negligible. Therefore the constant of integration of Eq. (7.5.8) is equal to the bottom shear stress.

7.5.2 Turbulent Flow in a Smooth Channel

Figure 7.15 shows a schematic description of the velocity profile. At the level z, the fluid particle fluctuations and turbulent vortices are characterized by a mixing length, L. If, due to a positive fluctuation w', the fluid particle jumps from level z to level $z + L$, it causes a local negative longitudinal fluctuation $(-u')$, as its momentum is lower than the typical momentum of fluid particles located at level $z + L$. Due to mass conservation, the absolute values of u' and w' should be close to each other. Also, the fluctuations u' should be similar to the difference between longitudinal velocities at levels $z + L$ and z. Therefore

$$|u'| \approx |w'| \approx L\frac{dU}{dz} \tag{7.5.9}$$

The mixing length value depends on the distance from the solid wall, namely the channel bottom. For most practical cases of open channel flow, it is possible to assume that the mixing length is proportional to z, with the coefficient of proportionality equal to the *Von Karman constant*, κ, with a nominal value of about 0.4. Introducing this relationship into Eq. (7.5.9) and then substituting into Eq. (7.5.8), we obtain

$$\tau_0 = \mu\frac{dU}{dz} + \rho\kappa^2 z^2 \left(\frac{dU}{dz}\right)^2 \tag{7.5.10}$$

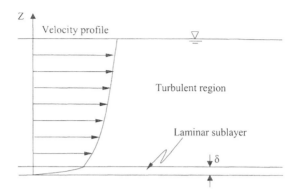

Figure 7.16   Schematic description of the laminar sublayer and the turbulent region.

If the channel bottom is very smooth, and Re is not extremely high, then the velocity profile can be considered to comprise two regions, as shown in Fig. 7.16. One region is located at the channel bottom and is called the laminar sublayer. The thickness of this region is δ and its value is of the order of magnitude 0.1 mm. In the laminar sublayer, the turbulent vortices are very small and velocity fluctuations are negligible. Therefore in the laminar sublayer, the last term of Eq. (7.5.10) can be neglected. Overlaying the laminar sublayer is the turbulent region. In the turbulent region, the effect of viscous forces is negligible. Therefore, in this region, the first term on the right-hand side of Eq. (7.5.10) can be neglected.

First consider the laminar sublayer. For this layer, Eq. (7.5.10) can be written in nondimensional form as (neglecting the turbulence term)

$$u^+ = z^+ \tag{7.5.11}$$

where

$$u^+ = \frac{U}{u_*} \qquad z^+ = \frac{zu_*}{v} \qquad u_* = \sqrt{\frac{\tau_0}{\rho}} \tag{7.5.12}$$

Here, u_* is the *shear velocity*, u^+ is the dimensionless velocity, and z^+ is the dimensionless distance from the bottom of the channel. Experimental data have shown that the dimensionless thickness of the laminar sublayer is

$$\delta^+ = \frac{\delta u_*}{v} = 11.6 \tag{7.5.13}$$

According to this result, Eq. (7.5.11) is thus applicable for z^+ values in the range

$$0 \le z^+ \le 11.6 \tag{7.5.14}$$

In the turbulent region, the first term on the right-hand side of Eq. (7.5.10) is ignored and, in dimensionless form, we have

$$du^+ = \frac{1}{\kappa} \frac{dz^+}{z^+}$$

(7.5.15)

where the dimensionless variables have the same meanings as before. Considering that $\kappa = 0.4$, and integrating Eq. (7.5.15), we obtain

$$u^+ = 2.5 \ln z^+ + C$$

(7.5.16)

where C is an integration constant.

At the top of the laminar sublayer, where $z^+ = \delta^+ = 11.6$ (Eq. 7.5.14), the velocity distributions of Eqs. (7.5.11) and (7.5.16) should match, with identical values of u^+. Therefore the value of the constant C of Eq. (7.5.15) must be

$$C = \delta^+ - 2.5 \ln \delta^+$$

(7.5.17)

By introducing Eqs. (7.5.17) and (7.5.13) into Eq. (7.5.16), we obtain

$$u^+ = 2.5 \ln z^+ + 5.5$$

(7.5.18)

Now, the dimensionless discharge per unit width of the channel is given by

$$q^+ = \frac{q}{\nu} = \int_0^{H^+} u^+ dz^+$$

(7.5.19)

where H^+ is the dimensionless water depth. This equation can be integrated, by substituting Eq. (7.5.18) for u^+. Then the dimensionless average flow velocity is obtained by dividing q^+ by H^+:

$$V^+ = 2.5 \ln(H^+) + 3$$

(7.5.20)

Recalling that the hydraulic radius of a wide rectangular channel is equal to the water depth, Eqs. (7.2.23) and (7.2.21) are rewritten as

$$S_f = \frac{f}{4H} \frac{V^2}{2g} \qquad \tau_0 = \rho g(4H)S_f$$

(7.5.21)

where H is the dimensional depth. Substituting in terms of V^+ then gives

$$V^+ = \sqrt{\frac{8}{f}}$$

(7.5.22)

In addition, the dimensionless water depth can be expressed using the Reynolds number and the friction coefficient, as

$$H^+ = \frac{H u_*}{\nu} = \frac{4HV}{\nu} \left(\frac{1}{4}\right)\left(\frac{u_*}{V}\right) = \mathrm{Re}\sqrt{f}\frac{1}{8\sqrt{2}} \tag{7.5.23}$$

By introducing Eqs. (7.5.22) and (7.5.23) into Eq. (7.5.20), we obtain

$$\frac{1}{\sqrt{f}} = 2.035 \log_{10}(\mathrm{Re}\sqrt{f}) - 1.084 \tag{7.5.24}$$

However, observations concerning smooth turbulent flow in open channels lead to a minor adjustment of the constants in this result, to allow closer fits to observations, giving

$$\frac{1}{\sqrt{f}} = 2 \log_{10}(\mathrm{Re}\sqrt{f}) - 0.796 \tag{7.5.25}$$

This expression is identical to a similar result that has been derived for turbulent pipe flow, except the constant at the end is 0.8 instead of 0.796.

The criterion for *smooth* turbulent flow in an open channel is that the roughness of the channel should be smaller than about 43% of the thickness of the laminar sublayer, or

$$\frac{k}{\delta} \le 0.43 \tag{7.5.26}$$

Using the definition of the sublayer thickness from Eq. (7.5.13), this criterion also can be written as

$$\varepsilon^+ \le 5 \tag{7.5.27}$$

7.5.3 Transition and Rough Turbulent Flow in Open Channels

If the roughness projections at the solid boundary are larger than the laminar sublayer thickness, the flow is no longer characterized as smooth flow. A transition regime exists between smooth and rough flow, until the roughness is larger than about seven times the possible thickness of the laminar sublayer, at which point the flow is fully rough and all effects of the laminar sublayer vanish. The flow is then controlled by the roughness of the channel. Equation (7.2.29) provides the flow definition in terms of ε^+.

In cases of $\varepsilon^+ > 5$, when the flow is either in a transition state or fully turbulent, the exact structure of the velocity profile at the channel wetted perimeter is not known exactly. However, in the turbulent region, due to the dominant inertial forces, the logarithmic velocity profile is preserved. We may apply Eq. (7.5.18) to the free water surface, to obtain a relationship between U_0^+ and H^+, where U_0 is the velocity at the free water surface. From this expression we subtract Eq. (7.5.18), to obtain

$$U_0^+ - u^+ = 2.5 \ln \left(\frac{H}{z} \right) \tag{7.5.28}$$

This is modified by dividing both the numerator and the denominator in the logarithmic term by k:

$$U_0^+ - u^+ = 2.5 \ln \left(\frac{H/k}{z/k} \right) \tag{7.5.29}$$

or

$$u^+ = U_0^+ - 2.5 \ln \left(\frac{H}{k} \right) + 2.5 \ln \left(\frac{z}{k} \right) \tag{7.5.30}$$

In the transition range between smooth and rough turbulent flows, the sum of the first and second terms on the right-hand side of Eq. (7.5.30) is subject to variations with regard to the value of ε^+. Under conditions of *rough turbulent flow*, that sum becomes constant, with an experimentally determined value of 8.5, so that Eq. (7.5.30) is modified as

$$u^+ = 2.5 \ln \left(\frac{z}{k} \right) + 8.5 \tag{7.5.31}$$

The average flow velocity is then found by integrating over the depth,

$$V^+ = 2.5 \ln \left(\frac{H}{k} \right) + 6.0 \tag{7.5.32}$$

Using Eq. (7.5.21), with minor modifications made in the coefficient values to comply with empirical observations, Eq. (7.5.32) leads to

$$\frac{1}{\sqrt{f}} = 2 \log_{10} \left(\frac{12 R_h}{\varepsilon} \right) \tag{7.5.33}$$

This expression is quite well correlated with the approximation used by the Manning equation.

The *Colebrook equation* for open channel flow gives an expression for the friction factor and is obtained by combining Eqs. (7.5.25) and (7.5.33):

$$\frac{1}{\sqrt{f}} = -2 \log_{10} \left(\frac{\varepsilon}{12 R_h} + \frac{2.5}{Re \sqrt{f}} \right) \tag{7.5.34}$$

This result provides an interpolating function between smooth and fully rough conditions, and is applicable to the entire range of smooth and rough turbulent flow, as well as for the transition region between these two regimes.

7.6   GRADUALLY VARIED FLOW

7.6.1   Control Sections

Steady flow in open channels is generally uniform, provided that the relevant portion of the channel is far from particular sections, where flow conditions are not connected with parameters of the open channel flow along the major portion of the channel. In those particular cross sections, called *control sections*, the water depth is determined either solely by the channel discharge or by external conditions. As an example, consider the water depth in the control section just upstream of a waterfall, where the water depth is at its critical value. Upstream of the waterfall, the channel flow is subcritical. The particular cross section, at which the water depth is determined only by the discharge, is a cross section of critical flow, where Fr = 1. Such a cross section is often called a *choke*. However, as shown in Fig. 7.17, there are cases of control sections in which flow is not necessarily critical.

In the portion of the open channel that is close to a control section, the flow is not uniform, and the water depth is subject to gradual variation. As shown in the following, calculations of water depth variations always start with a known water depth at the control section. Such calculations lead to the *hydraulic profile* or *backwater curve* of the flow. The analysis is most often concerned with steady-state conditions. If the flow is subcritical in the portion of the channel located upstream of the control, calculations of the hydraulic profile proceed from the control section in the upstream direction. If the flow downstream of the control section is supercritical, then the calculation of the hydraulic profile proceeds from the control section in the downstream direction.

7.6.2   The Differential Equations of Gradually Varied Flow

In steady, gradually varied flow, due to the gradual variation of the water depth and flow velocity, the friction slope, S_f, also is subject to gradual variations. From Eq. (7.2.15), the friction slope is related to parameters of the flow by

$$S_f = -\frac{dH}{dx} = -\frac{d}{dx}\left(z + y + \frac{V^2}{2g}\right) = S_0 - \frac{dE}{dx} \qquad (7.6.1)$$

Normal differentials are used in Eq. (7.6.1), since the variables are not considered to be functions of time here.

For a prismatic channel, the rate of change of E with x is

$$\frac{dE}{dx} = \frac{d}{dx}\left(y + \frac{Q^2}{2gA^2}\right) = \frac{dy}{dx}(1 - \mathrm{Fr}^2) \qquad (7.6.2)$$

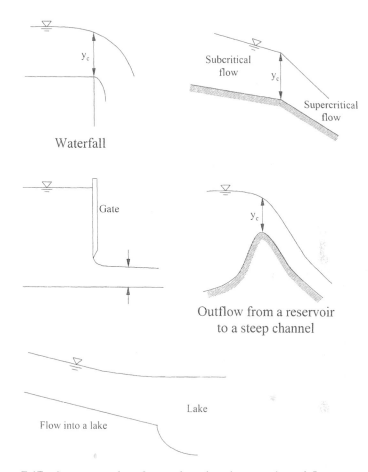

Figure 7.17 Some examples of control sections in open channel flow.

Thus Eq. (7.6.1) can be rewritten as

$$\frac{dy}{dx} = \frac{S_0 - S_f}{1 - \text{Fr}^2} \tag{7.6.3}$$

Expressions for S_0 and S_f can be obtained using the Manning equation (7.2.34),

$$S_f = \left(\frac{Qn}{AR_h^{2/3}} \right)^2 \tag{7.6.4a}$$

$$S_0 = \left(\frac{Qn}{A_n R_{hn}^{2/3}} \right)^2 \tag{7.6.4b}$$

where the subscript n refers to *normal conditions*, or *uniform flow*. Equation (7.6.3) then becomes

$$\frac{dy}{dx} = S_0 \frac{1 - (A_n R_{hn}^{2/3}/AR^{2/3})^2}{1 - \text{Fr}^2} \tag{7.6.5}$$

which may be integrated to obtain the hydraulic profile (y as a function of x). Since the Froude number for critical flow, $\text{Fr}_c = 1$, we can write

$$\text{Fr}^2 = \frac{\text{Fr}^2}{\text{Fr}_c^2} = \frac{Q^2 y/gA^3}{Q^2 y_c/gA_c^3} = \frac{A_c^3 y}{A^3 y_c} \tag{7.6.6}$$

Substituting this into Eq. (7.6.5) then gives an alternative form for the differential equation of the hydraulic profile,

$$\frac{dy}{dx} = S_0 \frac{1 - (A_n R_{hn}^{2/3}/AR^{2/3})^2}{1 - (A_c/A)^3 (y/y_c)} = f(x, y) \tag{7.6.7}$$

In the general case, values of basic geometrical parameters of the channel cross section are not necessarily constant along the channel. As an example, consider changes in the width of the channel bottom. In this case, the derivative of y with respect to x is in general a function of x and y. If geometric variables of the channel can be expressed as functions of y alone, then the channel is called a *prismatic channel*, as previously noted. Equation (7.6.7) indicates that, in cases of prismatic channels, the derivative of y with respect to x is a function of y only.

Equation (7.6.5) or (7.6.7) is a first-order nonlinear differential equation. Such a differential equation represents an initial value problem. It requires a single boundary condition at $x = 0$. As noted previously, the calculation of the hydraulic profile starts at the boundary of a control section, where the water depth is determined prior to the calculation of the profile. The calculation itself is commonly carried out using a numerical code, which usually provides a sufficiently accurate solution and description of the profile.

It should be noted that, in cases of prismatic channels, Eq. (7.6.7) can be represented as

$$\frac{dx}{dy} = f_1(y) \tag{7.6.8}$$

This equation can be integrated numerically to provide

$$\Delta x = \int_{y_0}^{y_1} f_1 \, dy \tag{7.6.9}$$

This provides a sort of inverted approach, where changes in distance along the channel are calculated as a function of changes in water depth, and in some applications this may be preferable (see Sec. 7.6.5).

7.6.3 General Forms of the Differential Equation for a Wide Rectangular Channel

In order to study the basic types of hydraulic profiles, without losing the general features of Eq. (7.6.7), we consider a wide rectangular channel. Recall that q = discharge per unit width is used, rather than the total discharge, and that the hydraulic radius is taken as the water depth.

Referring to Eqs. (7.6.3)–(7.6.7), the basic differential equation of the hydraulic profile can be expressed in several different formats for a wide rectangular channel:

$$\frac{dy}{dx} = \frac{S_0 - S_f}{1 - \text{Fr}^2} \tag{7.6.10a}$$

$$\frac{dy}{dx} = \frac{S_0 - \left(\dfrac{qn}{y^{5/3}}\right)^2}{1 - \text{Fr}^2} \tag{7.6.10b}$$

$$\frac{dy}{dx} = S_0 \frac{1 - \left(\dfrac{y_n}{y}\right)^{10/3}}{1 - \text{Fr}^2} \tag{7.6.10c}$$

$$\frac{dy}{dx} = S_0 \frac{1 - \left(\dfrac{y_n}{y}\right)^{10/3}}{1 - \left(\dfrac{y_c}{y}\right)^{3}} \tag{7.6.10d}$$

Either of the different forms of Eq. (7.6.10) can be used for the basic analysis of hydraulic profiles.

7.6.4 Various Types of Hydraulic Profiles

The major groups of hydraulic profiles are classified according to slope as follows:

> Mild slope, in which $y_n > y_c$ (*type M profiles*)
> Steep slope, in which $y_n < y_c$ (*type S profiles*)
> Horizontal slope, in which $S_0 = 0$ and $y_n \to \infty$ (*type H profiles*)
> Critical slope, in which $y_n = y_c$ (*type C profiles*)
> Adverse slope, in which, $S_0 < 0$ (*type A profiles*)

Each of these profiles is illustrated in Fig. 7.18 and can be analyzed using Eq. (7.6.10).

As shown in Fig. 7.18a, there are three types of M profiles. The, *type M_1* profile is associated with

$$y \geq y_n \tag{7.6.11}$$

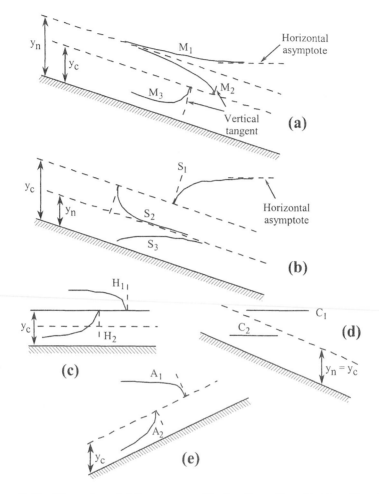

Figure 7.18 Illustration of different types of hydraulic profiles: (a) *M* profiles; (b) *S* profiles; (c) *H* profiles; (d) *C* profiles; and (e) *A* profiles.

Equation (7.6.11) indicates that the numerator and denominator of Eq. (7.6.10d) are positive. Therefore the water depth increases with *x*. There are two extreme cases to consider: (1) what happens when values of *y* approach the normal depth, and (2) what happens when *y* approaches infinity. By applying Eq. (7.6.10d), it can be shown that

$$\frac{dy}{dx} \to 0 \text{ as } y \to y_n \tag{7.6.12a}$$

and

$$\frac{dy}{dx} \to S_0 \quad \text{as} \quad y \to \infty \qquad (7.6.12b)$$

Equation (7.6.12a) indicates that the free surface asymptotically approaches y_n, at which point the flow becomes uniform. Equation (7.6.12b) indicates that the water surface asymptotically approaches a horizontal orientation when y becomes very large.

The *type* M_2 profile is associated with

$$y_c \le y \le y_n \qquad (7.6.13)$$

This indicates that the numerator of Eq. (7.6.10d) is negative and the denominator is positive. Therefore the water depth decreases with x. As with *type* M_1, there are two extreme cases to consider: (1) what happens when y approaches the normal depth, and (2) what happens when y approaches the critical depth. Again applying Eq. (7.6.10d), we find

$$\frac{dy}{dx} \to 0 \quad \text{as} \quad y \to y_n \qquad (7.6.14a)$$

and

$$\frac{dy}{dx} \to \infty \quad \text{as} \quad y \to y_c \qquad (7.6.14b)$$

Equation (7.6.14a) indicates that the free surface asymptotically approaches y_n, at which point the flow becomes uniform. Equation (7.6.14b) indicates that the water surface becomes vertical where the water depth approaches the critical depth.

The *type* M_3 profile is associated with

$$y \le y_c \qquad (7.6.15)$$

In this case, both the numerator and denominator of Eq. (7.6.10d) are negative. Therefore the water depth increases with x. There is a single extreme case to consider here, concerning the behavior as y approaches the critical depth. By applying Eq. (7.6.10d), we find

$$\frac{dy}{dx} \to \infty \quad \text{at} \quad y \to y_c \qquad (7.6.16)$$

Thus the free water surface becomes vertical as y approaches y_c.

As shown in Fig. 18b there are three types of S profiles. The *type* S_1 profile is associated with

$$y_c \le y \qquad (7.6.17)$$

In this case both the numerator and denominator of Eq. (7.6.10d) are positive. Therefore the water depth increases with x. There are two extreme cases to consider: (1) what happens when y approaches the critical depth, and (2) what happens when y becomes very large. By applying Eq. (7.6.10d), it can be shown that

$$\frac{dy}{dx} \to \infty \qquad \text{as} \qquad y \to y_c \tag{7.6.18a}$$

and

$$\frac{dy}{dx} \to S_0 \qquad \text{as} \qquad y \to \infty \tag{7.6.18b}$$

The *type S_2* profile is associated with

$$y_n \leq y \leq y_c \tag{7.6.19}$$

Then the numerator of Eq. (7.6.10d) is positive and the denominator is negative. Therefore the water depth decreases with x. Again, there are two extreme cases to consider: (1) what happens when y approaches the normal depth, and (2) what happens when values of y approach the critical depth. From Eq. (7.6.10d) we find

$$\frac{dy}{dx} \to 0 \qquad \text{at} \qquad y \to y_n \tag{7.6.20a}$$

and

$$\frac{dy}{dx} \to \infty \qquad \text{at} \qquad y \to y_c \tag{7.6.20b}$$

Equation (7.6.20a) indicates that the free surface asymptotically approaches the value of y_n, where the flow becomes uniform. Equation (7.6.20b) indicates that the water surface becomes vertical where the water depth approaches the critical depth.

The *type S_3* profile is associated with

$$y \leq y_n \tag{7.6.21}$$

This is associated with both the numerator and the denominator of Eq. (7.6.10d) being negative, so the water depth increases with x. There is a single extreme case to consider in this case: what happens when y approaches the normal depth. From Eq. (7.6.10d) we find

$$\frac{dy}{dx} \to 0 \qquad \text{at} \qquad y \to y_n \tag{7.6.22}$$

This indicates that the free water surface asymptotically approaches y_n, and at that value the flow becomes uniform.

Calculation of type H profiles should take into account that $S_0 = 0$. In this case, the numerator of Eq. (7.6.10b) is always negative. As shown in Fig. 7.18c, there are two types of H profiles.

The *type H_1* profile is associated with

$$y_c \leq y \qquad (7.6.23)$$

This result indicates that the denominator of Eq. (7.6.10b) is positive. Therefore the water depth decreases with x. There is a single extreme case to consider: what happens when y approaches the critical depth. By applying Eq. (7.6.10b) we find

$$\frac{dy}{dx} \to \infty \qquad \text{at} \qquad y \to y_c \qquad (7.6.24)$$

Thus when water depth approaches the critical depth, the water surface becomes vertical.

The *type H_2* profile is associated with

$$y \leq y_c \qquad (7.6.25)$$

In this case, the denominator of Eq. (7.6.10b) is negative. Therefore the water depth increases with x. There is a single extreme case to consider: what happens when values of y approach the critical depth. Again applying Eq. (7.6.10b), we find

$$\frac{dy}{dx} \to \infty \text{ at } y \to y_c \qquad (7.6.26)$$

This indicates that the free water surface approaches a vertical tangent as the value of y approaches y_c.

Calculation of type C profiles should take into account that $y_c = y_n$. Therefore the numerator and denominator of Eq. (7.6.10d) have almost identical values and, in any case of flow,

$$\frac{dy}{dx} \approx S_0 \qquad (7.6.27)$$

This suggests that the free water surface is almost horizontal, and water depth increases with x. As shown by Fig. 7.18c there are two types of C profiles:

The *type C_1* profile is associated with

$$y_c \leq y \qquad (7.6.28)$$

From Eqs. (7.6.27) and (7.6.28), the water depth is seen to increase with x until critical-normal water depth is obtained.

The *type C_2* profile is associated with

$$y \leq y_c \qquad (7.6.29)$$

Here, Eqs. (7.6.27) and (7.6.29) indicate that the water depth increases with x until the downstream end of the channel is reached.

Calculation of type A profiles should take into account that $S_0 < 0$ and there is no normal depth of flow in a channel with adverse slope. The numerator of Eq. (7.10d) is always negative. As shown in Fig. 7.18e, there are two types of A profiles.

The *type A_1* profile is associated with

$$y_c \leq y \tag{7.6.30}$$

When this is true, the denominator of Eq. (7.6.10b) is positive. Therefore the water depth decreases with x. There is a single extreme case to consider: what characterizes values of y approaching the critical depth. By applying Eq. (7.6.10b), we obtain

$$\frac{dy}{dx} \to \infty \quad \text{at} \quad y \to y_c \tag{7.6.31}$$

Thus the water surface becomes vertical where the water depth approaches the critical depth.

The *type A_2* profile is associated with

$$y \leq y_c \tag{7.6.32}$$

In this case, the denominator of Eq. (7.6.10b) is negative. Therefore the water depth increases with x. There is a single extreme case to consider: what happens when y approaches the critical depth. Using Eq. (7.6.10b), we find

$$\frac{dy}{dx} \to \infty \quad \text{at} \quad y \to y_c \tag{7.6.33}$$

This indicates that the free water surface becomes vertical as the value of y approaches y_c.

7.6.5 St. Venant Equations

As hinted at in the above discussion, in many problems of open channel flow it is sufficient to consider the mean flow velocity and depth, i.e., a one-dimensional (longitudinal) approach can supply the needed information. The St. Venant equations are commonly used for this purpose. They are based on continuity and momentum and represent a slight extension of previously described approaches in that possible changes in flow rate along the longitudinal direction are considered.

Consider a short section of a channel as shown in Fig. 7.19. Variables are defined at the center of the element as $Q = UA =$ flowrate, $U =$ mean velocity,

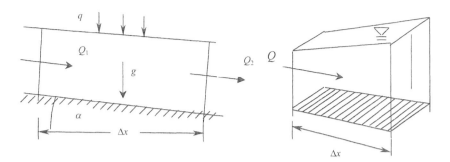

Figure 7.19 Control section used to develop St. Venant equations.

$A = BH$ = area, B = top width, y = depth, q = net inflow per unit length, Δx = length of the segment, $\forall$ = volume of the segment, α = channel slope angle (S_0 = bottom slope) and S_f = friction slope, as previously defined. Considering flows into and out of the section, a water mass balance may be written as

$$\frac{\partial \forall}{\partial t} = \Delta x \frac{\partial A}{\partial t} = \left(Q - \frac{\partial Q}{\partial x} \frac{\Delta x}{2}\right) - \left(Q + \frac{\partial Q}{\partial x} \frac{\Delta x}{2}\right) + q \, \Delta x$$

which results in

$$\frac{\partial A}{\partial t} = -\frac{\partial Q}{\partial x} + q \tag{7.6.34}$$

However, the rate of change of area can be expressed as

$$\frac{\partial A}{\partial t} = B \frac{\partial y}{\partial t} \tag{7.6.35}$$

which, when substituted into Eq. (7.6.34), gives

$$B \frac{\partial y}{\partial t} + \frac{\partial Q}{\partial x} = q \tag{7.6.36}$$

Using the definition of Q as the product of area and mean velocity, we have

$$B \frac{\partial y}{\partial t} + UB \frac{\partial y}{\partial x} + A \frac{\partial U}{\partial x} + y \frac{\partial B}{\partial x} = q \tag{7.6.37}$$

Finally, dividing by B and neglecting the last term on the left-hand side, which is equivalent to the usual wide channel assumption ($y/B \ll 1$),

$$\frac{\partial y}{\partial t} + U \frac{\partial y}{\partial x} + \frac{A}{B} \frac{\partial U}{\partial x} = \frac{q}{B} \tag{7.6.38}$$

This is the continuity equation for one-dimensional channel flow.

The momentum equation is developed using an integral approach that incorporates pressure forces acting on the side walls in the flow direction. Viscous effects are neglected. It is also assumed that the area centroid is approximately at depth $y/2$. The integral momentum equation for the channel segment in Fig. 7.19 is then

$$\sum F = \frac{\gamma}{2}\left(y - \frac{\partial y}{\partial x}\frac{\Delta x}{2}\right)\left(A - \frac{\partial A}{\partial x}\frac{\Delta x}{2}\right) - \frac{\gamma}{2}\left(y + \frac{\partial y}{\partial x}\frac{\Delta x}{2}\right)$$

$$\times \left(A + \frac{\partial A}{\partial x}\frac{\Delta x}{2}\right) + \frac{\gamma H^2}{2}\left[\left(B + \frac{\partial B}{\partial x}\frac{\Delta x}{2}\right) - \left(B - \frac{\partial B}{\partial x}\frac{\Delta x}{2}\right)\right]$$

$$+ \gamma \Delta x\, A\, \sin\alpha - \gamma \Delta x\, A S_f$$

$$= \frac{\partial}{\partial t}(\rho A\, \Delta x\, U) - \rho\left[\left(Q - \frac{\partial Q}{\partial x}\frac{\Delta x}{2}\right)\left(U - \frac{\partial U}{\partial x}\frac{\Delta x}{2}\right)\right.$$

$$\left. - \left(Q + \frac{\partial Q}{\partial x}\frac{\Delta x}{2}\right)\left(U + \frac{\partial U}{\partial x}\frac{\Delta x}{2}\right)\right]$$

which, when simplified, becomes

$$-\frac{g}{2}\left(y\frac{\partial A}{\partial x} + A\frac{\partial y}{\partial x} - y^2\frac{\partial B}{\partial x}\right) + Ag(S_0 - S_f) = \frac{\partial Q}{\partial t} + Q\frac{\partial U}{\partial x} + U\frac{\partial Q}{\partial x} \quad (7.6.39)$$

Using continuity, we substitute

$$\frac{\partial Q}{\partial t} = U\frac{\partial A}{\partial t} + A\frac{\partial U}{\partial t} = U\left(-\frac{\partial Q}{\partial x} + q\right) + A\frac{\partial U}{\partial t} \quad (7.6.40)$$

so that Eq. (7.6.39) becomes

$$A\frac{\partial U}{\partial t} + Q\frac{\partial U}{\partial x} + Uq = Ag(S_0 - S_f) - \frac{g}{2}\left(2A\frac{\partial y}{\partial x}\right) \quad (7.6.41)$$

Finally, dividing by A, we have

$$\frac{\partial U}{\partial t} + U\frac{\partial U}{\partial x} + \frac{U}{A}q = g\left(S_0 - S_f - \frac{\partial y}{\partial x}\right) \quad (7.6.42)$$

which is the desired momentum equation.

Simultaneous solution of Eqs. (7.6.38) and (7.6.42) provides a complete description of mean velocity, depth, and flow rate for the channel, at least within the one-dimensional framework. These are generally solved numerically, since variations in inflow rates and channel geometry can be incorporated directly (see below).

7.6.6 Numerical Calculation of the Hydraulic Profiles for Prismatic Channels

A general formulation of Eq. (7.6.10) for a prismatic channel can be written as

$$\frac{dy}{dx} = f(y) \tag{7.6.43}$$

With minor modification, this can be inverted in a form such as Eq. (7.6.8),

$$\frac{dx}{dy} = f_1(y) \tag{7.6.44}$$

where

$$f_1(y) = \frac{1}{f(y)} \tag{7.6.45}$$

Referring back to any of the various forms of Eq. (7.6.10), it is seen that, in general, $f(y)$ is a nonlinear function of y. Therefore Eq. (7.6.43) is described as a nonlinear ordinary first-order differential equation. This is thus an initial value problem, and a single boundary condition (an initial value) is required to obtain the distribution of y along the channel.

According to the required accuracy of the calculation of y, an appropriate method for the solution of Eq. (7.6.43) should be adopted. One of the most common methods is the *fourth-order Runge–Kutta method*. According to this method, the value of y_{i+1}, at the grid point x_{i+1}, is determined according to the value of y_i at the grid point x_i and an additional term depending on the grid interval Δx and several intermediate values of $f(y)$ evaluated between x_i and x_{i+1},

$$y_{i+1} = y_i + \tfrac{1}{6}[\Delta y_1 + 2\Delta y_2 + 2\Delta y_3 + \Delta y_4] \tag{7.6.46}$$

where

$$\Delta y_1 = \Delta x f(y_i) \qquad \Delta y_2 = \Delta x f\left(y_i + \frac{\Delta y_1}{2}\right) \tag{7.6.47}$$

$$\Delta y_3 = \Delta x f\left(y_i + \frac{\Delta y_2}{2}\right) \qquad \Delta y_4 = \Delta x f(y_i + \Delta y_3)$$

In the case of a prismatic channel, instead of solving Eq. (7.6.43), it is possible to solve Eq. (7.6.44). The solution of that equation requires a numerical integration. One possible method for accomplishing this is the *Simpson one-third method*. According to this method, the value of x_{i+1} is determined according to the value of x_i and an additional term depending on Δy and some intermediate values of $f_1(y)$, as

$$x_{i+1} = x_i + \frac{\Delta y}{3}[f_1(y_i) + 4f_1(y_i + \Delta y) + f_1(y_i + 2\Delta y)] \tag{7.6.48}$$

where Δy is a specified change in water depth, for which the appropriate value for Δx is sought. Specifically, it should be noted that at position x_i, the water depth is y_i, and at position x_{i+1}, the water depth is y_{i+1}, which is equal to $y_i + 2\,\Delta y$.

The method of calculation of y versus x by Eq. (7.6.48) is simpler and more accurate than that of Eq. (7.6.46). Also, it has some particular advantages with regard to the beginning of the hydraulic profile calculation. The value of $f(y)$ in Eq. (7.6.45) is infinite when the control section is subject to critical flow. Thus the calculation must be started at a point located close to the control section. Alternatively, at a control section of critical flow, $f_1(y)$ of Eq. (7.6.48) is equal to zero, and there is no difficulty in starting the calculation of the hydraulic profile at the control section itself. However, Eq. (7.6.48) should be used with care to make sure appropriate values of Δy are used, to avoid transition of the calculation from one type of hydraulic profile to another type of profile. In reality such a phenomenon may not occur in gradually varied flow.

7.6.7 Numerical Calculation of the Hydraulic Profiles for Nonprismatic channels

If the basic geometry of the channel cross section varies along the channel length, then the channel is nonprismatic and the equations of the previous section do not apply. As an example, we consider a trapezoidal channel with variable bottom width,

$$b = b(x) \tag{7.6.49}$$

The cross-sectional area of a nonprismatic channel is represented as

$$A = A[x,\, y(x)] \tag{7.6.50}$$

Since the basic geometry of the cross section varies along the channel, the value of y_c also varies along the channel, and there is no meaning of the term normal water depth.

Due to Eq. (7.6.49), the basic relationship of Eq. (7.6.2) is modified as

$$\frac{dE}{dx} = \frac{dy}{dx} + \frac{d}{dx}\left(\frac{Q^2}{2gA^2}\right) = \frac{dy}{dx} - \frac{Q^2}{gA^3}\left(\frac{\partial A}{\partial x} + \frac{\partial A}{\partial y}\frac{dy}{dx}\right)$$

$$= \frac{dy}{dx}\left(1 - \mathrm{Fr}^2\right) - \frac{\mathrm{Fr}^2}{B}\frac{\partial A}{\partial x} \tag{7.6.51}$$

By combining Eqs. (7.6.1) and (7.6.51), the differential equation describing the hydraulic profile is obtained as

$$\frac{dy}{dx} = \frac{S_0 - S_f - (\mathrm{Fr}^2/B)(\partial A/\partial x)}{1 - \mathrm{Fr}^2} = f(x,\, y) \tag{7.6.52}$$

From this result it may be seen that in nonprismatic channels the derivative of y with respect to x is a function of both x and y. Therefore this equation cannot be converted to a simple derivative of x with respect to y, although it can be converted to another differential equation that involves the derivative of x with respect to y.

The differential equation represented by Eq. (7.6.52) can be solved by an appropriate numerical procedure, provided that the water depth in a control section is given and used as an initial value for carrying out the calculation. The fourth-order Runge–Kutta method, as described in the previous section, can be used for these calculations. Similar to Eqs. (7.6.46) and (7.6.47), the general solution for y_{i+1} is given by

$$y_{i+1} = y_i + \tfrac{1}{6}[\Delta y_1 + 2\,\Delta y_2 + 2\,\Delta y_3 + \Delta y_4] \tag{7.6.53}$$

where

$$\Delta y_1 = \Delta x f(x_i, y_i) \quad \Delta y_2 = \Delta x f\left(x_i + \frac{\Delta x}{2},\, y_i + \frac{\Delta y_1}{2}\right)$$
$$\Delta y_3 = \Delta x f\left(x_i + \frac{\Delta x}{2},\, y_i + \frac{\Delta y_2}{2}\right) \quad \Delta y_4 = \Delta x f(x_i + \Delta x,\, y_i + \Delta y_3) \tag{7.6.54}$$

The basic format of Eqs. (7.6.53) and (7.6.54) also can be used for the solution of the differential equations given by Eqs. (7.3.9) and (7.4.4), namely, the equations concerning lateral flow out of or into a channel, respectively.

7.7   CIRCULATION IN LAKES AND RESERVOIRS

7.7.1   Introduction

Water motions in lakes and reservoirs are generally more complicated than in open channel flows. This is due to the much larger scales, both vertically and horizontally, as well as the greater variety of forces contributing to the velocity field. In addition to gravity, which is the main driving force in open channel flow, lakes are subject to wind shear stress, atmospheric pressure variations, river inflows and outflows, and convectively driven motions (due to buoyancy changes — see Chap. 13). The lake volume itself is variable, due to hydrologic factors controlling runoff, precipitation, and evaporation. In addition, if the lake is sufficiently large, Coriolis effects must be taken into consideration. Because of these factors, it is difficult or impossible to obtain analytical solutions to the equations governing the fluid motions, which in many cases should be written in full three-dimensional form for a complete description of the system. This necessitates the use of numerical solutions to model lakes and reservoirs, and previously described methods for solving the equations of motion (continuity and Navier–Stokes equations) may be applied, while accounting for appropriate boundary conditions.

For water quality modeling applications, there is usually interest in simulating the velocity and temperature fields using a hydrodynamic model. Results from the hydrodynamic model then provide the transport terms for a water quality model. The temperature field also is of interest since many environmental processes are temperature dependent. The hydrodynamic model is formulated by considering the various features that contribute to development of the velocity and temperature fields, as illustrated in Fig. 7.20. Wind exerts shear stress on the surface, which contributes to waves and currents. When the wind is relatively steady, wind setup and *seiche* motions may be generated (see Sec. 12.2), as the pressure distribution in the water body adjusts to balance the wind shear. Water surface variations also may arise in response to large-scale atmospheric pressure variations, if the lake has large enough horizontal extent. River inflows and outflows can strongly affect the local velocity field, and surface heat exchange can lead to convective motions that reach to the bottom of even very deep lakes. Ice cover also may be a factor in modeling a particular lake, as it affects the transfer of heat and momentum at the surface.

Because of the scale of most lakes, it is not practical to model directly the entire range of motions possible. A model must be chosen with a scale (i.e., a calculation spatial step) suitable to represent the motions of interest for a given application, and this must be balanced with available computer resources. For example, it would not be practical to model the local turbulence, with length scales on the order of several centimeters or smaller, in a lake that is several

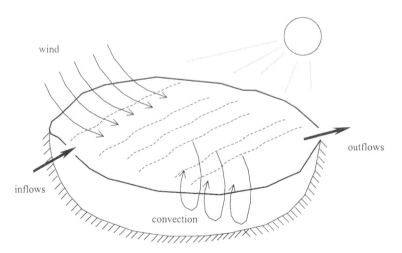

Figure 7.20   Forces contributing to development of the velocity field in a lake.

tens or hundreds of kilometers long. For any given model scale, it is then necessary to develop some sort of closure scheme to adequately represent processes that occur on scales smaller than the model scale. Alternatively, in many cases it is possible to assume that all flow variables are well mixed within the space represented by a model grid.

The simplest approach for modeling lakes is the "one-box" model, which considers the entire lake to be well mixed. In this case, force balances are applied to the lake as a whole to calculate setup, if required, and surface heat flux and inflows and outflows provide source and sink terms for whatever state variables are of interest (i.e., temperature, concentration, etc.). The basic water mass balance equation for this case is simply

$$\frac{d\forall}{dt} = \sum (Q_{\text{in}} - Q_{\text{out}}) + RO + (I - E - F)A \tag{7.7.1}$$

where $\forall$ is the lake volume, Q_{in} and Q_{out} are river inflow and outflow rates, respectively, RO is direct runoff volume rate, I is precipitation rate, E is evaporation rate, F is infiltration, or seepage rate, and A is the surface area of the lake. Each of the hydrologic variables I, E, and F are in units of length per time.

Equation (7.7.1) is easily modified to evaluate changes in other state variables, for example, mass concentration of a contaminant of interest (units of mass per unit volume). In that case, each of the terms in Eq. (7.7.1) would be multiplied by the appropriate value of concentration corresponding to that particular term. With the possible addition of biological/chemical reactions that may affect the contaminant mass, the equation is transformed into an expression for contaminant mass balance. This process is discussed further in Chaps. 10 and 16, with regard to surface water quality modeling and remediation of surface waters.

7.7.2 Horizontally Averaged Model for Temperature Distribution

Rather than the one-box model, the next higher level of complexity involves one-dimensional (vertical) models. These have provided much useful information, particularly with regard to reservoir operations, where vertical temperature distributions are of interest. In this case, as with the one-box models, the general velocity distribution is not of major interest, since all horizontal gradients are neglected. However, there is interest in the vertical extent of inflows and outflows, as they may affect the vertical temperature distribution. This is illustrated in Fig. 7.21, which also shows a discretization scheme that might be used in a finite difference model.

A one-dimensional temperature equation is obtained from Eq. (2.9.33) by integrating in the horizontal directions. First we assume, for simplicity,

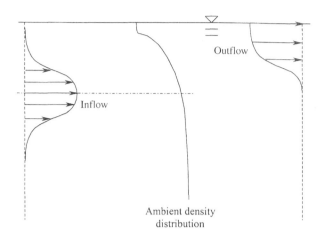

Figure 7.21 Side view of idealized one-dimensional (vertical) model, with inflow and outflow distributions.

that the coordinate system is oriented so that any flows into or out of the lake are in the x-direction, as shown in Fig. 7.21, and that the flow field is independent of x or y. This last assumption implies that we are considering the area-averaged velocity field. In principle, it is possible to account for variations in u and v, though this adds complexity in the integrations to follow that is unnecessary for the present discussion. Integrating Eq. (2.9.3) in the y-direction gives

$$\frac{\partial}{\partial T} \int_0^W T\,dy + u\frac{\partial}{\partial x} \int_0^W T\,dy + vT|_0^W + w\frac{\partial}{\partial z} \int_0^W T\,dy =$$
$$k_T \left[\frac{\partial^2}{\partial x^2} \int_0^W T\,dy + \frac{\partial^2}{\partial z^2} \int_0^W T\,dy \right] + k_T \frac{\partial T}{\partial y}\bigg|_0^W - \frac{1}{\rho c} \nabla \cdot \vec{\varphi}_r \int_0^W dy \tag{7.7.2}$$

where W is the width of the lake in the y-direction, c is the specific heat, k_T is the thermal diffusion coefficient (assumed constant), and φ_r is the radiation flux, which in this case is due to solar radiation (see Chap. 12 for further discussion). Only the vertical component of this term is relevant here. Now, the sum of the advective and diffusive fluxes in the y-direction,

$$\left(vT - k_T \frac{\partial T}{\partial y} \right)\bigg|_0^W$$

represents the total fluxes of temperature evaluated at the boundaries in the y-direction. These fluxes are zero, according to the assumed coordinate system orientation. Integrating Eq. (7.7.2) in the x-direction, we obtain

$$\frac{\partial}{\partial t}\int_0^L\int_0^W T\,dy\,dx + \left(u\int_0^W T\,dy\right)\Bigg|_0^L + w\frac{\partial}{\partial z}\int_0^L\int_0^W T\,dy\,dx =$$

$$\left(k_{\mathrm{T}}\frac{\partial}{\partial x}\int_0^W T\,dy\right)\Bigg|_0^L + K_{\mathrm{T}}\frac{\partial^2}{\partial z^2}\int_0^L\int_0^W T\,dy\,dx$$

$$-\frac{1}{\rho c}\frac{\partial\varphi_{\mathrm{rz}}}{\partial z}\int_0^L\int_0^W dy\,dx \qquad (7.7.3)$$

where φ_{rz} is the vertical component of φ_{r} and L is the length of the lake in the x-direction. Because the integral of T in the y-direction is simply the width-averaged temperature multiplied by W, the advective and diffusive terms in the x-direction can be combined as the total fluxes into and out of the lake at the two limits $x = 0$ and $x = L$:

$$\left[u\int_0^W T\,dy - k_{\mathrm{T}}\frac{\partial}{\partial x}\int_0^W T\,dy\right]\Bigg|_0^L = [(UT)_{\mathrm{in}} - (UT)_{\mathrm{out}}]W \qquad (7.7.4)$$

where U represents the total inflow or outflow velocity. Also, $(UT)_{\mathrm{out}}$ can be written as $U_{\mathrm{out}}T_{\mathrm{avg}}$, where T_{avg} is the area-averaged temperature,

$$T_{\mathrm{avg}} = \frac{1}{A}\int_0^L\int_0^W T\,dy\,dx \qquad (7.7.5)$$

Substituting Eqs. (7.7.4) and (7.7.5) into Eq. (7.7.3) then gives

$$\frac{\partial T_{\mathrm{avg}}}{\partial t} + w\frac{\partial T_{\mathrm{avg}}}{\partial z} = k_{\mathrm{T}}\frac{\partial^2 T_{\mathrm{avg}}}{\partial z^2} + [(UT)_{\mathrm{in}} - U_{\mathrm{out}}T_{\mathrm{avg}}]\frac{1}{L} - \frac{1}{\rho c}\frac{\partial\varphi_{\mathrm{rz}}}{\partial z}$$

$$+ \left\{\frac{k_{\mathrm{T}}}{A}\left(2\frac{\partial A}{\partial z}\frac{\partial T_{\mathrm{avg}}}{\partial z} + T_{\mathrm{avg}}\frac{\partial^2 A}{\partial z^2}\right) - \frac{wT_{\mathrm{avg}}}{A}\frac{\partial A}{\partial z}\right\} \qquad (7.7.6)$$

If W is not a function of z, then the term in the curly brackets of Eq. (7.7.6) becomes

$$+ \left\{\frac{1}{L}\left(k_{\mathrm{T}}\frac{\partial L}{\partial z}\frac{\partial T_{\mathrm{avg}}}{\partial z} - wT_{\mathrm{avg}}\frac{\partial L}{\partial z}\right)\right\} \qquad (7.7.7)$$

and if A is constant, the term in brackets is eliminated altogether.

The inflow is normally assumed to be centered around the vertical location at which the density (temperature) of the inflow is equal to the density in the lake or reservoir. Outflow is taken as near the surface in a natural lake, or at the withdrawal depth in a reservoir. Inflows and outflows are discussed further in Chap. 14.

7.7.3 Barotropic Models

A further level of complexity is obtained using a two-dimensional vertically averaged approach. Since variations in the vertical direction are not considered, this is usually known as a *shallow water approach*. This type of model may be used when vertical variations are not significant or not of interest. Because density variations are not incorporated, these types of models also are known as *barotropic models* (see also Chap. 9).

The vertically averaged continuity equation can most easily be developed by considering a control section as shown in Fig. 7.22, where H is the depth and U and V are vertically averaged velocities in the x- and y-directions, respectively:

$$U = \frac{1}{H} \int_0^H u \, dz; \qquad V = \frac{1}{H} \int_0^H v \, dz \tag{7.7.8}$$

Letting U, V, and H be defined at the center of the control section, which has dimensions dx and dy in the two horizontal directions, a water mass balance statement is written as

$$
\begin{aligned}
\rho \frac{\partial}{\partial t} (H \, dx \, dy) = {} & \rho \left(U - \frac{\partial U}{\partial x} \frac{dx}{2} \right) \left(H - \frac{\partial H}{\partial x} \frac{dx}{2} \right) dy \\
& - \rho \left(U + \frac{\partial U}{\partial x} \frac{dx}{2} \right) \left(H + \frac{\partial H}{\partial x} \frac{dx}{2} \right) dy \\
& + \rho \left(V - \frac{\partial V}{\partial y} \frac{dy}{2} \right) \left(H - \frac{\partial H}{\partial y} \frac{dy}{2} \right) dx \\
& - \rho \left(V + \frac{\partial V}{\partial y} \frac{dy}{2} \right) \left(H + \frac{\partial H}{\partial y} \frac{dy}{2} \right) dx \\
& + \rho (I - E - F) \, dx \, dy
\end{aligned}
\tag{7.7.9}
$$

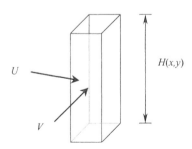

Figure 7.22 Control section used to develop vertically averaged model.

This is simplified and divided by $\rho\,dx\,dy$ to become

$$\frac{\partial H}{\partial t} + \frac{\partial(UH)}{\partial x} + \frac{\partial(VH)}{\partial y} = I - E - F \tag{7.7.10}$$

This same result can be obtained by integrating the continuity equation in the z-direction, incorporating the fact that $w = \partial H/\partial t$ at $z = H$, and accounting for I, E, and F, which may occur independently of U or V. In the case where H is constant and there is no precipitation, evaporation, or seepage, Eq. (7.7.10) reduces to the usual continuity expression,

$$\frac{\partial U}{\partial x} + \frac{\partial V}{\partial y} = 0 \tag{7.7.11}$$

The momentum equations are obtained in a similar manner, by vertically integrating. For purposes of illustration, we consider here the linearized versions of the Navier–Stokes equations, in which the advective acceleration terms are neglected:

$$\frac{\partial u}{\partial t} - fv = -\frac{1}{\rho}\frac{\partial p}{\partial x} + v_t\nabla^2 u \tag{7.7.12}$$

$$\frac{\partial v}{\partial t} + fu = -\frac{1}{\rho}\frac{\partial p}{\partial y} + v_t\nabla_2 v \tag{7.7.13}$$

where f is the *Coriolis parameter* and v_t is the horizontal eddy viscosity. Integrating Eqs. (7.7.12) and (7.7.13) over the depth H gives, respectively,

$$\frac{\partial(UH)}{\partial t} - fVH = -\frac{1}{\rho}\frac{\partial}{\partial x}\int_0^H p\,dz + v_t\left[\nabla_h^2(UH) + \frac{\partial u}{\partial z}\Big|_0^H\right] \tag{7.7.14}$$

$$\frac{\partial(VH)}{\partial t} + fUH = -\frac{1}{\rho}\frac{\partial}{\partial y}\int_0^H p\,dz + v_t\left[\nabla_h^2(VH) + \frac{\partial v}{\partial z}\Big|_0^H\right] \tag{7.7.15}$$

where ∇_h^2 is the horizontal Laplacian operator. If approximately hydrostatic conditions are assumed, then $p = \rho g(H - z)$. Also, the last terms on the right-hand sides of Eqs. (7.7.14) and (7.7.15), when multiplied by ρ, are the boundary shear stresses in each direction (at the surface and at the bottom). Thus

$$\frac{\partial(UH)}{\partial t} - fVH = -gH\frac{\partial H}{\partial x} + v_t\nabla_h^2(UH) + \frac{1}{\rho}(\tau_{sx} - \tau_{ox}) \tag{7.7.16}$$

$$\frac{\partial(VH)}{\partial t} + fUH = -gH\frac{\partial H}{\partial y} + v_t\nabla_h^2(VH) + \frac{1}{\rho}(\tau_{sy} - \tau_{oy}) \tag{7.7.17}$$

where τ_s is the surface stress, τ_o is the bottom stress, and subscripts x and y indicate in which direction the shear stress acts. For constant depth, both of these equations may be divided by H.

Eddy Viscosity

There are a number of models for the horizontal eddy viscosity, v_t. Experimental studies, in which the spreading of dye released at a point in the water body is monitored over time, are probably the most realistic, though also the most difficult and expensive, ways to obtain values for this parameter. The simplest analytical approach is to assume a constant value for v_t for the lake, usually as a function of wind stress. This has provided reasonable results for early lake and reservoir modeling, though it is not directly relevant in cases where the circulation and mixing are strongly affected by inflows and outflows. Most models express v_t as an increasing function of length.

From dimensional considerations, the eddy viscosity can be expressed as a product of appropriate length *(L)* and velocity *(v)* scales (also see discussion in Chap. 5). A simple *length scale model* for v_t is developed based on as assumption credited to Prandtl that the velocity scale should be proportional to the gradient of mean velocity as

$$V \approx L \frac{\partial U}{\partial x_i} \tag{7.7.18}$$

where x_i represents either y or z, depending on particular flow conditions, and the velocity gradient is understood to be a positive quantity. The eddy viscosity is then formulated as

$$v_t = c_L L^2 \frac{\partial U}{\partial x_i} \tag{7.7.19}$$

where c_L is a constant. A potential drawback to this formulation is in choosing appropriate values for c_L and L. These would often be chosen on the basis of fitting values of v_t to fit observations.

An alternative definition for the velocity scale is the square root of the turbulent kinetic energy (TKE), K, defined in Chap. 5. Then

$$v_t = c_k L \sqrt{K} \tag{7.7.20}$$

where c_k is a constant. This result is consistent with Eq. (5.5.14). Still, the problem of defining L remains and, as before, this may be considered as a fitting or calibration parameter. Using Eq. (7.7.20) also requires that the TKE equation (Eq. 5.5.21 or 5.5.23) be solved for K.

Still another approach is based on defining V in terms of the dissipation rate of TKE (ε) and results in the *4/3 power law* for eddy viscosity. Physically, this means that the inertial subrange eddies are responsible for the mixing. In this model, the diffusion process is characterized by ε and $L =$ the length scale of the diffusion process, usually taken as some proportion of the horizontal scale of the lake or, in the case of a dye release experiment, as distance from the dye release point. The velocity scale is defined by $(\varepsilon L)^{1/3}$, which when multiplied by L gives

$$v_{\mathrm{t}} \approx \varepsilon^{1/3} L^{4/3} \tag{7.7.21}$$

Equation (5.5.35) must be solved to obtain values for ε.

More complicated models have been developed for v_{t}, notably the two-equation $K-\varepsilon$ and Reynolds stress or *second-order closure models* discussed in Chap. 5. In particular, models based on solving the Reynolds stress terms (Eq. 5.5.28) have an advantage in that they can account for nonisotropic behavior. However, they also require solution of six additional equations, one for each of the stress components. The $K-\varepsilon$ models assume that the length scale is given, based on dimensional considerations, as

$$L = \frac{K^{3/2}}{\varepsilon} \tag{7.7.22}$$

Thus the eddy viscosity is given by

$$v_{\mathrm{t}} = c_{\mathrm{e}} \frac{K^2}{\varepsilon} \tag{7.7.23}$$

which is the same as eq. (5.5.16).

For large lake modeling v_{t} tends to be quite large, of order 100 m²/s, and in general it will be a function of wind and other forces generating motions in the water. Also, it is interesting to note that, as in Eq. (7.7.21), v_{t} increases with increasing length. Physically, this is explained because larger eddies participate in the diffusive process when larger horizontal extent is considered.

7.7.4 Three-Dimensional Models

As previously noted, three-dimensional models are needed for general applications. These involve solution of the full (nonaveraged) equations of motion. The problem of determining eddy viscosities remains, as with the barotropic models. In addition, vertical mixing must be accounted for, including shear-induced mixing as well as convectively driven motions. Coriolis terms also may be important in larger lakes and reservoirs.

Simpler three-dimensional models consider the lake as a series of layers, as illustrated in Fig. 7.23. At one extreme is the one-layer barotropic case described in the previous section. This can be generalized slightly to include two layers, which may be sufficient for cases in which there is a relatively thin well-defined thermocline that can represent the boundary between the two layers. Within each layer all fluid properties are vertically mixed, and each layer can be modeled as if it were a two-dimensional system. Interactions

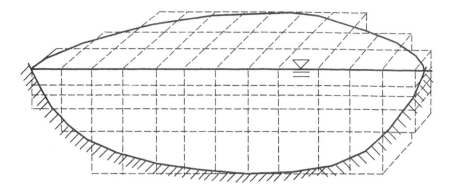

Figure 7.23 Layered three-dimensional discretization scheme (note that layers do not have to be of equal thickness).

between the layers also must be incorporated into the solution. Additional layers can be added to resolve finer variations in vertical structure, as needed for a given application. In the limit of a much larger number of layers, the problem is essentially fully three-dimensional. These problems generally require some care in designing and implementing an appropriate numerical solution, particularly since the horizontal scale may be much larger than the vertical scale. This introduces possible issues of numerical stability, for example. The reader is encouraged to review references listed at the back of this chapter for further information.

PROBLEMS

Solved Problems

Problem 7.1   Develop a version of the Manning equation appropriate for flow in a wide rectangular channel conveying a discharge per unit width q. Afterwards, use Eq. (7.2.35) for the development of the relationship between the discharge per unit width and the slope of the stable channel.

Solution

According to Eq. (7.2.4), the hydraulic radius of a wide rectangular channel is equal to the water depth. Therefore, from Eq. (7.2.34),

$$Q = qb = \frac{byy^{2/3}}{n}S_f^{1/2}$$

We divide this expression by the channel width to obtain

$$q = \frac{y^{5/3}}{n} S_f^{1/2} \tag{1}$$

Dividing both sides of Eq. (7.2.35) by the channel width also gives

$$y^{7/6} = \frac{qng^{1/2}}{u_{*c}} \tag{2}$$

According to Eq. (1),

$$S_f = \frac{q^2 n^2}{y^{10/3}} \tag{3}$$

We introduce the value of y from Eq. (2) into Eq. (3), to obtain

$$S_f = \frac{u_{*c}^{20/7}}{g^{10/7} q^{6/7} n^{6/7}}$$

This expression indicates that for a constant value of the critical shear stress and a constant value of Manning roughness coefficient, the slope of the channel is inversely proportional to the discharge raised to the power of 6/7.

Problem 7.2 Consider a wide rectangular channel of constant slope and constant Manning roughness coefficient. Develop an expression indicating how the bottom shear stress varies with the discharge per unit width in the channel.

Solution

Manning's equation for a wide rectangular channel is

$$y = \frac{q^{3/5} n^{3/5}}{S_f^{3/10}} \tag{1}$$

According to Eq. (7.2.21), for a wide rectangular channel,

$$\tau = \rho g y S_f \tag{2}$$

Introducing the value of y from Eq. (1) into Eq. (2),

$$\tau = \rho g q^{3/5} n^{3/5} S_f^{7/10}$$

This expression indicates that, for constant values of the slope and the roughness coefficient of Manning, the shear stress is proportional to the discharge per unit width raised to the power 3/5.

Problem 7.3 Consider a channel with trapezoidal cross section, that is to be designed to convey a discharge of $Q = 10$ m^3/s and to be free of erosion. The following geometrical variables are given: $b = 2$ m, $m = 3$. The roughness coefficient of Manning is $n = 0.025$. The critical shear stress is $\tau_c = 10$ Pa. Determine the water depth, the flow velocity, and the longitudinal slope of the channel.

Solution

In a channel free of erosion, according to Eq. (7.2.35),

$$AR_h^{1/6} = \frac{Qn}{\sqrt{\tau_c/(\rho g)}}$$

Substituting the numerical values of the parameters then leads to

$$AR_h^{1/6} = \frac{10 \times 0.025}{\sqrt{10/9810}} = 7.83$$

$$AR_h^{1/6} = \frac{(by + my^2)^{7/6}}{(b + 2y\sqrt{m^2 + 1})^{1/6}} = \frac{(2y + 3y^2)^{7/6}}{(2 + 2y\sqrt{10})^{1/6}}$$

Or

$$f(y) = \frac{(2y + 3y^2)^{7/6}}{(2 + 2y\sqrt{10})^{1/6}} - 7.83 = 0$$

This is solved iteratively to obtain $y = 1.35$ m. The cross-sectional area, wetted perimeter, and hydraulic radius are given, respectively, by

$$A = by + my^2 = 2 \times 1.35 + 3 \times 1.35^2 = 8.17 \text{ m}^2$$

$$P = b + 2y\sqrt{m^2 + 1} = 2 + 2 \times 1.35\sqrt{10} = 10.54 \text{ m}$$

$$R_h = \frac{R}{P} = \frac{8.17}{10.54} = 0.78 \text{ m}$$

The flow velocity and longitudinal slope of the channel are given by

$$V = \frac{Q}{A} = \frac{10}{8.17} = 1.22 \text{ m/s}$$

$$S_0 = \left(\frac{Vn}{R^{2/3}}\right)^2 = \left(\frac{1.22 \times 0.025}{0.78^{2/3}}\right)^2 = 1.3 \times 10^{-3}$$

Problem 7.4 Consider a channel with trapezoidal cross section, that should be designed to convey a discharge of $Q = 10$ m^3/s. It is required that the average flow velocity in the channel will be $V = 1.2$ m/s. The following geometrical variables are given: $b = 2$ m, $m = 3$. The roughness coefficient of Manning is $n = 0.025$. Determine the water depth, the longitudinal slope of the channel, and the bottom shear stress.

Solution

The cross-sectional area of the channel is given by

$$A = \frac{Q}{V} = \frac{10}{1.2} = 8.33 \text{ m}^2$$

The general equation of the cross-sectional area is

$$my^2 + by - A = 0$$

This is a second-order algebraic equation for the determination of y according to given values of b and A, which indicates that

$$y = \frac{-b + \sqrt{b^2 + 4mA}}{2m} = \frac{-2 + \sqrt{2^2 + 4 \times 3 \times 8.33}}{2 \times 3} = 1.37 \text{ m}$$

Therefore the wetted perimeter, the hydraulic radius, the channel slope, and the bottom shear stress are given, respectively, by

$$P = b + 2y\sqrt{m^2 + 1} = 2 + 2 \times 1.37 \times \sqrt{10} = 10.66 \text{ m}$$

$$R_h = \frac{A}{P} = \frac{8.33}{10.66} = 0.78 \text{ m}$$

$$S_0 = \left(\frac{Vn}{R_h^{2/3}}\right)^2 = \left(\frac{1.2 \times 0.025}{0.78^{2/3}}\right)^2 = 1.25 \times 10^{-3}$$

$$\tau = \rho g R_h S_0 = 9810 \times 0.78 \times 1.25 \times 10^{-3} = 9.56 \text{ Pa}$$

Problem 7.5 The width of a rectangular channel decreases from b_1 to b_2. The Froude number of the flow is smaller than 1. The discharge flowing through the channel is Q. Determine the minimum value of b_2, that does not cause an increase of the water depth upstream of the channel constriction.

Solution

In cross sections 1 and 2, the specific energy is identical. Its value is given by

$$E = y_1 + \frac{Q^2}{2gb_1^2 y_1^2} \tag{1}$$

In cross section 2, the flow is subject to critical flow conditions. Therefore

$$y_2 = \frac{2}{3}E \qquad \text{Fr}_2^2 = \frac{Q^2}{gb_2^2 y_2^3} = 1 \tag{2}$$

Equations (1) and (2) are then combined to give

$$b_2 = Q\sqrt{\frac{1}{g}\left(\frac{3}{2E}\right)^3}$$

Problem 7.6   The width of a rectangular channel decreases from b_1 to b_2. In cross section 1, the Froude number is smaller than 1, while for cross section 2, the Froude number of the flow is equal to 1. The discharge flowing through the channel is Q. Determine the value of y_1.

Solution

In cross section 2, the water depth and specific energy are given, respectively, by

$$y_2 = \sqrt[3]{\frac{Q^2}{gb_2^2}} \qquad E = \frac{3}{2}y_2$$

The specific energy in cross section 1 is equal to that of cross section 2. Therefore

$$E = y_1 + \frac{Q^2}{2gb_1^2}$$

The calculated value of E leads to the following third-order equation for the determination of y_1:

$$y_1^3 - Ey_1^2 + \frac{Q^2}{2gb_2^2} = 0$$

Problem 7.7   The width of a rectangular channel is b. Between two close cross sections, cross section 1 and cross section, 2, the bottom of the channel is elevated by the amount Δz. In cross section 1, the Froude number is smaller than 1. The discharge flowing through the channel is Q, and the water depth is y_1. Determine the maximum value of Δz that does not cause any change of the water depth in cross section 1.

Solution

The discharge per unit width of the channel is given by

$$q = \frac{Q}{b}$$

In cross section 2, the water depth and specific energy are given, respectively, by

$$y_2 = \sqrt[3]{\frac{q^2}{g}} \qquad E_2 = \frac{3}{2} y_2$$

In cross section 1, the specific energy is given by

$$E_1 = E_2 + \Delta z = y_1 + \frac{q^2}{2gy_1^2}$$

As values of E_1 and E_2 are determined, the value of Δz is given by

$$\Delta z = E_1 - E_2$$

Problem 7.8 The width of a rectangular channel is b_1 and the water depth is y_1. The discharge flowing through the channel is Q. In order to cause a local depression, Δy, of the water free surface, it is suggested to change locally the width of the channel to b_2. In cross section 1, where water depth is y_1, the Froude number is smaller than 1. Determine the value of b_2. What is the maximum value of Δy that can be obtained by using the arrangement described in this problem? If Δy obtains its maximum value, what is the value of b_2?

Solution

The specific energy at cross sections 1 (where the channel width is b_1) and 2 (where the channel width is b_2) is given by

$$E = y_1 + \frac{Q^2}{2gb_1^2 y_1^2} = y_2 + \frac{Q^2}{2gb_2^2 y_2^2} \tag{1}$$

We also know that

$$y_2 = y_1 - \Delta y \tag{2}$$

Thus

$$\frac{Q^2}{2gb_2^2(y_1 - \Delta y)^2} = E - y_1 + \Delta y \tag{3}$$

This expression yields

$$b_2 = \frac{Q}{(y_1 - \Delta y)\sqrt{2g(E - y_1 + \Delta y)}} \tag{4}$$

The maximum value of Δy is obtained when critical conditions prevail at cross section 2, i.e.,

$$y_2 = y_c = \tfrac{2}{3}E \tag{5}$$

By introducing Eq. (5) into Eqs. (1) and considering that the Froude number is equal to 1 at cross section 2, we find

$$\Delta y_{max} = \frac{1}{3}y_1 - \frac{Q^2}{3gb_1^2 y_1^2} \tag{6}$$

and

$$(b_2)_{\Delta y_{max}} = \frac{Q}{\sqrt{g[2/3E]^3}}$$

Problem 7.9   Water flows through a rectangular channel with a discharge q per unit width. The water depth is y_1. In order to cause a local depression, Δy, of the water free surface, it is suggested to change locally the elevation of the channel bottom. In cross section 1, of water depth y_1, the Froude number is smaller than 1. Determine the value of the elevation of the channel bottom. What is the maximum value of Δy that can be obtained by using the arrangement described in this problem? If Δy obtains its maximum value, what is the elevation of the channel bottom?

Solution

Referring to a datum $z = 0$, which is defined at the channel bottom at cross section 1 (where the water depth is y_1), the total head in cross sections 1 and 2 (where the bottom is elevated by the amount Δz) is given by

$$H = E_1 = E_2 + \Delta z = y_1 + \frac{q^2}{2gy_1^2} = y_2 + \frac{q^2}{2gy_2^2} + \Delta z \tag{1}$$

According to the particular conditions of the problem,

$$y_2 = y_1 - \Delta z - \Delta y \tag{2}$$

We introduce Eq. (2) into Eq. (1) to obtain

$$E_1 = y_1 - \Delta y + \frac{q^2}{2g(y_1 - \Delta z - \Delta y)^2} \tag{3}$$

A different arrangement of this expression yields

$$\Delta z = y_1 - \Delta y - \frac{q}{\sqrt{2g(E_1 + \Delta y)}} \tag{4}$$

The maximum value of y is obtained when critical conditions prevail at cross section 2. Then

$$y_2 = y_c = \sqrt[3]{\frac{q^2}{g}} \tag{5}$$

$$E_1 = \frac{3}{2} y_2 + \Delta z = \frac{3}{2} \sqrt[3]{\frac{q^2}{g}} + \Delta z \tag{6}$$

Therefore

$$\Delta z = (\Delta z)_{\Delta y_{max}} = E_1 - \frac{3}{2} \sqrt[3]{\frac{q^2}{g}} \tag{7}$$

According to Eqs. (2) and (6), we obtain

$$E_1 = \frac{3}{2}(y_1 - \Delta z - \Delta y) + \Delta z \tag{8}$$

This expression yields

$$\Delta y = (\Delta y)_{max} = y_1 - \tfrac{1}{3}\Delta z - \tfrac{2}{3}E_1 \tag{9}$$

By introducing Eqs. (1) and (7) into Eq. (9), we obtain

$$\Delta y = (\Delta y)_{max} = \frac{1}{2} \sqrt[3]{\frac{q^2}{g}} - \frac{q^2}{2g y_1^2}$$

This expression indicates that Δy is equal to the difference in velocity head between cross sections 1 and 2.

Problem 7.10   In Sec. 8.7 it is proved that the ratio between the water depths y_2 and y_1, downstream and upstream of a hydraulic jump, respectively, is given by

$$\frac{y_2}{y_1} = \tfrac{1}{2} \left(\sqrt{1 + 8\mathrm{Fr}_1^2} - 1 \right)$$

where Fr_1 is the Froude number upstream of the jump,

$$\mathrm{Fr}_1^2 = \frac{q^2}{g y_1^3}$$

Here, q is the discharge per unit width of the channel. Develop an expression for the ratio between y_1 and y_2 that depends on the Froude number *downstream* of the hydraulic jump.

Solution

From the definition of the momentum function downstream and upstream of the jump,

$$\frac{y_1^2}{2} + \frac{q^2}{g y_1} = \frac{y_2^2}{2} + \frac{q^2}{g y_2} \tag{1}$$

Multiplying this equation by 2 and rearranging results in

$$y_2^2 - y_1^2 = \frac{2q^2}{g} \left(\frac{y_1 - y_2}{y_1 y_2} \right) \tag{2}$$

We disregard the trivial solution that $y_1 = y_2$. Therefore we may divide Eq. (2) by the difference between y_2 and y_1 to obtain

$$y_2 + y_1 = \frac{2q^2}{g y_1 y_2} \tag{3}$$

We multiply this expression by y_1 and divide it by y_2^2 to obtain

$$\left(\frac{y_1}{y_2} \right)^2 + \left(\frac{y_1}{y_2} \right) - 2\mathrm{Fr}_2^2 = 0 \tag{4}$$

where

$$\mathrm{Fr}_2^2 = \frac{q^2}{g y_2^3} \tag{5}$$

Equation (4) is a second-order algebraic equation with regard to the ratio between y_1 and y_2; its solution is

$$\frac{y_1}{y_2} = \frac{1}{2} \left(\sqrt{1 + 8\mathrm{Fr}_2^2} - 1 \right)$$

Problem 7.11 Develop an expression for the head loss in a hydraulic jump in a rectangular channel.

Solution

The head loss in a hydraulic jump is equal to the difference between the specific energy upstream and downstream of the hydraulic jump,

$$\Delta H = \Delta E = \left(y_1 + \frac{q^2}{2g y_1^2} \right) - \left(y_2 + \frac{q^2}{2g y_2^2} \right) \tag{1}$$

Rearranging this expression yields

$$\Delta E = \frac{q^2}{2g} \left(\frac{y_2^2 - y_1^2}{y_1^2 y_2^2} \right) - (y_2 - y_1)$$

$$= (y_2 - y_1) \left[\frac{q^2}{2g y_1^2 y_2^2} (y_2 + y_1) - 1 \right] \tag{2}$$

We introduce Eq. (3) of problem 7.10 into Eq. (2), to obtain

$$\Delta E = (y_2 - y_1) \left[\frac{(y_2 + y_1)^2}{4 y_1^2 y_2^2} - 1 \right] = \frac{(y_2 - y_1)^3}{4 y_1^2 y_2^2}$$

This expression shows that the identity of the momentum function between two cross sections is always associated with energy loss, provided that the first one is subject to supercritical flow and the second is subject to subcritical flow.

Problem 7.12 Water flows through a wide rectangular channel. The shear stress at the bottom of the channel is $\tau = 10$ Pa. The water depth is $B = 1.5$ m. The channel bed is extremely smooth.

  (a) What is the average flow velocity?
  (b) What is the discharge per unit width?
  (c) What is the maximum flow velocity?
  (d) What is the slope of the channel?
  (e) What is the thickness of the laminar sublayer?

Solution

(a) According to Eq. (7.5.20),

$$V^+ = 2.5 \ln(B^+) + 3 \tag{1}$$

where B^+ is the dimensionless water depth and V^+ is the dimensionless average flow velocity. In general, we have

$$V^+ = \frac{V}{u_*} \qquad B^+ = \frac{u_* B}{v} \qquad u_* = \sqrt{\frac{\tau}{\rho}} \tag{2}$$

where V is the average flow velocity, u_* is the shear velocity, B is the water depth and v is the kinematic viscosity. For the present problem,

$$u_* = \sqrt{\frac{10}{1000}} = 0.1 \text{ m/s}$$

$$B^+ = \frac{0.1 \times 1.5}{10^{-6}} = 1.5 \times 10^5 \tag{3}$$

Introducing these values into Eq. (1), we obtain

$$\frac{V}{0.1} = 2.5 \ln(1.5 \times 10^5) + 3 = 32.79 \tag{4}$$

This expression yields

$$V = 3.28 \text{ m/s} \tag{5}$$

(b) From the definition,

$$q = VB = 3.28 \times 1.5 = 4.92 \text{ m/s}$$

(c) According to Eq. (7.5.18) the velocity profile is given by

$$u^+ = 2.5 \ln z^+ + 5.5 \tag{6}$$

where y is the distance from the bottom of the channel. Equations (1), (4), and (6) indicate that the maximum flow velocity takes place at the water free surface,

$$\frac{U}{0.1} = \frac{V}{0.1} + 2.5 = 32.79 + 2.5 = 35.29 \tag{7}$$

This expression yields

$$U = 3.53 \text{ m/s} \tag{8}$$

(d) According to Eq. (7.5.22), the Darcy–Weissbach friction coefficient is given by

$$f = \frac{8}{(V^+)^2} = \frac{8}{32.79^2} = 0.0074$$

According to the Darcy–Weissbach Eq. (7.5.21), the slope of the channel is

$$S_f = \frac{f}{4B} \frac{V^2}{2g} = \frac{0.0074}{4 \times 1.5} \times \frac{3.28^2}{2 \times 9.81} = 6.76 \times 10^{-5}$$

(e) From Eq. (7.5.13), the thickness of the laminar sublayer is

$$\delta = \frac{11.6\upsilon}{u_*} = \frac{11.6 \times 10^{-6}}{0.1} = 1.16 \times 10^{-4} \text{ m}$$

Problem 7.13   Water flows through a wide rectangular channel. The shear stress at the bottom of the channel is $\tau = 10$ Pa. The water depth is $B = 1.5$ m. The flow through the channel takes place in the rough turbulent regime. The channel roughness coefficient of Manning is 0.025.

(a) What is the average flow velocity? Compare results obtained using the Manning equation and using a logarithmic velocity distribution.

(b) What is the discharge per unit width?

(c) What is the maximum flow velocity?

(d) What is the slope of the channel?

Solution

(a) According to Eq. (7.5.34),

$$V^+ = 2.5 \ln \left(\frac{B}{\varepsilon} \right) + 6 \tag{1}$$

where V^+ is the dimensionless velocity, B is the water depth, and ε is the roughness of the channel bed. From Eqs. (7.5.12) and (7.2.33),

$$V^+ = \frac{V}{u_*} \qquad n = 0.038 \varepsilon^{1/6} \qquad u_* = \sqrt{\frac{\tau}{\rho}} \tag{2}$$

Here, V is the average flow velocity, u_* is the shear velocity, and n is the Manning roughness coefficient. For the present conditions,

$$u_* = \sqrt{\frac{10}{1000}} = 0.1 \text{ m/s}$$
$$\varepsilon = \left(\frac{0.025}{0.038} \right)^6 = 0.081 \text{ m} \tag{3}$$

Introducing these values into Eq. (1), we obtain

$$\frac{V}{0.1} = 2.5 \ln \left(\frac{1.5}{0.081} \right) + 6 = 13.30 \tag{4}$$

This gives

$$V = 1.33 \text{ m/s} \tag{5}$$

(b) According to Eq. (7.2.21),

$$S_f = \frac{\tau}{\rho g B} = \frac{10}{1000 \times 9.81 \times 1.5} = 6.8 \times 10^{-4}$$

Substituting into the Manning equation,

$$V = \frac{B^{2/3}}{n} S_f^{1/2} = \frac{1.5^{2/3}}{0.025} \sqrt{6.8 \times 10^{-4}} = 1.36 \text{ m/s}$$
$$q = VB = 1.33 \times 1.5 = 2.00 \text{ m/s}$$

(c) According to Eq. (7.5.33) the velocity profile is given by

$$u^+ = 2.5 \ln \left(\frac{y}{\varepsilon} \right) + 8.5 \tag{6}$$

where y is the distance from the bottom of the channel. Equations (1), (4), and (6) indicate that the maximum flow velocity takes place at the water free surface,

$$\frac{U}{0.1} = \frac{V}{0.1} + 2.5 = 13.30 + 2.5 = 15.80 \tag{7}$$

This expression yields

$$U = 1.58 \text{ m/s} \tag{8}$$

(d) According to Eq. (7.5.22), the Darcy–Weissbach friction coefficient is

$$f = \frac{8}{(V^+)^2} = \frac{8}{13.30^2} = 0.045$$

Then, from Eq. (7.5.21), the slope of the channel is

$$S_f = \frac{f}{4B} \frac{V^2}{2g} = \frac{0.045}{4 \times 1.5} \times \frac{1.33^2}{2 \times 9.81} = 6.76 \times 10^{-4}$$

This value is almost identical to that obtained in part (a) by using the expression for the shear stress.

Problem 7.14   The velocity distribution in a wide rectangular channel is given by

$$u = U \left[1 - \left(\frac{z}{B} \right)^2 \right]$$

where U is the velocity at the free surface and B is the water depth. The coefficient of diffusion D is given, and it has a constant value.

(a) Find the average flow velocity.
(b) Find the profile of u', namely, the deviation from the average flow velocity.
(c) Find an expression for C', namely, the deviation from the average concentration, in terms of the expected concentration profile, $\partial C / \partial x$.
(d) What is the value of the longitudinal dispersion coefficient?

Solution

(a) The average flow velocity is given by

$$V = \frac{1}{B} \int_0^B u \, dz = \frac{U}{B} \left[z - \frac{z^3}{3B^2} \right]_0^B = \frac{2}{3} U$$

(b) The profile of u' is

$$u' = u - V = U\left[\frac{1}{3} - \left(\frac{z}{B}\right)^2\right]$$

(c) According to Eq. (7.5.45),

$$C'(y) = \frac{1}{D}\frac{\partial C}{\partial x}\int_0^z\int_0^z u'\,dz\,dz + C'(0)$$

$$= \frac{U}{D}\frac{\partial C}{\partial x}\int_0^z\int_{00}^z\left[\frac{1}{3} - \left(\frac{z}{B}\right)^2\right]dz\,dz + C'(0)$$

$$C'(y) = \frac{U}{D}\frac{\partial C}{\partial x}\left[\frac{z^2}{6} - \frac{z^4}{12B^2}\right] + C'(0)$$

(d) From Eqs. (7.5.46) and (7.5.47), we have

$$K = -\frac{1}{B\partial C/\partial x}\int_0^B u'C'\,dz = -\frac{U^2}{BD}\int_0^B\left\{\left[\frac{1}{3} - \left(\frac{z}{B}\right)^2\right]\left[\frac{z^2}{6} - \frac{z^4}{12B^2}\right]\right\}dz$$

$$= \frac{8}{945}\frac{U^2B^2}{D} = 0.0085\frac{U^2B^2}{D}$$

Problem 7.15 A very long channel incorporates three long segments, as shown in Fig. 7.24. It delivers water from a reservoir and ends in an overfall. The discharge per unit width is $q = 3.6$ m^2/s. The Manning roughness coefficient is $n = 0.025$. The overfall is constructed on a bottom elevation of

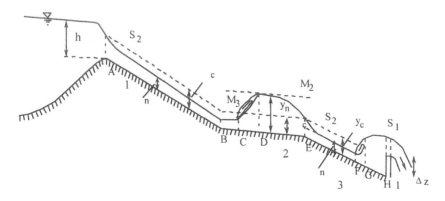

Figure 7.24 Flow in variable channel, Problem 7.15.

$\Delta z = 1.0$ m. The slopes of the different sections of the channel are (Section 1) $S_0 = 0.035$, (Section 2) $S_0 = 1.7 \times 10^{-3}$, (Section 3) $S_0 = 0.027$.

(a) Determine the value of y_c and values of y_n for each of the channel sections.

(b) Explain how the hydraulic profiles shown in Fig. 7.24 are obtained. How are the locations of the two hydraulic jumps shown in the figure determined?

(c) What is the water depth at cross sections A, B, C, D, E, F, G, and H? Which of these sections is a control section? What is the head loss in each of the hydraulic jumps? What is the elevation h of the free surface of the reservoir above the channel entrance?

(d) Determine the hydraulic profile of segment 1 of the channel, by using the Simpson one-third method of integration.

(e) Determine the hydraulic profiles of segment 2 of the channel, by using the Simpson one-third method of integration for the upstream portion of the segment and using a fourth-order Runge–Kutta method for the downstream portion of the channel segment.

(f) Determine the hydraulic profile at the upstream and downstream portions of segment 3 of the channel. Apply the Simpson and Runge–Kutta methods to the upstream portion of the segment. Compare the results obtained by the two methods. Apply the Runge–Kutta method for the downstream portion of the channel segment.

Solution

(a) As the channel is rectangular, we may apply Eq. (7.2.42) to calculate the critical water depth,

$$y_c = 3\sqrt{\frac{q^2}{g}} = 3\sqrt{\frac{3.6^2}{9.81}} = 1.10 \text{ m} \tag{1}$$

As the channel is wide, its hydraulic radius is equal to the water depth. Therefore, Manning's equation is given by (from problem 7.1)

$$q = \frac{y^{5/3}}{n} S_f^{1/2} \tag{2}$$

In the case of uniform flow, Eq. (2) yields

$$y = \left(\frac{qn}{S_0^{1/2}} \right)^{3/5} \tag{3}$$

Equation (3) yields for segments 1, 2, and 3 the following normal water depths:

$$(y_n)_1 = \left(\frac{qn}{S_0^{1/2}}\right)^{3/5} = \left(\frac{3.6 \times 0.025}{0.035^{1/2}}\right)^{3/5} = 0.65 \text{ m}$$

$$(y_n)_1 = \left(\frac{qn}{S_0^{1/2}}\right)^{3/5} = \left(\frac{3.6 \times 0.025}{\sqrt{1.7 \times 10^{-3}}}\right)^{3/5} = 1.60 \text{ m}$$

$$(y_n)_3 = \left(\frac{qn}{S_0^{1/2}}\right)^{3/5} = \left(\frac{3.6 \times 0.025}{0.027^{1/2}}\right)^{3/5} = 0.70 \text{ m}$$

(b) The hydraulic profile of segment 1 is obtained due to the steep slope of that segment. The discharge flowing from the reservoir into the channel is controlled by the capability of cross section A to convey water. Therefore the water depth in that cross section is the critical depth, and downstream of that cross section the hydraulic profile is S_1. At point B the water depth is the normal depth of segment 1, provided that the momentum function at normal depth of segment 1 is larger than that of normal depth of segment 2. As in segment 2, the normal depth is larger than the critical depth, downstream of cross section B, the hydraulic profile is of type M_3. This profile is extended until the momentum function at cross section C is equal to the momentum function at cross section D. Cross section E represents the border point between mild and steep slopes. Therefore the water depth at point E is the critical depth. Upstream of that cross section, the hydraulic profile is of type M_2. Downstream of that cross section, the hydraulic profile is of type S_2. At the overfall, at point I, the water depth is the critical depth. The type of the control section determines it. At cross sections I and H, the total head is identical. Upstream of cross section H, the hydraulic profile is required to be of the S_1 type. The location of the hydraulic jump is determined by the identity of the momentum function at cross sections F and G.

(c) *Cross section A*: This cross section controls the flow from the reservoir into the channel. Therefore it is a control section. The water depth in this cross section is the critical depth: $y_A = y_C = 1.1$ m.

Cross section B: The water depth at this cross section is the normal depth of channel segment 1. This cross section is not defined as a control section. However, calculation of the water depth downstream of point B is done by considering y_B as an initial value. The value of the water depth is $y_B = y_n = 0.65$ m.

Cross section C: In this cross section the flow is supercritical and the water depth is the alternate depth of the normal depth at point D. Therefore we may use the equation of the hydraulic jump to calculate the water depth at point C:

$$\frac{y_C}{y_D} = \frac{1}{2}\left(\sqrt{1 + 8\mathrm{Fr}_D^2} - 1\right) = \frac{1}{2}\left(\sqrt{1 + 8\frac{q^2}{g\,y_D^3}} - 1\right)$$

$$= \frac{1}{2}\left(\sqrt{1 + \frac{8 \times 3.6^2}{9.81 \times 1.6^3}} - 1\right) = 0.446$$

$$y_C = 0.446 y_D = 0.446 \times 1.60 = 0.71 \text{ m}$$

The head loss through the hydraulic jump located between points C and D is given by

$$(\Delta H)_{CD} = \frac{(y_D - y_C)^3}{4 y_C y_D} = \frac{(1.60 - 0.71)^3}{4 \times 1.60 \times 0.71} = 0.155 \text{ m}$$

Cross section D: The water depth at this cross section is the normal depth. Therefore, $y_D = y_n = 1.60$ m.

Cross section E: This cross section is located between segments of mild and steep slopes. Such a section is a control section, and the water depth is the critical depth: $y_E = y_c = 1.1$ m.

Cross section F: The water depth at this cross section is the normal depth, $y_F = y_n = 0.70$ m.

Cross section G: The water depth at this cross section is alternate to the normal water depth at cross section F. Therefore we apply the hydraulic jump equation to obtain

$$\frac{y_G}{y_F} = \frac{1}{2}\left(\sqrt{1 + 8\mathrm{Fr}_F^2} - 1\right) = \frac{1}{2}\left(\sqrt{1 + 8\frac{q^2}{g\,y_F^3}} - 1\right)$$

$$= \frac{1}{2}\left(\sqrt{1 + \frac{8 \times 3.6^2}{9.81 \times 0.70^3}} - 1\right) = 2.32$$

$$y_G = 2.32 y_F = 2.32 \times 0.70 = 1.62 \text{ m}$$

The head loss associated with this jump is given by

$$(\Delta H)_{FG} = \frac{(y_G - y_F)^3}{4 y_G y_F} = \frac{(1.62 - 0.70)^3}{4 \times 1.62 \times 0.70} = 0.172 \text{ m}$$

Cross section H: The total head at cross sections H and I is identical. Cross section I is a control section, in which the discharge alone determines the water depth. Therefore at cross section I the water depth is the critical depth. Thus

$$H_H = E_H = E_I + \Delta Z = \tfrac{3}{2} y_c + \Delta z = \tfrac{3}{2} \times 1.10 + 1.00 = 2.65 \text{ m}$$

where E is the specific energy. By applying the equation for the specific energy at point H we obtain

$$E_H = y_H + \frac{q^2}{2g y_H^2} = 2.65$$

We multiply this expression by the square of y_H, giving

$$y_H^3 - 2.65 y_H^2 + \frac{3.6^2}{19.62} = y_H^3 - 2.65 y_H^2 + 0.66 = 0$$

By trial and error, this results in $y_H = 2.55$ m.

Cross section I: This cross section is a control section, so the water depth is the critical depth, $y_I = y_c = 1.1$ m.

The value of h: Due to the equality of total head at the reservoir water free surface and cross section A,

$$h = E_A = \frac{3}{2} y_c = \frac{3}{2} \times 1.10 = 1.65 \text{ m}$$

(d) We represent the expression for the derivative of x with regard to y in segment 1, according to Eq. (7.6.10c),

$$\frac{dx}{dy} = \frac{1}{S_0} \frac{1 - (y_c/y)^3}{1 - (y_n/y)^{10/3}} = \frac{1}{0.035} \times \frac{1 - (1.10/y)^3}{1 - (0.65/y)^{10/3}}$$

$$= 28.57 \times \frac{1 - (1.10/y)^3}{1 - (0.65/y)^{10/3}} = f_1(y)$$

The hydraulic profile of type S_2 is extended between the water depth at cross section A, $y_A = y_c = 1.1$ m, and the normal water depth $y_n = 0.65$ m. We choose several intervals of the water depth Δy. The numerical integration rule according to Eq. (7.6.38) is given by

$$x_{i+1} = x_i + \Delta x_i = x_i + \frac{\Delta y}{3} [f_1(y_i) + 4f_1(y_i + \Delta y) + f_1(y + 2\Delta y)]$$

The following table was developed using a spreadsheet for the calculation of the hydraulic profile between $y = 1.10$ m and $y = 0.652$ m:

x	y	dy	f(y)	f(y + dy)	f(y + 2dy)	dx	
0	1.1	−0.05	0	−5.36323	−12.4086	0.564358	Row 1
0.564358	1	−0.05	−2.4086	−21.9887	−35.6381	2.266693	Row 2
2.83105	0.9	−0.05	−35.6381	−56.4232	−91.4947	5.880423	Row 3
8.711473	0.8	−0.05	−91.4947	162.293	−375.982	18.6108	Row 4
27.32227	0.7	−0.01	−375.982	−483.014	−661.482	49.49201	Row 5
76.81428	0.68	−0.01	−661.482	−1018.54	−2089.93	113.7591	Row 6
190.5734	0.66	−0.004	−2089.93	−3518.55	−10661.9	447.0997	Row 7
637.6731	0.652						Row 8

Col. 1   Col. 2 Col. 3   Col. 4   Col. 5   Col. 6   Col. 7

We work along the row elements. In row 1, column 1 we introduce the value of x in the control section at point A. In row 1 column 2, we introduce the value of y at that location. The value in column 3 represents Δy. Elements of columns 4, 5, and 6 are calculated according to the expressions for $f_1(y)$, $f_1(y + \Delta y)$, and $f_1(y + 2\Delta y)$, respectively. Column 7 is calculated according to the equation of Simpson's rule and specifies the value of Δx. In row 2, column 1 represents the sum of the elements of row 1, column 1 and row 1, column 7. Column 2 of row 2 is the sum of row 1, column 2 and two times the element of column 3, row 1. The following row elements are calculated by the copy–paste procedures of the relevant elements of row 1. After completing the copying and pasting of elements of row 2, all elements of row 3 are filled in by applying the copy–paste procedure for all elements of row 2. As required by the calculation, values of Δy are changed.

(e) We represent the expression for the derivative of x with regard to y in the upstream portion of segment 1, according to Eq. (7.6.10c), as

$$\frac{dx}{dy} = \frac{1}{S_0}\frac{1 - (y_c/y)^3}{1 - (y_n/y)^{10/3}} = \frac{1000}{1.7} \times \frac{1 - (1.10/y)^3}{1 - (1.60/y)^{10/3}}$$

$$= 588.2 \times \frac{1 - (1.10/y)^3}{1 - (1.60/y)^{10/3}} = f_1(y)$$

The hydraulic profile of type M_3 is extended between the water depth at cross section B, $y_B = y_n = 0.65$ m, and the water depth at point C, $y_C = 0.71$ m. We choose several intervals of the water depth, Δy. The numerical integration rule according to Eq. (7.6.38) gives

$$x_{i+1} = x_i + \Delta x_i = x_i + \frac{\Delta y}{3}[f_1(y_i) + 4f_1(y_i + \Delta y) + f_1(y + 2\Delta y)]$$

A spreadsheet is again used for the calculation of the hydraulic profile between $y = y_n = 0.65$ m and $y = 0.70$ m. The water depth of $y = 0.70$ m is the alternate depth of $y_n = 1.60$ m. This water depth takes place at the upstream side of the hydraulic jump.

x	y	dy	f(y)	f(y+dy)	f(y+2dy)	dx	
0	0.65	0.005	118.2226	117.9679	5.716869	0.993018	Row 1
0.993018	0.66	0.005	117.6991	117.416	5.68867	0.98842	Row 2
1.981438	0.67	0.005	117.1185	116.8064	5.65763	0.983336	Row 3
2.964774	0.68	0.005	116.4795	116.1376	5.623679	0.977756	Row 4
3.942529	0.69	0.005	115.7805	115.408	5.58674	0.971665	Row 5
4.914195	0.7						Row 6

Col. 1   Col. 2 Col. 3   Col. 4   Col. 5   Col. 6   Col. 7

Work with the spreadsheet is done as was done with the upstream portion of channel segment 1. However, here the calculation starts with point B and proceeds in the downstream direction. The downstream end of segment 2 represents a control section of critical depth. As the slope of the channel is mild, that portion of the segment incorporates an M_2 hydraulic profile, as noted previously. We apply the fourth-order Runge–Kutta method to calculate the shape of the hydraulic profile, starting our calculation from the control section in the upstream direction. A spreadsheet is used to perform the calculations with several values of Δx. The expression for the derivative of y with respect to x, according to Eq. (7.6.10d), is given by

$$\frac{dy}{dx} = S_0 \frac{1 - (y_n/y)^{10/3}}{1 - (y_c/y)^3} = 1.7 \times 10^{-3} \times \frac{1 - (1.60/y)^{10/3}}{1 - (1.1/y)^3} = f(y)$$

The hydraulic profile of type M_2 is extended between the water depth at cross section E, namely $y_E = y_c = 1.10$ m, and the water depth $y_n = 1.60$ m. We choose several values of the longitudinal interval Δx. The numerical solution according to Eq. (7.6.36) gives

$$y_{i+1} = y_i + \Delta y_i = y_i + \frac{1}{6}[\Delta y_1 + 2\Delta y_2 + 2\Delta y_3 + \Delta y_4]$$

where

$$\Delta y_1 = \Delta x f(y_i) \qquad \Delta y_2 = \Delta x f\left(y_i + \frac{\Delta y_1}{2}\right)$$

$$\Delta y_3 = \Delta x f\left(y_i + \frac{\Delta y_2}{2}\right) \qquad \Delta y_4 = \Delta x f(y_i + \Delta y_3)$$

Using a spreadsheet for the calculation of the hydraulic profile between $y_E = y_c \approx 1.105$ m and $y \approx y_n \approx 1.60$ m results in the following table:

x	y	dx	dy1	dy2	dy3	dy4	dy	
0	1.105	−1	0.30626	0.006296	0.186486	0.004637	0.116077	Row 1
−1	1.221077	−4	0.03696	0.030292	0.03136	0.026616	0.031147	Row 2
−5	1.252224	−15	0.100018	0.063324	0.074409	0.051445	0.071155	Row 3
−20	1.323378	−30	0.105748	0.06797	0.079603	0.054111	0.075834	Row 4
−50	1.399213	−50	0.093177	0.061041	0.071029	0.04763	0.067491	Row 5
−100	1.466704	−100	0.098904	0.05405	0.072869	0.036642	0.064898	Row 6
−200	1.531601	−200	0.084669	0.02905	0.064263	0.004381	0.045946	Row 7
−400	1.577547	−200	0.024812	0.010786	0.018615	0.004062	0.014613	Row 8
−600	1.59216	−200	0.008375	0.003865	0.006283	0.00164	0.005052	Row 9
−800	1.597211							Row 10

Col. 1   Col. 2   Col. 3   Col. 4   Col. 5   Col. 6   Col. 7   Col. 8

We work along the row elements. In row 1, column 1 we introduce the value of x in the control section, namely at point E. In row 1, column 2, we introduce the value of y at that location. The element of column 3 specifies the value of Δx. Elements of columns 4, 5, 6, and 7 of row 1 are then calculated according to the expressions for Δy_1, Δy_2, Δy_3, and Δy_4, respectively. Column 8 is calculated according to the fourth-order Runge–Kutta equation. Row 2, column 1 is the sum of the elements of row 1, columns 1 and 3. Row 1, column 2 is the sum of the elements of row 1, columns 2 and 8. The following row elements are calculated by applying the copy-and-paste procedures. At each row the value of Δx is changed if a change seems to be reasonable. The value of Δx should not be too large, to avoid negative values of any of the quantities Δy_1, Δy_2, Δy_3, or Δy_4.

(f) We apply first the numerical integration rule of the Simpson one-third method. With regard to the upstream portion of segment 3, where the hydraulic profile is of type S_2, the expression for the derivative of x with regard to y in segment 1, according to Eq. (7.6.10c), is

$$\frac{dx}{dy} = \frac{1}{S_0}\frac{1 - (y_c/y)^3}{1 - (y_n/y)^{10/3}} = \frac{1}{0.027} \times \frac{1 - (1.10/y)^3}{1 - (0.70/y)^{10/3}}$$

$$= 37.04 \times \frac{1 - (1.10/y)^3}{1 - (0.70/y)^{10/3}} = f_1(y)$$

The hydraulic profile of type S_2 is extended between the water depth at cross section E, where $y_E = y_c = 1.1$ m, and the normal water depth $y_n = 0.70$ m. We choose several intervals of the water depth Δy. The numerical integration rule according to Eq. (7.6.38) is given by

$$x_{i+1} = x_i + \Delta x_i = x_i + \frac{\Delta y}{3}[f_1(y_i) + 4f_1(y_i + \Delta y) + f_1(y + 2\Delta y)]$$

In the following table are results from a spreadsheet calculation of the hydraulic profile between $y = 1.10$ m and $y = 0.71$ m:

x	y	dy	f(y)	f(y + dy)	f(y + 2dy)	dx	
0	1.1	−0.05	0	−7.48474	−17.6292	0.792804	Row 1
0.792804	1	−0.05	−17.6292	−32.0379	−53.917	3.328299	Row 2
4.121103	0.9	−0.05	−53.917	−90.7423	−164.929	9.696919	Row 3
13.81802	0.8	−0.01	−164.929	−189.728	−220.751	19.07651	Row 4
32.89453	0.78	−0.01	−220.751	−260.668	−313.922	26.28906	Row 5
59.1836	0.76	−0.01	−313.922	−388.517	−500.456	39.47409	Row 6
98.65769	0.74	−0.01	−500.456	−687.082	−1060.42	71.82009	Row 7
170.4778	0.72	−0.005	−1060.42	−1433.8	−2180.61	149.6042	Row 8
320.082	0.71						Row 9
Col. 1	Col. 2	Col. 3	Col. 4	Col. 5	Col. 6	Col. 7	

Work with the spreadsheet is done as it was performed with regard to the upstream portion of the channel segment 1.

The calculation of the hydraulic profile of the upstream portion of segment 3, where the S_2 hydraulic profile takes place, is done by applying the fourth-order Runge–Kutta method, in which the expression for $f(y)$ is given by

$$\frac{dy}{dx} = S_0 \frac{1 - (y_n/y)^{10/3}}{1 - (y_c/y)^3} = 0.027 \times \frac{1 - (0.70/y)^{10/3}}{1 - (1.1/y)^3} = f(y)$$

The hydraulic profile of type S_2 is extended between the water depth at cross section E, $y_E = y_c = 1.10$ m, and the water depth $y_n = 0.70$ m. We choose several values of the longitudinal interval Δx. The numerical solution according to Eq. (7.6.36) yields

$$y_{i+1} = y_i + \Delta y_i = yi + \frac{1}{6}[\Delta y_1 + 2\Delta y_2 + 2\Delta y_3 + \Delta y_4]$$

where

$$\Delta y_1 = \Delta x f(y_i) \qquad \Delta y_2 = \Delta x f\left(y_i + \frac{\Delta y_1}{2}\right)$$

$$\Delta y_3 = \Delta x f\left(y_i + \frac{\Delta y_2}{2}\right) \qquad \Delta y_4 = \Delta x f(y_i + \Delta y_3)$$

The following table shows results from a spread sheet calculation of the hydraulic profile between $y = y_c \approx 1.095$ m and $y_n \approx 0.70$ m:

x	y	dx	dy1	dy2	dy3	dy4	dy	
0	1.09	0.5	−0.37497	−0.00952	−0.25076	−0.00489	−0.15007	Row 1
0.5	0.93993	1	−0.02802	−0.02419	−0.02468	−0.02168	−0.02457	Row 2
1.5	0.915357	5.5	−0.11934	−0.06445	−0.08586	−0.04817	−0.07802	Row 3
7	0.837334	13	−0.12454	−0.05437	−0.09005	−0.03139	−0.07413	Row 4
20	0.763207	20	−0.0678	−0.02838	−05032	−0.01192	−0.03952	Row 5
40	0.723688	20	−0.02257	−0.01145	−0.01684	−0.00622	−0.01423	Row 6
60	0.709457	20	−0.00866	−0.00464	−0.0065	−0.00266	−0.0056	Row 7
80	0.703857	20	−0.00348	−0.0019	−0.00262	−0.00111	−0.00227	Row 8
100	0.701587	20	−0.00142	−0.00078	−0.00107	−0.00046	−0.00093	Row 9
120	0.700655							Row 10
Col. 1	Col. 2	Col. 3	Col. 4	Col. 5	Col. 6	Col. 7	Col. 8	

Work with the spreadsheet is done as before with regard to the downstream portion of channel segment 2. However, in the present case the calculation starts with point E and proceeds in the downward direction. Here we start the calculation with a y value somehow smaller than y_c. The Runge–Kutta method requires an initial depth to be specified that is different from the critical depth. Values of Δx also were required to be comparatively small to avoid positive values of Δy_2 and Δy_4. The predicted length of the hydraulic profile is somehow different from that predicted by the Simpson method. However, these differences are not very significant.

For the hydraulic profile of the downstream portion of segment 3, where the S_1 hydraulic profile takes place, the fourth-order Runge–Kutta method is again applied. The expression for $f(y)$ is given by

$$\frac{dy}{dx} = S_0 \frac{1 - (y_n/y)^{10/3}}{1 - (y_c/y)^3} = 0.027 \times \frac{1 - (0.70/y)^{10/3}}{1 - (1.1/y)^3} = f(y)$$

The hydraulic profile of type S_1 is extended between the water depth at cross section G, where $y_G = 1.62$ m, and the water depth at section H, where $y_n = 2.55$ m. The calculation is done from point H moving in the upstream direction. We choose several values of the longitudinal interval Δx. The numerical solution according to Eq. (7.6.36) yields

$$y_{i+1} = y_i + \Delta y_i = y_i + \tfrac{1}{6}[\Delta y_1 + 2\Delta y_2 + 2\Delta y_3 + \Delta y_4]$$

where

$$\Delta y_1 = \Delta x f(y_i) \qquad \Delta y_2 = \Delta x f\left(y_i + \frac{\Delta y_1}{2}\right)$$

$$\Delta y_3 = \Delta x f\left(y_i + \frac{\Delta y_2}{2}\right) \qquad \Delta y_4 = \Delta x f(y_i + \Delta y_3)$$

The following table shows results of a spreadsheet for the calculation of the hydraulic profile between $y_H = 2.55$ m and $y_G = 1.62$ m:

x	y	dx	dy1	dy2	dy3	dy4	dy	
0	2.55		−5	−0.14481	−0.14576	−0.14576	−0.14686	−0.14579 Row 1
−5	2.404215		−5	−0.14686	−0.14812	−0.14813	−0.14961	−0.14816 Row 2
−10	2.256052		−5	−0.14961	−0.15136	−0.15138	−0.15347	−0.15143 Row 3
−15	2.104626		−5	−0.15347	−0.156	−0.15605	−0.15919	−0.15613 Row 4
−20	1.948499		−5	−0.15919	−0.16315	−0.16326	−0.16845	−0.16341 Row 5
−25	1.785086		−2	−0.06738	−0.06841	−0.06843	−0.0696	−0.06844 Row 6
−27	1.716644		−2	−0.0696	−0.07094	−0.07097	−0.07254	−0.07099 Row 7
−29	1.645652		−0.5	−0.01813	−0.01824	−0.01824	−0.01836	−0.01824 Row 8
−29.5	1.627407							Row 9
Col. 1	Col. 2	Col. 3	Col. 4	Col. 5	Col. 6	Col. 7	Col. 8	

Work with spreadsheet was done as with the calculation of the hydraulic profile taking place at the downstream portion of channel segment 2.

Unsolved Problems

Problem 7.16 A channel of trapezoidal cross section, with bottom width $b = 2$ m, side slope $m = 3$, and Manning roughness coefficient $n = 0.025$, conveys a water discharge of $Q = 15$ m^3/s. The average flow velocity is $V = 1.2$ m/s.

    (a) What is the water depth?
    (b) Calculate the wetted perimeter and the hydraulic radius.
    (c) Find the longitudinal slope of the channel.
    (d) What is the shear stress along the channel bed?
    (e) What are the values of the Reynolds and Froude numbers?
    (f) What are the values of the specific energy and the momentum function?

Problem 7.17 A channel of trapezoidal cross section, with bottom width $b = 2$ m, side slope $m = 3$, and Manning roughness coefficient $n = 0.025$, conveys a water discharge of $Q = 15$ m^3/s. The shear stress applied to the channel bed is $\tau = 10$ Pa.

    (a) What is the water depth?
    (b) Find the cross-sectional area, wetted perimeter, and hydraulic radius.
    (c) What is the average flow velocity?
    (d) Determine the longitudinal slope of the channel.
    (e) What are the values of the Reynolds and Froude numbers?
    (f) What are the values of the specific energy and the momentum function?

Problem 7.18   In a channel of trapezoidal cross section, the side slope is $m = 3$, the width of the channel bottom is $b = 2$ m, and the Manning roughness coefficient is $n = 0.025$. The shear stress at the channel bed is $\tau = 9.81$ Pa, and the average flow velocity is $V = 1.35$ m/s.

  (a)   Find the hydraulic radius of the channel.
  (b)   Find the longitudinal slope of the channel.
  (c)   What are the water depth and the cross-sectional area of the flow?
  (d)   What is the discharge flowing through the channel?

Problem 7.19   A rectangular channel has width $b_1 = 2$ m and water depth $y_1 = 1.5$ m. The Manning roughness coefficient is $n = 0.025$ and the discharge is $Q = 3.6$ m^3/s. In a short section of the channel, the width decreases to $b_2 = 1.2$ m.

  (a)   What is the hydraulic radius in the main channel section?
  (b)   What is the longitudinal slope of the channel?
  (c)   Find the water depth at cross section 2 (where the channel width is b_2).
  (d)   What is the minimum value of b_2, that does not affect the water depth upstream of the channel constriction?
  (e)   What is the water depth in cross section 2, under the condition of part (d)?

Problem 7.20   Figure 7.25 shows a wide rectangular channel in which the water depth is $y = 2$ m. The longitudinal slope is $S_0 = 5 \times 10^{-4}$. The Manning roughness coefficient is $n = 0.025$. At a minor portion of the channel the free water surface is subject to a local depression of $\Delta y = 0.05$ m due to a local elevation of the channel bottom. Head loss may be neglected.

  (a)   What are the discharge per unit width of the channel and the average flow velocity?

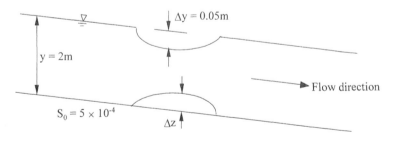

Figure 7.25   Definition sketch, Problem 7.20.

(b) What are the values of the Reynolds and Froude numbers upstream of the water depression?

(c) What is the specific energy upstream of the water depression?

(d) What is the shear stress along the channel bottom upstream of the water depression?

(e) What is the amount of local elevation of the channel bottom (Δz)?

(f) What is the critical depth of the water flow?

(g) Consider that the amount of bottom elevation is increased, resulting in a change of the value of y upstream of the bottom elevation. Under such conditions, what is the water depth above the elevated bottom?

Problem 7.21   A rectangular channel, shown in Fig. 7.26, has a width $b_1 = 2.0$ m and delivers a water discharge of $Q = 5$ m^3/s. Initially the water flow is uniform along the entire length of the channel, with $y_n = 1.90$ m. The Manning roughness coefficient is $n = 0.025$. Between points B and C of the channel, the channel bottom is elevated by $\Delta Z = 0.5$ m, and the width of the channel is increased to b_2, so that the free water surface is not changed.

(a) What is the hydraulic radius of the flow upstream of point A?

(b) What is the longitudinal slope of the channel?

(c) What is the water depth at point B?

(d) What is the width b_2?

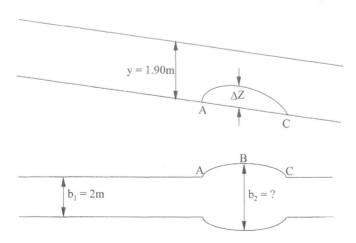

Figure 7.26   Flow over a hump in open channel flow, Problem 7.21.

(e) What is the maximum value of ΔZ possible so that the effect of raising the bottom can be compensated by increasing the width of the channel?

(f) Provide a schematic description of the curve showing the dependence of the ratio b_2/b_1 on the value of the ratio $\Delta Z/y_1$, under the condition that the free surface level remains unchanged.

Problem 7.22 Considering the same conditions as in problem 7.19, calculate the force acting on the channel constriction under all circumstances considered in the different parts of that problem.

Problem 7.23 Considering the same conditions as in problem 7.20, calculate the force acting on the channel constriction under all circumstances considered in the different parts of that problem.

Problem 7.24 Water flows through a rectangular channel. The water depth is $y_1 = 1.5$ m and the discharge per unit width is $q = 3$ m^2/s. At a minor section of the channel, the bottom is elevated by $\Delta z = 0.20$ m. Find the resulting change in free surface level.

Problem 7.25 A rectangular channel of width $b = 2.0$ m delivers a water discharge of $Q = 5.30$ m^3/s. Initially, the water flow is uniform along the entire length of the channel, with water depth $y_n = 0.75$ m. The Manning roughness coefficient is $n = 0.020$. Between points B and C of the channel, a constriction of the channel cross section is constructed and the width of the channel is reduced, as shown in Fig. 7.27. The channel width has a minimum value b_2 at point C. The constriction of the channel leads to the formation of

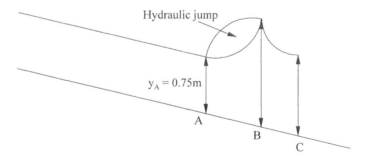

Figure 7.27 Definition sketch, Problem 7.25.

a hydraulic jump between points A and B. The water depth at point A is not affected by the channel constriction.

    (a)   What is the hydraulic radius of the flow upstream of point A?
    (b)   What is the longitudinal slope of the channel?
    (c)   What is the water depth at point B?
    (d)   What is the water depth at point C?
    (e)   What is the head loss between points A and B?
    (f)   What is the width of the channel at point C?

Problem 7.26   Water flows through a wide rectangular channel with a discharge per unit width of $q = 2.5$ m^2/s. The friction slope is $S_0 = 10^{-3}$ and the water depth is $B = 1.5$ m.

    (a)   What is the Manning roughness coefficient and what is the roughness of the channel?
    (b)   Develop an expression for the velocity profile in the channel.
    (c)   What is the maximum flow velocity?
    (d)   What is the value of the longitudinal dispersion coefficient?

Problem 7.27   Consider uniform free surface flow of water down a slope at angle α relative to horizontal, as shown in Fig. 7.28. Shear stress at the surface may be neglected. Set up and solve the equations for the velocity profile $u(y)$ and verify that the forces acting on a length L of the fluid layer are in equilibrium.

Problem 7.28   Water flows underneath a sluice gate in a wide rectangular channel with a water depth of $y_0 = 0.3$ m. This depth is one-third of the critical depth. Downstream of the sluice gate, the water flow creates a hydraulic

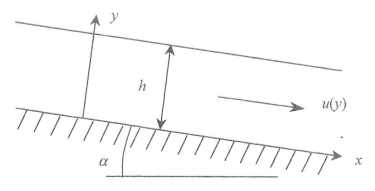

Figure 7.28   Open channel flow on a slope, Problem 7.27.

profile, which ends in a hydraulic jump. The alternate depth at the downstream side of the jump is $y_2 = 1.15$ m. The Manning roughness coefficient is $n = 0.022$. The channel is long and ends at an overfall.

(a)   What are the critical depth and the discharge per unit width in the channel?
(b)   What are the normal depth and the longitudinal slope of the channel?
(c)   Provide a schematic of the channel, in which you specify the various hydraulic profiles and their types.
(d)   What is the water depth upstream of the hydraulic jump?
(e)   What is the head loss associated with the hydraulic jump?
(f)   Calculate the shape of the hydraulic profile by applying the Simpson one-third rule of integration.

Problem 7.29   Figure 7.29 shows water flowing through a wide rectangular channel under uniform flow conditions. The average velocity is $V = 1.4$ m/s, the water depth is $y = 1.2$ m, and the Manning roughness coefficient is $n = 0.025$. In a short section of the channel, between points A and C, the bottom of the channel is elevated as a heap with a maximum elevation at point B. At point C, which is located at the downstream end of the bottom elevation, the water depth is $y_C = 0.30$ m. At a certain distance downstream of point C, between points D and E, a hydraulic jump develops.

(a)   What is the discharge per unit width?
(b)   What is the longitudinal slope of the channel?
(c)   What is the water depth at point D, and what is the head loss of the hydraulic jump?
(d)   What are the water depth and amount of elevation Δz at point B?
(e)   What is the amount of elevation Δy of the water depth at point A?

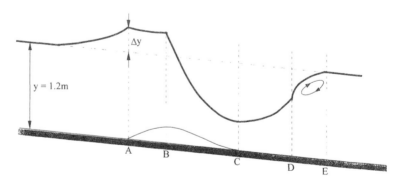

Figure 7.29   Definition sketch, Problem 7.29.

(f) What types of hydraulic profiles are created in the channel? Calcu-late the shape of the hydraulic profiles, upstream of point A and downstream of point C, by applying the Simpson one-third rule of integration.

(g) What is the gain in total head at point A, which is obtained by the elevated bottom of the channel?

(h) The gain of total head at point A is larger than the head loss of the hydraulic jump. Where is the head difference between these two quantities dissipated? Prove your answer using appropriate calcu-lations.

Problem 7.30 Figure 7.30 shows a wide rectangular channel with two seg-ments. Both segments are very long. The Manning roughness coefficient of segment I is $n = 0.025$. The roughness coefficient of segment II is $n = 0.020$. The first segment (segment I) comprises an outlet of a reservoir and its slope is $S_0 = 1.7 \times 10^{-3}$. The slope of segment II is $S_0 = 0.025$. Segment II ends at an overfall. However, right before the overfall the bottom of the channel is elevated by an amount ΔZ. The water depth above that elevation is $y_F = 0.95$ m. Upstream of that elevation, the water depth is $y_E = 2.50$ m. The free surface of the reservoir water above the bottom of the channel entrance is H.

(a) What is the discharge per unit width of the channel?

(b) What are the normal depths of segments I and II?

(c) What are the water depths at points A and B, and what are the values of ΔZ and H?

(d) How many hydraulic jumps are in the channel? Where are they and why do they occur?

(e) What are the water depths upstream and downstream of the hydraulic jumps? What is the head loss of the hydraulic jump?

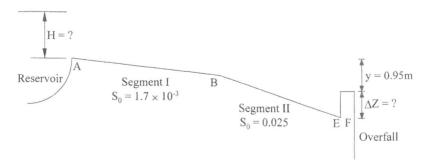

Figure 7.30 Definition sketch, Problem 7.30.

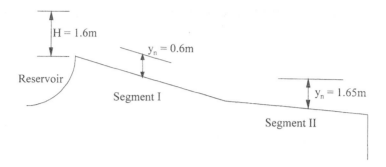

Figure 7.31   Definition sketch, Problem 7.31.

> (f)   What types of channel segments and hydraulic profiles are present at each of the channel segments?
>
> (g)   Provide a schematic sketch of the channel with the critical and normal depths labeled, as well as the hydraulic profiles and hydraulic jumps along the channel.
>
> (h)   Using spreadsheets, calculate the hydraulic profiles. Apply appropriate vertical and longitudinal geometrical scales, and provide a graphical description of each hydraulic profile.

Problem 7.31   Figure 7.31 shows a wide rectangular channel incorporating two segments. Both segments are very long. The Manning roughness coefficient for both segments is $n = 0.020$. The first segment (segment I) comprises an outlet of a reservoir. The reservoir free water surface elevation is $H = 1.6$ m above the bottom of the channel bottom entrance. The normal water depth of segment I is $y_{n1} = 0.60$ m. Segment II ends with an overfall. The normal water depth of segment II is $y_{n2} = 1.65$ m.

> (a)   What is the discharge per unit width of the channel?
>
> (b)   What are the slopes of segments I and II?
>
> (c)   What are the water depths at points A, B and C?
>
> (d)   In which channel segment is a hydraulic jump located? Why?
>
> (e)   What are the water depths upstream and downstream of the hydraulic jump? What is the head loss of the hydraulic jump?
>
> (f)   What types of hydraulic profiles are present in each of the channel segments?
>
> (g)   Provide a schematic diagram of the channel with the critical and normal depths labeled, as well as the hydraulic profiles along the channel.

(h) Using spreadsheets, calculate the hydraulic profiles. Apply appropriate vertical and longitudinal geometrical scales, and provide a graphical description of each hydraulic profile.

Problem 7.32 Over a one-month period a lake with surface area of 2 ha experiences an average inflow of 250 m^3/s and an average outflow of 220 m^3/s. During the month there are two storms, with a total precipitation of 5 cm. Average runoff into the lake is estimated as 10 m^3/s and seepage is negligible. What is the average evaporation rate during this month?

Problem 7.33 In a very large lake changes in area with depth are usually neglected. Under this condition and also assuming there are no inflows or outflows and no vertical velocities, demonstrate mathematically that under steady-state conditions the diffusive flux of temperature is exactly balanced by the radiation flux. If, in addition, a deep ice and snow layer also covers the lake, so that there is essentially zero radiation flux into the water, what is the temperature distribution in the water?

Problem 7.34 Simplify Eqs. (7.7.16) and (7.7.17) for the case of constant depth. Using these equations and also neglecting shear stresses and diffusive transport, what acceleration would be experienced by a flow initially traveling at 1 m/s northward (y-direction) at 40°N latitude? Provide a simple sketch of the flow.

SUPPLEMENTAL READING

Blumberg, A. F. and Mellor, G. L., 1987. A description of a three-dimensional coastal ocean circulation model, in *Three-Dimensional Coastal Ocean Models*. Coastal and Estuarine Sciences, v. 4, ed. N. S. Heaps. American Geophysical Union, Washington, D.C., pp. 1–16.

Csanady, G. T., 1984. *Circulation in the Coastal Ocean*. D. Reidel, Boston.

Henderson, F. M., 1966. *Open Channel Flow*. Macmillan, New York. (Contains a thorough coverage of many topics connected with open channel flow.)

Hofman, J. D., 1992. *Numerical Methods for Engineers and Scientists*. McGraw-Hill, New York. (Many books concerning numerical methods can be used to provide the reader with basic knowledge of numerical methods for the solution of initial value problems. However, we find that this book incorporates a comprehensive presentation of the topic and the different applicable methods.)

Fischer, H. B., List, E. J., Koh, R. C. Y., Imberger, J., and Brooks, N. H., 1979. *Mixing in Inland and Coastal Waters*. Academic Press, New York. (Contains a comprehensive presentation of issues associated with contaminant mixing and transport in surface water flows.)

Gray, W. G., ed., 1986. *Physics-Based Modeling of Lakes, Reservoirs and Impound-ments*. ASCE, New York.

Hutter, K., ed., 1984. *Hydrodynamics of Lakes*. CISM Lecture Series. Springer-Verlag, Wien.

Imberger, J. and Hamblin, P. F., 1982. Dynamics of lakes, reservoirs and cooling ponds. *Ann. Rev. Fluid Mech.* **14**:153–187.

Simons, T. J., 1980. Circulation models of lakes and inland seas. *Canad. Bull. Fish Aquatic Sci.* **203**.

8

Surface Water Waves

8.1 INTRODUCTION

Mechanical systems may vibrate if their displacement from a state of equilibrium is subject to a restoring force. Some examples of such vibrations are represented by water hammer in pipes, sound waves, surface gravity waves, surface capillary waves, and internal waves. The restoring force in these particular cases may be the pipe elasticity, fluid compressibility, gravity, surface tension, or Coriolis force in cases of rotating fluids. In most cases wave motions decay by viscous shear stresses unless there is a constant supply of energy to maintain the motion.

Water quality issues of the marine environment, as well as of lakes, channels, and rivers, are often closely related to topics of wave formation and propagation. For example, waves carry mechanical energy, which can lead to the destruction of maritime structures. With regard to environmental issues, wave energy leads to mixing in the water column and also along the bottom, causing movement of sediments along coastlines, affecting various current patterns in a water body and aiding the transport of solutes and floating materials in the environment.

The present chapter discusses basic features and properties of waves in such environments, focusing on surface waves. These waves, on a horizontal water surface of a marine or lake environment, are confined to two-dimensional propagation in the horizontal plane and are subject to a vertical external gravity restoring force. An initial presentation of the wave equation is given, along with a discussion of its application to several specific issues of environmental fluid mechanics. Special attention also is given to the development and propagation of waves in open channels and rivers. Internal waves are presented in Sec. 13.2.

8.2 THE WAVE EQUATION

Neglecting viscous and Coriolis forces, the general governing equations for fluid motion are the equations of motion and mass conservation as developed in Chap. 2,

$$\rho\left(\frac{\partial \vec{V}}{\partial t} + \vec{V}\cdot\nabla\vec{V}\right) = -\nabla(p + \rho g Z) \tag{8.2.1}$$

$$\frac{\partial \rho}{\partial t} + \vec{V}\cdot\nabla\rho + \rho\nabla\cdot\vec{V} = 0 \tag{8.2.2}$$

where V is the velocity, p is pressure, t is time, g is gravitational acceleration, ρ is fluid density, and Z is the elevation above an arbitrary datum. These equations are considered to apply to a flow field that represents a small deviation from an initial state of fluid at rest and uniform fluid density, ρ_0. Thus velocities, pressure variations from hydrostatic values, and density variations all are assumed to be small, so that any products of these quantities become even smaller (i.e., in nondimensional terms, these quantities are less than one). Equations (8.2.1) and (8.2.2) may then be linearized by neglecting all terms involving products of these small quantities, resulting in

$$\rho_0\frac{\partial \vec{V}}{\partial t} = -\nabla(p + \rho_0 g Z) \tag{8.2.3}$$

$$\frac{\partial \rho}{\partial t} = -\rho_0\nabla\cdot\vec{V} \tag{8.2.4}$$

The velocity vector may comprise two parts. One part is rotational and is associated with the vorticity. An equation for vorticity can be obtained by taking the curl of Eq. (8.2.3). Then, since the curl of the right-hand side of that equation vanishes, it is seen that the vorticity cannot be a time dependent variable. Therefore it is noted that *linear wave theory* (based on the linearized equations of motion) neglects movements of vortex lines with the fluid. Other properties may propagate in the domain.

In addition to the rotational part of the velocity, which is independent of time, the velocity incorporates another, irrotational part that is time dependent. This part can be considered as originating from a potential function Φ. Therefore we may write (see Sec. 4.2)

$$\vec{V} = \nabla\Phi \tag{8.2.5}$$

This expression implies that the velocity of interest for wave propagation stems from a potential function. According to linear wave theory, any steady rotational velocity field does not affect this velocity.

In the usual case of waves developing on a homogeneous water environment, we consider a fluid with constant density and apply a coordinate system in which z is the vertical upward coordinate, with $z = 0$ corresponding to the elevation of the free surface, where the value of p vanishes. For such a case, introducing Eq. (8.2.5) into Eq. (8.2.3) and integrating both sides results in

$$\frac{\partial \Phi}{\partial t} + \frac{p}{\rho} + gz = 0 \qquad (8.2.6)$$

If the density of the fluid is not constant, then combining Eqs. (8.2.5) and (8.2.4) produces

$$\frac{\partial \rho}{\partial t} = -\rho_0 \nabla^2 \Phi \qquad (8.2.7)$$

If there is no change in the fluid density, as with surface water waves, then the left-hand side of this last equation vanishes and the potential function satisfies *Laplace's equation*.

To summarize the results so far, Eqs. (8.2.1)–(8.2.7) indicate that with regard to small amplitude wave motion, the equations of motion can be linearized and the velocity of interest originates from a potential function. However, the phenomenon of surface water waves is associated with the propagation of surface disturbances. The equations representing the free surface of the water, as shown below, also can be linearized to represent the boundary condition of the water free surface by a linear differential equation with regard to the potential function.

Simple wave motions of small amplitude are described by the *wave equation*,

$$\frac{\partial^2 \eta}{\partial t^2} = c^2 \nabla^2 \eta \qquad (8.2.8)$$

where η is the displacement of the free surface and c is the wave velocity. This is a linear hyperbolic differential equation in a two-dimensional space. For the analysis and calculation of many wave phenomena, it is sufficient to consider a one-dimensional space. For this case, Eq. (8.2.8) collapses to

$$\frac{\partial^2 \eta}{\partial t^2} = c^2 \frac{\partial^2 \eta}{\partial x^2} \qquad (8.2.9)$$

where it has been assumed that the wave propagates along the x-direction. The general form of the solution of Eq. (8.2.9) is given by

$$\eta = f(x - ct) + F(x + ct) \qquad (8.2.10)$$

where f and F represent arbitrary functions. Specific forms of these functions are discussed in the following sections of this chapter.

8.3 GRAVITY SURFACE WAVES

Surface water waves develop at the water free surface, which is the interface between the water and air phases. In general, this interface is considered as a discontinuity in the overall distribution of density in the domain. The state of stable equilibrium of the system is represented by water occupying the lowest portions of the domain. Disturbances to the state of equilibrium are represented by surface gravity waves. Such waves propagate only in the horizontal direction, while the restoring gravity force acts in the vertical direction. Therefore there is no preferred horizontal direction of the disturbance propagation, and the waves are *isotropic* (they may move equally in any horizontal direction). However, waves of different wavelengths penetrate to different depths into the water phase. This phenomenon has an implication with regard to the inertia of the fluid particles that are directly affected by the waves. Therefore waves of different wavelengths have different wave speed. The dependence of the wave speed on the wavelength causes *dispersion* of the waves.

To develop a solution of the wave equation, we adopt a coordinate system as in Sec. 8.2, using z as the vertical upward coordinate, with its origin at the water free surface. We also drop the subscript 0 for ρ, with the understanding that water density is constant (not stratified). The undisturbed absolute pressure, p_0 is distributed hydrostatically,

$$p_0 = p_a - \rho g z \qquad (8.3.1)$$

where p_a is the atmospheric pressure. The pressure disturbance, p_e, originating from the wave disturbance, is defined by

$$p_e = p - p_0 \qquad (8.3.2)$$

The linearized equation of motion, from Eq. (8.2.3), is given by

$$\rho \frac{\partial \vec{V}}{\partial t} = -\nabla p_e \qquad (8.3.3)$$

where p_a has been assumed to be constant, so that its gradient is zero. Due to the incompressibility of the fluid, the continuity Eq. (8.2.7) collapses to Laplace's equation,

$$\nabla^2 \Phi = 0 \qquad (8.3.4)$$

Laplace's equation cannot describe wave propagation in a fluid that is completely bounded by stationary surfaces, but it can describe wave propagation by the employment of the boundary condition at the original free water surface. At the original free water surface, the pressure value is associated with the wave displacement, namely the disturbed free surface elevation, η, according to

$$p_e = \rho g \eta \qquad (8.3.5)$$

where

$$\eta = \eta(x, y, t) \tag{8.3.6}$$

Equation (8.3.5), along with Eqs. (8.3.1) and (8.3.2), indicates that at the disturbed free water surface, the pressure is identical to the atmospheric pressure.

According to Eqs. (8.2.6) and (8.3.3) for the irrotational part of the velocity, which is associated with the wave propagation,

$$p_e = -\rho \frac{\partial \Phi}{\partial t} \tag{8.3.7}$$

Therefore at the disturbed free water surface, Eqs. (8.3.5)–(8.3.7) yield the surface boundary condition,

$$\left. \frac{\partial \Phi}{\partial t} \right|_{z=\eta} = -g\eta \tag{8.3.8}$$

In practice, this represents a very complicated boundary condition, since the value of η is not known (in fact, it is part of the desired solution). However, according to linear theory, we may consider that η is a small quantity. Therefore Eq. (8.3.8) is approximated by

$$\left. \frac{\partial \Phi}{\partial t} \right|_{z=0} = -g\eta \tag{8.3.9}$$

According to the mean value theorem, the difference between the values of the left-hand sides of Eqs. (8.3.8) and (8.3.9) is equal to the product of the disturbance η with the derivative of the left-hand-side expression with respect to z, evaluated at a point intermediate to the disturbed and undisturbed water surfaces. This gives a means of checking the degree to which Eq. (8.3.9) provides a good approximation to Eq. (8.3.8).

The rate of change of η is equal to the vertical fluid velocity at the surface, namely,

$$\frac{\partial \eta}{\partial t} + \vec{V} \cdot \nabla \eta = \left. \frac{\partial \Phi}{\partial z} \right|_{z=\eta} \tag{8.3.10}$$

According to linear theory, this expression is simplified by neglecting the advective rate of change of η, since it is a product of two small quantities. Furthermore, the right-hand side of Eq. (8.3.10) is evaluated for $z = 0$, instead of $z = \eta$. Therefore

$$\frac{\partial \eta}{\partial t} = \left. \frac{\partial \Phi}{\partial z} \right|_{z=0} \tag{8.3.11}$$

By differentiating Eq. (8.3.9) with respect to t and applying Eq. (8.3.11), we obtain

$$\frac{\partial^2 \Phi}{\partial t^2} + g \frac{\partial \Phi}{\partial z} = 0 \qquad \text{at} \qquad z = 0 \tag{8.3.12}$$

The wave propagation in the domain is then fully determined by solving the Laplace equation (8.3.4), subject to the boundary condition given by Eq. (8.3.12).

8.4 SINUSOIDAL SURFACE WAVES ON DEEP WATER

8.4.1 The General Wave Propagation Equation

Recall that a general form of the solution for the wave equation was Eq. (8.2.10), which gives the surface displacement, η, as a function of time and position. An alternative function to describe wave movement is

$$F = F(\xi) \tag{8.4.1}$$

where $\xi = \omega t - kx$, ω is the *angular velocity*, or *radian frequency*, and k is the *wave number*. The parameters ω and k are connected with the wave velocity, c, as

$$c = \frac{\omega}{k} \tag{8.4.2}$$

In addition, the wave number is related to the wave length, λ, and the angular velocity is related to the wave period, t_p, as

$$\lambda = \frac{2\pi}{k}; \qquad t_p = \frac{2\pi}{\omega} \tag{8.4.3}$$

By combining Eqs. (8.4.2) and (8.4.3), it is seen that the wave propagates a distance of a single wavelength during one period, namely,

$$\lambda = c t_p \tag{8.4.4}$$

The function $F(\xi)$ of Eq. (8.4.1) has a constant value for a constant value of the variable ξ. Therefore this equation represents constant values of the function F, moving in the positive x-direction. Such a function may refer to waves propagating in that direction. Differentiating Eq. (8.4.1) twice with respect to t and twice with respect to x gives

$$\frac{\partial^2 F}{\partial t^2} = \omega^2 F'' \qquad \frac{\partial^2 F}{\partial x^2} = k^2 F'' \tag{8.4.5}$$

where the double prime represents differentiation with respect to the variable ξ. Using the wave velocity from Eq. (8.4.2) then gives

$$\frac{\partial^2 F}{\partial t^2} = c^2 \frac{\partial^2 F}{\partial x^2} \qquad (8.4.6)$$

This is the same wave equation as was obtained earlier in Eq. (8.2.9).

The potential function (Eq. 8.2.5), which represents motions in the entire water domain subject to surface water waves, usually consists of a product of a function similar to that given by Eq. (8.4.1) and another function, which describes an attenuation of the fluid motion with the water depth. In the case of periodic surface waves, $F(\xi)$ is usually specified as a sine or cosine function. Therefore the potential function can be represented using a sine, a cosine, or the real part of a complex function, such as

$$\Phi = f(z)\sin(\omega t - kx) = f(z)\sin\left[\omega\left(t - \frac{x}{c}\right)\right] \qquad (8.4.7a)$$

$$\Phi = f(z)\cos(\omega t - kx) = f(z)\cos\left[\omega\left(t - \frac{x}{c}\right)\right] \qquad (8.4.7b)$$

$$\Phi = f(z)\exp[i(\omega t - kx)] = f(z)\exp\left[i\omega\left(t - \frac{x}{c}\right)\right] \qquad (8.4.7c)$$

where $f(z)$ is a function that represents the variation in the vertical direction. Since we know that the potential function is governed by the Laplace equation (8.3.5), it follows that

$$f''(z) - k^2 f(z) = 0 \qquad (8.4.8)$$

8.4.2 The Potential Function for Deep Water Waves

The general solution of Eq. (8.4.8) can be obtained by a linear combination of an exponentially decaying term and an exponentially growing term. However, if the water depth is very large, then the exponential growing term should vanish. Therefore a solution of Eq. (8.4.8) that is consistent with a vanishing value of Φ with increasing depth (where $z \to -\infty$) is given by

$$f(z) = \Phi_0 e^{kz} \qquad (8.4.9)$$

where Φ_0 is a constant equal to the value of f at the surface, and $z = 0$.

Combining Eqs. (8.4.7a) and (8.4.9), we obtain

$$\frac{\partial \Phi}{\partial t} = -\omega\Phi_0 e^{kz}\sin(\omega t - kc) \qquad \frac{\partial \Phi}{\partial z} = k\Phi \qquad \frac{\partial^2 \Phi}{\partial t^2} = -\omega^2\Phi \quad (8.4.10)$$

Then, using the first part of this result with Eq. (8.3.10), a solution for the surface displacement is obtained as

$$\eta = \frac{\omega}{g}\Phi_0\sin(\omega t - kx) = a\sin(\omega t - kx) \qquad a = \frac{\omega}{g}\Phi_0 \qquad (8.4.11)$$

This is the relationship between the wave amplitude (a) and the maximum amplitude of the potential function.

By introducing Eq. (8.4.10) into Eq. (8.3.13), a relationship between the frequency and the wave number for gravity waves on a deep-water environment is found as

$$\omega^2 = gk \tag{8.4.12}$$

This is known as a *dispersion relationship*, giving the dependence of the wave propagation on the wavelength (or wave number). Using the definition for wave speed (Eq. 8.4.2), this last result shows that waves of different wavelengths propagate with different velocities:

$$c = \frac{\omega}{k} = \sqrt{\frac{g}{k}} = \sqrt{\frac{g\lambda}{2\pi}} \tag{8.4.13}$$

Considering that $g = 9.81$ m/s^2, we apply Eqs. (8.4.13) and (8.4.3) to obtain

$$c = 1.25\sqrt{\lambda} \qquad t_p = 0.80\sqrt{\lambda} \tag{8.4.14}$$

where λ is measured in meters, t_p is measured in seconds, and c in m/s. Then, considering that the range of typical wavelengths for surface water waves is between 1 m and 100 m, ranges of typical wave velocity and period are

$$1.25 \text{ m/s} \leq c \leq 12.5 \text{ m/s} \qquad 0.8 \text{ s} \leq t_p \leq 8.0 \text{ s} \tag{8.4.15}$$

It should be noted that although Eq. (8.4.15) represents typical values, extreme cases may exist, with wavelengths as low as 0.1 m or as large as 1000 m. Near the sea shore the wavelength is generally much less, and the waves should be described with alternative theories, since the deep-water assumption is no longer valid.

8.4.3 Pathlines of the Fluid Particles

Velocity components in a wavy flow field are obtained by differentiating the potential function. For example, taking Eq. (8.4.7b), the velocity components are found as

$$u = \frac{\partial \Phi}{\partial x} = k\Phi_0 e^{kz} \sin(\omega t - kx) \tag{8.4.16}$$

and

$$w = \frac{\partial \Phi}{\partial z} = k\Phi_0 e^{kz} \cos(\omega t - kx) \tag{8.4.17}$$

Thus both velocity components vary sinusoidally with time and have the same amplitude, which decays exponentially with depth. However, it should be noted that at a fixed position, the horizontal velocity lags the vertical velocity by 90° (this result holds when using either of the expressions of Eq. (8.4.7) to describe the potential function). Equations (8.4.16) and (8.4.17) also indicate that at a fixed position, the velocity vector maintains a constant absolute value and rotates in the clockwise direction.

Based on this oscillating velocity field, it is seen that over a long period of time, there is no net movement of a fluid particle. If it is assumed that the deviations of a fluid particle from an initial position (x_0, z_0) are relatively small, then the differential equations of the fluid particle pathline are given approximately by

$$\frac{dx}{dt} = u \cong k\Phi_0 e^{kz_0} \sin(\omega t - kx_0) \tag{8.4.18}$$

and

$$\frac{dz}{dt} = w \cong k\Phi_0 e^{kz_0} \cos(\omega t - kx_0) \tag{8.4.19}$$

Direct integration of these results gives an approximation for the instantaneous position of the fluid particle,

$$x = \frac{k}{\omega}\Phi_0 e^{kz_0} \cos(\omega t - kx_0) + C_1 \tag{8.4.20a}$$

$$z = \frac{k}{\omega}\Phi_0 e^{kz_0} \sin(\omega t - kx_0) + C_2 \tag{8.4.20b}$$

where C_1 and C_2 are constants for each particular fluid particle.

By eliminating time from Eqs. (8.4.20a) and (8.4.20b), we obtain

$$(x - C_1)^2 + (z - C_2)^2 = \left(\frac{k}{\omega}\Phi_0 e^{kz_0}\right)^2 \tag{8.4.21}$$

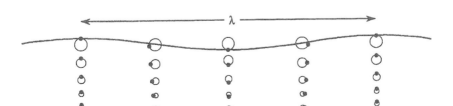

Figure 8.1 Schematic description of pathlines and instantaneous positions of fluid particles subject to motion due to sinusoidal surface wave on deep water.

This result shows that the fluid particles move in circular pathlines, which also is evident from the previous conclusion that the velocity components have equal amplitudes (Eqs. 8.4.16 and 8.4.17). The radius of the pathline decays exponentially with water depth and does not depend on the horizontal coordinate. According to Eqs. (8.4.20) and (8.4.21), the phase of the fluid particle location does not depend on the z coordinate. Figure 8.1 provides a schematic description of various pathlines of different fluid particles and their instantaneous positions.

8.4.4 The Shape of the Streamlines

The shape of the streamlines is calculated by considering the following relationships between the stream function and the real parts of the velocity components given by Eqs. (8.4.15) and (8.4.16):

$$\frac{\partial \Psi}{\partial z} = u = k\Phi_0 e^{kz} \sin(\omega t - kx) \tag{8.4.22}$$

$$\frac{\partial \Psi}{\partial x} = -w = -k\Phi_0 e^{kz} \cos(\omega t - kx) \tag{8.4.23}$$

Direct integration of these expressions yields

$$\Psi = \Phi_0 e^{kz} \sin(\omega t - kx) + C \tag{8.4.24}$$

where C is an arbitrary constant. This shows that streamlines, like the propagating surface waves, are sinusoidal and the amplitude of the streamlines decays exponentially with the water depth.

8.4.5 The Wave Energy

The excess energy in surface water waves is divided between kinetic and potential energy. The excess potential wave energy incorporated in a surface area of unit width, with length equal to one wavelength, and where the water depth, h, is large but finite, is

$$WE_p = \int_0^\lambda \left[\int_{-h}^\eta \rho g z \, dz - \int_{-h}^0 \rho g z \, dz \right] dx$$

$$= \int_0^\lambda \left[\frac{1}{2} \rho g (\eta^2 - h^2) + \frac{1}{2} \rho g h^2 \right] dx = \frac{1}{2} \rho g \int_0^\lambda \eta^2 \, dx \tag{8.4.25}$$

It should be noted that wave displacements above the level $z = 0$, as well as below $z = 0$, carry positive potential energy. The raised free surface adds potential energy by adding fluid above the level $z = 0$, while the depressed free surface adds potential energy by the removal of fluid from below the level $z = 0$.

The kinetic energy due to the wave motion can be obtained by a volumetric integral performed over a volume of a prism incorporating the entire water depth and a water surface area of unit width and a single wavelength in the longitudinal direction,

$$WE_k = \int \frac{1}{2}\rho(\nabla\Phi)^2 dU \qquad (8.4.26)$$

where dU is an elementary volume. Considering that Laplace's equation, given by Eq. (8.3.4), is satisfied, and applying the *divergence theorem*, Eq. (8.4.26) can be modified as

$$WE_k = \int \frac{1}{2}\rho\nabla\cdot(\Phi\nabla\Phi)\,dU = \int \frac{1}{2}\rho\left(\Phi\frac{\partial\Phi}{\partial n}\right)dS \qquad (8.4.27)$$

where dS is an elementary surface area and n is an outward unit normal to the surface of the prism.

There is no contribution of kinetic energy from the bottom of the water environment, where the derivative of Φ normal to the bottom vanishes. Contributions of kinetic energy at the left and right side walls of the prism cancel each other. Therefore the only surface of the prism that should be considered for the calculation of E_k is at the water surface. Therefore Eq. (8.4.27) can be replaced with

$$WE_k = \int_0^\lambda \frac{1}{2}\left[\Phi\frac{\partial\Phi}{\partial z}\right]_{z=0} dx \qquad (8.4.28)$$

This result is further modified by introducing an expression for the velocity potential. From Eq. (8.4.10),

$$\Phi = \frac{1}{k}\frac{\partial\Phi}{\partial z} \qquad (8.4.29)$$

Substituting Eqs. (8.4.29) and (8.3.12) into Eq. (8.4.28), we obtain

$$WE_k = \int_0^\lambda \frac{1}{2k}\rho\left(\frac{\partial\eta}{\partial t}\right)^2 dx \qquad (8.4.30)$$

The appearance of the wave number k in this result is related to the fact that there is a layer, with thickness proportional to the wavelength, though much smaller, which is subject to motion due to the surface wave movement upwards and downwards.

By rearranging Eq. (8.4.12) and substituting into Eq. (8.4.11), we obtain

$$\eta = \frac{k}{\omega} \Phi_0 \sin(\omega t - kx) = a \sin(\omega t - kx) \qquad (8.4.31)$$

Thus

$$\frac{\partial \eta}{\partial t} = k\Phi_0 \cos(\omega t - kx) = \omega a \cos(\omega t - kx) \qquad (8.4.32)$$

This result is then substituted into Eqs. (8.4.25) and (8.4.30), also using Eqs. (8.4.31) and (8.4.12). After dividing by the wave length, λ, to obtain the total wave energy per unit area of the water surface, the total energy is

$$E = E_p + E_k = \tfrac{1}{2}\rho g a^2 \qquad (8.4.33)$$

Note that this result refers to energies per unit area, whereas previous equations refer to the energies per an area of unit width and a single wavelength in length.

Equation (8.4.33) indicates that although the amount of potential and kinetic energies per unit surface area of the water varies from point to point, the total sum of potential and kinetic energies per unit water surface is constant over the entire water surface.

8.5 SINUSOIDAL SURFACE WAVES FOR SHALLOW WATER DEPTH

8.5.1 The Potential Function of Shallow Water Waves

Waves in deep water were considered in the previous section. In practice, deep water is defined when depth is greater than the wavelength, λ. If the water depth is uniform, but smaller than λ, then the water environment is considered to be shallow and the solution of the differential Eq. (8.4.8) should satisfy the condition of zero normal velocity at the bottom, or

$$\frac{\partial \Phi}{\partial z} = 0 \qquad \text{at} \qquad z = -h \qquad (8.5.1)$$

Therefore it is necessary to use the more general form of solution,

$$f(z) = C_1 e^{kz} + C_2 e^{-kz} \qquad (8.5.2)$$

(recall that in Sec. 8.4.2 the second term was unnecessary for deep water conditions — see Eq. 8.4.9). As the function $f(z)$ must vanish when $z = -h$, it follows that

$$C_1 e^{-kh} = -C_2 e^{kh} = \tfrac{1}{2}\Phi_0 \qquad (8.5.3)$$

where Φ_0 is a constant. Then Eq. (8.5.2) becomes

$$f(z) = \Phi_0 \cosh[k(z+h)] \tag{8.5.4}$$

where $\cosh(x)$ is the *hyperbolic cosine* function and

$$\cosh x = \frac{1}{2}(e^x + e^{-x}) \qquad \sinh x = \frac{d}{dx}\cosh x = \frac{1}{2}(e^x - e^{-x}) \tag{8.5.5}$$

It should be noted that at $z = -h$, the derivative of $f(z)$ vanishes, so that the condition of zero normal velocity at the bottom is satisfied.

Using Eq. (8.5.4) in Eq. (8.4.7a), we find

$$\frac{\partial \Phi}{\partial z} = \left. \frac{f'}{f} \right|_{z=0} \Phi = [k\tanh(kh)]\Phi \tag{8.5.6}$$

Where $\tanh(x) = \sinh(x)/\cosh(x)$ is the *hyperbolic tangent* function. Equations (8.3.12) and (8.5.6) then give

$$\omega^2 = gk\tanh(kh) \tag{8.5.7}$$

which is the *dispersion relation* for shallow-water waves.

8.5.2   The Wave Velocity of Propagation

Wave velocity is given by Eq. (8.4.2), which, when used with Eq. (8.5.7), gives

$$c = \frac{\omega}{k} = \sqrt{\frac{g}{k}\tanh(kh)} \tag{8.5.8}$$

In order to illustrate the effect of the wavelength on the velocity of wave propagation in a water layer of constant depth, Fig. 8.2 demonstrates several curves showing the variation of the hyperbolic functions. Using the relationships of Eq. (8.4.2), an expression relating wave velocity of propagation to the wavelength is obtained:

$$c = \sqrt{\frac{g\lambda}{2\pi}\tanh\left(\frac{2\pi h}{\lambda}\right)} \tag{8.5.9}$$

This expression and Fig. 8.2 indicate that for large values of λ, the wave velocity is represented by the square root of a product of a very large term and a very small term. The limit of this product for a very large value of λ can be obtained by a series expansion of $\tanh(x)$, for small values of x. For small values of x, using only to the first term of the series expansion,

$$\tanh(x) \longrightarrow x \qquad \text{if} \qquad x \to 0 \tag{8.5.10}$$

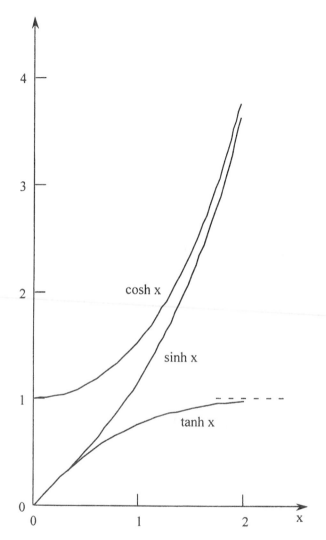

Figure 8.2 Graphs of cosh(x), sinh(x), and tanh(x).

Using this relation in Eq. (8.5.10) gives

$$c = \sqrt{gh} \qquad \text{if} \qquad \lambda \to \infty \tag{8.5.11}$$

For waves on deep water, the value of the tanh term of Eq. (8.5.9) is approximately unity (see Eq. 8.4.12). Figure 8.3 shows the variability of the wave velocity as a function of the relationship between the wavelength and the

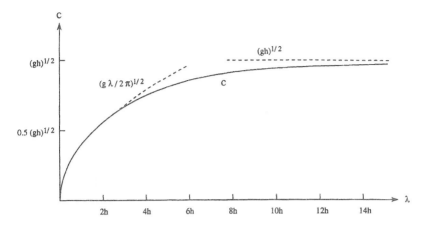

Figure 8.3 The wave speed versus the wavelength for water environment of constant depth.

depth of the water environment. This figure provides some guidance about the possible definition for deep and shallow water, as well as for long waves.

Using Eqs. (8.4.2), (8.5.7), and (8.5.9), it can be shown that

$$\frac{c\omega}{g} = \tanh\left(\frac{\omega h}{c}\right) \tag{8.5.12}$$

This result is useful for the evaluation of the variability of the wave velocity of propagation due to gradual decrease of the water depth, for a constant frequency. This is shown in Fig. 8.4, where Eq. (8.5.12) is applied to depict the variability of the wave velocity versus water depth, for constant wave frequency. Sinusoidal waves approaching the coastline pass through water of gradually decreasing depth, while their frequency is kept unchanged. Therefore the number of wave crests reaching the beach per unit time is equal to the number approaching the coastline. Figure 8.4 shows how the wave speed of such waves gradually decreases with the water depth. In addition, wavelength decreases and, in fact, is much more significant than the decrease of the wave velocity.

In general, the original wave crests approach the coastline with some orientation angle. Due to the decrease of the water depth, such wave crests tend to align with the coastline, and their orientation angle decreases. Figure 8.5 illustrates this phenomenon, which is associated with the decrease of the wave velocity in the region of shallow water. In other words, as a wave approaches the shore at some angle, the portion of the wave closest to the shoreline experiences a decrease in velocity sooner than portions further away. This

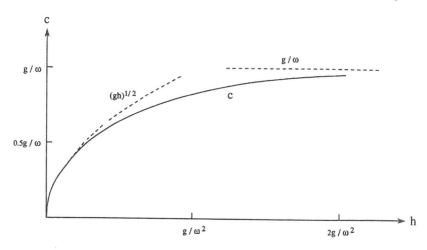

Figure 8.4 Wave speed as a function of water depth.

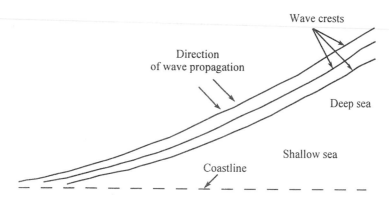

Figure 8.5 Alignment of wave crest approaching the coastline.

causes the wave crest to effectively change its alignment with respect to the shoreline.

8.5.3 Pathlines of the Fluid Particles

Following the same approach as in Sec. 8.4.3, expressions for the velocity components of fluid particles are obtained by differentiating the potential function Eq. (8.4.7b), using Eq. (8.5.4) for the depth variation, giving

$$u = \frac{\partial \Phi}{\partial x} = \{k\Phi_0 \cosh[k(z + h)]\} \sin(\omega t - kx) \qquad (8.5.13a)$$

and

$$w = \frac{\partial \Phi}{\partial z} = \{k\Phi_0 \sinh[k(z+h)]\} \cos(\omega t - kx) \tag{8.5.13b}$$

Using Eq. (8.3.12), w is found from Eq. (8.5.13) for $z = 0$, and the resulting expression is integrated over time to obtain the relationship between the potential function amplitude and the wave amplitude,

$$\eta = \frac{k\Phi_0}{\omega} \cosh(kh) \sin(\omega t - kx) = a \sin(\omega t - kx) \qquad a = \frac{k\Phi_0}{\omega} \cosh(kh) \tag{8.5.14}$$

Using this definition for the amplitude a allows alternative expressions for the velocity components, by substituting into Eq. (8.5.13):

$$u = \frac{dx}{dt} = \frac{a\omega}{\cosh(kh)} \cosh[k(z+h)] \sin(\omega t - kx) \tag{8.5.15a}$$

$$w = \frac{dz}{dt} = \frac{a\omega}{\cosh(kh)} \sinh[k(z+h)] \cos(\omega t - kx) \tag{8.5.15a}$$

In contrast to the situation for waves on deep water, in the case of water of finite depth, the amplitude of the fluid particle horizontal velocity is different from that in the vertical direction. As shown in Fig. 8.2, for waves on finite water depth the amplitude of the horizontal velocity is larger than that of the vertical velocity, but they become almost identical for large water depth.

Using the same approach as was used for deep water waves (i.e., integrating Eq. 8.5.15 over time), the following expressions for the pathlines of the fluid particles can be obtained:

$$x = -a \frac{\cosh[k(z_0 + h)]}{\cosh(kh)} \cos(\omega t - kx_0) + C_1 \tag{8.5.16a}$$

$$z = a \frac{\sinh[k(z_0 + h)]}{\cosh(kh)} \sin(\omega t - kx_0) + C_2 \tag{8.5.16b}$$

where C_1 and C_2 are constants connected with the initial location of the fluid particle.

We eliminate the time-dependent expression from the two parts of Eq. (8.5.16) to obtain the expression for the curve representing the particle pathline,

$$\frac{(x - C_1)^2}{\left\{ a \dfrac{\cosh[k(z_0 + h)]}{\cosh(kh)} \right\}^2} + \frac{(z - C_2)^2}{\left\{ a \dfrac{\sinh[k(z_0 + h)]}{\cosh(kh)} \right\}^2} = 1 \tag{8.5.17}$$

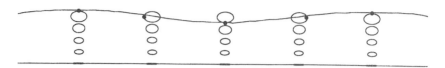

Figure 8.6   Pathlines of fluid particles in a sinusoidal wave on water with finite depth.

This indicates that particle pathlines are ellipses, as shown in Fig. 8.6. The major and minor axes of the elliptical pathlines are represented by the terms of the denominators of Eq. (8.5.17). In deeper water the pathlines tend to be more circular since, except near the bottom, the cosh and sinh terms of Eq. (8.5.17) become practically identical, as was the result found in Eq. (8.4.21). In very shallow water, the pathlines are flattened, as the minor axis of the elliptical pathlines decreases.

With regard to the calculation of the streamlines, we follow a similar procedure as before, to find

$$\frac{\partial \Psi}{\partial z} = u = \frac{a\omega}{\cosh(kh)} \cosh[k(z+h)] \sin(\omega t - kx) \tag{8.5.18a}$$

$$\frac{\partial \Psi}{\partial x} = -w = -\frac{a\omega}{\cosh(kh)} \sinh[k(z+h)] \cos(\omega t - kx) \tag{8.5.18b}$$

By direct integration of Eq. (8.5.18), the stream function is found as

$$\Psi = \frac{a\omega}{k \cosh(kh)} \sinh[k(z+h)] \sin(\omega t - kx) \tag{8.5.19}$$

This result indicates that the streamlines, namely lines of constant value of Ψ, have the shape of a sinusoidal wave whose amplitude decays with depth as sinh. At the bottom of the water environment $z = -h$, so it is represented by the streamline with $\Psi = 0$.

The excess potential energy of the waves on water of uniform finite depth takes the same form as for waves on deep water, given by Eq. (8.4.25). The expression for the kinetic energy of the wave also is identical to that of waves on deep water, given by Eqs. (8.4.28)–(8.4.30), with Eq. (8.5.6) used to specify the vertical gradient of the potential function (i.e., in Eq. 8.4.29). However, the total energy per unit area of the water surface has the same relationship typical of waves on deep water, namely Eq. (8.4.33).

8.6   THE GROUP VELOCITY

Previous sections of the present chapter refer to sinusoidal waves on deep water, as well as water of finite depth. A variety of properties of surface water waves were introduced and analyzed while assuming that the surface water waves are sinusoidal. It should be noted that the initial reference to sinusoidal surface water waves is justified for two main reasons: (1) surface waves are most commonly observed to be roughly sinusoidal, and (2) waves of more complicated shape can be analyzed by *Fourier analysis*, in which waves are considered as a linear combination of different sinusoidal disturbances. Due to the linearity of Laplace's equation and the linear boundary condition at the water surface as given by Eq. (8.3.12), a linear combination of various potential functions describing different sinusoidal waves can also be a potential function for a more general wave field.

When surface waves are represented as a linear combination of sinusoidal waves, it is important to keep in mind the dispersive property characterizing waves of different wavelengths. As shown previously, waves of different wavelengths have different wave speeds (refer to Eqs. 8.4.13 and 8.5.9). Therefore if a disturbance at the water surface is created at one location of the water environment, then at a later time different sinusoidal components of the water surface disturbance will be found at other locations, due to the original disturbance. For example, consider large disturbances created by a storm. In general, this causes the boundary conditions to be complicated and nonlinear, so that at the time and place of the storm, linear wave theory may not be appropriate. However, storm waves are reduced to groups of smaller size waves, which obtain energy from the high waves due to wave dispersion and are called *swell*. These reduced size waves can be analyzed by use of linear wave theory. Consider a small group of waves whose wave speed is c. After a time interval, t, this group of waves will be found at a distance Ut from the origin of the disturbance, where U is called the *group velocity* and is defined below (Eq. 8.6.10). In deep water, the group velocity is approximately equal to half of the wave speed. Also, it can be shown that the wave energy propagates at the group velocity.

Consider that the wavelength gradually varies from one wave to the next in the group of waves. A local phase, α, can be defined, and the value of α at every wave crest can be expressed as an even multiple of π, namely,

$$\alpha = 2\pi n \tag{8.6.1}$$

where n is an integer. The value of n increases by one for each successive wave crest that passes a particular point. Between the wave crests, the value of α varies smoothly. In the wave troughs, the value of α is an odd multiple of π. The rate of change of α with time is given in radians per second by

$$\frac{\partial \alpha}{\partial t} = 2\pi f = \omega \tag{8.6.2}$$

The phase also is a function of the longitudinal distance between successive wave crests. At a given time, the value of α decreases with x, at a rate equal to the wavenumber k in radians per meter,

$$\frac{\partial \alpha}{\partial x} = -k \tag{8.6.3}$$

Equations (8.6.2) and (8.6.3) imply that the water surface displacement, η, can be represented as a wave of slowly variable amplitude η_1, by either

$$\eta = \eta_1(x, t) \exp[i\alpha(x, t)] \tag{8.6.4a}$$

or

$$\eta = \eta_1(x, t) \cos[\alpha(x, t)] \tag{8.6.4b}$$

Around a particular location x_0, and particular time t_0, the value of α can be expressed by a Taylor series, from which the following linear combination is considered:

$$\alpha(x, t) = \alpha(x_0, t_0) - k_0(x - x_0) + \omega(t - t_0) \tag{8.6.5}$$

Then, differentiating Eq. (8.6.2) with respect to x and Eq. (8.6.3) with respect to t and subtracting one from the other, we obtain

$$\frac{\partial k}{\partial t} + \frac{\partial \omega}{\partial x} = 0 \tag{8.6.7}$$

We consider that ω is a function of k, as implied by Eq. (8.4.2), namely,

$$\omega = \omega(k) \tag{8.6.8}$$

Introducing this relationship into Eq. (8.6.7) gives

$$\frac{\partial k}{\partial t} + U \frac{\partial k}{\partial x} = 0 \tag{8.6.9}$$

where U is the *group velocity*, defined as

$$U = U(k) = \frac{d\omega}{dk} \tag{8.6.10}$$

Equation (8.6.9) indicates that k is constant along paths in the $(x–t)$ plane, which satisfy the condition that

$$x - Ut = const \tag{8.6.11}$$

In the case of waves on deep water, we introduce Eq. (8.6.11) into Eq. (8.6.10) and use Eq. (8.4.2) to obtain

$$U = \frac{d}{dk}(\sqrt{gk}) = \frac{1}{2}\sqrt{\frac{g}{k}} = \frac{1}{2}\frac{\omega}{k} = \frac{1}{2}c \tag{8.6.12}$$

This result indicates that, in the case of waves on deep water, the group velocity is equal to one-half of the wave speed.

In the case of waves on water of finite uniform depth, differentiation of Eq. (8.5.7) gives

$$2\omega\frac{d\omega}{dk} = g\left[\tanh(kh) + \frac{kh}{\cosh^2(kh)}\right] \tag{8.6.13}$$

By introducing Eq. (8.6.13) and Eq. (8.5.8) into Eq. (8.6.10), we obtain

$$U = c\frac{k}{\omega}\frac{d\omega}{dk} = c\frac{1}{2}\left[1 + \frac{2kh}{\sinh(2kh)}\right] \tag{8.6.14}$$

For large values of h this result converges to that of Eq. (8.6.12). For very small values of h, Eq. (8.6.14) indicates that the group velocity approaches c. Such a result is typical of *nondispersive waves*, whose wave speed is independent of k. In the case of nondispersive waves, the wave speed and the group velocity are identical. As an example of nondispersive waves we may consider long water waves, whose wave speed is given by Eq. (8.5.11). In this case the frequency ω is a multiple of a constant, namely the wave speed, and the wave number, k.

In order to demonstrate the group velocity concept, consider a sinusoidal wave represented by either of the following expressions:

$$\eta = a\exp[i(\omega t - kx)] \tag{8.6.15}$$
$$\eta = a\cos(\omega t - kx) \tag{8.6.16}$$

Equation (8.6.15) represents a sum of a real and an imaginary value. Each part has the form of a sinusoidal wave, but there is a phase lag of 90° between these two waves. Therefore it is possible to consider the sinusoidal wave represented by Eq. (8.4.31) or by Eq. (8.6.16). For the present example, we consider two waves with the format of Eq. (8.6.16), having the same amplitude and with wave numbers that are almost identical. The superposition of such waves is given by

$$\begin{aligned}a\cos(\omega_1 t - k_1 x) &+ a\cos(\omega_2 t - k_2 x)\\ &= \left\{2a\cos\left[\tfrac{1}{2}(\omega_2 - \omega_1)t - \tfrac{1}{2}(k_2 - k_1)x\right]\right\}\\ &\times \cos\left[\tfrac{1}{2}(\omega_2 + \omega_1)t - \tfrac{1}{2}(k_2 + k_1)x\right]\end{aligned} \tag{8.6.17}$$

This expression incorporates a slowly varying amplitude, represented by the term in the curly brackets. This term varies with a small wave number, equal to half of the difference between the wave numbers of the two waves. This slowly variable amplitude applies to a term oscillating with a much larger wave number equal to the average wave number of the superposed waves. The

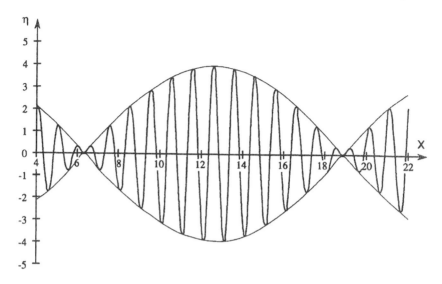

Figure 8.7 The linear combination of two sinusoidal waves.

combination of the two waves takes the form (shown in Fig. 8.7) of a series of wave packets traveling with the group velocity, U. Provided that k_2 and k_1 are close in magnitude, U is represented by

$$U = \frac{\omega_2 - \omega_1}{k_2 - k_1} = \frac{d\omega}{dk} \tag{8.6.18}$$

8.7 WAVES IN OPEN CHANNELS

8.7.1 General Aspects

Previous sections of this chapter refer to the development of sinusoidal waves on the water surface. It was considered that the wave motion can be described as originating from a potential function, under specific conditions that approximately take place at the water surface. The topics presented in those previous sections are relevant to the marine environment, as well as to open channels, but open channels usually represent a shallow water environment where flow, as well as wave propagation, are unidirectional phenomena. The present section considers a variety of nonsinusoidal types of surface waves that may develop in open channels.

Equation (8.5.11) provides the value of the sinusoidal wave speed in a shallow environment. As shown below, this expression is applicable with

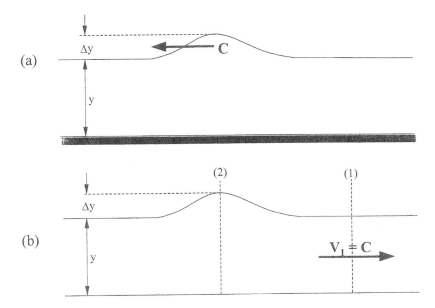

Figure 8.8 Propagation of a small disturbance on the channel water surface: (a) stagnant coordinate system; and (b) coordinate system moving with the disturbance speed c.

regard to any small disturbance developed on the surface of the channel water. Figure 8.8 shows the development of a small disturbance Δy on the water surface, assuming the channel water is stagnant. As shown in Fig. 8.8a, the disturbance propagates from left to right with a wave speed, c. It is assumed that the disturbance is small and its propagation is associated with negligible energy loss. Using a coordinate system moving with the speed of the small disturbance, the domain is subject to steady state, as indicated in Fig. 8.8.

Owing to the steady-state conditions and no energy loss, Bernoulli's equation may be applied between cross section 1 and cross section 2 in Fig. 8.8. Along with conservation of mass, basic equations for this flow situation can be written as

$$cy = V_2(y + \Delta y) \tag{8.7.1a}$$

$$y + \frac{c^2}{2g} = y + \Delta y + \frac{V_2^2}{2g} \tag{8.7.1b}$$

where y is the water depth in regions unaffected by the disturbance and V_2 is the velocity at the location of the small disturbance, measured by an observer moving with the disturbance propagation speed.

Introducing Eq. (8.7.1a) into Eq. (8.7.1b) and rearranging, we obtain

$$c^2 = \frac{2g\Delta y}{[1 - (y/y + \Delta y)^2]} = \frac{2g\Delta y}{[1 - 1/1 + 2\Delta y/y + (\Delta y/y)^2]}$$

$$\approx \frac{2g\Delta y}{1 - (1 - 2\Delta y/y)} = gy \tag{8.7.2}$$

This result indicates that if the water level at the downstream end of the channel gradually decreases, a series of small disturbances with decreasing propagation speed is created and propagates in the upstream direction, causing an oblique shape of the channel water surface. On the other hand, if the water surface increases at the downstream end of the channel, a series of small disturbances with increasing propagation speed is created and forms a *surge*.

Calculations of wave propagation in open channels are based on conservation of mass and the equation of motion. Referring to Fig. 8.9, the conservation of mass principle leads to (see also section 7.6.5)

$$\frac{\partial Q}{\partial x} + B\frac{\partial y}{\partial t} = 0 \tag{8.7.3}$$

where Q is the channel discharge and B is the width of the channel surface. For a rectangular channel, this simplifies to

$$\frac{\partial q}{\partial x} + \frac{\partial y}{\partial t} = 0 \tag{8.7.4}$$

where q is the channel discharge per unit width (applicable for rectangular channels — see Chap. 7). The discharge per unit width is calculated as the

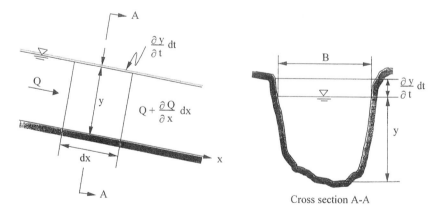

Figure 8.9 Definition sketch for the equation of mass conservation in open channel flow.

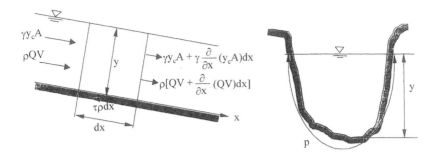

Figure 8.10 Definition sketch for the equation of motion in open channel flow.

product of mean channel velocity and depth,

$$q = Vy \tag{8.7.5}$$

Therefore Eq. (8.7.4) can be written as

$$\frac{\partial y}{\partial t} + \frac{\partial}{\partial x}(Vy) = 0 \tag{8.7.6}$$

This is a first-order partial differential equation, similar to the *advection equation*, and is a first-order hyperbolic partial differential equation.

As shown in Fig. 8.10, the forces acting on an elementary volume of the channel water are due to pressure and shear stress,

$$\sum \Delta F = -\frac{\partial}{\partial x}(\gamma y_c A)\Delta x - \frac{\partial z}{\partial x}(\gamma A)\Delta x - \tau_0 P \Delta x \tag{8.7.7}$$

where $\gamma = \rho g$ is the specific weight of the water, ρ is the water density, y_c is the depth of the center of gravity of the cross-sectional area of the channel from the water surface, z is the elevation of the channel bed with regard to an arbitrary datum, A is the cross-sectional area of the channel, P is the wetted perimeter, and τ_0 is the shear stress along the wetted perimeter. From conservation of momentum, this sum of forces is equal to the rate of change of momentum of the volume,

$$\sum \Delta F = \rho \Delta x \left[\frac{\partial Q}{\partial t} + \frac{\partial (QV)}{\partial x} \right] \tag{8.7.8}$$

Combining Eqs. (8.7.7) and (8.7.8) yields

$$\rho \left[\frac{\partial Q}{\partial t} + \frac{\partial}{\partial x}(QV) \right] = -\gamma A \frac{\partial h}{\partial x} - \tau_0 P \tag{8.7.9}$$

where h is the elevation of the water surface, given by

$$h = z + y \tag{8.7.10}$$

For a wide rectangular channel, P is approximately equal to the width, and Eq. (8.7.9) may be divided by the channel width to obtain

$$\rho \left[\frac{\partial(yV)}{\partial t} + \frac{\partial}{\partial x} \left(yV^2 + g\frac{y^2}{2} \right) \right] = \gamma y S_0 - \tau_0 \tag{8.7.11}$$

where S_0 is the slope of the channel bed,

$$S_0 = -\frac{dz}{dx} \tag{8.7.12}$$

We subtract the continuity result, Eq. (8.7.6), from Eq. (8.7.11) and divide the result by the water depth and the specific weight of the water, to obtain

$$\frac{1}{g} \left(\frac{\partial V}{\partial t} \right) + \frac{\partial}{\partial x} \left(\frac{V^2}{2g} + y + z \right) = -\frac{\tau_0}{\gamma y} \tag{8.7.13}$$

This also can be represented in another form, as

$$\frac{\partial H}{\partial x} + \frac{1}{g} \frac{\partial V}{\partial t} + \frac{\tau_0}{\gamma y} = 0 \tag{8.7.14}$$

where H is the water head,

$$H = h + \frac{V^2}{2g} \tag{8.7.15}$$

The last term on the left-hand side of Eq. (8.7.14) represents friction losses for the flow. This term is usually referred to as the *friction slope*, introduced earlier as Eq. (7.2.21),

$$S_f = \frac{\tau_0}{\gamma y} \tag{8.7.16}$$

By introducing Eq. (8.7.2) into Eq. (8.7.13), it can be shown that

$$2c\frac{\partial c}{\partial x} + V\frac{\partial V}{\partial x} + \frac{\partial V}{\partial t} = g(S_0 - S_f) \tag{8.7.17}$$

Also, introducing Eq. (8.7.2) into Eq. (8.7.6), we obtain

$$2V\frac{\partial c}{\partial x} + c\frac{\partial V}{\partial x} + 2\frac{\partial c}{\partial t} = 0 \tag{8.7.18}$$

Adding Eqs. (8.7.17) and (8.7.18) results in

$$\frac{\partial(V + 2c)}{\partial t} + (V + c)\frac{\partial(V + 2c)}{\partial x} = g(S_0 - S_f) \tag{8.7.19}$$

and subtracting them gives

$$\frac{\partial(V - 2c)}{\partial t} + (V - c)\frac{\partial(V - 2c)}{\partial x} = g(S_0 - S_f) \tag{8.7.20}$$

Equations (8.7.19) and (8.7.20) are two first-order hyperbolic partial differential equations. We may refer to the $(x-t)$ plane and identify the characteristics of each of these equations. The *family of characteristics* of Eqs. (8.7.19) and (8.7.20) are given, respectively, by

$$\frac{dx}{dt} = V + c \qquad \frac{dx}{dt} = V - c \tag{8.7.21}$$

If the terms on the right-hand sides of Eqs. (8.7.19) and (8.7.20) are small, then Eq. (8.7.19) indicates that an observer moving with velocity $(V + c)$ observes a constant fluid property $(V + 2c)$. This property incorporates a given relationship between the flow velocity and the water depth, which is seen from Eq. (8.7.2),

$$V + c = V + \sqrt{gy} \tag{8.7.22}$$

With regard to Eq. (8.7.20), vanishing values of the terms on the right-hand side indicate that an observer moving with velocity $(V - c)$ observes a constant fluid property $(V - c)$, and

$$V - c = V - \sqrt{gy} \tag{8.7.23}$$

The paths of the moving observers can be traced on the $(x-t)$ plane, as shown in Fig. 8.11.

The first family of characteristics shown in Fig. 8.11, referring to the observer moving with velocity $(V + c)$, has an inverse slope $(V + c)$ in the $(x-t)$ plane. A second family of characteristics refers to the observer moving with velocity $(V - c)$. The particular case represented in Fig. 8.11 is associated with the first characteristic family, given as straight lines. If the right-hand side of Eqs. (8.7.19) and (8.7.20) vanishes, and a straight line represents a particular characteristic in the $(x-t)$ plane, then all characteristics of the same type are straight lines. This can be shown by considering two characteristics of the first type, namely curves AB and DE, and two characteristics of the second type, namely AD and BE. Since DE is a straight line, both quantities, $(V + c)$ and $(V + 2c)$, are constant along that line. Therefore their difference (c) must be constant and V also should be constant. By referring to the second type of characteristics the following relationships may be obtained:

$$c_D = c_E \qquad V_D = V_E \tag{8.7.24a}$$
$$V_A - 2c_A = V_D - 2c_D \tag{8.7.24b}$$
$$V_B - 2c_B = V_E - 2c_E \tag{8.7.24c}$$
$$V_A - 2c_A = V_B - 2c_B \tag{8.7.24d}$$

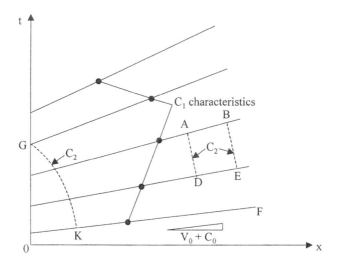

Figure 8.11   Characteristic curves for two types of moving observers.

Since AB is a characteristic of the first type, we also find

$$V_A + 2c_A = V_B + 2c_B \tag{8.7.24e}$$

The various expressions of Eq. (8.7.24) can be satisfied only if

$$V_A = V_B \qquad c_A = c_B \tag{8.7.25}$$

This indicates that the curve AB is a straight line, and therefore all characteristic curves of Fig. 8.11 are straight lines. Along each of these characteristics, values of V and c, and thereby y, are constant.

8.7.2   The Simple Negative Wave Problem

The set of Eqs. (8.7.19) and (8.7.20) has an analytical solution in the case of the simple negative wave problem. This problem is represented by the requirement to calculate the variation of flow velocity and water depth upstream of a very long river mouth, where the water depth is subject to a given rate of decrease. We may consider that Fig. 8.11 shows a set of characteristics of the first type issued from the river mouth, which is located at $x = 0$. The line OF, which is the first characteristic of the first type, has a constant inverse slope, $(V_0 + c_0)$. It is the line of separation between the "zone of quiet" and the domain of propagating disturbances. Because the line OF is a straight line, all characteristics of the first type, shown in Fig. 8.11, are straight lines. If values of the velocity (V and c) and the water depth y are given at $x = 0$, we may consider any arbitrary point G with coordinates $x = 0$ and an arbitrary t value on the x–t plane. The characteristic of the first type issued from G has the following inverse slope:

$$\frac{dx}{dt} = V(t) + c(t) \tag{8.7.26}$$

We also consider the characteristic of second type which is issued from G. This characteristic is shown by the dashed line between G and K. Owing to Eq. (8.7.20),

$$V(t) - 2c(t) = V_0 - 2c_0 \tag{8.7.27}$$

We introduce the value of $c(t)$ or $V(t)$ into Eq. (8.7.26), to obtain

$$\frac{dx}{dt} = \frac{3}{2}V(t) - \frac{1}{2}V_0 + c_0 \tag{8.7.28a}$$

$$\frac{dx}{dt} = 3c(t) + V_0 - 2c_0 \tag{8.7.28b}$$

According to this last result, values of V and c can be obtained for any point in the $(x$–$t)$ plane.

As an example, consider a channel of rectangular cross section, through which water flows with velocity $V_0 = 1$ m/s and uniform depth $y_0 = 2$ m. The channel slope and friction loss are neglected during the calculation of changes of the water flow velocity and depth due to the decrease of the water depth at the downstream end of the channel. The rate of decrease of the water depth at the downstream end of the channel is assumed to be uniform, with a value of 0.2 m/h. We calculate the time required for the channel level to fall by 0.6 m at a section located 2 km upstream from the end cross section of the channel.

The positive x direction is assumed in the upstream direction, measured from the end cross section of the channel. Therefore

$$V_0 = -1 \text{ m/s} \qquad c_0 = \sqrt{gy_0} = \sqrt{9.81 \times 2} = 4.43 \text{ m/s}$$
$$V_0 + c_0 = 3.43 \text{ m/s}$$

At the end cross section of the channel, the water depth falls by 0.6 m, at $t = 3\text{h} = 10{,}800$ s. At that time and location,

$$(c)_{x=0}^{t=10800} = \sqrt{gy} = \sqrt{9.81 \times 1.4} = 3.71 \text{ m/s}$$

According to Eq. (8.7.27b), the inverse slope of the first characteristic issued from the end cross section, at $t = 10800$ s, is given by

$$\frac{dx}{dt} = V(t) + c(t) = 3c(t) + V_0 - 2c_0$$
$$= 3 \times 3.71 - 1 - 2 \times 4.43 = 1.27 \text{ m/s}$$

Hence the time interval between the appearance of 1.4 m water depth at the end cross section and at the point located 2 km upstream from that point is given by

$$\Delta t = \frac{\Delta L}{V + c} = \frac{2000}{1.27} = 1575 \text{ s}$$

Therefore the water depth of 1.4 m is measured at a point located at a distance of 2 km upstream from the end cross section 12375 s = 3.44 h after the water depth starts falling at the end cross section of the channel.

8.7.3 Positive Waves and Surge Formation

If the water depth at the end cross section of the channel, as referred to in the previous section, is subject to an increasing water depth, then each new small disturbance propagating in the upstream direction moves with a greater speed than the previous disturbance. Therefore there is an accumulation of disturbances, which moves as a finite wave on the water surface. Such a wave is called a *surge* or a *bore*. Under specific upstream and downstream conditions, the surge may be stationary, in which case it is called a *hydraulic jump*. The movement of the surge, as well as the formation of the hydraulic jump, are analyzed using conservation of momentum, as presented by Eq. (8.7.10), neglecting the right-hand side of that equation. It also should be noted that, when analyzing the hydraulic jump, use of the momentum equation is preferable because of large energy losses associated with the jump.

Figure 8.12 shows a schematic of the formation of a hydraulic jump in a rectangular channel. Neglecting the right-hand side of Eq. (8.7.11), we have

$$\frac{q^2}{gy_1} + \frac{y_1^2}{2} = \frac{q^2}{gy_2} + \frac{y_2^2}{2} \tag{8.7.29}$$

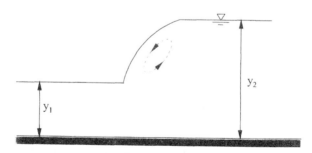

Figure 8.12 Hydraulic jump in a rectangular channel.

Rearrangement of this expression gives

$$\frac{q^2}{g}\left(\frac{y_2 - y_1}{y_1 y_2}\right) = \frac{1}{2}(y_2^2 - y_1^2) \tag{8.7.30}$$

Assuming that y_1 and y_2 are not equal, this is rewritten as

$$\frac{q^2}{g y_1^3} = \text{Fr}_1^2 = \frac{1}{2}\left(\frac{y_2}{y_1}\right)\left(\frac{y_2}{y_1} + 1\right) \tag{8.7.31}$$

where Fr_1 is the Froude number of the cross section located just upstream of the jump. Writing Eq. (8.7.31) as a second-order equation with regard to (y_2/y_1) results in

$$\frac{y_2}{y_1} = \frac{1}{2}\left(\sqrt{1 + 8\text{Fr}_1^2} - 1\right) \tag{8.7.32}$$

It is also possible to arrange Eq. (8.7.30) to represent the ratio between y_1 and y_2 as a function of Froude number of the cross section located downstream of the jump (Fr_2), giving

$$\frac{y_1}{y_2} = \frac{1}{2}\left(\sqrt{1 + 8\text{Fr}_2^2} - 1\right) \tag{8.7.33}$$

Equations (8.7.32) and (8.7.33) determine the relationships between y_1 and y_2 that lead to the formation of a stationary surge, or hydraulic jump. However, if y_1 is different from y_2 according to either Eq. (8.7.31) or (8.7.33), then a propagating surge is developed. Conservation of momentum also is employed in the case of a propagating surge. For this case, consider a surge propagating with velocity V_s, as shown in Fig. 8.13. By referring to a coordinate system moving with the surge velocity, we obtain a domain subject to steady-state conditions. Employment of momentum conservation in such a domain gives the result, analogous to Eq. (8.7.31),

$$\frac{(V_s + V_1)^2}{g y_1} = \frac{1}{2}\left(\frac{y_2}{y_1}\right)\left(\frac{y_2}{y_1} + 1\right) \tag{8.7.34}$$

The continuity equation yields

$$(V_s + V_1)y_1 = (V_s + V_2)y_2 \tag{8.7.35}$$

Equations (8.7.34) and (8.7.35) determine the relationships between the surge velocity and the water flow velocities and water depths upstream and downstream of the surge. Thus the total number of variables is five. Of these

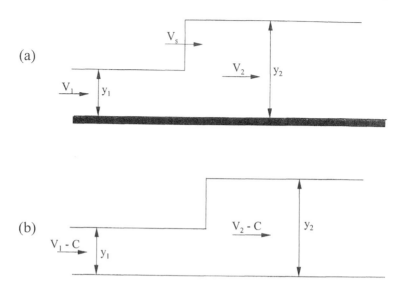

Figure 8.13 Surge propagating in a rectangular channel: (a) stationary coordinate system; and (b) coordinate system moving with the velocity V_s.

variables, three should be given and then, by applying Eqs. (8.7.34) and (8.7.35), the other two unknowns can be determined.

The term $(V_s + V_1)$ represents the velocity of the surge relative to the water upstream of the surge. Considering that y_2 is larger than y_1, Eq. (8.7.35) indicates that

$$(V_s + V_1) > \sqrt{gy_1} \qquad \text{if} \qquad y_2 > y_1 \qquad\qquad (8.7.36a)$$

and

$$(V_s + V_1) \longrightarrow \sqrt{gy_1} \qquad \text{if} \qquad y_2 \to y_1 \qquad\qquad (8.7.36b)$$

Equation (8.7.36a) shows that the surge moves with a velocity greater than that of small disturbances that develop on the water surface. Therefore such disturbances are absorbed by the propagating surge.

8.7.4 The Kinematic Wave

The concept of the kinematic wave is associated with estimates of phenomena connected with flood wave propagation in rivers and channels. A flood wave is a positive wave, namely, a rise of the water depth that propagates downstream. Since the rise of the water depth is only moderate, it is possible to consider that the channel bed slope and the friction slope of the channel flow are almost identical, and effects of friction on the wave propagation cannot be neglected. Therefore the channel discharge, Q, is solely a function of the water depth, y, and the features of the propagating wave are determined only by the equation of mass conservation. From Eq. (8.7.3), this is written as

$$\frac{\partial Q}{\partial x} + B\frac{\partial y}{\partial t} = 0 \qquad (8.7.37)$$

Since the channel discharge depends only on the water depth, this can be rewritten as

$$\frac{\partial y}{\partial t} + \left(\frac{1}{B}\right)\frac{dQ}{dy}\frac{\partial y}{\partial x} = 0 \qquad (8.7.38)$$

This equation represents an advection of the water depth, y, with the propagation velocity of the kinematic wave, given by

$$V_k = \frac{1}{B}\frac{dQ}{dy} \qquad (8.7.39)$$

According to Eqs. (8.7.38) and (8.7.39), the kinematic wave is featured with a single characteristic on the $(x-t)$ plane. For a wide rectangular channel, the result is

$$V_k = \frac{d}{dy}(Vy) = V + y\frac{dV}{dy} \qquad (8.7.40)$$

The equation of motion is identical to that of steady state flow in a channel. We may assume that the Chezy friction coefficient is constant during the propagation of the kinematic wave. Therefore (see Eq. 7.2.24),

$$V = C_c\sqrt{yS_0} \qquad (8.7.41)$$

where C_c is the Chezy friction coefficient. Introducing this result into Eq. (8.7.40) gives

$$V_k = \tfrac{3}{2}V \qquad (8.7.42)$$

Though the kinematic wave theory cannot be applied to the description of all wave phenomena associated with flood propagation in a channel, the speed of the main flood wave may be expected to be approximately that of the kinematic wave.

8.8 NUMERICAL ASPECTS

Previous sections of this chapter refer to several typical cases of environmental water waves. The basic concepts are based on the appropriate employment of the equations of mass and momentum conservation. Using each of these conservation principles leads to an advection equation, which is a first-order hyperbolic differential equation. Since two basic equations are used, two first-order hyperbolic equations should be solved simultaneously to provide the solution to the wave problem. In the particular case where only the principle of mass conservation is used, such as the kinematic wave, a single first-order partial differential equation is needed.

In general, problems of wave propagation on the water surface incorporate two unknown variables in two simultaneous first-order differential equations. The unknown variables are the flow velocity and the water depth. In cases of the marine environment or lakes, potential flow theory may be applied. Then the two unknown variables are the potential function and the wave height.

Here we present two general categories of methods of numerical simulations that can be applied to provide numerical solutions for problems of wave propagation on the water surface: (1) methods of characteristics, and (2) finite difference methods. It should be noted that usually the physical problem is formulated in conjunction with some possible boundary conditions. Wave propagation problems may refer to infinite domains. However, in many cases, major interest is in effects associated with the presence of specific types of boundary conditions in the domain. Therefore the simultaneous numerical solution of the two basic differential equations should satisfy the initial and boundary conditions specified for the domain. In some cases, implementation of the boundary conditions may be quite complicated.

8.8.1 The Method of Characteristics

A common category of methods is the *method of characteristics*. Discussion of the two types of characteristics typical of wave propagation in open channels was presented in Sec. 8.7. Use of the characteristics originates from the property that, in the solution domain, along the characteristics, linear combinations of the governing equations become total differential equations, called *compatibility equations*. The characteristics therefore can be used directly to solve the partial differential equations by numerically constructing the characteristic curves. According to the method of characteristics, the partial differential equations are replaced by the equivalent set of the corresponding characteristic equations and the compatibility equations.

The numerical simulation process starts with the identification of the characteristic equations. There are two, or in some particular cases, single families of characteristic equations. The compatibility equations are applied to the calculation of the unknown quantities in the $(x - t)$ plane at some points of the two different types of characteristics. Essentially, the method of characteristics is involved with transforming the partial differential equations from the physical coordinates to the characteristic coordinates, and calculation of the unknown variables along the characteristic curves.

It should be noted that a second-order hyperbolic equation can be represented by a set of two first-order advection equations. Each advection equation has a single characteristic. Therefore the characteristics of the two advection equations can be applied to solve the problem of wave propagation.

Simulation methods based on using the characteristics of the hyperbolic partial differential equation are usually stable and have good convergence. However, they require complicated programming, and calculation of the domain variables at points that are not necessarily of major interest. Therefore often it is more practical to use finite difference approximations.

8.8.2 Finite Difference Solutions

Finite difference methods are based on the replacement of the time and space derivatives by finite difference approximations on a fixed finite difference grid. The basic procedure for doing this is discussed in Sec. 3.5. The characteristics of the differential equation, however, determine the allowable step size in time and determine the allowable boundary conditions. Stability and convergence of the finite difference scheme used to solve the differential equation should be evaluated before its adoption.

PROBLEMS

Solved Problems

Problem 8.1 Simple unidirectional motion of small-amplitude surface water waves obeys the wave equation, which is a hyperbolic partial differential equation,

$$\frac{\partial^2 \eta}{\partial t^2} = c^2 \frac{\partial^2 \eta}{\partial x^2} \tag{1}$$

where η is the elevation of the free surface of the water above $z = 0$. The general solution of Eq. (1) can be represented by the following so-called D'Alembert's solution:

$$\eta = f(x - ct) + F(x + ct) \tag{2}$$

The first term on the right-hand side of Eq. (2) describes a wave propagating in the positive x-direction with velocity c. The second term describes a wave propagating in the negative x-direction with velocity c.

The substantial derivative of the free surface elevation is approximately equal to the vertical velocity at $z = 0$,

$$\left[\frac{\partial \eta}{\partial t} + \vec{V} \cdot \nabla \eta \right]_{z=0} \approx \left(\frac{\partial \eta}{\partial t} \right)_{z=0} \approx \left(\frac{\partial \Phi}{\partial z} \right)_{z=0} \tag{3}$$

where Φ is the potential function. Basically, the boundary represented by Eq. (3) is a nonlinear boundary in the sense that η itself is an unknown. However, the nonlinear terms are small, and this boundary is usually approximated by linear expressions.

(a) What is the complete set of boundary conditions for the function Φ, which is associated with surface water waves?
(b) How are the free surface boundary conditions represented by a differential equation for Φ?
(c) What type of differential equation represents the surface boundary?

Solution

(a) One boundary condition is represented by Eq. (3). For a fluid with constant density,

$$p - p_0 = \rho g \eta \tag{4}$$

where p_0 is the hydrostatic pressure. We introduce this relationship into Eq. (8.2.6), to obtain

$$\left(\frac{\partial \Phi}{\partial t} \right)_{z=0} = -g \eta \tag{5}$$

If the water depth is h, then at $z = -h$ the vertical velocity vanishes. Therefore

$$\left(\frac{\partial \Phi}{\partial z} \right)_{z=-h} = 0 \tag{6}$$

Equations (3), (5), and (6) represent the set of boundary conditions that should be satisfied by the wave potential function.

(b) We differentiate Eq. (5) with regard to time and apply Eq. (3), to obtain

$$\frac{\partial^2 \Phi}{\partial t^2} + g \frac{\partial \Phi}{\partial z} = 0 \tag{7}$$

(c) Equation (7) indicates that the water free surface boundary is represented by a parabolic differential equation.

Problem 8.2   What are the ranges of values of c, ω, k and t_p for waves of wavelength between $\lambda = 1$ m and $\lambda = 100$ m?

Solution

According to Eq. (8.4.12),

$$c = \sqrt{\frac{g\lambda}{2\pi}} = \sqrt{\frac{9.81}{2 \times 3.14}} \times \sqrt{\lambda} = 1.25\sqrt{\lambda}$$

Considering the given range of wavelengths, we obtain

$$1.25 \text{ m/s} \le c \le 12.5 \text{ m/s}$$

According to Eq. (8.4.2),

$$k = \frac{2\pi}{\lambda} = \frac{6.28}{\lambda}$$

This leads to

$$0.0628 \text{ m}^{-1} \le k \le 6.28$$

For frequency, Eq. (8.4.5) is

$$\omega = ck$$

which then gives

$$0.785 \text{ m/s} \le \omega \le 7.85 \text{ m/s}$$

Finally, from Eq. (8.4.6),

$$t_p = \frac{\lambda}{c} = 0.8\sqrt{\lambda}$$

Therefore

$$0.8 \text{ s} \le t_p \le 8 \text{ s}$$

Problem 8.3   Calculate the average power per unit width that is transported by a sinusoidal wave on deep water.

Solution

The average power per unit width transported by the waves is given by

$$N = \frac{1}{t_p} \int_0^{t_p} \int_{-h}^0 pu \, dz = \frac{1}{t_p} \int_0^{t_p} \int_{-\infty}^0 \left\{ -\rho \frac{\partial \Phi}{\partial t} u - \rho g u \right\} dz \, dt$$
$$\times \frac{1}{t_p} \int_0^{t_p} \int_{-h}^0 \left\{ -\rho \frac{\partial \Phi}{\partial t} u \right\} dz \, dt$$

We apply Eqs. (8.4.7) and (8.4.15), to obtain

$$N = \frac{1}{t_p} \int_0^{t_p} \int_{-\infty}^0 \rho \omega k \Phi_0^2 e^{2kz} \sin^2(\omega t - kx) \, dz \, dt$$

Letting $x = 0$ and performing the time and space integrations gives

$$N = \rho \frac{\omega \Phi_0^2}{4} = \left[\frac{1}{2} \rho g a^2 \right] \frac{c}{2}$$

This expression indicates that the wave energy propagates with the group velocity.

Problem 8.4 A standing wave in a lake of depth h is obtained by the superposition of two propagating waves of the same amplitude and wavelength, but moving in opposite directions. The resulting surface displacement is given by

$$\eta = a \sin(\omega t + kx) - a \sin(\omega t - kx) = 2a \sin(kx) \cos(\omega t)$$

Standing waves may be found in a limited water body, like a lake, due to reflection from its sides.

    (a)   What is the relevant potential function of the standing wave?
    (b)   Develop the expressions for the velocity components.
    (c)   Develop the expression for the stream function.
    (d)   Assuming that the standing waves develop in a one-dimensional lake of length L, develop expressions for possible values of λ and ω.

Solution

(a) According to Eq. (8.3.9), the potential functions of the waves produced by superposition are given by

$$(\Phi)_{z=0} = -2ga \sin(kx) \int \cos(\omega t) \, dt = -2 \frac{ga}{\omega} \sin(kx) \sin(\omega t)$$

According to Eq. (8.5.14), the amplitude of the propagating wave is associated with the potential function amplitude by

$$a = \frac{k\Phi_0}{\omega} \cosh(kh)$$

Therefore the potential function of the standing wave is given by

$$\Phi = -2\Phi_0 \cosh[k(z + h)] \sin(kx) \sin(\omega t)$$

(b) From the potential function, we obtain

$$u = \frac{\partial \Phi}{\partial x} = -2k\Phi_0 \cosh[k(z+h)][\cos(kx)\sin(\omega t)]$$

$$w = \frac{\partial \Phi}{\partial z} = -2k\Phi_0 \sinh[k(z+h)][\sin(kx)\sin(\omega t)]$$

(c) The velocity components and the stream function are related by

$$\frac{\partial \Psi}{\partial z} = u = -2k\Phi_0 \cosh[k(z+h)][\cos(kx)\sin(\omega t)]$$

$$\frac{\partial \Psi}{\partial x} = -w = 2k\Phi_0 \sinh[k(z+h)][\sin(kx)\sin(\omega t)]$$

By direct integration of these expressions, we obtain

$$\Psi = -2\Phi_0 \sinh[k(z+h)]\cos(kx)\sin(\omega t)$$

(d) Assuming that the walls are located at $x = -L/2$ and $x = L/2$, the condition of no flow through these walls implies that

$$\frac{kL}{2} = (2n+1)\frac{\pi}{2} \qquad n = 0, 1, 2, \ldots$$

This expression leads to

$$k = \frac{(2n+1)\pi}{L}$$

$$\lambda = \frac{2\pi}{k} = \frac{2L}{2n+1}$$

From this last expression, it is seen that the longest wavelength is $\lambda = 2L$. By applying Eqs. (8.5.8) and (8.5.9), the wave frequency is obtained as

$$\omega = \sqrt{\frac{\pi g(2n+1)}{L}} \tanh\left[\frac{(2n+1)\pi h}{L}\right]$$

Unsolved Problems

Problem 8.5   Find the progressive wave equations, whose superposition leads to the following standing wave equations:

   (a) $2a\cos(kx)\cos(\omega t)$
   (b) $2a\cos(kx)\sin(\omega t)$
   (c) $2a\sin(kx)\sin(\omega t)$

Problem 8.6   For each of the cases of standing waves given by problem (8.5.1), let the water depth be h. Determine the potential function and the stream function for each case.

Problem 8.7   For each case of standing waves, represented in problems (8.5.1) and (8.5.2), consider that the standing waves develop in a lake of width L. Where is the coordinate origin located in each case?

Problem 8.8   Develop the expression for the energy flux (power delivery by the waves) through $x = 0$, in the case of waves in a shallow water environment of depth h, and show that the wave energy propagates with the group velocity.

Problem 8.9   Analyze and describe the rate of propagation of a small wave, originating from a rock that falls into the centerline of a wide rectangular channel, through which water flows with velocity V. Consider cases of (a) subcritical flow, (b) critical flow, and (c) supercritical flow.

SUPPLEMENTAL READING

Abott, M. B., 1979. *Computational Hydraulics*. Pitman, London. (Contains a thorough coverage of numerical methods for the calculation of unsteady flow and wave propagation in open channels.)

Henderson, F. M., 1966. *Open Channel Flow*. Macmillan, New York. (Contains material in a format of a textbook referring to waves in open channels.)

Hofman, J. D., 1992. *Numerical Methods for Engineers and Scientists*. McGraw-Hill, New York. (Many books concerning numerical methods can be used to provide the reader with basic knowledge of numerical methods for the solution of hyperbolic partial differential equations. This textbook incorporates a comprehensive presentation of the topic and the different applicable methods of solution.)

Lighthill, J. L., 1978. *Waves in Fluids*. Cambridge University Press, Cambridge, England. (Contains a comprehensive presentation of general topics of water waves in open channels, lakes, and marine environments.)

Phillips, O. M., 1966. *The Dynamics of the Upper Ocean*. Cambridge at the University Press, London. (Includes a comprehensive reference to many theoretical aspects of water surface waves.)

Stokker, J. J., 1957. *Water Waves*. John Wiley, New York. (Presents the basic theory of wave propagation in open channels.)

9

Geophysical Fluid Motions

9.1 INTRODUCTION

Geophysical fluid mechanics generally deals with flows of large spatial extent. There are many subjects that fall within this category, including ocean currents, tides, estuaries, coastal flows, and others. In the present chapter we focus on motions for which the rotation of the earth, in particular the Coriolis term, is important in the equations of motion. This condition arises in problems with large spatial scales, which result in Rossby numbers approaching one or less.

Recall from Chap. 2 that the Rossby number is defined as the ratio of characteristic velocity to the product of characteristic length and rotation rate. This is derived from the ratio of the relative magnitude of the nonlinear acceleration terms to the Coriolis terms in the equations of motion, as shown in Sec. 2.9. Here we modify this definition slightly to be more consistent with the literature in this field, using

$$\text{Ro} = \frac{U}{fL} \tag{9.1.1}$$

where Ro = Rossby number, U = characteristic velocity, L = characteristic length, and f = *Coriolis parameter*, or *planetary vorticity* = $2\Omega_0 \sin \psi$, where Ω_0 is the angular rotation rate of the earth and ψ is the latitude. The magnitude of f varies between 1.45×10^{-4} s^{-1} at the poles to zero at the equator. Referring back to Table 2.1, Rossby numbers that are sufficiently small that rotation effects become important are associated with large lakes, estuaries, coastal regions, and oceanic currents. Atmospheric motions also are subject to Coriolis effects, but the present discussion focuses on aqueous systems.

In addition to the inclusion of the Coriolis terms, an interesting feature of the analysis of fluid motions in very large systems is the relative unimportance of solid boundaries. This sometimes poses difficulties in specifying boundary conditions, since the location of the boundaries is not well-defined.

The only clear boundary in the deep oceans, for instance, is the air/water interface. Boundaries do become important, however, in developing descriptions of general circulation.

9.2 GENERAL CONCEPTS

The general equations of motion for geophysical flows consist of the continuity and Navier–Stokes equations for incompressible flow introduced in Chap. 2. For convenience, these are repeated here:

$$\vec{\nabla} \cdot \vec{V} = \frac{\partial u}{\partial x} + \frac{\partial v}{\partial y} + \frac{\partial w}{\partial z} = 0 \tag{9.2.1}$$

$$\frac{\partial \vec{V}}{\partial t} + \vec{V} \cdot \nabla \vec{V} + 2\vec{\Omega} \times \vec{V} = \vec{g} - \frac{1}{\rho_0}\vec{\nabla}p + \nu\nabla^2\vec{V} \tag{9.2.2}$$

where $\vec{V} = (u, v, w)$ is the velocity vector, $\vec{\Omega}$ is the rotation rate of the earth, g is gravity, p is pressure, ρ_0 is a reference density, and ν is kinematic viscosity. Usually the main concern is with horizontal or two-dimensional motions, so that w will usually be assumed to be zero for the present discussion. With $w = 0$, Eq. (9.2.2) in component form appears as

$$\frac{\partial u}{\partial t} + u\frac{\partial u}{\partial x} + v\frac{\partial u}{\partial y} + w\frac{\partial u}{\partial z} - fv = -\frac{1}{\rho_0}\frac{\partial p}{\partial x} + \nu\nabla^2 u \tag{9.2.3}$$

$$\frac{\partial v}{\partial t} + u\frac{\partial v}{\partial x} + v\frac{\partial v}{\partial y} + w\frac{\partial v}{\partial z} + fu = -\frac{1}{\rho_0}\frac{\partial p}{\partial y} + \nu\nabla^2 v \tag{9.2.4}$$

$$0 = -g - \frac{1}{\rho_0}\frac{\partial p}{\partial z} \tag{9.2.5}$$

In the following, the viscous stress term will be replaced with a turbulent stress, with a possibly nonhomogeneous and nonisotropic turbulent eddy diffusivity (see Chap. 5). However, this does not change the basic form of the equation.

9.2.1 Geostrophic Balance

Geostrophic flow, as introduced in Eq. (2.9.26), involves a balance between the pressure and Coriolis terms in the equations of motion, that is,

$$-fv = -\frac{1}{\rho_0}\frac{\partial p}{\partial x} \tag{9.2.6a}$$

$$fu = -\frac{1}{\rho_0}\frac{\partial p}{\partial y} \tag{9.2.6b}$$

This result is obtained by assuming steady conditions and neglecting the nonlinear acceleration and friction terms in the equations of motion. Models that neglect the nonlinear accelerations are sometimes referred to as *Ekman models*. They are mostly appropriate for relatively small values of Ro, which as noted previously expresses the ratio of the magnitudes of the acceleration and Coriolis terms. It should be noted that the geostrophic balance is not valid near the equator, within a latitude of about $\pm 3°$, where f becomes very small.

An interesting result of Eq. (9.2.6) is that the flow direction is perpendicular to the pressure gradient. Therefore, on a weather map, isobars are approximately the same as streamlines of the flow, and the streamlines are lines of constant pressure. Also, the quantity $p/(f \rho_0)$ can be regarded as a stream function. Figure 9.1 shows a schematic description of flow along an isobar in the northern hemisphere, around centers of high and low pressure. The Coriolis force and the pressure gradient are colinear, with opposite directions. The velocity vector is perpendicular to those vectors and creates a counterclockwise angle of 90° with the Coriolis force. In the southern hemisphere, the velocity acts 90° to the left of the Coriolis force, due to the opposite sense of rotation.

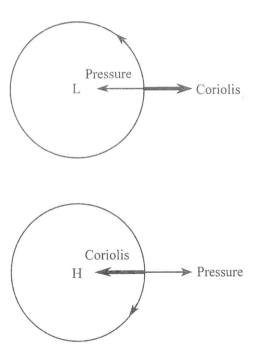

Figure 9.1   Relationship between isobars and streamlines in atmospheric flows (northern hemisphere).

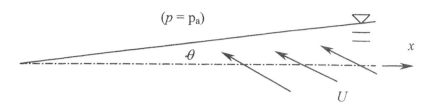

Figure 9.2 Surface tilt as a result of geostrophic balance (flow is into the page).

Another way of looking at the geostrophic balance is to consider a homogeneous fluid with a surface at an angle θ to the horizontal direction, as shown in Fig. 9.2. The pressure along the surface is the atmospheric pressure, p_a. The acceleration generated as a result of this angle is

$$a_x = -g \tan \theta \tag{9.2.7}$$

where the negative sign arises because x is positive to the right. If the fluid is moving at velocity U into the plane of Fig. 9.2, there will be a horizontal component of the Coriolis acceleration with magnitude fU in the positive x direction (northern hemishere). If these two accelerations are in balance, then

$$g \tan \theta = fU \Rightarrow U = \frac{g \tan \theta}{f} \tag{9.2.8}$$

This gives the expected geostrophic velocity, for a given surface slope or, conversely, the expected surface slope for a given flow velocity. Note that the flow is in a direction normal to the pressure gradient, as was shown in Eq. (9.2.6). Recall that in order for the balance that leads to Eq. (9.2.8) to exist, the flow must be uniform and in a straight direction, since no other accelerations are assumed than the pressure gradient and Coriolis terms.

It is somewhat surprising to consider the magnitude of the sea surface tilt angle that corresponds to expected velocities in the ocean. For example, if $U = 1$ m/s, and we assume a latitude of 45°, then $\tan \theta \cong 10^{-5}$, or about 1 cm/km. This is much too small to be measureable. However, measurements in a stratified ocean are much easier.

There is almost always some density stratification in the oceans, due to temperature or salinity variations or both. Issues related to stratification are discussed in Chap. 13, but for now consider that the stratification can be idealized as a two-layer system as sketched in Fig. 9.3, which shows a fluid of density ρ_1 flowing over a stagnant layer of density ρ_2. Since fluid 2 is at rest, its free surface must be horizontal, while the free surface of fluid 1 is tilted due to its motion, making an angle θ with the horizontal direction, as previously described. Consider that the interface between the two fluids lies at some angle θ_i, as shown in the figure.

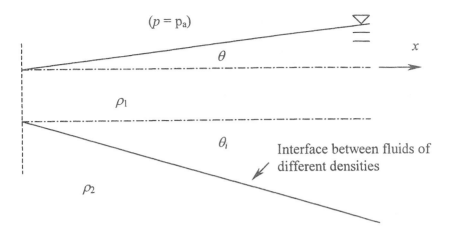

Figure 9.3 Surface and interface tilt for geostrophic flow in a two-layer ocean.

Along any horizontal line drawn in fluid 2, the pressure must be constant. For such a line drawn at depth h, the pressure is given by the hydrostatic relation

$$p_h = p_a + \rho_2 gh \tag{9.2.9}$$

Now, with x drawn so that $x = 0$ at the point at which the interface between the two fluids meets the surface, the pressure along the line drawn at depth h can be written as

$$p_h = p_a + g\{\rho_1(\tan\theta - \tan\theta_i)x + \rho_2(h + x\tan\theta_i)\} \tag{9.2.10}$$

Equating Eqs. (9.2.9) and (9.2.10) then results in

$$\tan\theta_i = -\tan\theta \frac{\rho_1}{\rho_2 - \rho_1} \tag{9.2.11}$$

This shows that the slope of the interface can have a much greater magnitude than the surface slope, depending on the relative values of ρ_1 and ρ_2. For example, if fresh water (specific gravity $= 1$) flows over sea water (specific gravity $= 1.025$), then the interface slope is approximately 40 times as great as the surface slope. In most cases the density difference is less than this, so that the interface slope would likely be even greater.

When surfaces of constant density are parallel to surfaces of constant pressure, the system is said to be in a *barotropic* state. When these surfaces intersect, the field is *baroclinic*. It is possible for a barotropic system to be statically stable, but in a baroclinic system there must be motion. The two-layer ocean considered above is an example of a baroclinic field, since motion was required to generate the Coriolis force to oppose the pressure gradient force.

The geostrophic flow that results from a horizontal density gradient is also called a *thermal wind*. This terminology stems from the usual situation in which the density differences are generated as a result of temperature variations. When there is a horizontal density gradient, the geostrophic flow also develops a vertical shear. This can be seen by considering the system shown in Fig. 9.4, which shows several contours of constant density and contours of contant pressure. Assuming that $\partial \rho / \partial x < 0$, then the density along section 1 is greater than that along section 2. In order to maintain hydrostatic equilibrium, the weight of columns δz_1 and δz_2 must be equal. Therefore the interval between the two isobars increases with x, or $\delta z_1 < \delta z_2$. The isobars, as shown in Fig. 9.4, then must be consistent with $\partial p / \partial x > 0$, and their slope increases with increasing z. Following the same arguments as before (coming from the geostrophic balance), the thermal wind is thus seen to be into the plane of Fig. 9.4, and its magnitude increases with z.

This phenomenon is clearly demonstrated using Eq. (9.2.6), along with the hydrostatic balance equation in the z-direction. Differentiating equation

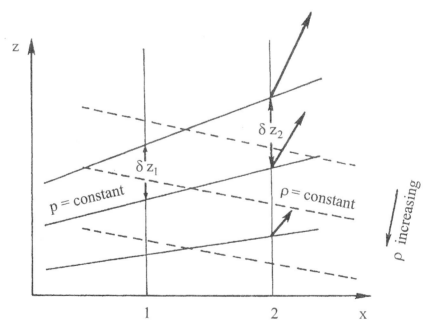

Figure 9.4 Baroclinic field, showing several contours of constant density and pressure.

(9.2.6a) with respect to z and using Eq. (9.2.5) to substitute for the vertical pressure gradient, we obtain

$$\frac{\partial v}{\partial z} = -\frac{g}{\rho_0 f}\frac{\partial \rho}{\partial x} \tag{9.2.12a}$$

Performing a similar operation by differentiating Eq. (9.2.6b) with respect to z gives

$$\frac{\partial u}{\partial z} = \frac{g}{\rho_0 f}\frac{\partial \rho}{\partial y} \tag{9.2.12b}$$

These equations are called the *thermal wind equations*. They provide the vertical variation of velocities from measurements of the horizontal temperature (density) gradients. The thermal wind, as indicated in Fig. 9.4, is associated with systems in which surfaces of constant pressure and constant density intersect, i.e., the baroclinic case.

9.2.2 Potential Vorticity

Another concept useful in the study of large-scale flows is that of conservation of potential vorticity. This is demonstrated by first writing the momentum equations for horizontal motion, neglecting the friction terms. From Eq. (9.2.3),

$$\frac{Du}{Dt} = \frac{\partial u}{\partial t} + u\frac{\partial u}{\partial x} + v\frac{\partial u}{\partial y} = fv - \frac{1}{\rho_0}\frac{\partial p}{\partial x} \tag{9.2.13}$$

and from Eq. (9.2.4),

$$\frac{Dv}{Dt} = \frac{\partial v}{\partial t} + u\frac{\partial v}{\partial x} + v\frac{\partial v}{\partial y} = -fu - \frac{1}{\rho_0}\frac{\partial p}{\partial y} \tag{9.2.14}$$

where (D/Dt) is taken here as the two-dimensional or horizontal material derivative operator. We now differentiate Eq. (9.2.13) with respect to y and subtract the result from the derivative of Eq. (9.2.14) with respect to x, giving

$$\frac{D}{Dt}\left(\frac{\partial v}{\partial x} - \frac{\partial u}{\partial y}\right) = \left(\frac{\partial v}{\partial x} - \frac{\partial u}{\partial y} - f\right)\left(\frac{\partial u}{\partial x} + \frac{\partial v}{\partial y}\right) - u\frac{\partial f}{\partial x} - v\frac{\partial f}{\partial y} \tag{9.2.15}$$

where horizontal density variations have been neglected. Now, f is not a function of time or longitude (x), so the last term on the right-hand side of Eq. (9.2.15) may be rewritten as

$$v\frac{\partial f}{\partial y} = \frac{Df}{Dt} \tag{9.2.16}$$

Substituting Eq. (9.2.16), and rearranging the terms of Eq. (9.2.15) then leads to

$$\frac{D}{Dt}\left[\left(\frac{\partial v}{\partial x}-\frac{\partial u}{\partial y}\right)+f\right]=-\left(\frac{\partial u}{\partial x}+\frac{\partial v}{\partial y}\right)\left[\left(\frac{\partial v}{\partial x}-\frac{\partial u}{\partial y}\right)+f\right] \quad (9.2.17)$$

Recall that the vertical component of vorticity, relative to the chosen coordinate system, is defined by Eq. (2.3.12),

$$\zeta=\omega_z=\frac{\partial v}{\partial x}-\frac{\partial u}{\partial y} \quad (9.2.18)$$

Also, if the flow satisfies the two-dimensional continuity equation, then the first term on the right-hand side of Eq. (9.2.17) (in parentheses) is zero, resulting in

$$\frac{D}{Dt}(\zeta+f)=0 \quad (9.2.19)$$

The sum of the relative vorticity (ζ) and the planetary vorticity (f) is called the *absolute vorticity*. Equation (9.2.19) states that the absolute vorticity is conserved, following a fluid particle along its path line.

If the flow field is required to satisfy the full continuity constraint, then Eq. (9.2.1) gives

$$\frac{\partial u}{\partial x}+\frac{\partial v}{\partial y}=-\frac{\partial w}{\partial z} \quad (9.2.20)$$

The right-hand side of this equation can be related to the rate of stretching of a column of fluid of thickness H, by

$$\frac{\partial w}{\partial z}=\frac{1}{H}\frac{DH}{Dt} \quad (9.2.21)$$

Introducing this result into Eq. (9.2.17) then gives

$$\frac{D}{Dt}(\zeta+f)=\frac{1}{H}\frac{DH}{Dt}(\zeta+f) \quad (9.2.22)$$

Dividing both sides by H, we obtain

$$\frac{1}{H}\frac{D}{Dt}(\zeta+f)-\frac{1}{H^2}\frac{DH}{Dt}(\zeta+f)=\frac{D}{Dt}\left(\frac{\zeta+f}{H}\right)=0 \quad (9.2.23)$$

where the quantity $(\zeta+f)/H$ is called the *potential vorticity*. This last result shows that, for frictionless incompressible flow, potential vorticity is conserved following a fluid particle. For steady flows the particle paths are the same as the streamlines, so potential vorticity is thus conserved along streamlines. Use of this concept, along with geostrophic flow assumptions and other results such as the Bernoulli equation (Chap. 2) has formed the basis for a number of theoretical models of ocean currents.

9.3 THE TAYLOR–PROUDMAN THEOREM

A number of laboratory experiments have been performed, usually using rotating tables, to simulate different aspects of low-Rossby-number flows. One of the more interesting experiments of this type involves simulation of geostrophic flow of a homogeneous fluid and produces direct observations of Taylor columns, as explained below. This experiment involves a tank of fluid that is rotated at a steady angular speed Ω. The rotation speed is sufficiently high that the Coriolis force is much larger than the acceleration terms, and conditions of geostrophic equilibrium may be assumed.

In regions that are not affected by the friction induced by the boundaries, the equations for geostrophic equilibrium in the horizontal directions, and hydrostatic conditions in the vertical direction, are given by Eqs. (9.2.6a), (9.2.6b), and (9.2.5), respectively. Note, however, that $f = 2\Omega$ for the conditions of the experiment. By differentiating Eq. (9.2.6a) with respect to y and Eq. (9.2.6b) with respect to x and subtracting, we find

$$2\Omega \left(\frac{\partial u}{\partial x} + \frac{\partial v}{\partial y} \right) = 0 \tag{9.3.1}$$

Using Eq. (9.2.20), and since $\Omega \neq 0$, Eq. (9.3.1) implies

$$\frac{\partial w}{\partial z} = 0 \tag{9.3.2}$$

Also, by differentiating each of Eqs. (9.2.6a) and (9.2.6b) with respect to z and substituting Eq. (9.2.5) for the vertical pressure gradient results in

$$\frac{\partial u}{\partial z} = \frac{\partial v}{\partial z} = 0 \tag{9.3.3}$$

Since there is no vertical motion, the angular velocity vector is oriented in the z-direction. Equations (9.3.2) and (9.3.3) indicate that the velocity vector does not vary with z, so we may conclude that steady, slow motions in a rotating, homogeneous, inviscid fluid are two-dimensional. This result is called the *Taylor–Proudman theorem*. It was obtained theoretically by Proudman in 1916. Soon afterwards, Taylor proved this theorem using an experimental setup as sketched in Fig. 9.5. A tank full of fluid was rotating as a solid body. A small cylinder was slowly dragged along the bottom of the tank and dye was released at point A, above the cylinder and slightly ahead of it. The thread

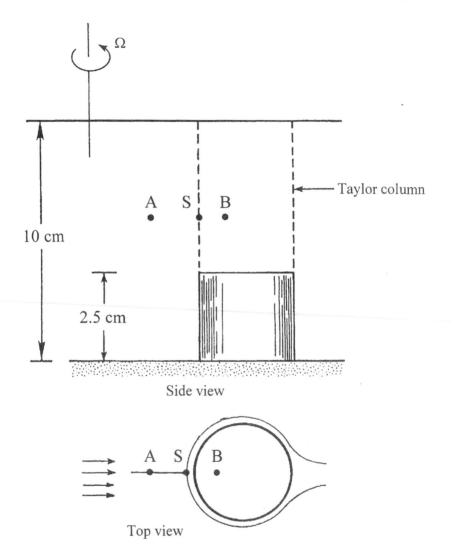

Figure 9.5 Schematic diagram of Taylor's experiment.

of dye divided at the point S, while appearing to belong to a column of fluid extending over the depth of the cylinder. This column of fluid is called a *Taylor column*. Taylor's experiment indicated that bodies moving slowly in a strongly rotating system of homogeneous fluid carry along their motion in a two-dimensional column of fluid.

9.4 WIND-DRIVEN CURRENTS (EKMAN LAYER)

We have already considered the special case of geostrophic flow, resulting
from a balance between the Coriolis and pressure terms in the momentum
equation. In that case, it was shown that the velocity vector is perpendicular
to the pressure gradient, so that the velocity follows the isobars. This balance
can only exist in situations where other factors such as the acceleration and
friction terms are negligible, and therefore these flows generally occur in the
upper atmosphere or deep oceans.

Another interesting effect of rotation can be seen on wind-driven currents, where the velocity field is formed into the so-called *Ekman spiral*.
Consider a flow far from any boundaries, with a wind stress acting on the
surface and with negligible pressure gradients. Here we again consider steady
horizontal flow, in a system with $z = 0$ at the surface and increasing downward. Neglecting also the acceleration terms (i.e., the Ekman model assumption) and assuming that the only friction effect is from vertical shear stresses,
the momentum equations for the two horizontal velocity components are

$$-fv = A\frac{\partial^2 u}{\partial z^2} \tag{9.4.1}$$

and

$$fu = A\frac{\partial^2 v}{\partial z^2} \tag{9.4.2}$$

where A is a horizontal eddy diffusivity, which is assumed to be constant. By
combining the constants into one term, we define

$$\lambda^2 = \frac{f}{A} \tag{9.4.3}$$

Note that λ has units of length^{-1}. Using Eq. (9.4.3), Eqs. (9.4.1) and (9.4.2)
are rewritten as

$$i^2\lambda^2 v = \frac{d^2 u}{dz^2} \tag{9.4.4}$$

and

$$i\lambda^2 u = i\frac{d^2 v}{dz^2} \tag{9.4.5}$$

where $i = \sqrt{-1}$ and ordinary derivatives are substituted for the partial derivatives, since the velocities are considered to be functions of z only. Now, define
$\xi = u + iv$, so that by adding Eqs. (9.4.4) and (9.4.5), we obtain

$$\frac{d^2\xi}{dz^2} = i\lambda^2\xi \tag{9.4.6}$$

This equation has a solution,

$$\xi = K_1 \exp(\lambda\sqrt{iz}) + K_2 \exp(-\lambda\sqrt{iz}) \tag{9.4.7}$$

where K_1 and K_2 are complex constants given by

$$K_1 = C_1 \exp(iC_1), \quad K_2 = C_2 \exp(-iC_2) \tag{9.4.8}$$

where C_1 and C_2 are constants that must be determined from known boundary conditions.

By separating the real and imaginary parts (according to the definition of ξ), the velocity components are found from

$$u = C_1 \exp\left(\frac{\lambda z}{\sqrt{2}}\right) \cos\left(\frac{\lambda z}{\sqrt{2}} + C_3\right)$$
$$+ C_2 \exp\left(-\frac{\lambda z}{\sqrt{2}}\right) \cos\left(\frac{\lambda z}{\sqrt{2}} + C_4\right) \tag{9.4.9}$$

and

$$v = C_1 \exp\left(\frac{\lambda z}{\sqrt{2}}\right) \sin\left(\frac{\lambda z}{\sqrt{2}} + C_3\right)$$
$$+ C_2 \exp\left(-\frac{\lambda z}{\sqrt{2}}\right) \sin\left(\frac{\lambda z}{\sqrt{2}} + C_4\right) \tag{9.4.10}$$

where C_3 and C_4 are constant phase shifts. For illustration, it is assumed that the x–y plane is oriented so that wind blows in the positive y direction. The boundary conditions include vanishing velocities when z becomes very large ($z \to \infty$), shear stress in the x-direction is zero at the surface, i.e., $\tau_x = 0$, at $z = 0$, and shear stress in the y-direction at the surface is given by

$$\tau_{y0} = \tau_y|_{z=0} = -\left(\rho A \frac{dv}{dz}\right)\bigg|_{z=0} \tag{9.4.11}$$

Note that $\tau_x = 0$ implies that $(du/dx) = 0$ at $z = 0$, since A is constant and $A \neq 0$.

Since both velocities vanish for large z, we conclude that $C_1 = 0$. Then, by differentiating Eqs. (9.4.9) and (9.4.10) with respect to z, using the shear boundary conditions, we obtain

$$\frac{du}{dz}\bigg|_{z=0} = -C_2 \frac{\lambda}{\sqrt{2}}(\sin C_4 + \cos C_4) = 0 \Rightarrow C_4 = -\frac{\pi}{4} \tag{9.4.12}$$

and

$$\frac{dv}{dz}\bigg|_{z=0} = -C_2 \frac{\lambda}{\sqrt{2}}\left(\cos\frac{\pi}{4} + \sin\frac{\pi}{4}\right) = -\frac{\tau_y}{A} \Rightarrow C_2 = \frac{\tau_y}{A\lambda} \tag{9.4.13}$$

To simplify the notation, define a length scale L, so that

$$L = \frac{\pi\sqrt{2}}{\lambda} = \pi \left(\frac{2A}{f}\right)^{1/2} \qquad (9.4.14)$$

Using this definition, the final results for the velocity components are written by substituting for C_2 and C_4 into Eqs. (9.4.9) and (9.4.10), respectively, giving

$$u = V_0 \exp\left(-\frac{\pi}{L}z\right) \cos\left(\frac{\pi}{4} - \frac{\pi}{L}z\right) \qquad (9.4.15)$$

and

$$v = V_0 \exp\left(-\frac{\pi}{L}z\right) \sin\left(\frac{\pi}{4} - \frac{\pi}{L}z\right) \qquad (9.4.16)$$

where V_0 is the magnitude of the surface drift current (at $z = 0$). It is obtained by differentiating Eq. (9.4.16) with respect to z and substituting into Eq. (9.4.11), to obtain

$$\left.\frac{dv}{dz}\right|_{z=0} = -\frac{\tau_{y0}}{\rho A} = -V_0 \frac{\pi\sqrt{2}}{L} \Rightarrow V_0 = \frac{\tau_{y0}L}{\rho A\pi\sqrt{2}} = \frac{\tau_{y0}}{\rho A\lambda} \qquad (9.4.17)$$

From this solution, it may be noted that at the surface, $u^2 + v^2 = V_0^2$, but the surface velocity is oriented at an angle 45° from the mean wind direction, as shown in Fig. 9.6. When $z = L$, the velocities are

$$u = V_0 e^{-\pi} \cos\left(-\frac{3\pi}{4}\right), \qquad v = V_0 e^{-\pi} \sin\left(-\frac{3\pi}{4}\right) \qquad (9.4.18)$$

and the velocity magnitude is $V = (u^2 + v^2)^{1/2} = V_0 e^{-\pi}$, or about $(1/23)V_0$. Also, the direction of the velocity at this depth is exactly opposite that of the surface velocity. For intermediate depths, the magnitude of the velocity decreases exponentially with increasing depth and its direction turns clockwise, according to Eqs. (9.4.15) and (9.4.16). Figure 9.6 illustrates this result, which is known as the *Ekman spiral*. The depth L is considered to represent the layer depth for which frictional force driven by surface wind shear has an influence on the motions. It is called the *Ekman depth*. As defined in Eq. (9.4.14), L is not a function of surface shear, but it does depend on latitude, approaching ∞ at the equator (where $f = 0$). This presents a problem, for instance, in defining V_0. However, Coriolis effects are not important at the equator, so this is not an issue of practical interest (i.e., the above derivation is not valid at the equator).

One further observation of interest concerns the mean mass transport associated with the wind-driven drift currents. The mass flux is calculated as

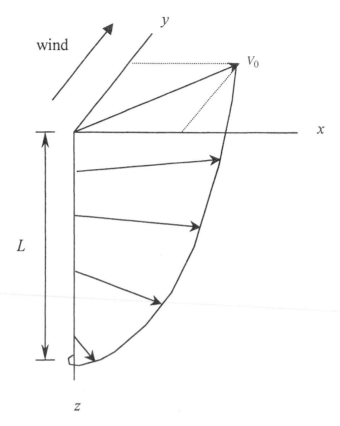

Figure 9.6 Ekman spiral; V_0 is the surface velocity, with other lines showing velocities at successively greater depths, ending with $V \cong 0$ and pointed in the opposite direction to V_0 at $z = L$.

the product of velocity and density. Therefore we multiply Eqs. (9.4.15) and (9.4.16) by ρ, assuming constant ρ, and integrate from $z = 0$ to $z = \infty$, so that the total mass transport rates in the x and y directions are, respectively,

$$M_x = \int_0^\infty \rho u \, dz = \rho \frac{V_0 L}{\pi \sqrt{2}} \tag{9.4.19}$$

and

$$M_y = \int_0^\infty \rho v \, dz = 0 \tag{9.4.20}$$

The mass flux in the x-direction also can be written using Eq. (9.4.17) for V_0, giving

$$M_x = \frac{\tau_{y0}}{f} \qquad (9.4.21)$$

Thus, curiously, there is zero net mass transport in the direction of the wind, and all of the transport is at 90° to the right of the wind direction. This *Ekman transport* process is of interest for a number of problems in physical and coastal oceanography.

Although mathematically sound, verificatioin of the Ekman spiral and transport calculations is difficult in practice, primarily because it is difficult to find a set of measured field conditions that matches all the required simplifications and assumptions used in the derivation. There have been observations of iceberg drift in directions normal to the wind direction, which is at least some verification of the mass flux calculations of Eqs. (9.4.20) and (9.4.21). One of the most difficult problems in checking Ekman spiral predictions is with the values for τ and A, which are usually not well known. Also, the ocean is usually stratified, so density gradients act to restrict the vertical transport of momentum. Observations have shown there is little deviation from the above results for the case of continuous stratification, but strong density interfaces such as might be found at the base of the upper mixed layer (see Chap. 13) tend to limit the vertical extent of the wind effect. Under this condition, the velocities do not necessarily vanish at the interface. Instead, a zero shear (zero gradient) boundary condition is usually used, and it can be shown that the deflection angle increases for small depths, relative to the infinitely deep layer.

For a layer of finite depth H, the boundary condition is modified so that $u, v \to 0$ for $z = H$. For this case, the deflection angle depends on H/L, and for sufficiently small H/L, the drift is nearly parallel to the wind, increasing to 45° only when H/L is greater than about 0.5.

It should also be noted that the above theory assumes no boundaries and no horizontal pressure gradients. In the real ocean, eventually the flow should encounter a boundary, where water would build up to create a pressure gradient opposing the flow. Alternatively, a wind blowing from the north along the western coast of a land mass in the northern hemisphere would cause transport of water mass westward, and this could result in upwelling of deeper waters along the coast. For a southerly wind, the flow may be turned in the direction of the wind stress and a geostrophic balance may be obtained.

9.5 VERTICALLY INTEGRATED EQUATIONS OF MOTION

Because of the complexity of motions possible in large lakes and the seas, involving a wide range of interacting scales, mathematical formulation of the problem often leads to complicated systems of equations that can be solved only after certain simplifying assumptions have been made. The geostrophic assumption is one such approach, and in this section we introduce a more formal approach to studying systems in which the motions are primarily horizontal. In fact, this assumption has been made in the previous sections of this chapter, but here we derive more complete equations by vertically integrating the governing equations. This is similar to the development in Sec. 7.7.3, except here we consider deep water, incorporate eddy viscosities, and consider only steady solutions.

The steady Ekman-type model for horizontal velocity components is

$$-fv = -\frac{1}{\rho}\frac{\partial p}{\partial x} + \frac{\partial}{\partial x}\left(A_x\frac{\partial u}{\partial x}\right) + \frac{\partial}{\partial y}\left(A_y\frac{\partial u}{\partial y}\right) + \frac{\partial}{\partial z}\left(A_z\frac{\partial u}{\partial z}\right) \qquad (9.5.1)$$

and

$$fu = -\frac{1}{\rho}\frac{\partial p}{\partial y} + \frac{\partial}{\partial x}\left(A_x\frac{\partial v}{\partial x}\right) + \frac{\partial}{\partial y}\left(A_y\frac{\partial v}{\partial y}\right) + \frac{\partial}{\partial z}\left(A_z\frac{\partial v}{\partial z}\right) \qquad (9.5.2)$$

From Chap. 2, these can be rewritten in terms of shear stresses as

$$-f\rho v = -\frac{\partial p}{\partial x} + \frac{\partial \tau_{xx}}{\partial x} + \frac{\partial \tau_{xy}}{\partial y} + \frac{\partial \tau_{xz}}{\partial z} \qquad (9.5.3)$$

and

$$f\rho u = -\frac{\partial p}{\partial y} + \frac{\partial \tau_{yx}}{\partial x} + \frac{\partial \tau_{yy}}{\partial y} + \frac{\partial \tau_{yz}}{\partial z} \qquad (9.5.4)$$

Integrating each of Eqs. (9.5.3) and (9.5.4) vertically from $z = 0$ to $z = L$ (Ekman layer depth) gives

$$-fM_y = -\frac{\partial P}{\partial x} + \frac{\partial T_{xx}}{\partial x} + \frac{\partial T_{xy}}{\partial y} + \tau_{xz}\Big|_{z=0} \qquad (9.5.6)$$

and

$$fM_x = -\frac{\partial P}{\partial y} + \frac{\partial T_{yx}}{\partial x} + \frac{\partial T_{yy}}{\partial y} + \tau_{yz}\Big|_{z=0} \qquad (9.5.7)$$

where M_x and M_y are the mass transport rates in the x- and y-directions, respectively, P represents the integral of pressure between 0 and L, and T is the integrated shear stress between 0 and L. Note that the shear stresses (τ_{xz} and τ_{yz}) disappear at $z = L$. Equation (9.5.6) is now differentiated with respect to y and the result is subtracted from the derivative of Eq. (9.5.7) with respect to x, resulting in

$$
f\left(\frac{\partial M_x}{\partial x} + \frac{\partial M_y}{\partial y}\right) + M_y \frac{\partial f}{\partial y} = \left[\frac{\partial}{\partial x}(\tau_{yz}|_{z=0}) - \frac{\partial}{\partial y}(\tau_{xz}|_{z=0})\right]
$$
$$
+ \left(\frac{\partial^2 T_{yx}}{\partial x^2} - \frac{\partial^2 T_{xy}}{\partial y^2}\right) + \left(\frac{\partial^2 T_{yy}}{\partial x \partial y} - \frac{\partial^2 T_{xx}}{\partial y \partial x}\right) \tag{9.5.8}
$$

where the derivative of f with respect to x has been set to zero.

Equation (9.5.8) is the vertically integrated vorticity equation. Consistent with continuity, the first term on the left-hand side is normally zero, unless the model allows some vertical flow into the layer. The second term on the left-hand side is sometimes referred to as the *planetary vorticity tendency* and is a measure of the torque imposed upon the water column as it moves to regions of different f. The rate of change of f with latitude is usually written as

$$
\frac{\partial f}{\partial y} = \beta \tag{9.5.9}
$$

and models that consider constant values of β are called *beta-plane models*. The first term on the right-hand side of Eq. (9.5.8) is the vertical component of the curl of the wind stress vector and gives a measure of the torque exerted about a vertical axis by surface wind. The remaining two terms are contributions to torque by the action of viscous and turbulent stresses within the water column. There are a number of ways of dealing with these terms, from neglecting them completely to making detailed formulations for eddy viscosities, usually as functions of velocities. In any case, formulation of the two-dimensional vertically integrated problem has led to some important results in understanding general circulation in the oceans.

PROBLEMS

Solved Problems

Problem 9.1 (Demonstration of nondimensional formulation of Ekman-type model.) The so-called Ekman model was developed as a simplification of the general momentum equations, in which the horizontal diffusivities are neglected, as well as the nonlinear acceleration terms. In addition, a large width-to-depth ratio is assumed, with small Rossby number, $\mathrm{Ro} = U/fL$, where U and L are the characteristic velocity and length scales, respectively. Hydrostatic pressure in the vertical direction is then justified and the horizontal momentum equation (x-direction) has the form

$$
\frac{\partial u}{\partial t} \cong -\frac{1}{\rho}\frac{\partial p}{\partial x} + fv \tag{1}
$$

A characteristic time scale is also defined as $T = L/U$. Then, writing each term in Eq. (1) in nondimensional form, using $u' = u/U$, $x' = x/L$, $t' = t/T$ (primes indicate nondimensional quantities), we have

$$\frac{U^2}{L}\frac{\partial u'}{\partial t'} \cong -\frac{1}{\rho L}\frac{\partial p}{\partial x'} + f U v' \Rightarrow \frac{\partial u'}{\partial t'} \cong -\frac{1}{\rho U^2}\frac{\partial p}{\partial x'} + \frac{fL}{U}v' \tag{2}$$

Thus it is seen that the behavior of this model is still at least partly controlled by Ro, which appears in the last term on the right-hand side (note that it is actually the inverse of Ro that appears; since Ro is assumed to be small, this term is large). One of the great advantages of models such as this is the neglect of the nonlinear terms, which greatly simplifies the solution for Eq. (2).

Problem 9.2 Express the dependence of the Coriolis parameter f on the latitude angle θ, and provide a table of f values versus θ, at intervals of $10°$.

Solution

As defined in Eq. (9.1.1),

$$f = 2\Omega \sin\theta$$

where Ω is the angular velocity of the earth,

$$\Omega = \frac{2\pi}{24 \times 3600} = 7.27 \times 10^{-5} \text{ s}^{-1}$$

By applying this value, we obtain the following table:

θ	$10°$	$20°$	$30°$	$40°$	$50°$	$60°$	$70°$	$80°$	$90°$
$f \times 10^4$	0.252	0.497	0.727	0.935	1.114	1.259	1.366	1.432	1.454

Unsolved Problems

Problem 9.3 The usual equation used to estimate geostrophic velocities in ocean currents or in the atmosphere is

$$\frac{1}{\rho}\frac{\delta p}{\delta n} = f c$$

where δp is the pressure difference measured along a line in the n direction and c is the geostrophic velocity. Explain how this equation is related to Eqs. (9.2.6) and describe the relative directions of orientation of n and c, in (a) the northern hemisphere and (b) the southern hemisphere.

Problem 9.4 Consider the general equations of motion for an incompressible Newtonian fluid,

$$\nabla \cdot \vec{V} = 0$$

and

$$\frac{D\vec{V}}{Dt} + 2\vec{\Omega} \times \vec{V} = -\frac{1}{\rho}\nabla p - \vec{g} + v\nabla^2\vec{V}$$

where $\vec{V} = (u, v, w)$ is the velocity vector, $\vec{\Omega}$ is the background rotation rate, ρ = density, p = pressure, g = gravity, v = kinematic viscosity, ∇ is the vector gradient operator, and ∇^2 is the Laplacian operator.

(a) Describe the physical meaning of each of the five terms in the momentum equation.

(b) Consider a system such as a large but relatively shallow lake, which has primarily two-dimensional horizontal motions (i.e., w = vertical velocity component can be neglected). Write the equations of motion for this situation in component form, i.e., for each of the three coordinate directions.

(c) Rewrite the equations from part (b) by making the additional assumptions of steady state and negligible viscous effects, and also neglect the nonlinear acceleration terms. This should result in the geostrophic balance equations.

(d) Consider the situation sketched in Fig. 9.7, which shows the water surface tilted at an angle θ to the horizontal, in the positive x-direction. Is this a physically possible situation? If not, why not? If so, explain what else must be happening to allow the surface to remain tilted. (Note: there is no wind.)

Problem 9.5 Find the horizontal pressure gradient across the surface of the Gulf Stream, where the current is 100 km wide and the difference in water surface elevation is 1.0 m (higher on the right than on the left, looking in the

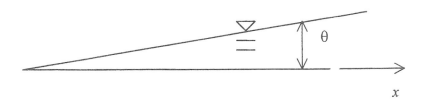

Figure 9.7 Definition sketch, Problem 9.4.

direction of motion). Assuming a latitude of 42°N, what is the corresponding geostrophic velocity?

Problem 9.6 *Cyclostrophic motion* develops from a balance between the pressure gradient and centrifugal force terms in the equations of motion (i.e., it is independent of the earth's rotation) and produces small-scale disturbances such as tornados and whirlpools.

    (a) Show that the mathematical equation that expresses a balance between pressure gradient and centrifugal acceleration can be written as

$$\frac{1}{\rho_0}\frac{dp}{dr} = \omega^2 r$$

        where r is radial distance and ω is angular velocity.

    (b) Sketch the general corresponding flow patterns for positive and negative pressure gradients.

Problem 9.7 Derive an equation for potential vorticity in which friction along the bottom cannot be neglected.

Problem 9.8 Why, in Taylor's experiment, was it important to move the cylinder very slowly across the bottom?

Problem 9.9 Describe the methods and instrumentation you might use in an attempt to observe directly the Ekman spiral in the deep ocean.

Problem 9.10 *Upwelling* is a process in which deeper water is brought to the surface, usually near a coastline, as a result of surface water being driven offshore. If upwelling is driven by wind, in what direction must the wind be blowing for the upwelling to occur on the western coast of the United States?

Problem 9.11
    (a) If the depth of significant wind-driven flow in the ocean at latitude 45°N is found to be 100 m, what is the average value of the vertical diffusivity?
    (b) If the magnitude of the surface velocity is 1 m/s, what is the magnitude of the surface wind shear stress?
    (c) What is the total mass flux rate induced by the wind and in what direction is it, relative to the wind direction?

Problem 9.12 Explain why it is reasonable to set the derivative of f with respect to x equal to zero in Eq. (9.5.8). Also, why do the shear stresses vanish at $z = L$?

Problem 9.13 Calculate different values of β that would be applicable over different ranges of latitudes, from the equator to the poles.

Problem 9.14 Formulate a finite difference expression that could be used to solve Eq. (9.5.8). Assume the first term on the left-hand side is zero and that β is constant. Suggest the approach you would use to solve this expression.

SUPPLEMENTAL READING

Defant, Albert, 1961. *Physical Oceanography*. Pergamon Press, Oxford.

Eckart, Carl, 1960. *Hydrodynamics of Oceans and Atmospheres*. Pergamon Press, Oxford.

Ekman, V. W., 1905. On the influence of the earth's rotation on ocean currents. *Ark. F. Mat. Astr. Och Fysik. K. Sv. Vet. Ak. Stockholm* **2**:11.

Fofonoff, N. P., 1954. Steady flow in a frictionless homogeneous ocean. *J. Mar. Res.* **13**(3):245–262.

Fofonoff, N. P., 1962. Dynamics of ocean currents, in *The Sea*, Vol. 1. Interscience, New York.

McLellan, Hugh J., 1965. *Elements of Physical Oceanography*. Pergamon Press, Oxford.

Munk, W. H., 1950. On the wind driven ocean circulation. *J. Meteorol.* **7**:79–93.

Neuman, Gerhard and Piersol Willard, J., 1966. *Principles of Physical Oceanography*. Prentice Hall, Englewood Cliffs, NJ.

Pedlosky, Joseph, 1987. *Geophysical Fluid Dynamics*, 2ND Ed. Springer-Verlag, New York.

Rossby, C. G., 1932. A generalization of the theory of the mixing length with application to atmospheric and oceanic turbulence. *Papers Phys. Oceanog. Meterol.* **1**(4).

Rossby, C. G. and Montgomery, R. B., 1935. The layer of frictional influence in wind and ocean currents. *Papers Phys. Oceanog. Meterol.* **3**(3).

Stommel, H., 1956. On the determination of the depth of no meridional motion. *Deep Sea Res.* **3**(4):273–278.

Sverdrup, H. U., Johnson, M. W. and Fleming, R. M., 1942. *The Oceans*. Prentice Hall, New York.

Sverdrup, H. U., 1947. Wind driven currents in a baroclinic ocean, with application to the equatorial currents of the eastern Pacific. *Proc. Nat. Acad. Sci.* **33**:318–326.

Veronis, G. and Stommel, H., 1956. The action of variable wind stresses on a stratified ocean. *J. Mar. Res.* **15**(1):43–75.

von Arx, William S., 1962. *An Introduction to Physical Oceanography*. Addison Wesley, Reading, MA.

10
Environmental Transport Processes

10.1 INTRODUCTION

The field of environmental fluid mechanics spans a broad range of topics. Though based primarily on fluid mechanics and hydraulics concepts, as described in Part 1 of this text, the area has grown in the past few decades to encompass many applications in water quality modeling. A major connection between this latter field and pure fluid mechanics lies in the determination of terms needed to specify the transport and mixing rates for a given parameter of interest. This will be seen in the present chapter, in which the classic advection–diffusion equation is derived to express a mass balance statement for a dissolved chemical species distributed in a fluid flow field. In later chapters this idea is expanded to include transport of suspended sediment particles, and several important classes of environmental flows are discussed. Typical parameters of interest might include concentrations of dissolved gases (particularly oxygen), nutrients such as phosphorus or nitrogen, various chemical contaminants, both organic and inorganic, salinity, suspended solids, temperature, biological species, and others. In order to fully describe the fate and transport of a particular species, a knowledge of specific source and sink terms, including interactions with other species, must be incorporated in the general conservation equation. The present text, with several exceptions, generally does not deal directly with these terms, but rather concentrates on the physical transport mechanisms.

10.1.1 Water, Heat, and Solute Transport

Earlier we saw that certain quantities control the rate at which different properties of a flow are transported, either by mean motions or by diffusion. For example, kinematic viscosity may be thought of as a molecular diffusivity for momentum. Thermal diffusivity represents a similar transport term for heat energy, and solute diffusivity represents a corresponding transport mechanism for a dissolved species. Mean motions generally carry all properties of a flow at the same rate, but molecular diffusivities can vary widely. For instance, kinematic viscosity of water is about 10^{-2} cm²/s, thermal diffusivity is of order 10^{-3} cm²/s, and salt diffusivity is of order 10^{-5} cm²/s. These differences are associated with molecular activity, as shown below, and can be related by values of the Prandtl and Schmidt numbers,

$$\mathrm{Pr} = \nu/k_\theta \quad \mathrm{Sc} = \nu/k \tag{10.1.1}$$

where ν is kinematic viscosity, k_θ is thermal diffusivity, and k is molecular diffusivity of a dissolved material. Given the above values, Pr for water is around 10 and Sc is around 10^3. However, when turbulent diffusion is considered, it is normally assumed that the eddies responsible for transporting the properties of the flow are effective in transporting all properties at about the same rate (this is generally referred to as the *Reynolds analogy* — see Chap. 5). Of course, the net transport depends on mean gradients, as described previously, but the diffusion coefficients or diffusivities are the same. This implies that the *turbulent* Prandtl and Schmidt numbers (defined similar to Eqn. 10.1.1, but using turbulent or eddy diffusivities) are both around 1, which is a basic result of the Reynolds analogy.

Another transport term of interest is that of *dispersion*. Many authors have used the terms diffusion and dispersion interchangeably, since the net results of these processes are similar in causing spreading or mixing of material fluid properties. In fact, both terms are often represented mathematically in the same way in the conservation equations. However, dispersion arises from a completely different process than diffusion, as described hereinafter. The effect of dispersion on transport of different properties of a flow is similar to that of turbulent diffusion, in the sense that dispersion causes mixing of a fluid property about a mean position, and dispersion coefficients for momentum, heat and mass all tend to be similar.

Various transport processes of interest may be summarized as follows:

Advection. These motions are associated with mean flow or currents, such as rivers, streams, or tidal motions. They are normally driven by gravity or pressure forces and are usually thought of as primarily horizontal motions.

Convection. This term usually refers to vertical motions induced by hydrostatic instability, i.e., they are buoyancy driven. Examples of this type of motion include heating a pot of water on a stove, or fall and spring lake overturns occurring when the surface temperature on a lake passes through 4°C (temperature of maximum density).

Molecular diffusion. Molecules of a fluid are naturally in random motion, relative to other molecules (Brownian motion), and this leads to a mixing or spreading of fluid properties, consistent with the second law of thermodynamics.

Turbulent diffusion. This is a type of mixing similar to molecular diffusion but with a much stronger effect. Mixing in this case derives from the larger scale movement of packets of fluid (rather than individual molecules) by turbulent eddies.

Shear. Shear exists when there is a variation of advection (mean flow velocity) at different locations in a flow field, so that a gradient exists for flow velocity. This produces a variation in the rate of advective transport of a fluid property, with associated spreading of the average concentration of that property, as illustrated in Fig. 10.1.

Dispersion. This is spreading of a fluid property by the combined effects of shear and transverse diffusion.

The main distinction between advection (or convection) and diffusion or dispersion is that advection represents a net movement of the center of mass of a packet of fluid or fluid property, while diffusion and dispersion represent a spreading about the center of mass. This is illustrated in Fig. 10.2. As will be seen later, dispersion is normally included in the conservation equations in a similar manner as diffusion, but the effects of dispersion are normally much greater than those of diffusion. In this text the term mixing will refer to either diffusion or dispersion.

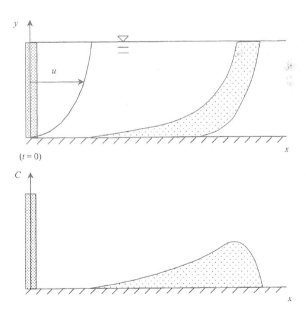

Figure 10.1 Effect of shear on spreading of mean (here, depth-averaged) concentration.

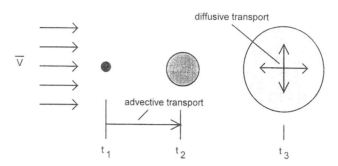

Figure 10.2 Illustration of advective and diffusive transport; advection moves the center of mass while diffusion spreads mass relative to the center, whether it is moving or not.

10.2 BASIC DEFINITIONS, ADVECTIVE TRANSPORT

The basic balance equation for dissolved mass may be derived following procedures similar to those used in the derivations for water mass, momentum, and energy balances from Chap. 2. However, it will first be important to define the various transport properties involved, especially diffusion, and this will lead to a derivation of the classic *advection–diffusion equation*.

It will be important to keep in mind the definition of a *flux*, which is the transport of a given property across a surface, per unit time and per unit area of that surface, and the definition of *mass concentration*, or density, C or ρ_1 (see below), which is the amount of mass of dissolved solids per unit volume of fluid. In the following discussion, it is also helpful to define a *pure concentration*, C^*, which is a dimensionless ratio of dissolved solid mass in a given volume of fluid to total mass in that volume. C is related to C^* by

$$C = \rho C^* \tag{10.2.1}$$

where ρ is the total density of the solution. C^* also is referred to as *mass fraction* or *relative concentration*. Concentrations are often listed in units of ppm (parts per million), referring to C^*, or as mg/L (milligrams per liter), referring to C or ρ_1. In aqueous systems, the numerical values for both C_1^* and ρ_1 turn out to be the same, since the density of water is typically $\rho_2 \cong 1$ gm/cm$^3 = 10^6$ mg/L. Thus for chemical concentrations on the order of 1 mg/L, both C_1^* and ρ_1 would have values of about 10^{-6} (1 mg/L or 1 ppm).

In addition, we will be concerned primarily with *binary systems*, i.e., water plus one other component (the extension to conditions with more than one additional component is straightforward). Thus let $\rho_1 = $ mass of species

1 per unit volume of solution and ρ_2 = mass of species 2 per unit volume of solution. The total density of the solution is $\rho = \rho_1 + \rho_2$. Normally there is a relatively small amount of one species compared with the other, such as with a pollutant dissolved in water. If species 2 is water, then $\rho_2 \gg \rho_1$ and $\rho \cong \rho_2$. In other words, the addition of the pollutant, or tracer, does not significantly affect the water (solution) density. The (nondimensional) concentration of species 1 is

$$C_1^* = \frac{\rho_1}{\rho} \qquad (10.2.2)$$

and in binary systems, $C_1^* + C_2^* = 1$.

The *advective flux* of species 1, for a velocity field $\vec{V} = (u, v, w)$ is defined as

$$\vec{A} = \rho_1 \vec{V} \qquad (10.2.3)$$

Thus the advective fluxes in each Cartesian coordinate direction are $A_x = u\rho_1$, $A_y = v\rho_1$, and $A_z = w\rho_1$.

10.3 DIFFUSION

10.3.1 Molecular Diffusion, Fick's Law

The basic form of diffusion is *molecular diffusion*, which is due to the random motions all molecules undergo (Brownian motion). From considerations of nonequilibrium thermodynamics, the simplest assumption about diffusion of mass that is consistent with the second law of thermodynamics (increasing entropy), is that this diffusion is proportional to the gradient of the chemical potential of the system (μ_c). This statement is analogous to Fourier's law for heat conduction and can be written as

$$\vec{F}_1 = -k\vec{\nabla}\mu_c \qquad (10.3.1)$$

where $\vec{F}_1$ is the diffusive flux of species 1, k is a constant, and the negative sign indicates the flux is in the opposite direction to the gradient of μ_c. Now, μ_c is generally a function of system properties, mostly temperature (θ), mass density (ρ_1), and pressure (p). Then (in one dimension for simplicity),

$$F_{1_x} = -k\left[\left(\frac{\partial\mu_c}{\partial\theta}\frac{\partial\theta}{\partial x}\right) + \left(\frac{\partial\mu_c}{\partial\rho_1}\frac{\partial\rho_1}{\partial x}\right) + \left(\frac{\partial\mu_c}{\partial p}\frac{\partial p}{\partial x}\right)\right] \qquad (10.3.2)$$

The main contribution to the flux is the middle term on the right-hand side. The first term is called the *Soret effect* and indicates the possible diffusion of mass due to a temperature gradient. This term may become important under some circumstances, usually involving high salinity concentrations, but is ignored in most applications. Not much is known about the third term, but it is assumed to be negligible. Considering only the density gradient term, then,

$$F_{1_x} = -\left(k\frac{\partial\mu_c}{\partial\rho_1}\right)\frac{\partial\rho_1}{\partial x} = -k_1\frac{\partial\rho_1}{\partial x} \tag{10.3.3}$$

where k_1 is defined as the *molecular diffusivity* for species 1. This result, extended to all three directions, becomes

$$\vec{F}_1 = -k_1\vec{\nabla}\rho_1 \tag{10.3.4}$$

which is known as *Fick's law* or *Fickean diffusion*.

The concept of diffusivity can also be derived using *mixing length* theory, illustrated in Fig. 10.3 for a simple one-dimensional stratification (z direction) of species 1. On the left-hand plot is shown a general density profile for species 1, while the right-hand plot is a magnified view of a small part of the profile. The average velocity of molecules in the z-direction, due to inherent Brownian motion, is w_1, and l_m is the mixing length, which is assumed to be related to the average distance the molecules travel before colliding with other molecules (molecular free path). Considering a small "window" in the fluid, perpendicular to the z axis, the flux of species 1 through the window in the positive z-direction is $(\rho_1 w_1)_1$, and in the negative z-direction it is $-(\rho_1 w_1)_2$, where the subscripts outside the parentheses indicate the levels at which the flux terms are evaluated. If w_1 is assumed constant, the net flux across this window is

$$F_{z_1} = [(\rho_1)_1 - (\rho_1)_2]w_1 \tag{10.3.5}$$

For small l_m, the density gradient is approximately constant, so

$$(\rho_1)_1 - (\rho_1)_2 \cong -l_m\frac{\partial\rho_1}{\partial z} \tag{10.3.6}$$

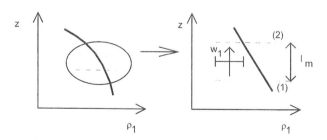

Figure 10.3   Mixing length concept.

where the minus sign is used since $(\partial \rho_1 / \partial z)$ is negative. Substituting Eq. (10.3.6) into Eq. (10.3.5),

$$F_{z_1} = -w_1 l_m \frac{\partial \rho_1}{\partial z} \tag{10.3.7}$$

The product of mean molecular velocity and mean free path $(w_1 l_m)$ is a property of the system and is given the symbol k_1 and defined as the *diffusivity* for species 1, similar to the term in Eq. (10.3.3). Similar analyses may be applied for the x and y directions, to arrive at the same result as before (Eq. 10.3.4). In general, k_1 is a function of temperature and density (ρ_1), and possibly pressure. It is interesting to note that k_1 is defined here as the product of a characteristic length and a characteristic velocity scale — this idea also is applied when discussing turbulent diffusion, though of course the velocity and length scales are different.

10.3.2   Turbulent Diffusion

The concept of turbulent diffusive transport is analogous to the turbulent transport of momentum, discussed in Chap. 5. In particular, the Reynolds transport terms are derived in exactly the same way as the Reynolds transport terms for momentum, (i.e., Eqs. 5.4.7 and 5.4.8).

Here, however, fluctuations of concentration are used instead of a second velocity component. In other words, the turbulent or Reynolds transport of dissolved mass in the i direction is

$$-\overline{u_i' c'} = E_i \frac{\partial \overline{C^*}}{\partial x_i} \tag{10.3.8}$$

where, as before, the overbar indicates a time-averaged quantity and the primes denote fluctuating terms. This equation serves to define the turbulent diffusivity, E_i, which in general is anisotropic and inhomogeneous. That is, in a general sense, turbulent diffusivity may depend on orientation and on location (as will be seen later, it is a function of stratification, for instance), so that E_x, E_y, and E_z could all have different values (anisotropic) and they might all be functions of (x, y, z) (inhomogeneous). Often turbulent diffusivities are not very well known and must be estimated, unless direct measurements of the terms in Eq. (10.3.8) are available. Instrumentation must be capable of measuring the fluctuating quantities making up the Reynolds transport term (left-hand side), and mean concentration gradient must be measured. In many cases, the turbulent diffusivities are treated as fitting parameters, chosen to optimize the results of a particular model, compared to observations.

10.3.3 Statistical Theory of Diffusion

One example of the use of observations to estimate diffusive-type spreading characteristics is illustrated by the use of dye release experiments in a natural system. This approach is based on the observation that average dye concentration distributions are often closely approximated by Gaussian profiles (i.e., they are normally distributed). Figure 10.4 shows a conceptual sketch of the outer extent of spreading of dye in the surface layer of a lake, following a concentrated "instantaneous" release over a small area. The dye patch is shown at three different times after this release. For simplicity, mean velocity is assumed to be in the x-direction. Figure 10.5 shows corresponding concentration profiles measured across the patch at one of these times. The ensemble average profile is approximately normally distributed. At earlier times it would be more peaked, and at later times it would be more flattened. In other words, the variance of the distribution increases with time.

Assuming the Gaussian distribution is appropriate, the concentration can be described in terms of the variances in each of the three coordinate

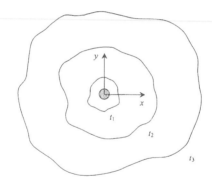

Figure 10.4   Spreading of a dye patch at three times: $t_3 > t_2 > t_1$.

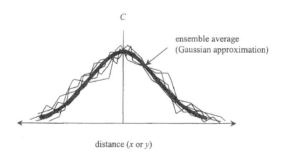

Figure 10.5   Measurements of concentration across the patch at one time, with ensemble average.

directions as

$$C^* = \frac{Me^{-Kt}}{\rho(2\pi)^{3/2}\sigma_x\sigma_y\sigma_z} \exp\left\{-\frac{(x-Ut)^2}{2\sigma_x^2} - \frac{y^2}{2\sigma_y^2} - \frac{z^2}{2\sigma_z^2}\right\} \quad (10.3.9)$$

where M is the mass of dye released, σ_x, σ_y, and σ_z are the standard deviations (square roots of variances) in each of the coordinate directions, ρ is density, K is a first-order decay rate (see Sec. 10.4), and t is time following the dye release. The exponential term with K incorporates any loss of the original mass M due to physical, biological, or chemical processes. In general, the mass remaining in the patch at any time t is given by

$$(Me^{-Kt}) = \rho \iiint C^* dx\, dy\, dz \quad (10.3.10)$$

The variances are measures of the degree of spreading along each of the coordinate directions. For example, variance in the x-direction is calculated from

$$\sigma_x^2 = \frac{\rho}{Me^{-Kt}} \iiint (x-Ut)^2 c\, dx\, dy\, dz \quad (10.3.11)$$

and similar expressions may be defined for the y and z directions. It will be shown later how the variances are related to the turbulent diffusivities.

10.4 THE ADVECTION–DIFFUSION EQUATION

Having developed the basic transport terms, we are now in position to write a conservation equation for dissolved mass of a tracer in an aqueous system. First, a total flux of dissolved mass of species 1 is defined as the sum of the advective (Eq. 10.2.2) and diffusive terms (Eq. 10.3.4),

$$\vec{N}_1 = \rho_1\vec{V} + \vec{F}_1 = \rho_1\vec{V} - k_1\vec{\nabla}\rho_1 \quad (10.4.1)$$

where $\vec{N}_1$ is the total flux of mass of species 1. For now, a molecular diffusion term is used (recall that k_1 is molecular diffusivity), though the following development holds equally well for turbulent diffusion, by substituting a turbulent diffusivity for k_1.

Sometimes it is convenient to define a representative velocity for species 1, $\vec{q}_1$, such that the total flux may be written as

$$\vec{N}_1 = \rho_1\vec{q}_1 \quad (10.4.2)$$

Although this velocity cannot be measured directly, it is defined by equating Eqs. (10.4.1) and (10.4.2). If $\vec{q}_2$ is a representative velocity for species 2, then the *total* momentum per unit volume is

$$\rho\vec{q} = \rho_1\vec{q}_1 + \rho_2\vec{q}_2 \tag{10.4.3}$$

where $\vec{q}$ is the bulk velocity, or total momentum per unit mass,

$$\vec{q} = \frac{\rho_1\vec{q}_1 + \rho_2\vec{q}_2}{\rho} = \frac{\vec{N}_1 + \vec{N}_2}{\rho} = \vec{V} \tag{10.4.4}$$

Note that this result indicates that diffusion does not contribute to the total momentum (both Eqs. (10.4.4) and (10.4.1) can be satisfied together only when $\vec{F}_1 = \vec{F}_2 = \vec{0}$), which is consistent with the idea that diffusion causes spreading about the center of mass and not net transport of the center of mass.

The mass balance for species 1 may be formulated in several ways, but the most direct way is to consider a differential fluid element and incorporate the transport of mass of species 1 across the boundaries of the element. The approach is similar to the development leading to the continuity equation (conservation of fluid mass — see Chap. 2), except diffusive flux must also be included here. A general statement of dissolved mass conservation for the fluid element (volume $\forall$) sketched in Fig. 10.6 is

    [rate of change of mass in $\forall$ per unit time] =

        [rate at which mass moves across the boundaries by net flux]

        $\pm$ [rate at which mass is produced $(+)$ or consumed $(-)$ by

          chemical and biological reactions]

For simplicity, only the fluxes in the x-direction are shown in Fig. 10.6, but similar fluxes may be defined for the y- and z-directions. Using a truncated

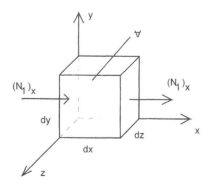

Figure 10.6 Fluid element, showing fluxes of species 1 in the x-direction.

Taylor series to evaluate differences in flux values across dx, dy, and dz, the general statement above is expressed mathematically by

$$\frac{\partial}{\partial t}(\rho_1 dx\, dy\, dz) = -\left[\frac{\partial}{\partial x}(N_1)_x dx\right](dy\, dz) - \left[\frac{\partial}{\partial y}(N_1)_y dy\right](dx\, dz)$$
$$-\left[\frac{\partial}{\partial z}(N_1)_z dz\right](dx\, dy) + R_1(dx\, dy\, dz) \qquad (10.4.5)$$

where R_1 is a rate constant, giving the net production rate of species 1 per unit volume. Then, dividing by the fluid element volume, $\forall = dx\, dy\, dz$,

$$\frac{\partial \rho_1}{\partial t} = -\frac{\partial(N_1)_x}{\partial x} - \frac{\partial(N_1)_y}{\partial y} - \frac{\partial(N_1)_z}{\partial z} + R_1$$

or

$$\frac{\partial \rho_1}{\partial t} + \vec{\nabla} \cdot \vec{N}_1 = R_1 \qquad (10.4.6)$$

A similar development for species 2 results in

$$\frac{\partial \rho_2}{\partial t} + \vec{\nabla} \cdot \vec{N}_2 = R_2 \qquad (10.4.7)$$

where each term has analogous meanings for species 2 as previously defined terms for species 1. Since the present development is for binary systems, any change in either of the species must correspond with an exact opposite change in the other species, so $R_1 = -R_2$. Adding Eqs. (10.4.6) and (10.4.7),

$$\frac{\partial(\rho_1 + \rho_2)}{\partial t} + \vec{\nabla} \cdot (\vec{N}_1 + \vec{N}_2) = 0 \implies \frac{\partial \rho}{\partial t} + \vec{\nabla} \cdot (\rho \vec{V}) = 0 \qquad (10.4.8)$$

which is the continuity equation. In other words, the mass balance statements for the binary system are consistent with an overall statement of conservation of mass.

A more common form of Eq. (10.4.6) can be derived by substituting Eq. (10.4.1) for total flux and dividing by total density (assuming $\rho =$ constant),

$$\frac{\partial C_1^*}{\partial t} + \frac{1}{\rho}\vec{\nabla} \cdot (\rho_1 \vec{V} - k_1 \vec{\nabla} \rho_1) = \frac{\partial C_1^*}{\partial t} + \vec{\nabla} \cdot (C_1^* - k_1 \vec{\nabla} \cdot C_1^*) = \frac{R_1}{\rho}$$
$$\implies \frac{\partial C_1^*}{\partial t} + C_1^*(\vec{\nabla} \cdot \vec{V}) + \vec{V} \cdot \vec{\nabla} C_1^* = \vec{\nabla} \cdot (k_1 \vec{\nabla} C_1^*) + \frac{R_1}{\rho} \qquad (10.4.9)$$

This result is equally valid in terms of C_1, simply by not dividing by ρ. For incompressible flow $(\vec{\nabla} \cdot \vec{V}) = 0$ and, if $k_1 =$ constant, Eq. (10.4.9) becomes

$$\frac{\partial C_1^*}{\partial t} + \vec{V} \cdot \vec{\nabla} C_1^* = \frac{\partial C_1^*}{\partial t} + u\frac{\partial C_1^*}{\partial x} + v\frac{\partial C_1^*}{\partial y} + w\frac{\partial C_1^*}{\partial z}$$

$$= k_1 \nabla^2 C_1^* + \frac{R_1}{\rho} \tag{10.4.10}$$

which is commonly known as the *advection–diffusion equation* for incompressible, dilute (and laminar) flow. As noted earlier, turbulent transport can be incorporated by defining a turbulent diffusivity as appropriate. If $\vec{V} = \vec{0}$ and $R_1 = 0$ (in which case the material is known as a *conservative substance*), Eq. (10.4.10) reduces to

$$\frac{\partial C_1^*}{\partial t} = k_1 \nabla^2 C_1^* \tag{10.4.11}$$

This result is a simple diffusion equation and is known as *Fick's second law*. It is analogous to Fourier's law of heat conduction.

For some problems it is useful to apply the advection–diffusion equation in cylindrical coordinates. The development of this equation is not presented here, since it follows the same general procedure as described above, but the final result is

$$\frac{\partial C_1^*}{\partial t} + v_r \frac{\partial C_1^*}{\partial r} + \frac{1}{r} v_t \frac{\partial C_1^*}{\partial \alpha} + v_z \frac{\partial C_1^*}{\partial z}$$
$$= k \left[\frac{1}{r} \frac{\partial}{\partial r} \left(r \frac{\partial C_1^*}{\partial r} \right) + \frac{1}{r^2} \frac{\partial^2 C_1^*}{\partial \alpha} + \frac{\partial^2 C_1^*}{\partial z^2} \right] + \frac{R_1}{\rho} \tag{10.4.12}$$

where subscript t indicates a tangential component and α is the angular coordinate.

10.4.1   Source and Sink Reaction Terms

Reactions leading to increases or decreases in species 1 are classified into two categories, depending on whether those reactions occur uniformly throughout the volume or at specific locations within the system, usually at a boundary. The former are called *homogeneous reactions* and are usually incorporated in the governing equation through the source/sink term R_1. The latter are called *heterogeneous reactions;* these are more appropriately included as boundary conditions. In some cases, usually depending on the number of physical dimensions being modeled, reactions that might appear as homogeneous in one situation may appear as an internal or boundary source in another. For example, gas transfer across an air/water interface is normally incorporated as a boundary condition when gas concentrations in the vertical direction are of interest, such as in lakes. In this case the advection–diffusion equation would be solved explicitly for the vertical direction (horizontal directions might also be modeled), and the air/water interface would represent a boundary along that direction. However, when considering gas modeling in a river, a common approach (see Chap. 12) is to use a one-dimensional (longitudinal) model, in which case any flux across the air/water surface would be considered to be instantaneously mixed over the entire depth. In other words, the model would have no capability for simulating a vertical distribution of concentration, and therefore the air/water surface has no meaning as a boundary condition. In this case, fluxes across the surface would be considered as an internal source term for the model.

Internal (homogeneous) reactions may be specified in a number of ways, and a full description of all possibilities, for all potential parameters of interest, would require a separate text. It is worthwhile to note here, at least, two common classes of reaction terms. In general, reaction terms may be grouped according to the assumed dependence of the reaction on concentration of the chemical species of interest. The most common reaction terms are either zero order or first order. A zero-order source/sink term does not depend on concentration at all but would be a constant (possibly time-varying) term added to the right-hand side of Eq. (10.4.10). An example of this type of reaction is a municipal waste stream discharging into a river. The loading of a contaminant of interest to the river through this waste stream would not depend on concentrations in the river, and would depend only on the characteristics of the discharge. A first-order reaction is one that depends linearly on concentration. In this case, a first-order reaction rate K would be defined so that

$$\frac{R_1}{\rho} = (\pm)KC_1 \tag{10.4.13}$$

where either plus $(+)$ or minus $(-)$ is used depending on whether concentration is growing or decaying. This form of reaction term is useful for many natural processes and also allows analytical solutions for the advection–diffusion equation, as discussed further in Section 10.7.

10.4.2 Boundary and Initial Conditions

Both boundary conditions and initial conditions are needed to obtain solutions to any differential equation, and the advection–diffusion equation is no exception. Boundary conditions apply to specific locations in the modeled physical domain, as noted above, and are usually specified in one of three ways (see also Sec. 4.6):

1. Specify concentration (e.g., $C = C_0$ at $x = 0$), possibly time-dependent; in combination with velocity, this gives advective flux.
2. Specify gradient (also possibly time-dependent), which, in combination with the diffusivity, gives diffusive flux — this is useful,

for example, at impermeable surfaces, where velocities go to zero; zero gradient implies a "perfectly insulating surface" (using heat conduction analogy).
3. Specify total flux, as a (linear) combination of both diffusive and advective fluxes.

Boundary conditions play an important role in determining the behavior of a particular solution, and care must be taken in specifying the correct conditions for any given problem. This is particularly true when solutions are desired near one of the boundaries of the system domain. In some cases, where numerical solutions are applied, it is useful to define additional grids or nodes outside of the actual system being modeled, and to apply the boundary conditions at the limits of these additional grids. That way, the boundary condition itself does not directly affect as strongly the solution at the point of interest. This and other considerations for numerical solutions are discussed in more detail in Sec. 10.8.

Initial conditions also are needed for time-dependent problems (i.e., $\partial C / \partial t \neq 0$). These are usually specified by setting all values of C throughout the domain to known values for $t = 0$. A useful function for specifying initial conditions, which allows analytical solutions, is the *Dirac delta* function, δ, used to indicate an instantaneous input of mass at a point, along a line or across a surface (corresponding to a three-dimensional, two-dimensional, or one-dimensional problem, respectively). For example, for a planar source parallel to the y–z plane and passing through $x = x_1$, the initial concentration can be described by $C(x_i, 0)dx = M\delta(x - x_1)$, where $M =$ input mass per unit area, dx is the thickness of the source and

$$\delta(x - x_1) = 0 \qquad \text{if} \qquad x \neq x_1$$
$$\delta(x - x_1) = 1 \qquad \text{if} \qquad x = x_1$$
$$\iiint\limits_{\forall} \delta(x - x_1)d\forall = 1 \qquad\qquad\qquad (10.4.14)$$

Application of this function to solve a simple example problem shows how it can be used. For this example, consider an infinitely long cylinder filled with quiescent water, as shown in Fig. 10.7. At time $t = 0$, an infinitely thin cylinder of dye with mass M is introduced at $x = 0$. The problem is then to calculate the concentration at any x and any $t > 0$. For this problem, note that C is constant across any cross section, so there is no diffusion in the radial $(y$–$z)$ plane. The initial concentration distribution is shown in Fig. 10.8, along with several distributions at later times. Note that $C \rightarrow \infty$ at $x = 0$, since a finite amount of mass (M) is injected into an infinitely small volume (if $dx \rightarrow 0$).

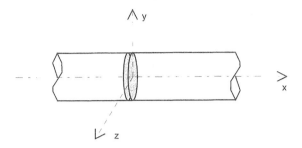

Figure 10.7 Infinite cylinder with initial injection of dye at $x = 0$.

The full advection–diffusion equation is given by Eq. (10.4.9) (the subscript 1 is dropped here and in most of the text to follow, since it will be understood that species 1 is being analyzed). Since the water is quiescent, $u = v = w = 0$, and C^* is a function of x only. The dye also is assumed to be conservative ($R = 0$), so the governing equation is the simple diffusion equation (10.4.11). The initial condition ($t = 0$) states that $C^* = 0$ when $x \neq 0$ and $C^* \to \infty$ when $x = 0$ (i.e., $\delta(x = 0)$ would be used to specify the initial injection of mass). The boundary conditions are $C^* \to 0$ when x is very large, i.e., $x \to \pm\infty$. The general solution is

$$C^* = \frac{B}{\sqrt{t}} \exp\left\{-\frac{x^2}{4kt}\right\} \tag{10.4.15}$$

where B is a constant. The value for B is determined by relating it to the total mass M, using the fact that M is constant (no decay) and using

$$M = \int_{-\infty}^{+\infty} \rho C^* A \, dx = \rho A \int_{-\infty}^{+\infty} C^* dx \tag{10.4.16}$$

Substituting for C^* from Eq. (10.4.15),

$$M = \rho A \int_{-\infty}^{+\infty} \frac{B}{\sqrt{t}} \exp\left\{-\frac{x^2}{4kt}\right\} dx = \rho A \frac{B}{\sqrt{t}} (2\sqrt{\pi k t})$$

$$= 2\rho A B \sqrt{\pi k} \tag{10.4.17}$$

Note that M is independent of time, as it should be for a conservative dye. Rearranging this last result to solve for B, and substituting into Eq. (10.4.15), the final solution is

$$C = \frac{M}{2\rho A \sqrt{\pi k t}} \exp\left\{-\frac{x^2}{4kt}\right\} \tag{10.4.18}$$

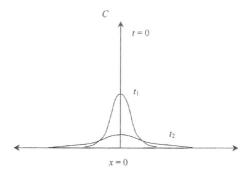

Figure 10.8 Concentration distributions at three different times $(t_1 < t_2)$.

This solution is sketched schematically in Fig. 10.8 for two times, $t > 0$. Note that the solution is in the form of a normal distribution, symmetric and centered around the dye center of mass at $x = 0$. This is consistent with the statistical approach to diffusion as discussed previously in Sec. 10.3.3.

10.5 DISPERSION

The term dispersion is often used interchangeably with the term diffusion, since both terms refer to processes that tend to spread a fluid property relative to mean transport (i.e., advective transport). Dispersion also is often characterized similarly as a diffusion-like term in the mass balance advection–diffusion equation, at least in surface water modeling. However, diffusion and dispersion really refer to two different processes. Diffusion results from temporal averaging of velocity and other property fluctuations, as used in defining turbulent diffusivities, for example. Dispersion, however, results from *spatial* averaging. In many ways, the idea of dispersion is analogous to that of turbulent diffusion. For instance, it would not be necessary to define turbulent closure schemes (Chap. 5) if models were solved with time and spatial steps small enough to resolve directly the turbulent fluctuations. In most applications this is impractical, so that turbulent diffusivities, or eddy viscosities, in the case of the momentum equations, are defined to account for processes occurring on time scales much less than the model time step.

Similarly, simplifications in spatial representation require that smaller-scale processes be represented in some way. The most common application of dispersion is when the spatial domain, particularly for the velocity field, is simplified by integrating (averaging) over one or more coordinate directions. For example, one-dimensional river models are by definition averaged over depth and width, and thus require a longitudinal dispersion term to account for variations or processes occurring in those directions, which are not directly included in the model formulation. Dispersion is often included in such models using a Fickean diffusion term, i.e., dispersive transport may be defined by the product of a *dispersion coefficient*, or dispersivity, with a mean gradient. This dispersion coefficient is in most cases much larger than the turbulent diffusivity, so that diffusion terms are often neglected when dispersion is included. In many applications dispersion is used as a sort of bulk adjustment to the model, to account for any processes (known or unknown) which are not directly represented. In fact, the advection–diffusion equation might be more aptly called the advection–dispersion equation in these cases. As long as the dispersion coefficient can be estimated, this usually allows for significant simplifications in model approach and structure, by reducing the number of spatial dimensions that must be considered.

10.5.1 Taylor analysis

Dispersion was originally formulated by Taylor in the context of one-dimensional transport in a tube (Fig. 10.9). For this analysis an instantaneous injection of a dye, uniformly over a cross section, is assumed at time $t = 0$ and at location $x = 0$. The tube has constant diameter and there is a steady turbulent flow in the positive x-direction. In general, the concentration is a function of x, t, and $r =$ radial position, so $C^* = C^*(x, r, t)$. The dependence on r arises because the velocity is a function of r. The governing equation for this problem is a two-dimensional (radially symmetric) advection–diffusion equation, written in cylindrical coordinates as

$$\frac{\partial C^*}{\partial t} + u(r)\frac{\partial C^*}{\partial x} = \frac{1}{r}\frac{\partial}{\partial r}\left[rE_r(r)\frac{\partial C^*}{\partial r}\right] + E_x(r)\frac{\partial^2 C^*}{\partial x^2} - kC^* \qquad (10.5.1)$$

where $u(r) =$ longitudinal velocity, $E_r(r) =$ radial diffusivity, $E_x(r) =$ longitudinal diffusivity, and $k =$ first-order decay rate. It is assumed that steady, fully developed flow exists in the tube, so that u, E_r, and E_x are not functions of x.

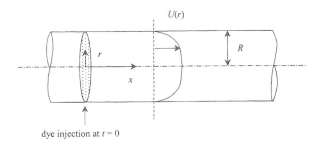

dye injection at $t = 0$

Figure 10.9 Problem definition for Taylor analysis of dispersion in a tube.

Although the velocity distribution is not necessarily known, it is assumed that the Reynolds analogy holds, and E_r is the same as the eddy viscosity in the radial direction, so

$$E_r(r) = -\frac{\tau}{\rho \partial u / \partial r} \tag{10.5.2}$$

where τ = shear stress, given by

$$\tau = \tau_0 \frac{r}{R} \tag{10.5.3}$$

and τ_0 = wall shear stress and R = tube radius. Further substituting $u* = (\tau_0/\rho)^{1/2}$ = friction velocity, Eq. (10.5.2) is rewritten as

$$E_r(r) = -\frac{r}{R} \frac{u_*^2}{(\partial u / \partial r)} \tag{10.5.4}$$

Also, following Taylor, the longitudinal diffusion term is neglected (this should be valid after a sufficient time has elapsed following the injection).

In principle, once $u(r)$ and $E_r(r)$ are specified, the original equation (10.5.1) may be integrated directly to obtain a solution. However, this information is not always available, and the integration may be difficult, possibly requiring a numerical solution. A simpler alternative approach was suggested by Taylor, involving use of a one-dimensional transport equation,

$$\frac{\partial C^*}{\partial t} + U \frac{\partial C^*}{\partial x} = E_L \frac{\partial^2 C^*}{\partial x^2} - kC^* \tag{10.5.5}$$

where $C^* = C^*(x, t)$ is the cross-sectional average concentration, $U = Q/A$ is the mean velocity, Q = flow rate, A = cross-sectional area, and E_L is a *longitudinal dispersion coefficient*. Note that both C^* and U are no longer functions of r. The solution to this problem is already known; it is the same as Eq. (10.3.9), rewritten here in terms of the present variables,

$$C^* = \frac{M}{2\rho A \sqrt{\pi E_L t}} \exp\left\{ -\frac{(x - Ut)^2}{4E_L t} - kt \right\} \tag{10.5.6}$$

where M is the original mass of dye injected. Taylor showed that this solution was a good approximation to the exact solution, using

$$E_L = 10.1 R u^* \tag{10.5.7}$$

This value may be compared with the radial or longitudinal diffusivities, estimated from

$$E_r \cong E_x \cong 0.07 R u^* \tag{10.5.8}$$

which are several orders of magnitude smaller.

The different effects of diffusion and dispersion are illustrated in Fig. 10.10, for an instantaneous injection of dye at time $t = 0$ and $x = 0$.

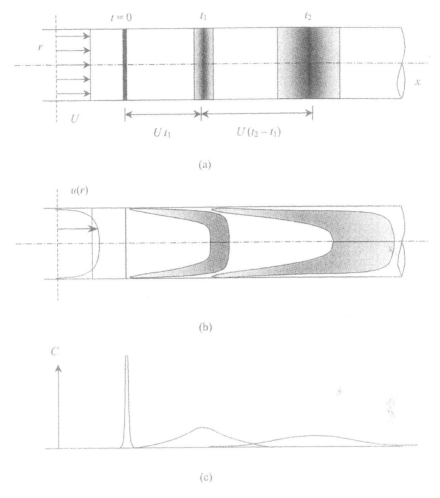

Figure 10.10 Effect of velocity shear on the spreading of dye, with a longitudinal dispersion approach: (a) spreading due to diffusion only; (b) spreading due to diffusion and velocity shear; and (c) resulting cross-sectional average concentration profiles, according to the dispersion model.

For purposes of demonstration, the dye is assumed to be conservative. In Fig. 10.10a a uniform velocity is assumed, so that only longitudinal diffusion occurs (because there are no gradients in the radial direction, radial diffusion is not important). In this case the dye spreads symmetrically upstream and downstream about the center of mass being advected downstream at velocity U. Figure 10.10b shows a more realistic situation that accounts for the no-slip conditions at the tube wall, although the mean velocity is unchanged. The variations in velocity cause a stretching of the dye distribution, effectively spreading the concentration more quickly than by diffusion alone, as shown in Fig. 10.10c. In this case, radial diffusion will occur because this stretching creates radial gradients. For the longitudinal dispersion approach, the same velocity field as in part (a) is assumed (i.e., uniform), but E_L is defined to account for the effects of nonuniform velocity, so that a better fit with the cross-sectional average concentration (C^*) is obtained, as seen in Fig. 10.10c. Although not shown here, this produces a response similar to that shown in part (a), but with a faster spreading rate.

10.5.2 Longitudinal Dispersion in Rivers

As previously noted, dispersion is commonly used when a model is simplified in the number of spatial dimensions or coordinate directions that it includes explicitly. In general, since almost all real processes are three-dimensional in nature, this implies that any model that does not solve for all three coordinate directions should include a dispersion term to account for those processes not directly included. In order to illustrate this procedure, and also to show how a dispersion coefficient may be estimated, the problem of two-dimensional open channel flow is considered here for modeling in a one-dimensional (longitu- dinal) framework. Two-dimensional flow is a common assumption for open channel modeling (see Chap. 7) and in principle should include a dispersion term to account for the fact that any processes occurring in the lateral direction (the y-direction, in the following discussion) are already averaged. The focus here is on the further simplifying step of going to a one-dimensional model.

A two-dimensional flow and concentration profile for a tracer, assumed to be conservative for simplicity of discussion, are shown in Fig. 10.11, where arbitrary distributions are assumed. The important point to note is the decom- position of both velocity u and concentration c into mean and fluctuating

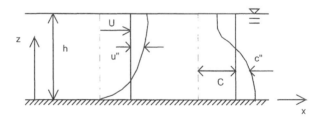

Figure 10.11 Two-dimensional velocity and concentration profiles, with depth-ave- raged values and spatial deviations shown.

components, i.e.,

$$u = U + u''(z) \qquad c = C + c''(z) \tag{10.5.9}$$

(note that a lower case c is used to denote a local value, while the upper case indicates an average, as defined below; also, the discussion here applies equally to a dimensional and a nondimensional concentration). The decomposition in Eq. (10.5.9) is similar to the Reynolds decomposition discussed in Sec. 5.3, except that the means and fluctuations here are defined over space instead of time, so

$$U = \frac{1}{h} \int_0^h u \, dz \qquad \text{and} \qquad C = \frac{1}{h} \int_0^h c \, dz \tag{10.5.10}$$

and by definition,

$$\int_0^h u'' \, dz = \int_0^h c'' \, dz = 0 \tag{10.5.11}$$

In general, u and c also may be functions of x and/or time t. For the present discussion it is assumed that the profiles are already averaged over a time period for which a dispersion coefficient value is desired. Similar calculations as shown below could be performed for other time periods or for other longitudinal positions, as needed.

The governing equation for transport of a conservative tracer in two-dimensional flow is

$$\frac{\partial c}{\partial t} + u \frac{\partial c}{\partial x} = \frac{\partial}{\partial x} \left(E_x \frac{\partial c}{\partial x} \right) + \frac{\partial}{\partial z} \left(E_z \frac{\partial c}{\partial z} \right) \tag{10.5.12}$$

where E_x and E_z are diffusivities in the x and z directions, respectively. Following the same approach as in Taylor's analysis of dispersion in a tube, the diffusive transport in the x-direction is assumed to be negligible. Substituting Eq. (10.5.9) into Eq. (10.5.12),

$$\frac{\partial}{\partial t}(C + c'') + (U + u'') \frac{\partial}{\partial x}(C + c'') \cong \frac{\partial}{\partial z} \left[E_z \frac{\partial}{\partial z}(C + c'') \right] \tag{10.5.13}$$

Again following a procedure similar to Reynolds averaging, these terms are depth-averaged to obtain

$$\frac{\partial C}{\partial t} + U \frac{\partial C}{\partial x} + \overline{\overline{u'' \frac{\partial c''}{\partial x}}} = \frac{1}{h} \left. \left(E_z \frac{\partial C}{\partial z} \right) \right|_0^h \tag{10.5.14}$$

where the double overbar indicates a spatial average and Eq. (10.5.11) has been used to set the average of any single fluctuating term to zero. The right-hand side (RHS) of Eq. (10.5.14) is the difference between the vertical transport at the upper and lower boundaries and, as long as there is no loss or gain of material at these boundaries (i.e., zero-flux boundary conditions), both of these transport rates are zero.

The third term on the left-hand side (LHS) of Eq. (10.5.14) is similar to a Reynolds stress term. It is rewritten as

$$\overline{u''\frac{\partial c''}{\partial x}} = \frac{\partial}{\partial x}\overline{(u''c'')} - \overline{c''\frac{\partial u''}{\partial x}} = \frac{\partial}{\partial x}\overline{u''c''} \qquad (10.5.15)$$

where the last result is due to continuity, since the fluctuating velocity field must satisfy the continuity equation just as the mean field does. Making this last substitution, the result is

$$\frac{\partial C}{\partial t} + U\frac{\partial C}{\partial x} = -\frac{\partial}{\partial x}\overline{u''c''} \qquad (10.5.16)$$

This last expression may be written using a dispersion term, by defining the dispersion coefficient as

$$E_L = \frac{-\overline{u''c''}}{\partial C/\partial x} \qquad (10.5.17)$$

so that

$$\frac{\partial C}{\partial t} + U\frac{\partial C}{\partial x} = E_L\frac{\partial^2 C}{\partial x^2} \qquad (10.5.18)$$

This result is the (one-dimensional) advection–dispersion equation for this flow. The equation is mathematically much simpler to solve than the original two-dimensional equation (10.5.12) and provides adequate calculations for the average concentration C.

While it is possible to calculate terms such as $\overline{u''c''}$ directly, as long as sufficient data are available, an alternate approach may be used that relies only on velocity data and an estimate for E_z. For this approach, the original two-dimensional equation, neglecting longitudinal diffusion, is used as a starting point,

$$\frac{\partial c}{\partial t} + u\frac{\partial c}{\partial x} = \frac{\partial}{\partial z}\left(E_z\frac{\partial c}{\partial z}\right) \qquad (10.5.19)$$

The coordinate system is then transformed by considering a new coordinate χ, which is moving with mean velocity U, so

$$\chi = x - Ut \qquad \frac{\partial \chi}{\partial t} = -U \qquad \left.\frac{\partial c}{\partial t}\right|_\chi = \left.\frac{\partial c}{\partial t}\right|_x + U\frac{\partial c}{\partial \chi} \qquad \frac{\partial c}{\partial x} = \frac{\partial c}{\partial \chi} \qquad (10.5.20)$$

Then

$$\frac{\partial c}{\partial t} - U\frac{\partial c}{\partial \chi} + u\frac{\partial c}{\partial \chi} = \frac{\partial}{\partial z}\left(E_z\frac{\partial c}{\partial z}\right)$$

$$\implies \frac{\partial c}{\partial t} + u''\frac{\partial c}{\partial \chi} = \frac{\partial}{\partial z}\left(E_z\frac{\partial c}{\partial z}\right) \tag{10.5.21}$$

This equation may be simplified for the case of steady state, also assuming $c'' \ll C$ and noting that C is not a function of z,

$$u''\frac{\partial C}{\partial \chi} = \frac{\partial}{\partial z}\left(E_z\frac{\partial c''}{\partial z}\right) \tag{10.5.22}$$

Then, integrating with respect to z, we have (z' and z'' are dummy integration variables in the following)

$$\int_0^z u''\frac{\partial C}{\partial \chi}\,dz' = \int_0^z \frac{\partial}{\partial z'}\left(E_z\frac{\partial c''}{\partial z'}\right)dz'$$

$$\implies \frac{\partial C}{\partial \chi}\int_0^z u''\,dz' = E_z\frac{\partial c''}{\partial z} \implies \frac{\partial c''}{\partial z} = \left[\frac{1}{E_z}\int_0^z u''\,dz'\right]\frac{\partial C}{\partial \chi}$$

where a zero-flux boundary at $z = 0$ has been assumed in the first integration step. Integrating once again,

$$\int_0^z \frac{\partial c''}{\partial z''}\,dz'' = c''(z) - c''(0) = \frac{\partial C}{\partial \chi}\int_0^z \left[\frac{1}{E_z}\int_0^z u''\,dz'\right]dz''$$

This last result is multiplied by u'', and the resulting product is depth-averaged, to obtain

$$u''c'' = u''\frac{\partial C}{\partial \chi}\int_0^z \left[\frac{1}{E_z}\int_0^z u''\,dz'\right]dz'' + u''c''(0)$$

$$\implies \overline{u''c''} = \frac{1}{h}\int_0^h u''c''\,dz = \frac{1}{h}\int_0^h \left\{u''\frac{\partial C}{\partial \chi}\int_0^z \left[\frac{1}{E_z}\int_0^z u''\,dz'\right]dz''\right\}dz$$

$$+ \frac{1}{h}\int_0^h u''c''(0)\,dz$$

Or, since the last term on the right-hand side is zero because $c''(0)$ is treated like a constant, the final result is

$$\overline{u''c''} = \frac{1}{h}\int_0^h \left\{u''\frac{\partial C}{\partial \chi}\int_0^z \left[\frac{1}{E_z}\int_0^z u''\,dz'\right]dz''\right\}dz \tag{10.5.23}$$

This gives an explicit expression for the dispersive transport term, which may be calculated as long as the velocity and diffusion profiles are known. This may be taken one step further, to develop an expression for the longitudinal dispersivity, by substituting Eq. (10.5.17),

$$E_L = -\frac{1}{h} \int_0^h \left\{ u'' \int_0^z \left[\frac{1}{E_z} \int_0^z u'' dz' \right] dz'' \right\} dz \qquad (10.5.24)$$

Consistent with the idea that the dispersive transport is related to the spatial fluctuations in velocity, this last result shows that E_L is larger when the velocity fluctuations u'' are larger. Estimates for open channel flow reported by Fischer et al. (1979) show that E_L is two to four orders of magnitude larger than longitudinal turbulent diffusivity, E_x, and may be approximated by $E_L \cong 20.2 \, hu*$. For practical applications, E_L is normally chosen as a fitting parameter, since it is difficult to know its precise value without detailed velocity profile information, which normally is not available. Statistical approaches can be used, as discussed previously in the context of turbulent diffusivities, or values can be chosen to allow the best fit of a particular model. The final value also can take into account numerical dispersion resulting from the solution procedure used to solve the partial differential equation of mass transport (advection–dispersion equation), as discussed further in Sec. 10.8.1.

10.6 DISPERSION IN POROUS MEDIA

Experiments of solute diffusion in the stagnant fluid phase that saturates sediments have indicated that the molecular diffusive transport in such a domain is subject to attenuation due to several factors, including electrical effects and *tortuosity*. Electrical effects originate from the gradients of other ions, which may be present in the solution or sorbed onto the particles. However, the electric effects are usually much smaller than the effect of tortuosity.

Tortuosity is associated with the ratio of the actual path of ions as they move around sediment particles to the straight distance of that path. Basically, it may be assumed that the tortuosity should depend on the porosity of the porous medium and a characteristic length of the sediment particles. It is common to define the tortuosity, τ, by

$$D_0 = \frac{D_m}{\tau^2} \qquad (10.6.1)$$

where D_m is the coefficient of molecular diffusion of the contaminant in free solution and D_0 is the diffusion coefficient in the fluid phase which saturates the porous medium. In the case of lake sediments of similar characteristic particle size, the tortuosity can be expressed as a function of the porosity, and experiments have indicated that the following approximation can often be useful:

$$D_0 = \phi^2 D_{\mathrm{m}} \qquad (10.6.2)$$

where ϕ is the porosity (see Sec. 4.3). In the case of a heterogeneous porous medium with an axis of symmetry, the tortuosity may be considered as a second-order tensor depending on the vector, which indicates the direction of preferred diffusion. Then the diffusion coefficient also should be represented by a tensor.

Generally, the fluid phase that saturates the porous medium is subject to flow. Then the solute, advected by the flowing fluid particles, follows the curved paths of the fluid particles, and some mixing between flow lines is inevitable, even though the general macroscale fluid motion occurs along straight lines. Again, the difference between straight-line advection and advection in curved lines through the porous medium is associated with the tortousity of the porous medium. Section 4.4 shows how the laminar flow through porous media can be represented by a model of flow through small capillaries. Then, the average flow rate per unit area, namely the *specific discharge*, which is a macroscale (scale much larger than the characteristic pore size) parameter, is shown to be proportional to the hydraulic gradient. Therefore, though the flow is basically laminar, its macroscale characteristics can be modeled and simulated by methods applied to inviscid flows. Hence, the specific discharge is shown to originate from a potential function that is proportional to the piezometric head. The macroscale average interstitial velocity through the porous medium is equal to the specific discharge divided by the porosity of the porous medium. However, contaminant advection in the domain is accomplished by the microscale flow of the fluid particles. The microscale flow velocity can be expressed, in Eulerian terms, by

$$\vec{V} = \overline{\vec{V}} + \vec{V}' \qquad (10.6.3)$$

where $\overline{\vec{V}}$ is the average macroscale local flow velocity, which originates from the gradient of the potential function, and $\vec{V}'$ is the local deviation of the microscale velocity, relative to the macroscale local flow velocity. The expression given by Eq. (10.6.3) is very similar to the expression represented by Eq. (5.2.1), with regard to turbulent flow. However, in turbulent flow, under steady-state conditions, the value of the local velocity deviation from the average value is still a time-dependent quantity. With regard to flow through porous media, under steady-state conditions, the local deviation from the macroscale velocity is a space-dependent variable. This, then, is more consistent with the definition of a dispersive transport, as defined in the previous section. It should be noted that by the employment of the Lagrangian approach, the deviation of the fluid particle velocity from the local macroscale velocity is always a time-dependent variable.

As with the discussion of dispersion in open channel flow in the previous section, the deviation of the microscale velocity from the macroscale velocity is associated with the dispersion of contaminant in the porous medium domain. On the other hand, the macroscale velocity is considered as the only parameter leading to contaminant advection in the domain. In addition, there are two major differences between dispersion in surface waters and dispersion in flow through porous media: (a) dispersion in free turbulent flow often tends to be nearly isotropic, while dispersion in flow through a porous medium, even in an homogeneous and isotropic porous medium, is a nonisotropic phenomenon, provided that the Peclet number is high (the definition of Peclet number is given hereinafter); and (b) for large-size domains the dispersion coefficients of flow through a porous medium are larger than for smaller size domains — in large domains, there is likely to be a greater degree of inhomogeneity in the properties of the porous medium, which intensifies the effect of contaminant dispersion, as discussed below.

The dispersion in flow through a porous medium depends on the properties of the porous medium and on the magnitude of the flow velocity. For larger macroscale velocity, the deviations of the microscale velocities from the macroscale velocity also become larger. Therefore the dispersion coefficient value increases with an increase of the macroscale velocity. If the fluid that saturates the porous medium is flowing, and the porous medium is isotropic, then dispersion coefficients are larger in the direction of the macroscale velocity vector.

Assuming that the porous medium is isotropic, the dispersion coefficient should depend on scalar properties of the porous medium, as well as the velocity vector and its invariant, namely, its absolute value or magnitude. Therefore the dispersion coefficient should be a second-rank tensor, which can be represented using a series approximation,

$$D_{ij} = D_0\delta_{ij} + a_1 V \delta_{ij} + a_2 \frac{V_i V_j}{V} + a_3 V^2 \delta_{ij} + a_4 V_i V_j \qquad (10.6.4)$$

where D_0 is the molecular diffusivity affected by the tortuosity of the domain, $a_i (i = 1 \ldots 4)$ are coefficients with constant values, V is the absolute value of the velocity, V_i is the ith component of the velocity vector, and δ_{ij} is the Kronecker delta.

The coefficients a_i of Eq. (10.6.4) depend on the structure of the porous medium, which can be represented by a characteristic advection length. In most cases, not all terms of the series given by Eq. (10.6.4) are considered in studies of contaminant dispersion in a porous medium. Basically, the number of significant terms of Eq. (10.6.4) depends on *Peclet number*, defined as

$$\text{Pe} = \frac{Vd}{D_m} \qquad (10.6.5)$$

where V is the large-scale interstitial flow velocity, d is the mean grain size or any other characteristic length of contaminant advection in the domain, and D_m is molecular diffusivity of the contaminant in the fluid phase. The Peclet number, in this case, represents the ratio between contaminant advection and molecular diffusion.

If Pe is extremely small, then only the first term on the right-hand side of Eq. (10.6.4) should be considered. If Pe is of order O(1), then the first three terms should be taken into account. In most cases relevant to contaminant transport in aquifers, Pe is quite high. Then the first and two last right-hand-side terms of Eq. (10.6.4) can be neglected, and an approximate relation is obtained,

$$D_{ij} = a_T V \delta_{ij} + (a_L - a_T) \frac{V_i V_j}{V} \tag{10.6.6}$$

where a_T is called the *transverse dispersivity*, and a_L is called the *longitudinal dispersivity*. The longitudinal dispersivity is normally about 20 times larger than the transverse dispersivity.

As indicated by Eq. (10.6.6), the principal directions of the dispersion tensor are parallel and perpendicular to the macroscale flow direction. Therefore by adopting a coordinate system with a coordinate parallel to the macroscale velocity, we obtain a matrix of dispersion coefficients whose entries are zero, except for those occupying the major diagonal. The values of the major diagonal dispersion coefficients are given by

$$D_L = a_L V \qquad D_T = a_T V \tag{10.6.7}$$

where D_L is the dispersion coefficient in the longitudinal direction (the direction parallel to the velocity vector) and D_T is the dispersion coefficient in an arbitrary direction perpendicular to the velocity vector.

It was noted previously that in large domains dispersion is usually more significant than in smaller domains. This phenomenon is commonly referred to as the *scale effect*. It is generally connected with some heterogeneity that characterizes common large-size porous domains. To exemplify the possible effect of the domain heterogeneity, consider the conceptual model shown in Fig. 10.12. The large-size domain shown in this figure incorporates porous blocks, which are permeable. In these blocks, two sets of equidistant fractures are embedded. There is laminar flow through the small-aperture fractures. Therefore these fractures may be considered as another type of porous medium. Through the large-size domain of Fig. 10.12, two types of flow are available for contaminant advection, the porous block flow and the fracture flow. Contaminant disposed at a certain point of the domain is subject to advection by both of these flows. However, the flow through the fractures is usually much quicker than that through the porous blocks. Therefore the fracture

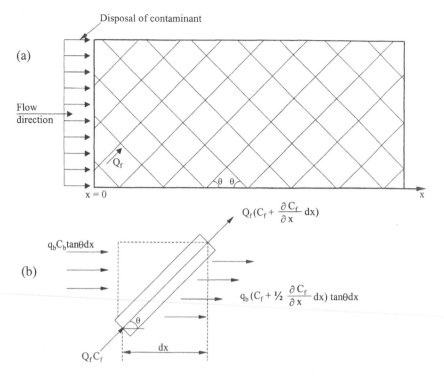

Figure 10.12 A conceptual model of a fractured permeable medium: (a) porous blocks embedding small aperture fractures; (b) mixing in the elementary fracture volume.

flow conveys contaminant into regions in which the porous block flow is not contaminated. Furthermore, there is mixing between the porous block flow and the fracture flow. The incorporation of contaminant advection in the blocks and the fractures and the mixing between the fracture flow and the porous block flow can be represented as advection and dispersion, respectively, in a homogeneous domain. However, the presentation of the domain of Fig. 10.12 as a homogeneous continuum porous matrix requires consideration of a sufficiently large volume of the blocks with the embedded fractures. Such a volume can be termed a representative elementary volume (REV) with regard to contaminant transport in the domain.

In Fig. 10.12b a schematic description of the mixing between the porous block flow and the fracture flow is given under steady-state flow. The mixing is assumed to take place in the elementary volume of the fracture. As an example, consider the case when there is complete mixing between the fracture flow and porous block flow in the elementary fracture volume, shown in Fig. 10.12b. By referring to the Figure we obtain

$$Q_f \frac{\partial C_f}{\partial x} + q_b \tan \theta \left(C_f + \frac{1}{2} \frac{\partial C_f}{\partial x} dx - C_b \right) = 0 \tag{10.6.8}$$

where Q_f is the fracture flow-rate, C_f is the contaminant concentration of the fracture flow, q_b is the specific discharge of the porous block flow, C_b is the contaminant concentration of the porous block flow that enters the fracture elementary volume, and θ is the orientation angle of the fractures.

As shown in the figure, initially (at $t = 0$), the value of C_b is zero for the entire domain, and the value of C_f is C_0 at the entrance of the first fracture segment (at $x = 0$). Therefore direct integration of Eq. (10.6.8) yields the initial distribution of C_f as

$$C_f = C_0 \exp \left(-\frac{q_b \tan \theta}{Q_f} x \right) \tag{10.6.9}$$

Equation (10.6.9) indicates that, due to the mixing between the fracture flow and the porous block flow, the contaminant distribution in the domain is subject to variations. The contaminant distribution in the porous blocks is not uniform downstream of the fracture segment. Therefore, in the segment following, the mixing between the porous block flow and the fracture flow produces another type of distribution of contaminant in the domain. By numerical experiments, it is possible to show that contaminant transport in the domain of Fig. 10.12 can be simulated as a combination of advection and dispersion in a domain composed of a continuum porous medium, provided that the simulated domain is sufficiently large.

10.7 ANALYTICAL SOLUTIONS TO THE ADVECTION–DIFFUSION EQUATION

In this section we consider analytical solutions for the advection–diffusion equation that have been developed for certain simplified conditions. These solutions can be applied directly to predict the behavior of a system under the stated conditions, and they are often useful for checking the results of a numerical solution that might be developed for more complicated situations. In other words, numerical model solutions are often used for simulations of more realistic conditions than are usually assumed for the analytical solutions presented here. In many cases the problems introduced in real applications are associated with the specification of initial and/or boundary conditions. The numerical model might be run under simplified conditions to compare with an analytical solution, as a verification that the model is properly formulated. Numerical modeling considerations are discussed further in Sec. 10.8.

Consider first the general three-dimensional form of the advection–diffusion equation, with first-order reaction,

$$\frac{\partial C}{\partial t} + u\frac{\partial C}{\partial x} + v\frac{\partial C}{\partial y} + w\frac{\partial C}{\partial z} = \frac{\partial}{\partial x}\left(E_x\frac{\partial C}{\partial x}\right) + \frac{\partial}{\partial y}\left(E_y\frac{\partial C}{\partial y}\right)$$

$$+ \frac{\partial}{\partial z}\left(E_z\frac{\partial C}{\partial z}\right) - KC \tag{10.7.1}$$

where, as before, C is concentration (C^* could be used just as well), (x, y, z) are the three spatial coordinates, (u, v, w) are the three corresponding velocity components, (E_x, E_y, E_z) are the corresponding diffusion coefficients, and K is the first-order decay constant. Specific solutions to Eq. (10.7.1) are outlined below for different domains, boundary conditions, and source conditions, which are commonly introduced through the boundary conditions.

10.7.1 Point Sources

Instantaneous Point Source

The instantaneous point source is perhaps the most fundamental situation to consider, since it forms a basis for most of the other solutions presented in this section. An instantaneous point source is one where a finite amount of mass is injected instantaneously at an infinitesimally small point. A solution is obtained first for the following conditions: (1) infinitely large domain with a source located at (x_1, y_1, z_1) — see Fig. 10.13; (2) homogeneous, anisotropic turbulence, constant in time; and (3) uniform and steady velocity field in the x-direction only.

The governing equation is Eq. (10.7.1), without the advection terms for v and w. A change of variables is used to eliminate the advection and decay terms, by defining

$$C = \varphi e^{-Kt} \tag{10.7.2}$$

and

$$x = \chi + Ut \tag{10.7.3}$$

This transforms the problem into one as viewed in a frame of reference moving with the mean velocity U (recall discussion in Sec. 10.5.2). By substituting for C and x into Eq. (10.7.1), with $v = w = 0$, the resulting equation is

$$\frac{\partial \varphi}{\partial t} = E_x\frac{\partial^2 \varphi}{\partial x^2} + E_y\frac{\partial^2 \varphi}{\partial y^2} + E_z\frac{\partial^2 \varphi}{\partial z^2} \tag{10.7.4}$$

The initial condition is assumed to be $C = 0$ everywhere and the boundary conditions are $C \rightarrow 0$ at large distances from the source, i.e., for (x, y, z) $\rightarrow$

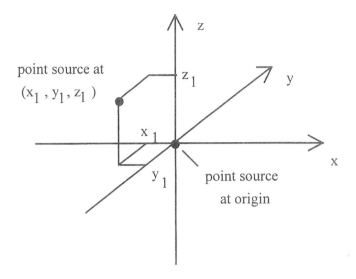

Figure 10.13 Instantaneous point sources in an infinite domain.

$\pm\infty$. The solution, in terms of the original variables, is

$$C = \frac{M \exp(-\beta)}{(4\pi t)^{3/2}(E_x E_y E_z)^{1/2}} \tag{10.7.5}$$

where M is the injected mass and

$$\beta = \frac{[x - x_1 - Ut]^2}{4E_x t} + \frac{(y - y_1)^2}{4E_y t} + \frac{(z - z_1)^2}{4E_z t} + Kt \tag{10.7.6}$$

It should be noted that this solution is invalid at $t = 0$, where C is infinite. Also, the maximum concentration at any time t occurs for $y = y_1$, $z = z_1$, and $(x - x_1) = (Ut)$. This is the position of the (moving) center of the injected mass. This maximum concentration is proportional to $t^{-3/2}$, which is a faster rate of decrease than for the line and plane sources, as will be seen below.

The solution given by Eqs. (10.7.5) and (10.7.6) can be generalized to consider a shear flow, with velocities

$$u = u_0(t) + \frac{\partial u}{\partial y}y + \frac{\partial u}{\partial z}z = u_0(t) + \lambda_y y + \lambda_z z \quad \text{and} \quad v = w = 0 \tag{10.7.7}$$

For simplicity in notation, a parameter φ also is defined as

$$\varphi^2 = \frac{1}{12}\left(\lambda_y^2 \frac{E_y}{E_x} + \lambda_z^2 \frac{E_z}{E_x}\right) \tag{10.7.8}$$

The inverse of φ is thought of as a time scale for the importance of velocity shear in causing mixing of concentration. The general solution under these conditions is

$$C = \frac{M \exp(-\beta)}{(4\pi t)^{3/2}(E_x E_y E_z)^{1/2} (1 + \varphi^2 t^2)^{1/2}} \tag{10.7.9}$$

where M is again the mass injected, but β is defined here as

$$\beta = \frac{\left[x - x_1 - \int_0^t u_0\, dt - \frac{1}{2}(\lambda_y y + \lambda_z z)t\right]^2}{4E_x t(1 + \varphi^2 t^2)} + \frac{(y - y_1)^2}{4E_y t}$$

$$+ \frac{(z - z_1)^2}{4E_z t} + Kt \tag{10.7.10}$$

This solution simplifies to Eqs. (10.7.5) and (10.7.6) for a steady, uniform velocity field (where $\lambda_y = \lambda_z = \varphi = 0$). This simplified velocity field will be assumed for most of the solutions derived below. Numerical approaches may be used to calculate the integral in the numerator of Eq. (10.7.10) in cases where analytical results are not possible. It is interesting to note that when $\varphi t \gg 1$, the maximum concentration decreases with $t^{-5/2}$, considerably faster than the $t^{-3/2}$ behavior determined for the uniform velocity field. This results from the additional effect of shear-induced mixing and implies that some adjustments would have to be made in the dispersion coefficients if a model using the simplified velocity field were to be used (also recall the earlier discussion of dispersion).

Continuous Point Source

In general, to develop solutions for continuous sources, the basic procedure is to integrate the corresponding instantaneous source solution over time. That is, a continuous source is considered to be a series of instantaneous sources acting over a given time interval. As will be seen, for continuous sources the concept of the steady state becomes of interest, when the source is acting over a very long time.

For a continuous point source the same basic assumptions are used as for the instantaneous source, except in this case the source occurs over a time interval t_1. The velocity field is steady and uniform and in the x-direction only, and the source is located at position (x_1, y_1, z_1), as before. Each instantaneous source (with solution given by Eq. 10.7.5) is associated with an amount of mass $(q\, dt)$, where $q = dM/dt =$ mass injection rate. The total concentration response is then obtained by integrating over time t_1,

$$C = \frac{d\sqrt{\pi}}{2\sqrt{a}}\{e^{2\sqrt{ab}}[\mathrm{erf}(F_1 + F_2) - \mathrm{erf}(F_3 + F_4)]$$

$$+ e^{-2\sqrt{ab}}[\mathrm{erf}(F_1 - F_2) - \mathrm{erf}(F_3 - F_4)]\} \tag{10.7.11}$$

where

$$a = \frac{(x - x_1)^2}{4E_x} + \frac{(y - y_1)^2}{4E_y} + \frac{(z - z_1)^2}{4E_z} \qquad (10.7.12)$$

$$b = \frac{U^2}{4E_x} + K \qquad (10.7.13)$$

$$d = \frac{q \exp\left[(x - x_1)/2E_x\right]}{(4\pi)^{3/2}(E_x E_y E_z)^{1/2}} \qquad (10.7.14)$$

and

$$F_1 = \left(\frac{a}{t - t_1}\right)^{1/2} \qquad F_2 = [b(t - t_1)]^{1/2} \qquad \text{erf}(\eta) = \frac{2}{\sqrt{\pi}} \int_0^{\eta} e^{-\xi^2} d\xi$$

$$F_3 = \left(\frac{a}{t}\right)^{1/2} \qquad F_4 = (bt)^{1/2} \qquad (10.7.15)$$

and erf is the *error function*. Values of the error function can be found in various statistic texts or from various web sites. A useful calculator may be found at http://ourworld.compuserve.com/homepages/MTE/gerr_e.htm.

For a *continuous injection*, the actual time is set equal to the injection time, $t = t_1$. In this case it should be noted that F_1 becomes unbounded (approaches ∞), but $F_2 = 0$ and the error function is still defined. For a *steady-state solution*, we let $t \to \infty$, and Eq. (10.7.11) becomes

$$C = \frac{d\sqrt{\pi}}{\sqrt{a}} \exp(-2\sqrt{ab}) \qquad (10.7.16)$$

An approximate solution for the steady state is also obtained using a *disk diffusion* approach, where advective transport is assumed to outweigh the effects of diffusion in the x-direction. The word "disk" refers to the fact that under this assumption the concentration spreads radially with respect to the x axis (assuming the source and the velocity field are aligned along the x axis) and that transport in the x-direction is dominated by advection, with diffusive transport neglected. This assumption is sometimes referred to as a *boundary layer approximation*, since longitudinal gradients are often neglected in boundary layer analyses, relative to transverse gradients. Figure 10.14 illustrates this situation, and the solution (for a source at $x_1 = y_1 = z_1 = 0$) is

$$C = \frac{q}{4\pi(E_y E_z)^{1/2}x} \exp\left(-\frac{y^2 U}{4xE_y} - \frac{z^2 U}{4xE_z} - \frac{Kx}{U}\right) \qquad (10.7.17)$$

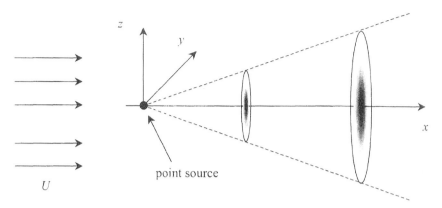

Figure 10.14 Illustration of the disk diffusion assumption; diffusion in the *x*-direction is neglected, relative to advective transport.

10.7.2 Line Sources

Instantaneous Line Source

An instantaneous line source is similar to an instantaneous point source, except the mass is injected evenly over a line (Fig. 10.15). Otherwise, the domain is again assumed to be infinite in all three coordinate directions. For now, a solution is developed for an infinitely long source, parallel to the z axis and passing through the coordinates (x_1, y_1). The result of this last assumption, along with the idea that the mass is evenly distributed over the source line, is that there is no gradient of concentration in the z-direction, and thus there is no transport in the z-direction (diffusion acts only to reduce gradients and does not generate them). With this conclusion in mind, it is easy to see that the solution developed here will apply equally well to a domain bounded in the z-direction, as long as the source stretches across the entire width (in the z-direction). Other assumptions used in the following are a homogeneous, anisotropic, and steady turbulence field; a steady velocity in the x-direction only (component U); the initial condition $C = 0$ every-where; and boundary conditions $C = 0$ at large distances from the source, $(x, y, z) \rightarrow \infty$.

For these conditions the governing equation (10.7.1) becomes

$$\frac{\partial C}{\partial t} + U \frac{\partial C}{\partial x} = E_x \frac{\partial^2 C}{\partial x^2} + E_y \frac{\partial^2 C}{\partial y^2} - KC \tag{10.7.18}$$

There are several ways of obtaining a solution, but perhaps the simplest way is to integrate the solution for an instantaneous point source (10.7.5) along

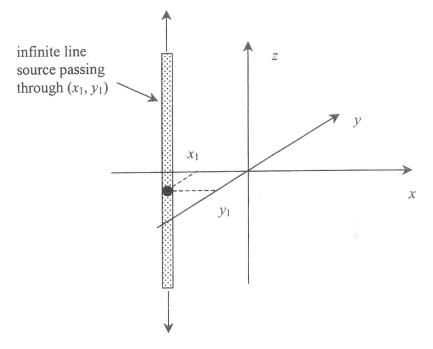

Figure 10.15 Infinite line source parallel to the z axis.

the source length. In other words, the line source is considered to consist of an infinitely long distribution of instantaneous point sources. This integration is similar to the approach used to obtain the solution for a continuous point source from the instantaneous source solution, except here the integration is performed over space rather than time. Each point source contributes an amount of mass ($M'\,dz$), where M' is the mass injected per unit length (this is known as the *source strength*). The resulting concentration due to one of these sources is denoted by C'. Making these substitutions in Eq. (10.7.5) and integrating over $z = \pm\infty$, we find

$$C = \frac{M'}{4\pi(E_xE_y)^{1/2}t}\,\exp\left\{-\frac{[(x-x_1)-Ut]^2}{4E_xt}-\frac{(y-y_1)^2}{4E_yt}-Kt\right\} \quad (10.7.19)$$

If the source has a finite length of $2Z$ (i.e., the source is bounded between $z = \pm Z$), and $x_1 = y_1 = 0$, the solution becomes

$$C = \frac{M'}{4\pi(E_xE_y)^{1/2}t}\,\exp\left\{-\frac{(x-Ut)^2}{4E_xt}-\frac{y^2}{4E_yt}-Kt\right\}$$
$$-\left[\mathrm{erf}\left(\frac{z+Z}{(4E_zt)^{1/2}}\right)-\mathrm{erf}\left(\frac{z-Z}{(4E_zt)^{1/2}}\right)\right] \qquad (10.7.20)$$

This last solution is obtained by integrating the point source solution between $z = \pm Z$, instead of $z = \pm\infty$. In comparison to the instantaneous point source solution, the maximum concentration for a line source decays with the inverse of time (t^{-1}). This is slower than for the point source solution because there is one less direction available for diffusion.

Continuous Line Source

The solution for this situation uses the same basic assumptions as for the instantaneous line source. For the following, we assume $x_1 = y_1 = 0$, for simplicity. The solution is obtained either by integrating the continuous point source solution over space, which is similar to the procedure used to generate an instantaneous line source solution from the instantaneous point source solution, or by integrating the instantaneous line source solution over time, which is similar to the approach used to obtain the continuous point source solution. In either case, the result is (for $t \geq t_1$)

$$C = \frac{q' \exp(xU/2E_x)}{4\pi(E_xE_y)^{1/2}} \left\{ \int_0^{(U^2t/4E_x+Kt)} \frac{1}{\zeta_1} \exp\left(-\zeta_1 - \frac{\beta_2^2}{\zeta_1}\right) d\zeta_1 \right\} \quad (10.7.21)$$

where

$$\zeta_1 = \left(\frac{U^2}{4E_x} + K\right)(t - t_1) \quad (10.7.22)$$

$$\beta_2 = \frac{[(E_yx^2 + E_xy^2)(U^2E_y + 4E_xE_yK)]^{1/2}}{4E_xE_y} \quad (10.7.23)$$

and q' is the injection rate (per unit length). The steady-state solution is

$$C = \frac{q' \exp(xU/2E_x)}{2\pi(E_xE_y)^{1/2}} \kappa_0 (2\beta_2) \quad (10.7.24)$$

where

$$\kappa_0(\eta) \cong \left(\frac{\pi}{2\eta}\right)^{1/2} e^{-\eta} \quad \text{for } \eta > 1 \quad (10.7.25)$$

10.7.3 Plane Source

Instantaneous Plane Source

Next we consider an extension of the line source to a plane source. This follows a similar development to that for the extension of the point source to

the line source. Here, we consider an infinite domain, or one bounded in the
y- and z-directions, with a planar source parallel to the $y-z$ plane and passing
through a point x_1 (Fig. 10.16). The injected mass is assumed to be uniformly
distributed over the area of the plane, so that there are no gradients in the y-
or z-directions (this is why it does not matter whether the source is infinite or
bounded in the y- and z-directions; in either case the transport in both these
directions is zero). A constant E_x is assumed, as well as constant velocity
U in the x-direction only. The initial mass distribution is M'' (mass per unit
area), over the area of the source, with $C = 0$ everywhere else initially. As
before, the boundary conditions are $C \to 0$ for $x \to \pm\infty$ (note that boundary
conditions do not apply for the y- and z-directions).

For these conditions the governing equation (10.7.1) becomes

$$\frac{\partial C}{\partial t} + U\frac{\partial C}{\partial x} = E_x \frac{\partial^2 C}{\partial x^2} - KC \tag{10.7.26}$$

It should be noted that the diffusion terms in the y- and z-directions do not
appear. The solution is obtained by integrating the line source solution in
the y-direction. In other words, the plane source is considered as an infinite
distribution of line sources. The injected mass per unit area for the source

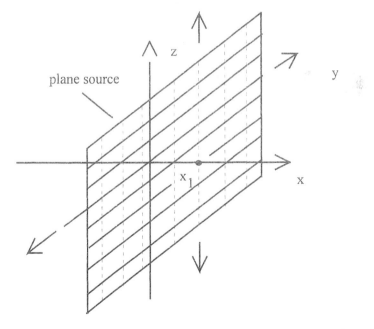

Figure 10.16 Instantaneous plane source parallel to the $y-z$ plane and passing
through point x_1.

is $M''(= source\ strength)$, and the corresponding source for each of the line sources is $M' = M''\,dy$, with resulting concentration response dc given by Eq. (10.7.19). Upon integrating, we find

$$C = \frac{M''}{(4\pi E_x t)^{1/2}} \exp\left\{-\frac{[(x - x_1) - Ut]^2}{4E_x t} - Kt\right\} \tag{10.7.27}$$

Again, with the reduction in the number of directions allowed for diffusion, there is a further reduction in the rate of decrease of maximum concentration with time, relative to the point or line sources. Here, the maximum concentration decreases with $t^{-1/2}$, which is considerably slower than for the point source, for instance.

Continuous Plane Source

The development for a continuous plane source solution follows similar procedures as used previously. With the same assumptions as for the instantaneous plane source, along with $x_1 = 0$, the general solution is (for $t \geq t_1$)

$$C = \frac{q'' \exp(xU/2E_x)}{2\phi}\{[\mathrm{erf}(G_1) - \mathrm{erf}(G_2)]\exp(G_3)$$
$$- [\mathrm{erf}(G_4) - \mathrm{erf}(G_5)]\exp(-G_3)\} \tag{10.7.28}$$

where

$$G_1 = \frac{x + \phi t}{\sqrt{4E_x t}} \tag{10.7.29}$$

$$G_2 = \frac{x + \phi(t - t_1)}{\sqrt{4E_x(t - t_1)}} \tag{10.7.30}$$

$$G_3 = \frac{x\phi}{2E_x} \tag{10.7.31}$$

$$G_4 = \frac{x - \phi t}{\sqrt{4E_x t}} \tag{10.7.32}$$

$$G_5 = \frac{x - \phi(t - t_1)}{\sqrt{4E_x(t - t_1)}} \tag{10.7.33}$$

$$\phi = (U^2 + 4KE_x)^{1/2} \tag{10.7.34}$$

and q'' is the rate of mass injection per unit area. For a continuous injection, $t = t_1$, and the solution becomes

$$C = \frac{q'' \exp(xU/2E_x)}{2\phi}\{[\mathrm{erf}(G_1) \mp 1]\exp(G_3) - [\mathrm{erf}(G_4) \mp 1]\exp(-G_3)\} \tag{10.7.35}$$

where the $-$ sign is used for $x > 0$ and the $+$ sign is used for $x < 0$.

The steady-state solution is found as before by letting $t \to \infty$:

$$C = \frac{q''}{\phi} \exp\left[\frac{x}{2E_x}(U \mp \phi)\right] \tag{10.7.36}$$

where the $-$ and $+$ signs are used as above. If C_0 is the steady-state concentration at the source ($x = 0$), then $C_0 = q''/U$, and an approximate solution is given by

$$\frac{C}{C_0} = \exp\left(\frac{xU}{E_x}\right) \qquad \text{for } x < 0 \tag{10.7.37}$$

and

$$\frac{C}{C_0} = \exp\left(-\frac{xK}{U}\right) \qquad \text{for } x > 0 \tag{10.7.38}$$

These results also are sketched in Fig. 10.17. They come from an analysis of the term ϕ. For example, from the definition (10.7.34), and using the binomial theorem,

$$\phi = U\left(1 + \frac{4KE_x}{U^2}\right)^{1/2} \cong U\left[1 + \frac{1}{2}\left(\frac{4KE_x}{U^2}\right)\right] \tag{10.7.39}$$

where it has been assumed that $(4KE_x/U^2) \ll 1$. Thus for $x < 0$ it is assumed that decay plays a minor role in the region close to the source, and substitution of Eq. (10.7.39) into Eq. (10.7.36) results in Eq. (10.7.37), where the second term in brackets in (10.7.39) is neglected compared with 1. For $x > 0$ this same assumption is used when substituting for ϕ in the denominator of Eq. (10.7.36),

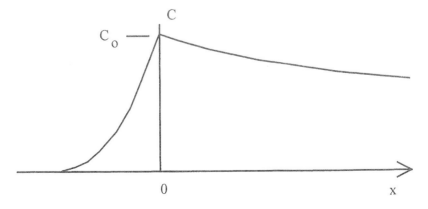

Figure 10.17 Approximate continuous plane source solutions, for a source located at $x = 0$.

but in the exponential the only term left after subtracting is $(-xK/U)$, as shown in Eq. (10.7.38). These results suggest that the extent of spreading upstream is governed primarily by a balance between advective downstream transport and diffusive transport upstream, with decay playing a relatively minor role in this region close to the source. On the other hand, for locations downstream $(x > 0)$, the diffusive transport is neglected in comparison with advection, and both decay and advection are important. In the case of a conservative substance, Eq. (10.7.38) shows that the downstream concentration would be constant and equal to the source value.

10.7.4 Instantaneous Volume Source

The next logical extension of these procedures is to consider an instantaneous volume source, consisting of a continuous distribution of instantaneous plane sources. The volume source is often useful in solving problems involving the concentration response to a spill occurring over a relatively short period of time. For this development we consider steady open channel flow (one-dimensional). The concentration response due to a distribution of instantaneous plane sources is found from integrating Eq. (10.7.27) over a coordinate in the x-direction,

$$C = \int_{-\infty}^{\infty} \frac{C_{\mathrm{i}}(\zeta_{\mathrm{i}})}{(4\pi E_{\mathrm{L}} t)^{1/2}} \exp\left\{-\frac{[(x - \zeta_{\mathrm{i}}) - Ut]^2}{4E_{\mathrm{L}} t} - Kt\right\} d\zeta_{\mathrm{i}} \qquad (10.7.40)$$

where E_{L} is the longitudinal dispersion coefficient, replacing E_x, ζ_{i} is a dummy variable indicating distance in the x-direction, and C_{i} is the initial concentration distribution for the source. Note that although the integration in Eq. (10.7.40) is over $(\pm\infty)$, the only regions contributing mass are those where $C_{\mathrm{i}} \neq 0$. As a special case, suppose that C_{i} is a constant for a source region located between $\zeta_{\mathrm{i}} = L_1$ and $\zeta_{\mathrm{i}} = L_2$ (Fig. 10.18). The dummy variable is seen to have a real physical interpretation, with $d\zeta_{\mathrm{i}}$ equal to the thickness of one of the plane sources and ζ_{i} equal to the distance from that source to the point at which C is being calculated. The total concentration C given in Eq. (10.7.40) is then just the sum of responses due to the distribution of plane sources.

The initial concentration for this simplified case is the mass injected, divided by the volume, or

$$C_{\mathrm{i}} = \frac{M}{A(L_2 - L_1)} \qquad (10.7.41)$$

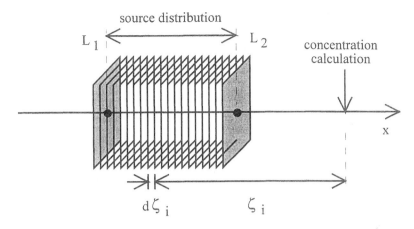

Figure 10.18 Distribution of plane sources along the x axis.

(an equivalent expression is easily written for the case where C_i is variable in x). Integration of Eq. (10.7.40) then gives

$$\frac{C}{C_i} = \frac{e^{-Kt}}{2} \left\{ \text{erf} \left[\frac{x - L_1 - Ut}{\sqrt{4E_L t}} \right] - \text{erf} \left[\frac{x - L_2 - Ut}{\sqrt{4E_L t}} \right] \right\} \qquad (10.7.42)$$

This solution may be used, for example, to check the accuracy of a finite difference solution (see Sec. 10.8), for a source centered about $x = 0$, with $L_1 = -a$ and $L_2 = a$, and with the finite difference grid size, $\Delta x = 2a$. Another special case is one in which there is a semi-infinite initial injection of a conservative substance, between $L_1 = -\infty$ and $L_2 = 0$. The solution is similar to the one in Eq. (10.7.42), except that one of the integration limits is $(-\infty)$,

$$\frac{C}{C_i} = \frac{1}{2} \text{erfc} \left[\frac{x - Ut}{\sqrt{4E_L t}} \right] \qquad (10.7.43)$$

where $\text{erfc}(\eta) = 1 - \text{erf}(\eta)$ is the *complementary error function*.

One of the interesting qualities of the volume source solutions is that, unlike other forms of sources considered previously, concentration is defined at $t = 0$. This means the initial condition can be prescribed by concentration instead of by amount of mass injected. This is sometimes more convenient in obtaining solutions for these problems.

10.7.5 Effect of Finite Domain

Although appealing, the solutions developed in the foregoing sections are restricted to infinite domains or, for the case of line and plane source solutions, at least to domains bounded in direction(s) normal to the source. One possibility for dealing with finite boundaries is to use the method of images, first introduced in Chap. 4 to develop potential flow solutions for problems involving solid walls. This method is illustrated here using the example of a point source, though image solutions can be applied for any of the sources discussed previously.

The method is first illustrated using the case of a contained spill of a tracer at the surface of a deep ocean. The ocean may be considered as a semi-infinite domain, with a boundary at $z = 0$, and the spill will be treated as an instantaneous point source (Fig. 10.19). Assuming the tracer does not volatilize, the ocean surface represents a no-flux boundary for the tracer. Flow in the water could be considered either as uniform or as shear flow, but it is assumed to be in the x-direction only, as before. The previous instantaneous point source solution (10.7.5) would allow diffusion above the air/water interface, effectively "losing" half the initial mass of the source to the atmosphere. In order to add back the mass that is lost, a second *image source* is superimposed on top of the original source (actually, the original source may be thought of as being just slightly below the surface and the image source just slightly above). This image source is identical to the original and, just like the original, half its mass diffuses into the atmosphere. However, half of the mass also diffuses downward into the water, thus replacing the mass that was lost by the original source. Although the solution produces atmospheric concentrations, these are ignored, since the primary interest is with the water domain. In this case, since the position and strength of the image are identical to the

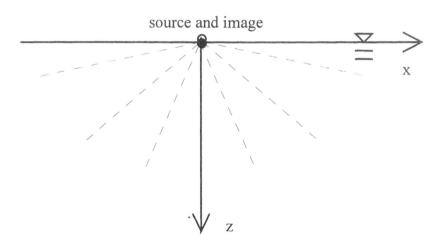

Figure 10.19 Point source at the edge of a semi-infinite domain.

original source, the solution is simply double the solution for one source,

$$C = \frac{2M \exp(-\beta)}{(4\pi t)^{3/2}(E_xE_yE_z)^{1/2} \; (1 + \varphi^2 t^2)^{1/2}} \tag{10.7.44}$$

where β is given by Eq. (10.7.6). A similar result would hold for a continuous source (i.e., double the solution for a single source).

The problem is slightly more complicated when the source is at some distance from the impermeable boundary. As an extension to the above example, consider a spill occurring at some depth H below the surface. In this case the mass diffuses for some distance before it might cross the air/water boundary, when it would be lost from the water. Following the same logic as before, this problem can be solved by positioning an image source so that it would add mass back to the water column to balance exactly the mass flux coming out of the water. This is done by adding an image source, with exactly the same strength as the original source, at exactly the same distance from the boundary, as shown in Fig. 10.20. Again, concentrations in the air are simply neglected, as they are not of interest for the present problem. Assuming an instantaneous point source, with uniform flow field (U), the solution is the superposition of the solutions for the two sources, taking into account their different positions (from Eq. 10.7.5),

$$C = \frac{M}{\rho(4\pi t)^{3/2}(E_xE_yE_z)^{1/2}} \left\{ \exp\left[-\frac{(x - Ut)^2}{4E_xt} - \frac{y^2}{4E_yt} - \frac{(z - H)^2}{4E_zt} - Kt \right] \right.$$
$$\left. + \exp\left[-\frac{(x - Ut)^2}{4E_xt} - \frac{y^2}{4E_yt} - \frac{(z + H)^2}{4E_zt} - Kt \right] \right\} \tag{10.7.45}$$

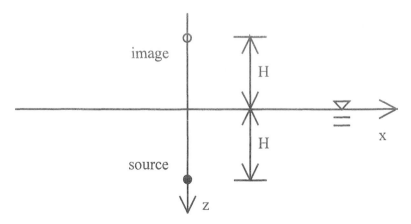

Figure 10.20 Source and image arranged to account for an impermeable boundary at some distance H from the original source.

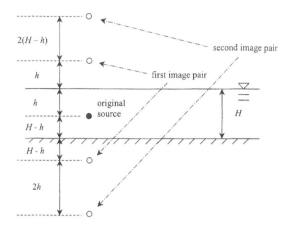

Figure 10.21 Point source and images in a system with two impermeable boundaries.

This same procedure can be generalized again to account for more than one boundary, for example, in a wide open channel flow of constant depth H, with impermeable boundaries at the bottom and surface. In this case, image sources must be added both above and below the respective surfaces. As shown in Fig. 10.21, a point source is located in a wide channel at a distance h below the surface. Two image sources, one at distance h above the surface and one at distance $(H-h)$ below the bottom are added to account for the no-flux boundaries at the top and bottom, respectively. In fact, additional image sources may be needed to account for mass from the first set of image sources lost through the boundaries. This process of adding image sources could, conceivably, continue until a large number of sources have been added. However, in practice, only one or two sets of images are needed, since the effect of additional sources becomes vanishingly small as distance from the boundary increases. This is particularly true when decay is a factor. It should also be noted that lateral boundaries may be treated in a similar fashion to the upper and lower boundaries. Of course, when the flow is bounded on four sides, it is probably simpler to use a plane or volume source solution.

10.8 NUMERICAL SOLUTIONS TO THE ADVECTION–DIFFUSION EQUATION

Many authors have written about the basic methodology used in numerical approaches for solving the advection–diffusion equation. Both finite difference and finite element models are possible, though finite difference representations for the derivatives are a much more common practice for this equation. Finite differences also are more directly related to bulk segmentation, or box models, common in water quality and mass balance modeling for contaminants in surface water systems. This approach has been used, for example, in recent large-scale mass balance modeling for Green Bay and Lake Michigan. The present discussion focuses on this class of solutions.

Finite difference methods are broadly classified as either explicit or implicit. Explicit methods express all derivatives in terms of known values (of concentration), while implicit methods use some of the unknown values (i.e., concentrations at a new time step), leading to the need for solving simultaneous equations, as will be evident in the following discussion. The finite difference time and space steps are denoted by Δt and Δx, respectively, as indicated in Fig. 10.22.

In an explicit method, the new function values (at time $i + 1$) are all calculated on the basis of values from the previous time step (time i), which are known. Initial conditions must be specified (for $i = 0$) in order to start the process. Programming for explicit methods is generally straightforward, but

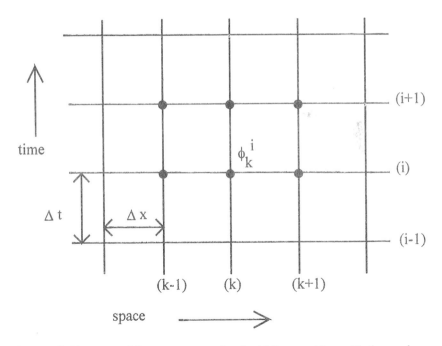

Figure 10.22 Finite difference computational grid for a problem with time and one dimension in space dependency; k is an index for spatial position, with corresponding spatial grid step size Δx; i is a similar parameter, but for time (temporal grid step size is Δt); ϕ_k^i is any function evaluated at position k and time i.

these methods tend to have stability problems. Fully implicit methods generally avoid stability problems but may require longer running times. They are based on a simultaneous calculation of function values at the new time step.

In addition to implicit and explicit considerations, there are a number of possibilities for expressing the derivatives, depending on the degree of accuracy desired. These issues are described in many textbooks on numerical methods for solving partial differential equations and were introduced earlier in Chap. 1. In general, the degree of accuracy depends on the number of terms included in a Taylor series expansion of the function to be integrated. For example, if it is assumed that the arbitrary (but continuous) function ϕ depends only on x (for simplicity), then it can be expanded about some point x_0 as

$$\phi(x_0 + h) = \phi(x_0) + h \frac{d\phi}{dx}\bigg|_{x_0} + \frac{h^2}{2} \frac{d^2\phi}{dx^2}\bigg|_{x_0} + \cdots \tag{10.8.1}$$

where h is the distance between x_0 and the new location where ϕ is to be evaluated. This is an infinite series, and the degree of accuracy of this expansion depends on the number of terms included. Substituting the nomenclature of Fig. 10.22, this expression is equivalent to

$$\phi_{i+1} = \phi_i + \Delta x \frac{d\phi_i}{dx} + \frac{\Delta x^2}{2} \frac{d^2\phi_i}{dx^2} + \cdots \tag{10.8.2}$$

A simple approximation for the first derivative of ϕ is then obtained by rearranging;

$$\frac{d\phi_i}{dx} \cong \frac{\phi_{i+1} - \phi_i}{\Delta x} \tag{10.8.3}$$

This is known as a *forward difference* approximation for evaluating the derivative of ϕ at position i. It is a first-order approximation, meaning that the error term, which is the first term in the series that is neglected, is first order in the spacing Δx (i.e., in rearranging the equation, dividing by Δx makes this neglected term linearly proportional to Δx).

As an example, consider a general finite difference formulation for the solution to the one-dimensional advection–diffusion equation, Eq. (10.7.26), assuming a conservative substance (no reactions) and constant U and E_L, using a forward difference for the time derivative:

$$\frac{C_k^{i+1} - C_k^i}{\Delta t} + \frac{U}{\Delta x}\{(1 - \omega)[\alpha(C_{k+1}^{i+1} - C_k^{i+1}) + (1 - \alpha)(C_k^{i+1} - C_{k-1}^{i+1})]$$

$$+ \omega[\alpha(C_{k+1}^i - C_k^i)] + (1 - \alpha)(C_k^i - C_{k-1}^i)\} = \frac{E_L}{\Delta x^2}\{(1 - \omega)$$

$$\times (C_{k+1}^{i+1} - 2C_k^{i+1} + C_{k-1}^{i+1}) + \omega(C_{k+1}^i - 2C_k^i + C_{k-1}^i)\} \tag{10.8.4}$$

where ω is a weighting factor that allows different weighting for implicit and explicit terms, and α is a weighting factor for the first spatial derivative (advection term), indicating whether a forward, backward, or central difference is being used. With $\omega = 1$ the method is fully explicit, and all values for C appearing in Eq. (10.8.4) correspond to time i, except in the time derivative. For $\omega = 0$ the method is fully implicit, and all concentration values correspond to time $(i + 1)$, again except for one of the values in the time derivative. When $\alpha = 1$ a forward difference is indicated, as in Eq. (10.8.3). A *backward difference* results when $\alpha = 0$, and $\alpha = 1/2$ refers to a *central difference* expression. A special case is when $\alpha = \omega = 1/2$, which is called the *Crank–Nicholson scheme* and represents a simple average between the approximations for the derivatives evaluated at the current and previous time steps. Both this and fully implicit methods have the advantage of much better stability characteristics, but they require solving a potentially large number of simultaneous equations. Fortunately, the coefficient matrices for these equations tend to be banded, so that banded matrix solvers can be used to save on memory requirements and run times.

The numerical solutions become more complicated when additional spatial dimensions are considered, but they are generally extensions of results such as in Eq. (10.8.4). A popular method for two-dimensional problems is the alternating direction implicit (ADI) method. In this approach the equation is solved using an implicit technique for the derivatives in one of the spatial directions, advancing the solution one-half of a time step. These new values are then used to apply the method in the remaining direction, to complete calculations for the full time step.

For water quality modeling applications, emphasis is usually placed on the reaction terms in the advection–diffusion equation. These terms result from internal reactions or interactions (for example, with suspended particles — see Chap. 15) that affect the concentration of the substance being modeled. Because of the degree of complexity involved in representing these processes, it is usually difficult also to include a large degree of complexity with regard to spatial or temporal resolution. Many water quality models are in fact formulated as "box"- or "segment"-type models, especially when applied to large systems or for long-term simulations (see, for example, Bierman et al., 1992; DePinto et al., 1995). On the other hand, hydrodynamic models are usually applied with smaller time and spatial steps, i.e., achieving better resolution while not modeling as many processes. This poses problems when it is desired to link these two types of models. For example, a hydrodynamic model might be run to generate the flow and dispersion fields for input to the water quality model, and output from the model must be converted in some way to produce appropriate values for input to a water quality model.

10.8.1 Numerical Dispersion

Most finite difference methods induce some additional spreading, referred to as *numerical diffusion* or *numerical dispersion*. Since the present discussion refers to one-dimensional models, we will use the term dispersion, consistent with earlier discussion (Sec. 10.5). The total dispersion actually present in the solution of any finite difference method is the sum of the input dispersivity, E_{in}, and the numerical dispersion, E_N, introduced by the numerical technique, i.e.,

$$E_T = E_{in} + E_N \,(= E_L)$$
(10.8.5)

On this basis, as long as E_N can be accurately predicted, then the value of E_{in} for the model may be modified accordingly. Many studies have shown that the advective term is the primary source of numerical dispersion in finite difference solutions to the advective–dispersion equation.

Numerical dispersion depends on the numerical discretization scheme used to approximate the partial derivatives in the transport equation. It results in an artificial spreading of material, much like a true, physical dispersion. To illustrate this process, consider the one-dimensional advection–diffusion equation, as before, without any *source/sink* terms. For purposes of illustration, an explicit central difference representation of the derivatives is used, although analogous results apply for other formulations. Then, with $\omega = 1$ and $\alpha = 1/2$, Eq. (10.8.4) becomes

$$\frac{C_k^{i+1}}{\Delta t} = C_{k-1}^i \left\{ \frac{U}{2\,\Delta x} + \frac{E_L}{\Delta x^2} \right\} + C_k^i \left\{ \frac{1}{\Delta t} - 2\frac{E_L}{\Delta x^2} \right\}$$
$$+ C_{k+1}^i \left\{ -\frac{U}{2\,\Delta x} + \frac{E_L}{\Delta x^2} \right\}$$
(10.8.6)

The solution is then obtained for arbitrary boundary conditions (for this example we consider the calculations only over several time steps, away from any boundaries), assuming an initial concentration distribution with $C = 0$ everywhere except for some value $C_K = C_0$ (Fig. 10.23a). The parameter E_L is supposed to represent any spreading of the concentration about the centroid of the distribution. However, it is interesting to see what happens in the numerical solution after only one time step, with $E_L = 0$, i.e., even when we expect there to be no spreading.

In this case the width of the distribution should remain constant at Δx. However, it is easy to verify that the concentrations at positions $(K - 1)$, (K), and $(K + 1)$ are all nonzero, by substituting $k = (K - 1)$, (K), or $(K + 1)$ in Eq. (10.8.6). The solution in fact produces negative concentrations, shown in Fig. 10.23b, an obviously undesirable result. Further calculations show that concentrations at other locations also may become negative at certain time

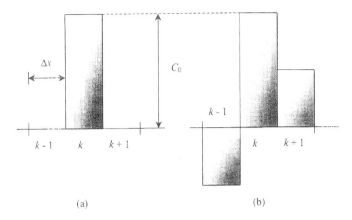

Figure 10.23 (a) Initial concentration distribution; (b) distribution after one time step, calculated from Eq. (10.8.6).

steps, and there are several places where the calculations oscillate between positive and negative values. Oscillations of this type are well known in advection-dominated problems. Overall mass is conserved, since the concentration $C_K = C_0$ is unchanged, while $C_{K-1} < 0$ and $C_{K+1} > 0$, with equal magnitudes. This result is shown in Fig. 10.23b, where the magnitudes have values of $(U \Delta t / 2 \Delta x) C_0$. Thus the width of the distribution has increased to $3 \Delta x$, which is considerably greater than the expected result. This is direct evidence of numerical dispersion.

In general, the relative effect of numerical dispersion can be characterized by looking at the ratio of the two terms in brackets, multiplying C_{K-1} and C_{K+1}, in Eq. (10.8.6). This gives

$$\frac{U/2 \, \Delta x}{E_L / \Delta x^2} = \frac{U \, \Delta x}{2 E_L} = \frac{1}{2} \text{Pe}_c \tag{10.8.7}$$

where Pe_c is called a *cell Peclet number*. This parameter plays a major role in determining the stability of explicit finite difference schemes, with $\text{Pe}_c < 2$ being a common stability criterion. In the context of Eq. (10.8.7), this means that the relative effect of advective transport is no greater than the effect of dispersive transport in any given time step, and the step size must be chosen sufficiently small to insure that this criterion is satisfied. When Pe is large, numerical dispersion also will be large. It can be shown that the numerical dispersion for an arbitrary spatial differencing scheme, with forward explicit time differencing (i.e., as in Eq. 10.8.6), can be expressed as

$$E_N = U \, \Delta x \left[(\alpha - 0.5) - \left(\frac{U \Delta t}{2 \Delta x} \right) (\omega - 0.5) \right] \tag{10.8.8}$$

where the weighting ratio, α, depends on the solution technique used (as above). This result is obtained by reintroducing the second-order terms in the Taylor series expansions for the first derivatives directly into the finite difference representation of the original differential equation.

Three dimensionless numbers may be defined when discussing the constraints of stability, solution positivity, and numerical dispersion in finite difference methods. These are (1) cell Peclet number, defined in Eq. (10.8.7); (2) *dispersion number*, $\lambda = E_L \Delta t / \Delta x^2$; and (3) *Courant number*, $\gamma = U \Delta t / \Delta x$. The input value for dispersivity, E_{in}, is understood to be E_L. Depending on the specific technique used, different constraints are placed on the values of these three parameters which allow stable positive solutions and estimates of E_N. In terms of these parameters, Eq. (10.8.8) can be rewritten in dimensionless form as

$$\frac{E_N}{E_L} = \frac{Pe}{2}[(2\alpha - 1) - \gamma(\omega - 0.5)] \tag{10.8.9}$$

To insure positivity and stability in an explicit scheme, the traditional criteria are given by Pe < 2 and $\lambda < 0.5$, which limit the time step that can be used for a given spatial step. An additional constraint, particularly for upwind or backward differencing schemes (for the advective term), is $\gamma < 1$. This prevents the mean flow from moving more than one grid space in one time step.

From Eq. (10.8.9), use of smaller Pe can help alleviate the problem of numerical dispersion and it is clear that numerical dispersion would be zero if there were no advection (Pe $= 0$). It also is worthwhile noting that the numerical dispersion for the Crank–Nicholson scheme is 0 (where $\alpha = \omega = 0.5$). Other solution techniques such as upwinding or higher order schemes can be used to reduce the problem further, depending on the approach used.

An alternative procedure that is capable of modeling the advection term more accurately, without introducing associated numerical dispersion, involves a *semi-Lagrangian approach*, in which the advection term is calculated with a simple routing scheme. In this approach the mean velocity, time, and spatial steps must be related by $\Delta x = U \Delta t$, so that each parcel of mass (contained in a numerical grid segment) moves exactly one grid space in one time step. The solution approach is then to separate the calculations for the advection and diffusion terms, using the routing calculation for the advection term and the normal finite difference solution for the diffusion term. This procedure is referred to as a split-operator approach and is illustrated for a one-dimensional calculation in Fig. 10.24. It is very effective in preventing numerical dispersion due to the advective term, but unfortunately the general application of this approach for two- or three-dimensional problems is more difficult, except for some simple cases with constant or unidirectional flow fields. Ultimately, the user must weigh a number of factors in choosing a numerical solution for any given problem.

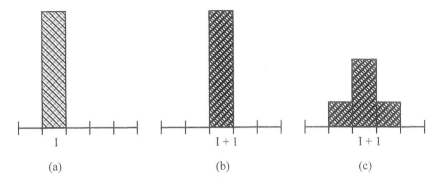

Figure 10.24 Illustration of "split-operator" approach for semi-Lagrangian model; (a) initial condition; (b) distribution after the advection term is calculated; (c) distribution after the diffusion term is calculated.

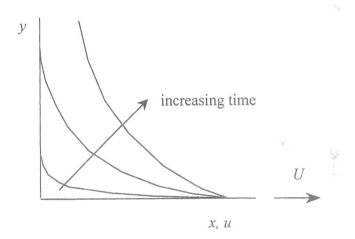

Figure 10.25   General solution, Problem 10.1.

PROBLEMS

Solved Problems

Problem 10.1   In some cases of calculating the spread of a chemical concentration it is possible to use a boundary layer–type solution (see also Section 3.4.2). For the present example, we review the problem of diffusion of momentum from a plate that suddenly starts moving in a fluid that is initially at rest. The problem is sketched in Fig. 10.25, which shows a flat plate along $y = 0$, which starts moving at speed U in the x-direction for time

$t > 0$ (also refer to Sec. 3.4). The problem is treated as two-dimensional flow, for simplicity. Calculate the resulting velocity profile.

Solution

The governing equations for this problem are the two-dimensional continuity and momentum equations for incompressible flow, from Chap. 2,

$$\frac{\partial u}{\partial x} + \frac{\partial v}{\partial y} = 0$$

$$\frac{\partial u}{\partial t} + u\frac{\partial u}{\partial x} + v\frac{\partial u}{\partial y} = -\frac{1}{\rho}\frac{\partial p}{\partial x} + v\left(\frac{\partial^2 u}{\partial x^2} + \frac{\partial^2 u}{\partial y^2}\right)$$

$$\frac{\partial u}{\partial t} + u\frac{\partial u}{\partial x} + v\frac{\partial u}{\partial y} = -\frac{1}{\rho}\frac{\partial p}{\partial x} - g_y + v\left(\frac{\partial^2 u}{\partial x^2} + \frac{\partial^2 u}{\partial y^2}\right)$$

where g_y is the y component of gravity. For this problem it is possible to use a boundary layer approach, since the vertical velocity can be neglected, as well as any gradients in the x-direction, relative to gradients in the y-direction. In fact, since $v = 0$, then the velocity gradient in the x-direction is exactly zero by the continuity equation. The momentum equation in the y-direction is not of further interest, and the x-momentum equation is written in simplified terms as

$$\frac{\partial u}{\partial t} = v\frac{\partial^2 u}{\partial y^2}$$

This is a simple diffusion equation and is the same as Eq. (3.4.1), where v is seen as the diffusion coefficient for momentum. The initial and boundary conditions are

$$u = 0 \quad \text{at} \quad t = 0$$
$$u = U \quad \text{at} \quad y = 0$$
$$u \longrightarrow 0 \quad \text{for} \quad y \to \infty$$

By substituting a dimensionless variable,

$$\eta = \frac{y}{2\sqrt{vt}}$$

and rewriting the derivatives in terms of η, the diffusion equation becomes

$$\frac{d^2 u}{d\eta^2} + 2\eta\frac{du}{d\eta} = 0$$

with boundary conditions $u = U$ for $\eta = 0$ and $u \to 0$ for $\eta \to \infty$. The solution to this problem is

$$u = U(1 - \text{erf}\,\eta) = U\text{erfc}\,\eta$$

This solution is sketched in Fig. 10.25. Note that this solution allows an estimate of the length of time it should take for the velocity at a certain distance y to reach a given fraction of U. A similar formulation can be developed for diffusion of concentration.

Problem 10.2   A river passes through a region of heavy industry. Each industry has its own (constant) waste stream discharging into the river, and it is desired to model the concentration response (C) in the river for a particular chemical species. The chemical is expected to be in each of the discharges, though at possibly different concentrations. The flow rates for the discharges are negligible compared with the river flow rate.

(a) Provide a simple sketch of this problem and label any appropriate points. In order to evaluate the distribution C in the river, would you treat this as a steady or an unsteady problem? Explain why.

(b) Assuming this to be treated as a one-dimensional (longitudinal) problem, how would you incorporate the waste discharges into your model (i.e., as initial condition, boundary condition, internal reaction or source, or other)?

(c) What processes should be considered for inclusion as a source or sink term for C?

Solution

(a) A general sketch of the problem is shown in Fig. 10.26 (for n industries). The cross sections indicate an initial or boundary location upstream of the industrial region (cross section 0), sections immediately downstream of each industrial discharge point, and a section at some point downstream, where

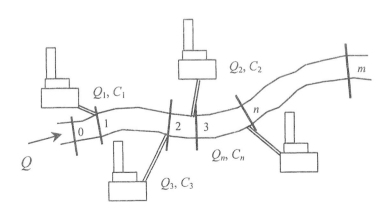

Figure 10.26   Definition sketch, Problem 10.2.

there is interest in knowing the concentration response (C) in the river. A one-dimensional approach is assumed, so that each discharge is fully mixed over the river cross section at each discharge point. Since the discharges are constant, with (assumed) constant concentrations C_i, then as long as the river flow (Q) is steady and any decay processes occur at constant rates, then the problem can be considered as steady.

(b) Since the discharges are assumed to be completely mixed, and also because they occur at locations along the river, and not at one of the boundaries (note that the only boundaries of concern in this one-dimensional, longitudinal, approach are at cross sections 0 and m), then the best way to incorporate them is as internal sources. It might be possible to consider these discharges as boundary conditions if the river were to be modeled as a series of connected reaches, starting at each of the cross sections identified in the sketch. In that case, a simple dilution calculation would have to be performed at each discharge point to obtain the boundary condition for the model for the associated river segment. Since from part (a) the problem is being treated as one of steady state, initial conditions are irrelevant and this is not an option in terms of incorporating the discharges.

(c) Possible processes that might be considered as source or sink terms include biological or chemical reactions (probably decay terms), adsorption of the chemical to sediment particles and subsequent settling of those particles (again, another sink term), volatilization across the air/water interface, degradation or transformation by sunlight, and possible secondary sources such as nonpoint sources (e.g., leaching from groundwater into the stream).

(d) The general governing equation for steady-state one-dimensional transport with constant flow rate is

$$U\frac{\partial C}{\partial x} = \frac{1}{A}\frac{\partial}{\partial x}\left(AE\frac{\partial C}{\partial x}\right) - kC$$

where a first-order single decay coefficient is used to represent the various possible internal sources and sinks. Note that the area A is assumed to possibly vary with x in this equation, although the flow rate Q remains constant. Since the equation is for steady state, there are no initial conditions. Boundary conditions must be specified at the upstream and downstream cross sections. Upstream, at cross section 0, the usual boundary condition would be of the Dirichlet type, so that C_0 should be specified. The downstream boundary condition is somewhat more difficult. In order not to overly constrain the model, the best condition is one of zero gradient. However, cross section m must be chosen to be far enough downstream so that this condition is likely to be realistic.

Unsolved Problems

Problem 10.3   For a delta function input at $x = 0$ into an infinite cylinder filled with a fluid, the mean square displacement (i.e., the variance) of the concentration can be defined as

$$
\sigma_x^2 = \frac{\displaystyle\int_{-\infty}^{\infty} x^2 c(x, t)\, dx}{\displaystyle\int_{-\infty}^{\infty} c(x, t)\, dx}
$$

Find the root-mean-square (rms) displacement as a function of time and diffusion coefficient E. Note that this relationship shows how you would calculate E based on an experiment in which the spreading of a dye was measured.

Problem 10.4   It is desired to evaluate changes in salt concentration (salinity) in a quiescent tank of water 10 m deep in which a salinity gradient has been established in the vertical direction. The tank is not stirred and is completely sealed from its surroundings (no water or salt is added or taken away).

(a) What is the governing equation you would use for this problem? Use z as the vertical coordinate.

(b) Based on your answer to part (a) (or using any other way you can think of), estimate the time required for the gradient to be smoothed by molecular diffusion (molecular diffusivity for salt in water is 10^{-5} cm^2/s). Does the magnitude of the initial gradient make any difference?

Problem 10.5   The turbulent spread of a conservative tracer in a diffusion experiment in the ocean (measured in terms of a standard deviation of the concentration distribution) is observed to be proportional to $t^{3/2}$ ($t = $ time). What can you say about the variation of turbulent diffusion coefficient as a function of time? (Hint: compare Eqs. 10.3.9 and 10.7.5). What is a possible physical reason for the diffusivity to increase with time?

Problem 10.6   Concentration measurements of a conservative tracer are made in a pipe flow with water ($\rho = 1$ g/cm^3). At a particular cross section at a distance x downstream, the concentration is observed to increase from 3 ppm to 13 ppm over a 10 minute period. The mean velocity in the pipe is 10 cm/s and the diffusion coefficient is 10 cm^2/s. The cross-sectional area of the pipe is 5 cm^2.

(a) Demonstrate (graphically is sufficient) that the following kinematic transformation may be applied:

$$
\left.\frac{\partial C}{\partial t}\right|_x = -U \left.\frac{\partial C}{\partial x}\right|_x
$$

(b) What is the total mass transport rate (in mg/s) across the section at x, at a time half way through the observation period? List any assumptions necessary.

Problem 10.7   Using an appropriately defined elemental volume, derive the equation used in part (d) of problem 10.2, for steady one-dimensional open channel transport of a dissolved chemical that is undergoing first-order decay. Note that the flow rate in the channel is constant, but that natural variations in the cross section lead to variations in area and thus velocity.

Problem 10.8   Figure 10.27 shows a shallow lake of depth H and large horizontal extent, which is initially quiescent (no motions). At time $t = 0$ a constant wind starts, generating a velocity U at the surface of the lake. There is some concern about resuspending contaminated sediments at the bottom of the lake. For this problem, it is desired to estimate the length of time the wind would have to blow (i.e., at a constant speed and direction) before water motion would be transferred down to the bottom and, assuming the wind acts for a sufficiently long time, the maximum shear stress expected at the bottom.

(a) List the governing equation and initial and boundary conditions for the problem. Assume a constant viscosity v.

(b) From problem 10.1, a boundary layer solution (in terms of error function) was presented for the problem of finding the velocity profile in a fluid where a flat plate was moving at constant velocity U. Can this solution be used here? Why or why not? Under what conditions might it be valid?

(c) Assuming it is valid, use the boundary layer solution to estimate the time required before motions would be seen in the vicinity of the bottom. Suppose $H = 10$ m, $v = 10^{-2}$ m^2/s, $U = 1$ m/s, and $\rho = 1000$ kg/m^3. How much time is needed?

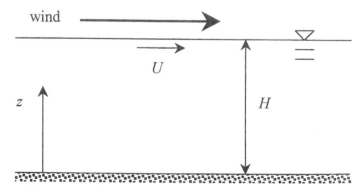

Figure 10.27   Definition sketch, Problem 10.8.

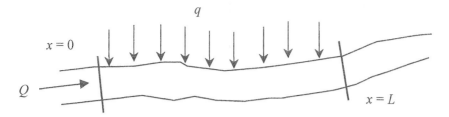

Figure 10.28 Definition sketch, Problem 10.9.

(d) What is the steady-state solution to this problem [i.e., the problem formulated in part (a)]? What is the maximum shear stress expected at the bottom [using parameter values from part (c)]?

Problem 10.9 Consider a small river, with average width W and depth H, passing by a large agricultural field as shown in Fig. 10.28. For simplicity, assume W and H are constants. During rain storms there is direct runoff (flow rate $= q$, per unit length along the stream) from the field into the stream, carrying a concentration of some parameter of interest (say, a pesticide) with concentration C_{r_0}. Upstream of this field the concentration is zero. This contaminant tends to adsorb strongly onto sediment particles, so that settling of particles in the stream represents a loss (sink) of the contaminant from the stream. Assume that particles settle at a rate u_s, so that the settling flux of the contaminant is $u_s C$. In addition, assume that there is an overall decay of the chemical due to biological activity, which may be modeled using a first-order reaction with rate constant k.

(a) Show that the rate of change of stream flow is given by

$$\frac{\partial Q}{\partial x} = q$$

(b) Develop a one-dimensional time-dependent advection–diffusion model, including initial and boundary conditions, to simulate concentration C in the stream from $x = 0$ to $x = L$.

(c) List at least one factor that makes this problem difficult (maybe impossible) to solve using an analytical solution.

Problem 10.10 In a certain treatment process water flows from a large reservoir into a thin and wide channel or duct as shown in Fig. 10.29. After some distance into the duct, boundary layer development is complete and a stable velocity profile is reached, with $U = U_{max}/(2by - y^2)/b^2$. A contaminant of concentration C is transported with the water flow, and the walls of the duct are such that they completely remove any contaminant that comes into contact with

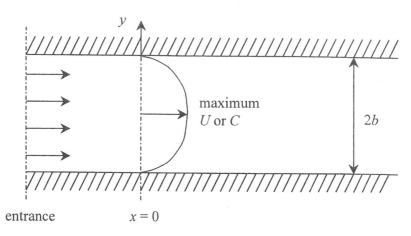

Figure 10.29 Definition sketch, Problem 10.10.

the wall, i.e., $C = 0$ at $y = 0, 2b$. The concentration profile may be assumed to have a similar shape as the velocity profile, $C = C_{\max} (2by - y^2)/b^2$. If $b = 10$ cm, $U_{\max} = 0.1$ m/s, and $C_{\max} = 100$ mg/L at $x = 0$, at what value of x will $C_{\max}$ fall to 1 mg/L? Assume the diffusion/dispersion coefficient is 10 cm^2/s.

Problem 10.11 A cooling "pond" is designed to provide cooling for the condensers of a power plant, as sketched in Fig.10.30. Water is taken from the canal at flow rate Q_{in} and passed through the heat exchangers of the plant, where it undergoes a temperature rise ΔT and is returned to the canal, where there is a general circulation flow rate of Q_0 (note: $Q_{\text{in}} < Q_0$). There are two common designs for the return flow: (1) surface discharge with minimal mixing, and (2) submerged discharge, with (assumed) full mixing. Considering heat engine efficiency, it is desired to maintain the lowest possible condenser temperature, and a major design consideration is to be able to predict these intake temperatures.

(a) In general, would you treat this as a steady or unsteady problem? Why?

(b) For each of the two discharge designs, describe the modeling approach you would use, including governing equation(s) and initial and boundary conditions (as appropriate).

(c) Comment briefly on any advantages or disadvantages you see with the two designs. Which one would you choose?

Problem 10.12 An estuary is defined as that portion of a river that is affected by the salinity and tidal motions of the ocean into which it discharges. With x

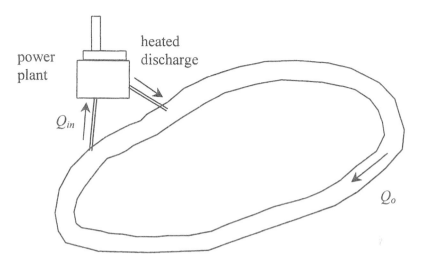

Figure 10.30 Definition sketch, Problem 10.11.

defined as the longitudinal coordinate, so that $x = 0$ at the boundary between the estuary and the ocean, x is negative moving upstream into the estuary, a possible governing equation has been proposed as

$$0 = -U\frac{ds}{dx} + E\frac{d^2s}{dx^2}$$

where U is the mean (tidal average) velocity, E is the dispersion coefficient, and s is salinity. The associated boundary conditions for this model are $s = 0$ for $x \to -\infty$ and $s = s_0$ at $x = 0$. List all assumptions that have been made in writing this as the governing equation and develop a solution for s as a function of x.

Problem 10.13 An experiment is conducted to measure the longitudinal dispersion coefficient in a river. The river is relatively wide and shallow, so that the flow may be approximated as two-dimensional. A line source is used to inject a conservative dye evenly and nearly instantaneously over the depth, near the middle of the river.

Velocity measurements in the river show that the velocity distribution can be described by a 1/7th law, which also describes the vertical distribution of dye concentration C, i.e.,

$$\frac{u}{u_s} = \left(\frac{z}{h}\right)^{1/7} \qquad \text{and} \qquad \frac{C}{C_s} = \left(\frac{z}{h}\right)^{1/7}$$

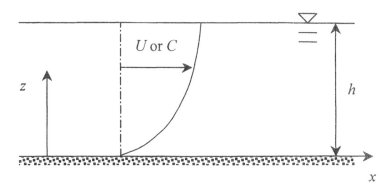

Figure 10.31 Definition sketch, Problem 10.13.

where z = height above bottom, h = total depth, and subscript s indicates the surface value (see Figure 10.31). Several measurements are made of the decay in mean (cross-sectional average) concentration with distance downstream to reveal

$$\frac{\partial \overline{\overline{C}}}{\partial x} \cong -0.1 \frac{C_s}{h} \qquad \text{where} \quad \overline{\overline{C}} = \frac{1}{h} \int_0^h C \, dz$$

where the double overbar indicates a spatial average; $\overline{\overline{u}}$ is defined similarly. Using this information, along with the formal definition, estimate a value for the dispersion coefficient in this river, in terms of u_s and h.

Problem 10.14   An experiment is conducted to evaluate mixing in a 60 km long section of a river. The river is 50 m wide, 3 m deep, and has a mean velocity of 1.0 m/s. In the experiment, a conservative dye is injected continuously at the upstream end of the river reach over a sufficiently long period of time. The dye is injected at a rate so that, if it is assumed that the dye is instantaneously mixed over the river cross section, the initial concentration would be C_0. At the downstream end of the reach (at x = 60 km), the area-averaged concentration is measured, producing the data plotted in Fig. 10.32 (note that t is time after starting the injection).

(a) List the governing equation you would use to solve this problem, including any assumptions made. Also list initial and boundary conditions you would use.

(b) Use the data shown in Fig. 10.32 to estimate the average longitudinal dispersion coefficient for the river.

Problem 10.15   An experiment is carried out to determine the longitudinal dispersion coefficient in a river. The experiment is started by injecting a

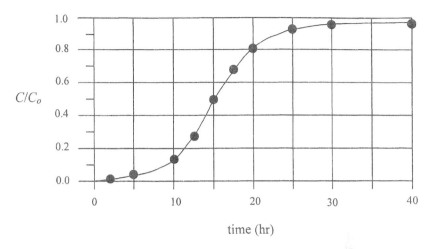

Figure 10.32 Data for river transport experiment, Problem 10.14.

conservative dye into the river at a certain point in such a way that full mixing across the width and depth is achieved quickly, resulting in an initial concentration C_0. The dye is added continuously over a period of 24 hours. This period of time is considered to be sufficiently long that a continuous plane source solution may be used (actually, an instantaneous volume source solution also is valid, and easier to work with). The proposed solution is

$$\frac{C}{C_0} = \frac{1}{2}\left[\text{erfc}\left(\frac{x - Ut}{2\sqrt{E_L t}}\right)\right]$$

where erfc is the complimentary error function. At a point $x = 25$ km downstream, the cross-sectional average dye concentration is measured as a function of time, resulting in the plot shown in Fig. 10.33

(a) What is the governing equation for this problem? List all assumptions and any initial and boundary conditions. Is the above solution consistent with these conditions?

(b) What is the mean river velocity (U)?

(c) Estimate the longitudinal dispersion coefficient E_L.

Problem 10.16 A cargo boat accidentally spills a hazardous waste into a river. There is a water supply intake a distance L downstream of the spill. Between the spill and the intake the stream depth H is approximately constant, but the stream widens at a rate that can be approximated with an exponential function, $B = B_0 e^{ax}$, where B_0 is the width at $x = 0$, x is distance downstream, and a is a constant. The stream flow rate Q is constant and steady, and the

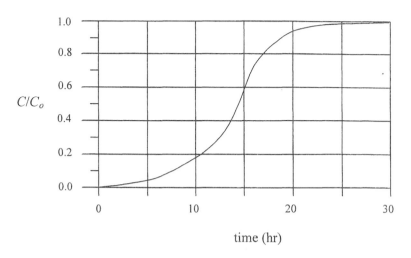

C/C_o

time (hr)

Figure 10.33 Data for river transport experiment, Problem 10.15.

mass of waste spilled is M. The waste decays with first-order rate constant K. The problem is then to evaluate the chemical concentration at the intake.

(a) What is the governing equation for this problem? This equation should be written in a form that directly incorporates the widening of the stream. List any assumptions used.

(b) What are the initial and boundary conditions?

(c) Is the concept of dispersion applicable for this problem? Why or why not?

Problem 10.17 Consider an aquifer in which fresh water of depth h flows over a salt water layer, as sketched in Fig. 10.34. The salt water layer is stagnant and has salinity S_0, while the inflowing freshwater has initial salinity $S = 0$. The freshwater flow is driven by a mean gradient of 10^{-3}, and the aquifer has hydraulic conductivity 50 m/day and porosity 0.2. The vertical dispersivity is 2 cm. Except near the entrance, horizontal dispersion may be neglected. Under steady state, calculate the isohaline lines corresponding to $S/S_0 = 0.25, 0.5$, and 0.75. Assume that h is large. (*Note:* this type of problem is developed further in Chap. 11.)

Problem 10.18 Emissions from the smokestack of a coal-fired power plant contain sulfur dioxide concentrations of 100 ppm at a constant discharge rate of 10,000 ft^3/s. The smokestack is 150' above ground level (ground is assumed to be flat). Average wind velocity (in the x-direction) is 10 mph, $E_x = E_y =$

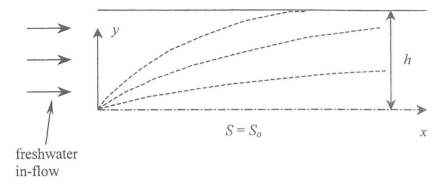

Figure 10.34 Definition sketch, Problem 10.17.

10 ft²/s, and $E_z = 3$ ft²/s (z is the vertical direction). The decay rate (first order) is K.

(a) Show that, as long as x is greater than about 250', diffusion in the x-direction can be neglected.

(b) What are the governing equation and initial and/or boundary conditions for this problem?

(c) A possible analytical solution for this problem is proposed as

$$C = \frac{q}{4\pi\rho(E_y E_z)^{1/2} x} \exp\left(-\frac{y^2 U}{4x E_y} - \frac{z^2 U}{4x E_z} - \frac{Kx}{U}\right)$$

where q is the rate of mass injection. Is this solution consistent with your answer to part (b)? Why or why not? State specific assumptions needed in using this solution.

(d) Assuming $K = 0$, using the solution from part (c), what is the maximum concentration expected at ground level ($z = 0$)?

(e) If SO_2 undergoes the following reaction (leading to acid rain),

$$SO_2 + H_2O \longrightarrow H_2SO_3$$
$$H_2SO_3 + \tfrac{1}{2}O_2 \longrightarrow H_2SO_4$$

$$\overline{\phantom{SO_2 + H_2O + \tfrac{1}{2}O_2 \longrightarrow H_2SO_4}}$$

$$SO_2 + H_2O + \tfrac{1}{2}O_2 \longrightarrow H_2SO_4$$

and the reaction is first order with respect to SO_2, with $K = 1,000$ day^{-1}, what is the maximum ground-level concentration? Use the solution from part (c).

Problem 10.19   A truck travels along a long, straight road at 35 mph. The exhaust pipe discharges a smoke with a particular constituent having an initial concentration of 500 ppm. The pipe exhaust is 12 ft above ground level and the discharge is 10 cfs. Neglecting decay and buoyancy effects, estimate the maximum ground level concentration and the time at which it occurs (measured from the time at which the truck passes a given point). List the governing equation and any assumptions used in your solution. Air density is 0.002 slug/ft^3 and there is no wind. Assume homogeneous, anisotropic diffusion with $E_x = E_y = 10$ ft^2/s and $E_z = 3$ ft^2/s, where z is the vertical coordinate.

Problem 10.20   A conservative pollutant is discharged through a multiport diffuser evenly across the bottom of a river that is 25' (8 m) wide. Assume constant uniform velocity U and homogeneous isotropic diffusion. Use $U = 2$ ft/s (0.6 m/s), depth = 10' (3 m), $E = 0.1$ ft^2/s(0.03 m^2/s), source flow $q = 10$ cfs (0.3 cms) and source concentration $C_0 = 100$ ppm.

(a) Use a line source solution to calculate the downstream concentration response.

(b) Repeat part (a), but using a plane source solution. At what distance downstream does the plane source solution become valid?

(c) Compare the results from each of the solutions from parts (a) and (b) and comment on the relative advantages and disadvantages of these two approaches.

Problem 10.21   Consider an infinite cylinder with dyed water of concentration C_0 enclosed between 2 partitions at $x = \pm L$, as shown in Fig. 10.35. The rest of the cylinder is filled with clear water ($C = 0$). Write the governing equation and initial and boundary conditions for this problem. If at time $t = 0$ the partitions are instantaneously removed, find the dye concentration at an arbitrary point $x(x \geq L)$ as a function of time.

Hint: Consider the source to be composed of a number of infinitesimal sources of thickness dx, and combine the responses for each of these sources. The solution will be expressed in terms of the error function.

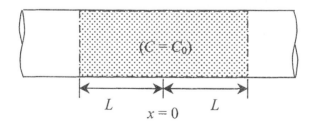

Figure 10.35   Initial dye distribution, Problem 10.21.

Problem 10.22   A city of 500,000 people is discharging sewage into the ocean from a submerged multiport diffuser as shown in Fig. 10.36. It is assumed that an initial dilution of 100 is achieved at the point where the discharge reaches the surface (i.e., at $x = 0$ the sewage concentration is 1/100 of its discharge value). Under certain wind conditions an on-shore current of 0.1 m/s is observed in this region, and there is concern for coliform concentrations at the beach, 3000 m away. The production of coliforms is estimated to be 400×10^9 coliforms per person per day for a sewage flow rate of 600 liters per person per day. The diffusivity corresponding to the initial plume width of $b_0 = 100$ m is estimated to be $E_0 = 0.1$ m^2/s, and it is assumed to increase with the 4/3 power of width, i.e., $(E/E_0) \approx (b/b_0)^{4/3}$, where $b =$ plume width and subscript zero indicates the initial value (at $x = 0$). Coliforms are assumed to decay with a first-order decay rate of $K = 10$ per day.

(a) Calculate the flow rate and concentration of coliforms at the discharge point and also at $x = 0$ (after initial dilution).

(b) Estimate the thickness of the discharge plume under the given flow condition, assuming an initial width of 100 m (note that this thickness does not change significantly with x, due to inhibition of vertical spreading by buoyancy).

(c) Assuming a steady state and neglecting diffusion in the longitudinal and vertical directions, what is the governing equation for this problem?

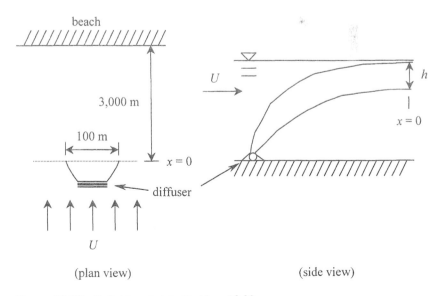

Figure 10.36   Definition sketch, Problem 10.22.

(d) A solution is proposed, given by

$$\frac{C_{\max}}{C_0} = e^{-\frac{kx}{U}} \operatorname{erf}\left\{\left[\frac{3/2}{(1 + 8xE_y/Ub_0^2)^3 - 1}\right]^{1/2}\right\}$$

Using this solution, what is the maximum concentration expected at the beach? Is this solution consistent with the governing equation you used in part (c)? Why or why not? Provide a qualitative plot of the shape of the concentration distribution resulting from this solution (show several plots at different values of x).

SUPPLEMENTAL READING

Bierman, V. J., DePinto, J. V., Young, T. C., Rodgers, P. W., Martin, S. C. and Raghunathan, R. K., 1992. Development and Validation of an Integrated Exposure Model for Toxic Chemicals in Green Bay, Lake Michigan. EPA Report, Office of Research and Development, Large Lakes Research Station, MI.

Chapra, Steven C., 1997. *Surface Water Quality Modeling.* McGraw-Hill, Boston.

Clark, Mark M., 1996. *Transport Modeling for Environmental Engineers and Scientists,* John Wiley and Sons, Inc., New York.

Csanady, G. T., 1973. *Turbulent Diffusion in the Environment.* D. Reidel Publishing Company, Dordrecht-Holland.

DePinto, Joseph V., Morgante, Michael, Zaraszczak, Joseph, Bajak, Tricia and Atkinson, Joseph, F., 1995. Application of Mass Balance Modeling to Assess Remediation Options for the Buffalo River. EPA 905-R-95-007, US Environmental Protection Agency, Great Lakes National Program Office, Chicago, IL.

Eskinazi, S., 1975. *Fluid Mechanics and Thermodynamics of Our Environment,* Academic Press, New York, NY.

Fischer, H. B., List, E. J., Koh, R. C. Y., Imberger, J., Brooks, N. H., 1979. *Mixing in Inland and Coastal Waters,* Academic Press, San Diego, CA.

Hemond, H. F. and Fechner, E. J., 1994. *Chemical Fate and Transport in the Environment,* Academic Press, San Diego, CA.

Logan, Bruce, 1999. *Environmental Transport Processes.* Wiley and Sons, Inc., New York.

Thomann, Robert, V. and Mueller, John, A., 1987. *Principles of Surface Water Quality Modeling and Control.* Harper Collins, New York.

11

Groundwater Flow and Quality Modeling

11.1 INTRODUCTION

In Sec. 4.4 it was shown how laminar flow through porous media can be represented by a model of flow through small capillaries. Then the average flow rate per unit area, namely the specific discharge, is proportional to the hydraulic gradient. Therefore, though the flow is basically laminar, it can be modeled and simulated by methods applied to inviscid flows, where the specific discharge originates from a potential function. In the present chapter we explore some further applications of fluid mechanics principles with regard to groundwater flow, its contamination, and its preservation.

Groundwater is always associated with the concept of an aquifer. An aquifer comprises a layer of soil that may store and convey groundwater. Therefore an aquifer is a layer of soil whose effective porosity and permeability (or hydraulic conductivity) are comparatively high. There are various types of aquifers, as illustrated in Fig. 11.1: (1) the *confined aquifer*, (2) the *phreatic aquifer*, and (3) the *leaky aquifer*. It should be noted that in addition to this classification, there are other properties of aquifers that are of interest, such as the presence and effects of fractures, etc. However, regarding the aquifers shown in Fig. 11.1, a confined aquifer is an aquifer whose top and bottom consist of impermeable layers. A phreatic aquifer has an impermeable bottom and a free surface, and a leaky aquifer is an aquifer whose boundaries are leaky, i.e., there is flow across its boundaries. Figure 11.1c shows a leaky phreatic aquifer, for example.

Considering length scales of aquifer flows, the thickness of the aquifer is usually quite small, of the order of several tens of meters, whereas the horizontal extent is of the order of kilometers. Therefore it can be assumed that in many cases the groundwater flow is approximately in the horizontal direction. Such an assumption leads to the *Dupuit approximation,* introduced

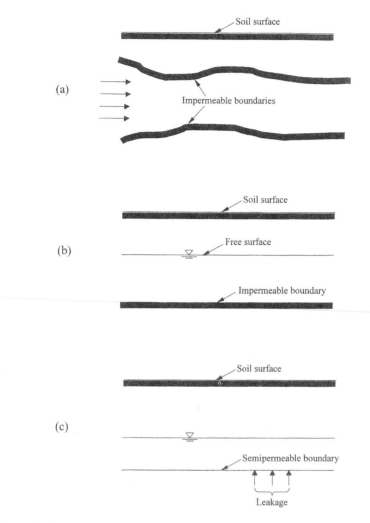

Figure 11.1 Typical aquifers: (a) confined aquifer; (b) phreatic aquifer; and (c) leaky phreatic aquifer.

in Chap. 4. The essence of this type of approximation is represented in the following section.

11.2 THE APPROXIMATION OF DUPUIT

The Dupuit approximation is based on several simplifying assumptions, which are generally quite well satisfied in groundwater systems. The major assumption is that streamlines in groundwater flows are almost horizontal. Such an assumption with regard to a free surface (phreatic) aquifer greatly simplifies the boundary condition at the free surface. The free surface of the phreatic aquifer, by definition, represents a streamline on which the pressure is equal to atmospheric pressure. A boundary condition of prescribed pressure, according to Bernoulli's equation, is a nonlinear boundary for the calculation of the potential function. Also, the exact location of the streamline of the free surface is not known prior to the performance of the calculations. Both of these difficulties are resolved by the employment of the Dupuit approximation.

Figure 11.2 shows the basic differences between the presentation of the groundwater flow according to potential flow theory and the modification of that presentation by the employment of the Dupuit approximation.

Basically, the Dupuit approximation does not consider the exact shape of the streamlines. The conservation of mass is considered with no reference to the stream function. The vertical component of the specific discharge is ignored, but the horizontal component of the specific discharge varies along the longitudinal x coordinate. It is assumed that due to the small curvature of the streamlines, the elevation of the free surface represents the piezometric head, which is constant along vertical lines, instead of along lines perpendicular to the free surface of the groundwater. Therefore the specific discharge vector is approximated as

$$|q| \approx q_x = \frac{\partial \Phi}{\partial x} \approx -K \frac{\partial h}{\partial x} \qquad (11.2.1)$$

where K is the hydraulic conductivity of the porous medium, q is the specific discharge, Φ is the potential function, and h is the elevation of the free surface, with regard to an arbitrary datum. In the particular case of Fig. 11.2, the bottom of the aquifer is horizontal. Therefore the thickness of the flowing water layer is adopted to represent the value of h. As shown in the following paragraph, such an adoption of h for Fig. 11.2 may provide a complete linearization of the equation of flow and the surface boundary condition.

The assumption of vertical lines of constant piezometric head implies that the specific discharge is uniformly distributed in a vertical cross section of the aquifer. Therefore the total discharge per unit width flowing through any vertical cross section of the aquifer, shown in Fig. 11.2b, is given by

$$Q = -Kh \frac{\partial h}{\partial x} = -\frac{K}{2} \frac{\partial}{\partial x}(h^2) \qquad (11.2.2)$$

If there are no sources of water in the domain of Fig. 11.2, and the domain of this figure is subject to steady-state conditions, then due to the conservation of mass, the value of Q is constant for all vertical cross sections shown in Fig. 11.2. Under such conditions, the value of h varies only with

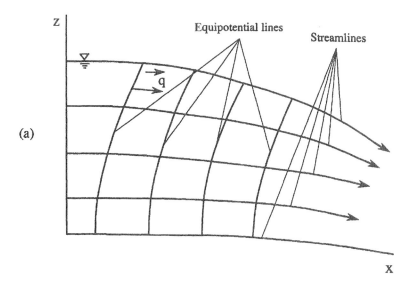

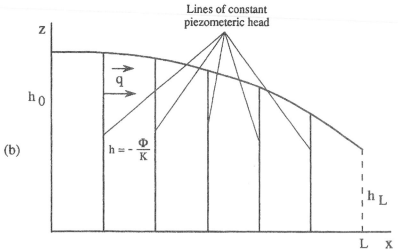

Figure 11.2 Differences between the potential flow theory and the Dupuit approximation: (a) the potential flow presentation; and (b) the Dupuit approximation.

the x coordinate, and the solution of Eq. (11.2.2) is given by

$$Qx = -\frac{K}{2}h^2 + C \tag{11.2.3}$$

where C is a constant of integration.

By applying two measured values of h at two arbitrary points, we can identify the value of the two constant coefficients of Eq. (11.2.3), namely Q and C. Consider that the measured value of h at $x = 0$ is h_0, and at $x = L$ the value of h is h_L. Introducing these values into Eq. (11.2.3), we obtain

$$Q = K\frac{h_0^2 - h_L^2}{2L} \tag{11.2.4}$$

As previously noted, the Dupuit approximation basically neglects the component of the specific discharge in the vertical direction. Therefore, in the most general case, that approximation allows two horizontal components of the specific discharge, namely,

$$q_x = -K\frac{\partial h}{\partial x} \qquad q_y = -K\frac{\partial h}{\partial y} \tag{11.2.5}$$

If the domain is subject to steady-state conditions, then by applying the procedure of Eq. (11.2.2), we obtain

$$Q_x = -Kh\frac{\partial h}{\partial x} \qquad Q_y = -Kh\frac{\partial h}{\partial y} \tag{11.2.6}$$

By considering the conservation of mass under steady-state conditions, Eq. (11.2.6) yields

$$\frac{\partial}{\partial x}\left(Kh\frac{\partial h}{\partial x}\right) + \frac{\partial}{\partial y}\left(Kh\frac{\partial h}{\partial y}\right) = 0 \tag{11.2.7}$$

This expression may look similar to Laplace's equation, but it refers only to steady-state conditions. In the case of a phreatic aquifer, some quantities of percolating runoff, called *accretion*, penetrate into the aquifer through its free surface. Under such conditions, the free surface of the aquifer is not a streamline. However, this case also can be completely linearized by the Dupuit approximation.

In cases of flow through a confined aquifer, the boundary conditions of the domain are linear, but their shape may lead to some difficulties in solving Laplace's equation. In such cases, the Dupuit approximation simplifies the calculations, as it leads to an assumption of unidirectional flow.

If the domain of Fig. 11.2 is subject to unsteady conditions, then the groundwater free surface is subject to variations, as shown in Fig. 11.3. Then, consideration of Eq. (11.2.2) and the mass conservation for the elementary volume of unit width shown in Fig. 11.3 yields

$$\frac{\partial}{\partial x}\left(Kh\frac{\partial h}{\partial x}\right) = \phi\frac{\partial h}{\partial t} \tag{11.2.8}$$

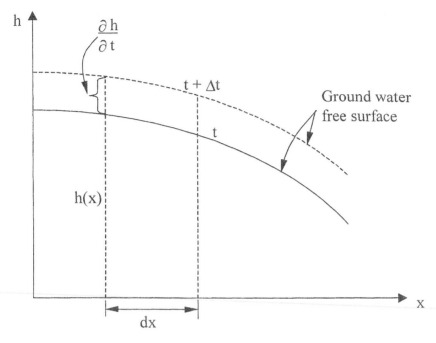

Figure 11.3   Variation of groundwater surface in a phreatic aquifer.

In the most general case of applying the Dupuit approximation, instead of Eq. (11.2.8), we obtain

$$\nabla \cdot (Kh\nabla h) = \phi \frac{\partial h}{\partial t} \qquad (11.2.9)$$

where ϕ is the portion of the porosity which takes part in the water flow, namely the effective porosity of the aquifer; and the gradient vector refers only to the horizontal directions.

   If the aquifer is confined, then the Dupuit approximation is useful to simplify the equations based on *Darcy's law* and mass conservation. In the case of a confined aquifer, the parameter h in Eq. (11.2.9) is still considered as the piezometric head with regard to calculation of the specific discharge, but it is replaced by the thickness, B, of the confined aquifer, with regard to the calculation of the flow rate per unit width of the aquifer, as indicated in Eq. (11.2.6). In a confined aquifer, contrary to a phreatic aquifer, the thickness of the region of flowing groundwater is kept almost constant. However, variations of flow in the aquifer are accompanied by some compression of the water phase, as well as restructuring of the solid skeleton and porosity of the porous medium. Such changes are characterized by the *strativity* or *coefficient of storage, S*. Therefore in the case of a confined aquifer, Eq. (11.2.9) is modified to yield

$$\nabla \cdot (T\nabla h) = S\frac{\partial h}{\partial t} \qquad (11.2.10)$$

where T is the *transmissivity* of the aquifer. Its value is given by

$$T = KB \qquad (11.2.11)$$

A common approach is to linearize Eq. (11.2.9) by considering that also in the case of a phreatic aquifer, the transmissivity can be defined by

$$T = Kh_{av} \qquad (11.2.12)$$

where h_{av} is an average value of the aquifer thickness. By introducing Eq. (11.2.12) into Eq. (11.2.9), we obtain

$$\nabla \cdot (T\nabla h) = \phi\frac{\partial h}{\partial t} \qquad (11.2.13)$$

Equations (11.2.10) and (11.2.13) are basically identical. Differences are between values of S and ϕ, where $S = O(10^{-2})$ and $\phi = O(10^{-1})$; and T of Eq. (11.2.13) results from the approximation given in Eq. (11.2.12).

Characterization of aquifers is usually obtained by the analysis of various types of field tests by reference to Eqs. (11.2.10) and (11.2.13). A common procedure is the use of pumping tests, or sludge tests. Laboratory tests of permeability of core samples can provide information with regard to the type of the soil but cannot be useful for the prediction of the response of the large-scale aquifer to various types of flow conditions. With regard to confined and phreatic aquifers, the analysis of field tests yields the storativity and transmissivity of the aquifer. Contemporary methods are sometimes used to characterize phreatic aquifers by reference to more sophisticated parametric analysis. Sometimes such approaches are needed, mainly in cases of large variations of the phreatic aquifer thickness. With regard to leaky aquifers, an additional parameter, the leakage factor, is required for the complete parametric presentation of the aquifer characteristics. Other types of aquifers, like fractured aquifers, require definitions of some other characteristic quantities. However, the topic of well hydraulics is based on practical uses of the Dupuit approximation, as exemplified by Eqs. (11.2.10) and (11.2.13), to characterize the capability of aquifers to supply required quantities of water to water supply systems.

The Dupuit approximation is very useful in obtaining simplified approximate solutions of groundwater flow problems. It simplifies problems of flow between impermeable layers, like confined aquifers, by simplifying the format of space-dependent coefficients. In cases of free surface flows, it linearizes the nonlinear surface boundary condition. In cases of immiscible fluid flows, it linearizes the nonlinear boundary of the interface between the immiscible fluids. In coastal aquifers, it is common to assume that the sea saltwater and freshwater of the aquifer are immiscible fluids. Then the Dupuit approximation can be useful for the calculation of the movement and location of the interface between salt and fresh waters.

It will be shown in the following sections of this chapter that the Dupuit approximation can often be useful for the solution of environmental problems associated with groundwater contamination and reclamation. Such topics are often defined as "contaminant hydrology." Such a definition is suggested to separate topics of pure hydraulics, which refer only to flow through porous media, from topics related to the quality of groundwater.

11.3 CONTAMINANT TRANSPORT

11.3.1 General Introduction

The basic equations of contaminant transport in any fluid system were introduced in Chap. 10. The same general approach, using elementary or finite control volumes, is applicable for modeling transport in porous media. However, in the case of a porous medium, we need to consider that a portion of the control volume is occupied by the solid matrix, and another portion incorporates the fluid or fluids. The elementary volume of reference in a porous medium system also must be much larger than the characteristic pore size. Such a representative elementary volume (REV) is much larger than is usually required, according to continuum mechanics of single-phase materials. In single-phase materials, continuum mechanics requires reference to an elementary volume significantly larger than the molecular size. In Chap. 10, we discussed some topics of dispersion in porous media. In the present section, we present the basic modeling approach to the analysis and calculation of contaminant transport in porous media.

11.3.2 Basic Equation of Contaminant Transport

Consider a constituent distributed in small concentrations in the water phase. The constituent concentration represents the mass of the constituent per unit volume of the water phase, consistent with the definition of mass concentration in Chap. 10. Also as in Chap. 10, a binary mass system of water and the constituent is assumed here. The total mass of the constituent is assumed to be very small in comparison to the quantity of water. Therefore the introduction of the minute quantity of constituent into the water phase does not affect the original volume of that water phase. The constituent may be present as a dissolved material in the water phase, it can be present as a material adsorbed to the solid skeleton of the porous medium, and it can be added, or taken away, in different forms to and from the control volume of the porous medium. Referring to an elementary representative volume of the porous medium, the basic equation of mass conservation of the dissolved constituent in the groundwater can be obtained as

$$\frac{\partial}{\partial t}(\phi C) + \nabla \cdot (\vec{q} C) = \nabla \cdot (\phi \tilde{D} \nabla C) - f + \phi \, \rho \Gamma - PC + RC_R$$

$$\quad (1) \qquad (2) \qquad\qquad (3) \qquad (4) \quad (5) \quad (6) \quad (7) \qquad (11.3.1)$$

It should be noted that this equation is valid for a porous medium saturated with water. In the unsaturated zone the porosity, ϕ, should be replaced by the water saturation. Each of the terms included in Eq. (11.3.1) requires some consideration, as presented in the following paragraphs, where the number of the paragraph corresponds to the number of the term in the equation.

(1) This term represents the rate of change of constituent mass per unit volume in the elementary control volume of the porous medium. It is usually assumed that variations of the porosity, ϕ, may be neglected.

(2) This term represents the difference between advective fluxes of contaminant leaving and entering the elementary control volume of the porous medium through its surfaces; q is the specific discharge of the flowing water phase.

(3) This term represents the effects of molecular diffusion and hydrodynamic dispersion on fluxes of contaminant entering and leaving the elementary control volume through its surfaces. Fluxes of diffusion and dispersion are proportional to the gradient of the constituent concentration. D is a second-order hydrodynamic dispersion tensor, which is represented by a matrix of the nine coefficients of dispersion. The values of the dispersion coefficients depend on the type of the porous medium, its isotropy and homogeneity, and its Peclet number. The Peclet number is defined by

$$\text{Pe} = \frac{VL}{D_d} \qquad\qquad (11.3.2)$$

where V is the interstitial average flow velocity, which is also the specific discharge divided by the porosity; L is a characteristic length of the pores, and D_d is the coefficient of molecular diffusion of the constituent in the water phase. Appropriate expressions for the hydrodynamic dispersion tensor are well presented in the scientific literature for cases of homogeneous and isotropic porous media. For an isotropic porous medium, i.e., one in which the permeability is identical in all directions, if $\text{Pe} \ll 1$, then the hydrodynamic dispersion tensor is an *isotropic tensor*, whose main diagonal components are smaller than the molecular diffusion coefficient. If $0.4 < \text{Pe} < 5$, then some anisotropy characterizes the hydrodynamic dispersion tensor, and it becomes a symmetric second-order tensor. The principal directions of this tensor are parallel and perpendicular to the velocity vector. If $\text{Pe} > 5$, then the effect of molecular diffusion is minor, and the dispersion tensor can be represented by

$$D_{ij} = a_{\text{T}}|V| + (a_{\text{L}} - a_{\text{T}})\frac{V_i V_j}{|V|} \qquad (11.3.3)$$

where a_{T} and a_{L} are the transverse and longitudinal dispersivity, respectively. Studies report that the longitudinal dispersivity is between 5 to even 100 times larger than the transverse dispersivity. The common ratio between the longitudinal and transverse dispersivity is considered to be between 20 and 40.

(4) This term represents phenomena of sorption–desorption. A positive value of f indicates larger quantities of the constituent adsorbed to the solid skeleton of the porous medium than those desorbed from the solid skeleton. It is common to analyze phenomena of sorption–desorption using linear isotherm models, such as developed by Langmuir or Freundlich. These models provide approximate linear relationships between the concentration of the constituent dissolved in the water phase and its mass quantity adsorbed to the solid skeleton of the porous medium. Such a presentation of the adsorption process leads to incorporation of the fourth term of Eq. (11.3.1) with the first term as

$$\frac{\partial C}{\partial t} + f = R\frac{\partial C}{\partial t} \qquad (11.3.4)$$

where R is called the *retardation factor*.

(5) This term refers to the constituent added to the water phase, as a result of chemical reactions inside the elementary control volume. It incorporates the decay of the constituent mass and possible microbial uptake. The value of this term represents the mass of the constituent added (or taken away) by the internal chemical reactions per unit time, per unit volume of the porous medium.

(6) This term represents the artificial removal of the constituent, which may consume water with the constituent. The consumed water leaves the system with the current concentration level of the constituent.

(7) This term represents the artificial recharge of the constituent, which supplies water with constituent. The constituent concentration of the recharged water is C_{R}.

11.3.3 Various Issues of Interest

Solutions of Eq. (11.3.1) can be developed, provided that the appropriate forms for each of the terms (4) through (7) are known, values of dispersivities are given, initial conditions are defined, and boundary conditions of the system are well presented. Various issues of contaminant transport in groundwater are then quantified by the appropriate solution of Eq. (11.3.1), depending on the relative magnitudes of each of these terms. Often, initial and boundary conditions of the saturated porous system are not very well defined. In these cases, it is common to study the sensitivity of the system to a set of different initial and boundary conditions. After performing a set of such simulations, it is usually possible to choose a set of initial and boundary conditions in such a way that conservative results can be assured. Time scales of the different phenomena represented in Eq. (11.3.1) are often very different. Therefore considering different time scale phenomena may allow significant simplification of Eq. (11.3.1), since some terms may be neglected.

Also, it is common to use a conservative approach for the quantification of contaminant transport, by considering transport of a *conservative contaminant*. With this assumption, terms (4) through (7) can be neglected in most portions of the domain. Then contaminant migration in the domain is affected only by advection and dispersion, and Eq. (11.3.1) becomes an advection–diffusion equation similar to the mass balance equation derived in the previous chapter,

$$\frac{\partial C}{\partial t} + \vec{V} \cdot \nabla C = \nabla \cdot (\tilde{D} \nabla C) \tag{11.3.5}$$

At this point, it is important to consider the difference between the time scale of processes associated with pumping of water for water supply purposes, and the time scale of contaminant transport in groundwater. Pumping tests usually require several days of pumping. During that time period, changes in groundwater table or piezometric head are measured and evaluated. The effect of pumping on the groundwater table is quick, and it depends on the availability of water in the aquifer. With regard to contaminant transport in the aquifer, processes are determined by the advection of the contaminant, which also is associated with the contaminant dispersion. Regarding natural flow in an aquifer, the magnitude of the hydraulic gradient is usually of order 10^{-3}, the hydraulic conductivity is of order 10 m/d, and the porosity is of order 0.2. Therefore the interstitial flow velocity, or the advection velocity, is of order 5 cm/d. Under such conditions, a pollutant discharged into the aquifer is advected a distance of less than 20 m in a year. Therefore contamination in aquifers can persist for many years before there is any indication about such a process. Close to a pumping well, the hydraulic gradient is large. Therefore a contaminant that for many years may have spread only a comparatively small distance in the aquifer by natural flow is subject to relatively quick advection after its penetration into the region of influence of the well.

Some important solutions of the advection–diffusion equation can be applied to determine basic characteristics of contaminant transport through

porous media. As an example, we consider here the Ogata and Banks solution
named after the scientists who developed it. This problem refers to contaminant
transport in a semi-infinite column, through which the water flows with a
constant velocity V. The value of this velocity is, as previously defined, equal
to the specific discharge divided by the porosity of the porous medium. At time
$t \leq 0$, there is no contaminant in the flowing water phase. At $t > 0$, at one end
of the column, where $x = 0$, the contaminant concentration is kept constant,
at $C = C_0$. This may be the case, for example, when the semi-infinite column
is connected to a large reservoir, in which the contaminant distribution is kept
uniform due to mixing. Under these conditions, if the dispersion coefficient is
constant, the differential equation (11.3.5) reduces to

$$\frac{\partial C}{\partial t} + V \frac{\partial C}{\partial x} = D \frac{\partial^2 C}{\partial x^2} \tag{11.3.6}$$

The *initial and boundary conditions* of the problem are given as:

$$C = 0 \qquad \text{at} \qquad t \leq 0, x \geq 0 \tag{11.3.7a}$$

$$C = C_0 \qquad \text{at} \qquad t > 0, x = 0 \tag{11.3.7b}$$

$$C = 0 \qquad \text{at} \qquad t > 0, x = \infty \tag{11.3.7c}$$

The *Laplace transform* of Eqs. (11.3.6), (11.3.7b), and (11.3.7c), respectively,
yields

$$\frac{\partial^2 \overline{C}}{\partial x^2} - \frac{V}{D} \frac{\partial \overline{C}}{\partial x} - p\overline{C} = 0 \qquad \text{where} \qquad \overline{C} = \overline{C}(x, p) \tag{11.3.8}$$

$$\overline{C} = \frac{C_0}{p} \qquad \text{at} \qquad x = 0 \tag{11.3.9a}$$

$$\overline{C} = 0 \qquad \text{at} \qquad x = \infty \tag{11.3.9b}$$

where p is the Laplace transform variable; and the Laplace transform of C is
defined by

$$\overline{C}(x, p) = \int_0^\infty C \exp(-pt) \, dt \tag{11.3.10}$$

The solution of the differential Eq. (11.3.8), subject to the boundary
conditions (11.3.9), is

$$\overline{C} = \frac{C_0}{p} \exp \left\{ x \left[\frac{V}{2D} - \sqrt{\left(\frac{V^2}{4D} + \frac{p}{D} \right)} \right] \right\} \tag{11.3.11}$$

By returning to the x, t coordinates, Eq. (11.3.11) yields

$$C(x, t) = \frac{1}{2} C_0 \left\{ \text{erfc} \frac{x - Vt}{2\sqrt{Dt}} + \exp \left[\frac{Vx}{D} \right] \text{erfc} \frac{x + Vt}{2\sqrt{Dt}} \right\} \tag{11.3.12}$$

where erfc is the complementary error function, defined by

$$\text{erfc}(\eta) = 1 - \text{erf}(\eta) = 1 - \frac{2}{\sqrt{\pi}} \int_0^\eta \exp(-\xi^2) d\xi \qquad (11.3.13)$$

This solution was developed by Ogata and Banks in 1961, and it is very useful for the identification of the dispersion coefficient, as explained in the following paragraphs. If Pe is high, then the dispersion coefficient in Eq. (11.3.12) is the longitudinal dispersion, which, according to Eq. (11.3.3), is given by

$$D = D_L = a_L V \qquad (11.3.14)$$

Introducing this expression into Eq. (11.3.12), we obtain

$$C(x, t) = \frac{1}{2} C_0 \left\{ \text{erfc} \frac{x - Vt}{2\sqrt{a_L Vt}} + \exp\left[\frac{x}{a_L} \right] \text{erfc} \frac{x + Vt}{2\sqrt{a_L Vt}} \right\} \qquad (11.3.15)$$

If x/a_L is sufficiently large, the second term of this expression can be neglected and Eq. (11.3.15) reduces to

$$C(x, t) = \frac{1}{2} C_0 \left\{ \text{erfc} \frac{x - Vt}{2\sqrt{a_L Vt}} \right\} \qquad (11.3.16)$$

A *breakthrough curve* is obtained by continuously measuring contaminant concentration at a point $x = x_0$, and plotting the results on the C, t plane. An example of a set of breakthrough curves is shown in Fig. 11.4, where each curve is associated with measurements made at several different locations along the column. Breakthrough curves similar to these can be obtained in field tests, where a tracer is injected into a strip of wells, and contaminant concentration measurements are performed at different points located downstream of the injection.

By applying Leibniz's theorem, we may obtain the rate of change of C at time $t_{1/2}$, which is the time at which the contaminant concentration at the measurement point is half of C_0. Thus

$$S_{1/2} = \left(\frac{\partial C}{\partial t} \right)_{t=t_{1/2}} = \frac{C_0 V}{2\sqrt{\pi D t_{1/2}}} \qquad (11.3.17)$$

where $S_{1/2}$ is the slope of the breakthrough curve at $t = t_{1/2}$. Therefore, by measuring $S_{1/2}$, the value of the longitudinal dispersion coefficient and dispersivity can be calculated, using Eqs. (11.3.14) and (11.3.17),

$$D_L = \frac{C_0^2 V^2}{4\pi t_{0.5} S^2} \qquad a_L = \frac{D_L}{V} \qquad (11.3.18)$$

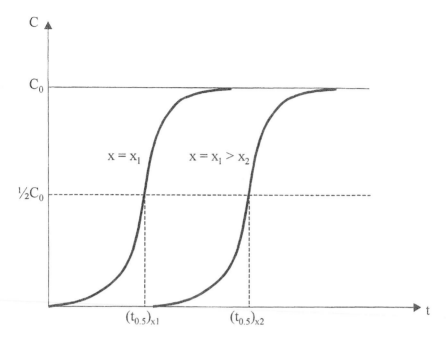

Figure 11.4 Breakthrough curves.

11.3.4 Application of the Boundary Layer Approximations

The boundary layer approach can be useful in obtaining approximate solutions of Eq. (11.3.5). For example, consider the one-dimensional contaminant transport problem represented by Eq. (11.3.6), subject to the boundary conditions given by Eq. (11.3.7). By adopting a moving longitudinal coordinate,

$$x_1 = x - Vt \tag{11.3.19}$$

Equation (11.3.6) becomes

$$\frac{\partial C}{\partial t} = D \frac{\partial^2 C}{\partial x_1^2} \tag{11.3.20}$$

Now, referring to the C, x plane shown in Fig. 11.5, consider the buildup of two boundary layers at the moving front of the advected contaminant. In this figure, it is assumed that advection in the x-direction is the dominant transport mechanism, so that contaminant dispersion is basically a second-order phenomenon, leading to the development of front and rear boundary layers. We assume that both boundary layers have an identical shape. Therefore

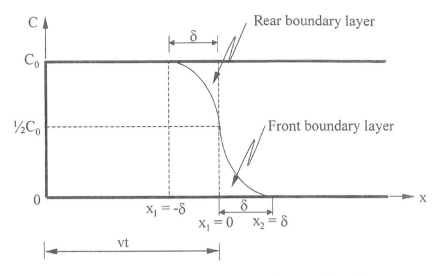

Figure 11.5 Boundary layers developed at the moving contaminant front.

the contaminant distribution at the moving front of the contaminant is given by

$$C = 0.5(1 - \eta)^n \qquad \eta = \frac{x_1}{\delta} \qquad \text{at} \qquad 0 \le x_1 \le \delta \qquad (11.3.22a)$$

$$C = 1 - 0.5(1 + \eta)^n \qquad \eta = \frac{x_1}{\delta} \qquad \text{at} \qquad -\delta \le x_1 \le 0 \qquad (11.3.22b)$$

where n is a power coefficient. The front of the contaminant is represented by

$$C = \frac{1}{2} \qquad \text{at} \qquad x_1 = 0 \qquad (11.3.21)$$

As the front and rear boundary layers are symmetrical with regard to $x_1 = 0$, we proceed with calculating the development of the front boundary layer. By introducing Eq. (11.3.22a) into Eq. (11.3.20), applying Leibnitz's theorem, and integrating the expressions over the boundary layer,

$$\frac{d}{dt} \left[\delta \int_0^1 (1 - \eta)^n \, d\eta \right] = \frac{nD}{\delta} \qquad (11.3.23)$$

where δ is the boundary layer thickness. After another integration with respect to time,

$$\delta^2 = 2Dn(n + 1)t \qquad (11.3.23)$$

As indicated in Fig. 11.5, the calculation, based on using the boundary layer approximation, can be applied provided that

$$Vt > 2\sqrt{2Dn(n + 1)t} \qquad (11.3.24)$$

By rearranging Eq. (11.3.24), it can be seen that the boundary layer approximation can be applied, provided that the point of 50% contaminant concentration is located at

$$x_{1/2} > \frac{8Dn(n + 1)}{V} \qquad (11.3.25)$$

The solution obtained by the boundary layer approach also may be applied to the determination of the dispersion coefficient. It leads to results identical to those of Eqs. (11.3.17) and (11.3.18), provided that $n = 2.05$.

As a further example of the application of the boundary layer approximation to different problems of contaminant hydrology, consider a quick evaluation of the migration of a NAPL (nonaqueous phase liquid — NAPLs in groundwater are discussed in more detail in Sec. 11.5). NAPL can be either lighter or denser than water (in the latter case it is referred to as a DNAPL, a dense nonaqueous phase liquid), but here we consider light NAPLs. NAPL migration arises due to the dissolution of the NAPL lens, which is created when a NAPL is released into the aquifer. Figure 11.6 shows how a NAPL

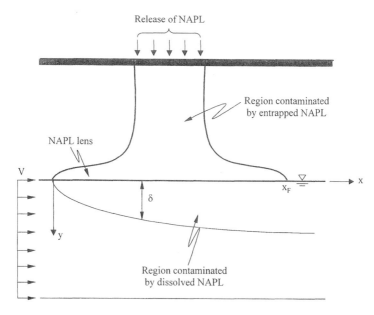

Figure 11.6   NAPL lens floating on top of the groundwater table.

released at the soil surface arrives at the groundwater table and creates a lens, which floats on top of the groundwater table. Due to the contact between the NAPL lens and the flowing groundwater, a region adjacent to the NAPL lens is contaminated by a dissolved or solubilized NAPL. Although a NAPL is considered as being immiscible in water, there is a small but finite miscibility, on the order of 0.1%. The advection velocity in the aquifer, V, is uniformly distributed, and the flow is only in the x-direction. Therefore Eq. (11.3.6) may be written as

$$\frac{\partial C}{\partial t} + V\frac{\partial C}{\partial x} = D_x\frac{\partial^2 C}{\partial x^2} + D_y\frac{\partial^2 C}{\partial y^2} \tag{11.3.26}$$

Although D_x is usually much larger than D_y, the first right-hand-side term in Eq. (11.3.26) is much smaller than the second one and may be neglected, due to the large gradient of the concentration profile in the y-direction. Furthermore, the contaminant distributions in vertical cross sections are similar, i.e.,

$$C = C_0\,(1 - \eta)^n \qquad \eta = \frac{y}{\delta} \tag{11.3.27}$$

where n is a power coefficient, δ is the boundary layer thickness, which represents the region of significant penetration of the dissolved NAPL, and C_0 is the dissolved NAPL concentration at the NAPL lens. We may consider that C_0 is the concentration of equilibrium.

Equation (11.3.27) complies with the following boundary conditions:

$$C = C_0 \qquad \text{at} \qquad y = 0 \tag{11.3.28a}$$
$$C = 0 \qquad \text{at} \qquad y = \delta \tag{11.3.28b}$$
$$\frac{\partial C}{\partial y} = 0 \qquad \text{at} \qquad y = \delta \tag{11.3.29c}$$

An additional boundary condition should be provided with regard to the x coordinate. This is usually done using a boundary condition for δ rather than for C, as

$$\delta = 0 \qquad \text{at} \qquad x = 0 \tag{11.3.30}$$

In general, an initial condition may be specified as

$$\delta = 0 \qquad \text{at} \qquad t = 0 \tag{11.3.31}$$

However, in the framework of this presentation, a steady state will be assumed, so that initial conditions are not needed.

Under steady-state conditions, and neglecting the first right-hand-side term of Eq. (11.3.26), we introduce Eq. (11.3.27) into Eq. (11.3.26), apply

Leibniz's theorem, and integrate over the boundary layer thickness to obtain

$$\frac{d}{dx}\left[V\delta\int_0^1 (1-\eta)^n d\eta\right] = \frac{D_y n}{\delta} \tag{11.3.32}$$

After performing the integral of Eq. (11.3.32), using separation of variables and integrating again, using Eq. (11.3.39), we obtain

$$\delta^2 = 2\frac{D_y}{V}n(n+1)x \tag{11.3.33}$$

As an example, consider $n = 2$, a NAPL lens with length 1000 m, a flow velocity in the aquifer of 10 cm/d, and transverse dispersivity of 1 cm. Then the transverse dispersion coefficient and the thickness of the region contaminated by dissolved NAPL are given by

$$D_y = a_T V = 1\,\text{cm} \times 10\,\text{cm/s} = 10\,\text{cm}^2/\text{s} \tag{11.3.34a}$$

and

$$(\delta)_{x=1000\,\text{m}} = \sqrt{2 \times 0.01\,\text{m} \times 2 \times 3 \times 1000\,\text{m}} \approx 11\,\text{m}$$

11.4 SALTWATER INTRUSION INTO AQUIFERS

Saltwater intrusion into coastal aquifers is a problem of interest in many places around the world. It can be quantified and analyzed by various methods. Some methods are based on the assumption that saltwater and freshwater are immiscible fluids, and there is a sharp interface that separates the two fluids. Such an approach allows a solution to be obtained simply by solving the equations of flow in the two portions of the domain. However, saltwater and freshwater do in fact mix, and several approaches have been developed that take into account the effect of salt diffusion and dispersion as a perturbation to the advective transport of the salt. Other approaches consider the domain saturated with groundwater, in which salinity is nonuniformly distributed. The effect of the salt on the density of the water is taken into account. These approaches require the simultaneous solution of the equations of flow and salt transport.

It should be noted that saltwater intrusion is not only typical of coastal aquifers. It represents an acute issue in many inland aquifers, whose partially permeable bottom may convey brine from deep formations into the overlying freshwater aquifer, due to natural or artificial causes.

11.4.1 The Sharp Interface Approximation

The sharp interface approximation considers that saltwater and freshwater are immiscible fluids. The sharp interface represents a streamline and a nonlinear boundary (since its position is not well known before solving the problem) with regard to the velocity vector. Such a boundary is similar to the free surface of a phreatic aquifer. As shown hereinafter, by using the Dupuit approximation, the equation of flow is completely linearized under steady-state conditions. Figure 11.7 shows two-dimensional steady-state freshwater flow in a phreatic coastal aquifer. According to the Dupuit approximation, lines of constant potential are vertical. Therefore the elevation of a point of the groundwater table represents the piezometric head of the freshwater in the vertical cross section that incorporates that point. Considering the small control volume of the freshwater portion of the aquifer, we obtain

$$\frac{dQ}{dx} = K \frac{d}{dx} \left[(h_f + B) \frac{dh_f}{dx} \right] - N \tag{11.4.1}$$

where Q is the discharge per unit width of the freshwater aquifer, K is the hydraulic conductivity of the aquifer to freshwater, h_f is the elevation of the groundwater table above the sea level — also the piezometric head, B is the depth of the interface below the sea level, N is the rate of accretion per unit width of the aquifer, and x is the longitudinal coordinate.

Both sides of the interface are subject to the same pressure. Therefore, according to Fig. 11.7,

$$B\gamma_s = (h_f + B)\gamma_f \tag{11.4.2}$$

where γ_f and γ_s are the specific weights of freshwater and saltwater, respectively. Rearrangement of Eq. (11.4.2) yields

$$B = \left(\frac{\gamma_f}{\gamma_s - \gamma_f} \right) h_f \tag{11.4.3}$$

Introducing Eq. (11.4.3) into Eq. (11.4.1), and performing a single integration with respect to x, we find

$$Q = K \left(\frac{\gamma_s}{\gamma_s - \gamma_f} \right) h_f \frac{dh_f}{dx} + \int N\, dx \tag{11.4.4}$$

Equation (11.4.4) is subject to two boundary conditions. According to the boundary condition at the seashore,

$$h_f = 0 \quad \text{at} \quad x = 0 \tag{11.4.5}$$

Another boundary condition can be given by a measured value of h_f at a distance $x = L$ from the seashore, or

$$h_f = h_L \quad \text{at} \quad x = L \tag{11.4.6}$$

In the particular case of no accretion, the aquifer discharge per unit width is kept constant in the domain of Fig. 11.7.

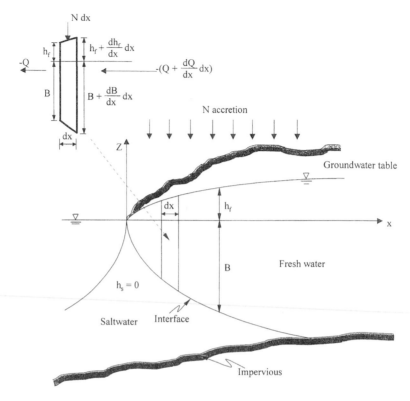

Figure 11.7   Saltwater intrusion into a coastal aquifer.

Equation (11.4.4) is a linear ordinary differential equation with respect to h_f^2. If there is no accretion, then direct integration yields

$$Qx = \frac{K}{2}\left(\frac{\gamma_s}{\gamma_s - \gamma_f}\right)h_h^2 + C \tag{11.4.7}$$

where C is an integration constant. The boundary condition of Eq. (11.4.5) indicates that $C = 0$. By applying the boundary condition of Eq. (11.4.6),

$$Q = \frac{K}{2L}\left(\frac{\gamma_s}{\gamma_s - \gamma_f}\right)h_L^2 \tag{11.4.8}$$

Also, according to Eqs. (11.4.7) and (11.4.8),

$$h_f = h_L\sqrt{\frac{x}{L}} \tag{11.4.9}$$

The results of the preceding paragraphs show that by a small number of piezometric head measurements, it is possible to obtain an estimate of the freshwater discharge of the aquifer and the location of the interface between the freshwater and the saltwater.

In order to obtain some quantitative information about coastal aquifers, we consider the following approximate values:

$$\rho_s = 1025 \, \text{kg/m}^3 \qquad \rho_f = 1000 \, \text{kg/m}^3 \tag{11.4.10}$$

where ρ_f and ρ_s are the density of fresh and saltwater, respectively. Introducing Eq. (11.4.10) into Eq. (11.4.3),

$$B = 40 h_f \tag{11.4.11}$$

This result is called the Ghyben–Herzberg relationship after the scientists who first developed this expression in the beginning of the 20th century by considering hydrostatic pressure distribution in the coastal aquifer.

As an example, consider a coastal aquifer with an impervious bottom located at an elevation of 40 m below sea level. Then, according to Eq. (11.4.11), the toe of the interface is located where $h_f = 1$ m. If the hydraulic conductivity of that aquifer is 40 m/d, and the toe of the interface is located at a distance of 1 km from the seashore, then, according to Eq. (11.4.8), the aquifer discharge per unit width is

$$Q = \frac{K}{2L}\left(\frac{\rho_s}{\rho_s - \rho_f}\right) h_L^2 = \frac{40}{2 \times 1000}\left(\frac{1025}{1025-1000}\right) \times 1^2 = 0.82 \, \text{m}^2/\text{d} \tag{11.4.12}$$

Under unsteady conditions, the interface is subject to movement, and additional terms representing the variation of the groundwater table and the displacement of the interface should be added to the differential Eq. (11.4.1). This renders the equation nonlinear. Therefore problems of saltwater intrusion into a coastal aquifer, in which the interface is subject to movement, are usually solved by numerical simulation.

11.4.2 Salinity Transport

The sharp interface assumption is often used, at least as a first approximation for the evaluation of saltwater intrusion into coastal aquifers. It also is used for the evaluation of saltwater intrusion into inland aquifers. However, as previously noted, freshwater and saltwater are miscible fluids. The main difference between the two fluids is simply the difference in salt concentrations. Therefore, logically, the appropriate method of simulation of saltwater intrusion into an aquifer should focus on the water flow associated with advection and dispersion of salt in the domain. Such an approach requires that the flow equation and salt transport equation must be solved simultaneously, as the increase of the salt concentration increases the density of the water phase.

There are various numerical models, some of them in the public domain, that can adequately provide such solutions. However, a simplified approach may apply the sharp interface approximation incorporating the assumption that the transition zone between freshwater and saltwater can be represented as a boundary layer. Using this approach, it is assumed that the interface represents a boundary of constant salt concentration, as shown in Fig. 11.8. The transition zone develops along that boundary, and it is similar to a boundary layer. Assuming that the curvature of the interface is small, we adopt a two-dimensional coordinate system x, y, where x is the longitudinal coordinate extended along the interface and y is perpendicular to the interface. Due to the small curvature of the interface, the equation of salt transport in the proximity of the interface is given by

$$\frac{\partial C}{\partial t} + V \frac{\partial C}{\partial x} = D_y \frac{\partial^2 C}{\partial y^2} \qquad (11.4.13)$$

For steady-state conditions, we apply the following approximations (refer to Eqs. 11.3.14 and 11.3.27):

$$D_y = a_T V \qquad (11.4.14a)$$

$$C = C_0 (1 - \eta)^n \qquad \eta = \frac{y}{\delta} \qquad (11.4.14b)$$

where δ is now the thickness of the transition zone, C_0 is salt concentration of the saltwater, and n is a power coefficient. We introduce Eq. (11.4.14) into Eq. (11.4.13) and integrate over the transition zone to obtain

$$\frac{d\delta^2}{dt} = 2a_T n(n + 1) \qquad (11.4.15)$$

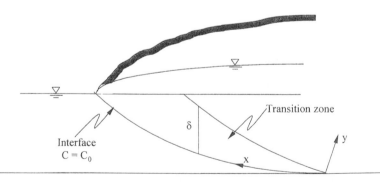

Figure 11.8 Development of the transition zone at the sharp interface.

By direct integration of Eq. (11.4.15), and assuming that $\delta = 0$ at $x = 0$,

$$\delta^2 = 2a_T n(n + 1)x \tag{11.4.16}$$

11.5 NON-AQUEOUS PHASE LIQUID (NAPL) IN GROUNDWATER

Due to the development and use of products originating from the oil industry, such as fuels, oil distillates, and other oil products, there have been many cases of groundwater contamination by such products. Such fluids are considered to be immiscible in water, and they are termed *nonaqueous phase liquids*, or NAPLs. When a NAPL arrives at an aquifer, small quantities may dissolve in groundwater and have a significant effect on the taste and odor of the water. Humans are sensitive to the presence of fuel in water, even when its concentration is smaller than 1 mg/L. Some of the hydrocarbon fractions in the fuel are poisonous at specific threshold levels. Various organic liquid compounds, many including chlorine, are suspected to be carcinogens.

Figure 11.6 describes a typical case of groundwater contamination by NAPL. NAPL quantities that are released at the ground surface percolate almost vertically through the unsaturated zone. Some quantities of the NAPL are entrapped in that zone. If the released NAPL amount is large, or the groundwater table is close to the soil surface, then the NAPL spill may arrive at the capillary fringe, where it migrates horizontally. If the density of the NAPL is smaller than that of water, it is called LNAPL (*light NAPL*), and a NAPL lens floating on top of the groundwater is created, as shown in Fig. 11.6. Such a NAPL lens is subject to gradual release of dissolved NAPL into the flowing groundwater. It also provides NAPL vapors into the gaseous phase.

Analysis of NAPL migration in the porous medium may be subject to various stages. In the first stage, possible simultaneous flow of NAPL, water, and air may be considered. Numerical models for such multiphase flow were developed by oil companies some time ago. In the 1980s such models were simplified and adopted for environmental flows. However, the stage of multiphase flow in porous media is relatively short and concerns minor issues of groundwater contamination and possible reclamation. Usually, even in cases of very significant NAPL spills, the volume of the NAPL released into the environment is small, compared to the volume of water contaminated by the NAPL. After a relatively short time the NAPL reaches a state of equilibrium, in which it is entrapped within the porous medium. The NAPL lens shown in Fig. 11.6 is almost stagnant. However, it is subject to oscillations in the upward and downward directions with the groundwater table. Such oscillations originate from natural as well as artificial reasons with seasonal and annual time scales. As a result of such oscillations, the NAPL may be trapped within the aquifer, usually in the form of *ganglia* (blobs), which are surrounded by the water phase. Figure 11.9 shows some typical shapes of entrapped NAPL blobs. Under such conditions, the flowing groundwater gradually dissolves the entrapped NAPL.

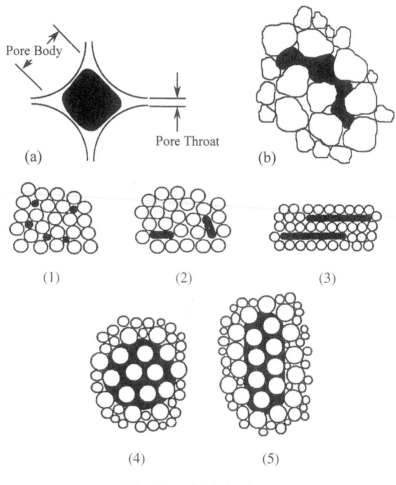

Idealized blob shapes

Figure 11.9 Typical shapes of entrapped NAPL blobs. (a) Singlet formed by "snap-off" in pore with high pore body-to-throat size ratio; and (b) blob formed by "bypassing" in course sand lens. (1) Singlets; (2) doublets; (3) fingers; (4) large sphere; and (5) cylinder.

The rate of mass transfer of the NAPL from the continuous entrapped NAPL phase into the flowing water phase is given by

$$F = k_f a_b (C_s - C) \tag{11.5.1}$$

where F is the rate of mass transfer between the NAPL and water phases, k_f is the effective mass transfer coefficient, a_b is the specific interfacial area of contact between the NAPL blob and the mobile water phase, C is the concentration of dissolved NAPL in the water phase, and C_s is the equilibrium concentration of the dissolved NAPL in the water phase. The specific interfacial area of contact between the NAPL blob and the mobile water is given by

$$a_b = S_n \left(\frac{A}{V} \right)_b f \tag{11.5.2}$$

where S_n is the NAPL saturation, namely the portion of pores saturated by NAPL, $(A/V)_b$ is the ratio between the surface area and volume of the NAPL blob, and f is the fraction of the blob surface area exposed to contact with the mobile water.

Figure 11.6 and a quantitative example connected with that figure provide an approximate description of groundwater contamination originating from a NAPL lens floating on top of the flowing groundwater. In the following paragraph, we consider how groundwater is contaminated by dissolved NAPL, which originates from the dissolution of entrapped NAPL blobs. This is done assuming one-dimensional flow of groundwater through a portion of the aquifer contaminated by entrapped NAPL. The equation of dissolved NAPL transport in the domain is thus given as

$$\frac{\partial C}{\partial t} + V \frac{\partial C}{\partial x} = D_x \frac{\partial^2 C}{\partial x^2} + \frac{k_f a_b (C_s - C)}{\phi S_w} \tag{11.5.3}$$

where D_x is the longitudinal dispersion coefficient, ϕ is the porosity, and S_w is the water saturation. The longitudinal dispersion coefficient is again assumed to be given by Eq. (11.3.14), or

$$D_x = a_L V \tag{11.5.4}$$

where a_L is the longitudinal dispersivity, as before.

The flow of groundwater through the site contaminated by entrapped NAPL leads to a gradual increase of the dissolved NAPL concentration in the water phase. It also decreases the saturation of the entrapped NAPL. However, there are two different time scales for these two phenomena. For the calculation of the increase of the dissolved NAPL concentration along the x coordinate, quasi-steady-state conditions may be assumed, with constant NAPL saturation.

Therefore, Eq. (11.5.3) is simplified as

$$\frac{dC}{dx} = a_L \frac{d^2C}{dx^2} + K_f(C_s - C) \tag{11.5.5}$$

where

$$K_f = \frac{k_f a_b}{\phi s_w V} \tag{11.5.6}$$

The denominator of Eq. (11.5.6) represents the specific discharge of the groundwater.

Equation (11.5.5) is an ordinary second-order differential equation subject to a single important boundary condition,

$$C = 0 \qquad \text{at} \qquad x = 0 \tag{11.5.7}$$

We may adopt another boundary condition, that is not necessarily materialized, as

$$C = C_s \qquad \text{at} \qquad x \to \infty \tag{11.5.8}$$

If the coefficient of effective mass transfer is large, or the portion of the aquifer contaminated by entrapped NAPL is large, then the boundary condition of Eq. (11.5.8) is materialized. Otherwise, downstream of that portion of the aquifer, the dissolved NAPL concentration is smaller than its equilibrium value.

The solution of Eq. (11.5.5), subject to the boundary conditions of Eqs. (11.5.7) and (11.5.8), is

$$C = C_s - C_s \exp\left\{\frac{x}{2a_L}\left[1 - \sqrt{1 + 4a_L K_f}\right]\right\} \tag{11.5.9}$$

In most cases, the effect of dispersion on the NAPL dissolution process is very minor. Therefore Eq. (11.5.5) can usually be approximated by

$$\frac{dC}{dx} = K_f(C_s - C) \tag{11.5.10}$$

Direct integration of Eq. (11.5.10), and use of the boundary condition given by Eq. (11.5.7), give

$$C = C_s \left[1 - \exp(-K_f x)\right] \tag{11.5.11}$$

This result also satisfies the boundary condition of Eq. (11.5.8).

Further discussion concerning NAPL in groundwater is provided in Chap. 16, which concerns the remediation of contaminated environments.

11.6 NUMERICAL MODELING ASPECTS

11.6.1 The Equations of Flow

Groundwater flow simulation by the employment of adequate numerical models is nowadays a common practice of hydrologists and engineers. Public domain computer programs, as well as commercial programs, are available to cover all aspects of groundwater hydrology. The oldest numerical method is the method of finite differences. Other approaches that are commonly used are the finite element method, boundary elements, and analytical elements.

Employment of the Dupuit approximation indicates that the basic aquifer flow equations, represented by Eqs. (11.2.1)–(11.2.12), originate from the basic differential equation

$$S\frac{\partial h}{\partial t} = \frac{\partial}{\partial x}\left(T\frac{\partial h}{\partial x}\right) + \frac{\partial}{\partial y}\left(T\frac{\partial h}{\partial y}\right) + q \tag{11.6.1}$$

where S is the storage coefficient, h is the piezometric head, T is the transmissivity, x and y are two horizontal coordinates, and q represents a source of groundwater. This equation indicates that simulation of steady-state flow in an aquifer is basically providing the solution to an elliptic partial differential equation. If the transmissivity is constant, then Eq. (11.6.1) reduces under steady-state conditions to the Laplace or Poisson equation. The numerical solution of these equations is discussed in Chap. 4.

Considering the common finite difference approach, the domain is represented by a finite difference grid. Each nodal point of that grid provides a linear equation, which incorporates several unknown values of the piezometric head (or potential function) at that particular point, and several other grid points in its proximity. The boundary conditions of the domain are introduced into that set of equations. Later, the set of linear equations is solved by an iterative method, like the *successive over-relaxation method* (SOR). As the coefficient matrix incorporates many zero elements, an iterative solution method is more efficient for steady-state cases than an elimination method. Common boundary conditions of the domain are either of the Dirichlet type, where the piezometric head is given at the boundary, or the Neumann type, where the spatial derivatives of the piezometric head are given at the boundary, or the mixed boundary condition incorporating both Dirichlet and Neumann expressions. If Eq. (11.6.1) is applied under steady-state conditions to a phreatic aquifer, then the transmissivity depends on the elevation of the groundwater table. Under such conditions, if the bottom of the aquifer is flat, Eq. (11.6.1) reduces to Eq. (11.2.7) with sources, which is a linear partial differential equation with regard to h^2. With such an equation, a solution for values of h^2 at the nodal points of the finite difference grid has a better convergence than the solution for values of h. The latter solution should incorporate an artificial linearization of the differential equation.

Under unsteady-state conditions, Eq. (11.6.1) is a parabolic partial differential equation, which is basically similar to the diffusion equation. Such an equation can be solved by explicit or implicit methods. If calculations refer to a confined aquifer, then Eq. (11.6.1) is a linear partial differential equation. If calculations refer to a phreatic aquifer, then Eq. (11.6.1) is a nonlinear partial differential equation, as the transmissivity depends on the elevation of the groundwater table, namely the piezometric head. The transmissivity can be proportional to the piezometric head, provided that the impervious bottom of the aquifer is flat. Therefore, in cases of a phreatic aquifer, Eq. (11.6.1) is nonlinear, and linearization methods should be applied during the numerical simulation process.

In general, two categories of numerical schemes can be developed for the solution of Eq. (11.6.1): (a) explicit schemes and (b) implicit schemes. As previously described (Chaps. 1, 4, and 10), an explicit scheme calculates values of the piezometric head at the time step $m + 1$, at each individual nodal point of the finite difference grid, by applying values of the piezometric head and the transmissivity at the time step m. An implicit numerical scheme creates a set of linear algebraic equations, in which values of the piezometric head, at the time step $m + 1$, at all nodal points of the finite difference grid, are the unknown variables. When forming the set of linear equations, the numerical scheme requires known values of the piezometric head at time step m (these are part of the right-hand sides of the equations). If the aquifer is a phreatic aquifer, then implicit methods are often associated with some iterative calculations, needed to determine the aquifer transmissivity at an intermediate time step, between times m and $m + 1$, before the values of the piezometric heads at time step $m + 1$ can be obtained.

There are various types of explicit numerical schemes, but as a basic illustration of the procedure, consider the finite difference approximations of the terms in Eq. (11.6.1),

$$S \frac{\partial h}{\partial t} \approx S_{i,j}^m \frac{h_{i,j}^{m+1} - h_{i,j}^m}{\Delta t} \tag{11.6.2a}$$

$$\frac{\partial}{\partial x} \left(T \frac{\partial h}{\partial x} \right) \approx \frac{1}{\Delta x} \left[\left(T \frac{\partial h}{\partial x} \right)_{i+1/2, j}^m - \left(T \frac{\partial h}{\partial x} \right)_{i-1/2, j}^m \right]$$

$$= \frac{1}{(\Delta x)^2} [T_{i+1/2, j}^m (h_{i+1, j}^m - h_{i,j}^m) - T_{i-1/2, j}^m (h_{i,j}^m - h_{i-1/2, j}^m)] \tag{11.6.2b}$$

$$\frac{\partial}{\partial y}\left(T\frac{\partial h}{\partial y}\right) \approx \frac{1}{(\Delta y)^2}[T^m_{i,j+1/2}(h^m_{i,j+1/2} - h^m_{i,j}) - T^m_{i,j-1/2}(h^m_{i,j} - h^m_{i,j-1/2})]$$

(11.6.2c)

where i and j are the subscripts for the x and y locations, respectively, of the grid nodal point, m is a superscript referring to the number of the time step, and Δx and Δy are increments in the x- and y-directions, respectively. By introducing these approximations into Eq. (11.6.1), we obtain the following explicit scheme for the calculation of values of the piezometric head at each individual nodal point of the finite difference grid:

$$h^{m+1}_{i,j} = h^m_{i,j} + \frac{\Delta t}{S_{i,j}(\Delta x)^2}[T^m_{i+1/2,j}(h^m_{i+1,j} - h^m_{i,j})$$

$$-T^m_{i-1/2,j}(h^m_{i,j} - h^m_{i-1/2,j})] + \frac{\Delta t}{S_{i,j}(\Delta y)^2}[T^m_{i,j+1/2}(h^m_{i,j+1} - h^m_{i,j})$$

$$-T^m_{i,j-1/2}(h^m_{i,j} - h^m_{i,j-1/2})] + \frac{\Delta t}{S_{i,j}}q^m_{i,j}$$

(11.6.3)

It should be noted that T varies with time only in a phreatic aquifer. In a confined aquifer, it is common to assume that T has the same value for significant portions of the aquifer. Usually, T and S do not vary substantially in an aquifer.

Explicit schemes are subject to convergence and stability limitations. Such limitations are associated with the maximum value of the time step that can be used during the simulation. Usually, a safe time step is proportional to the square of the space increments. If in Eq. (11.6.1) the values of T and S are constant in time and space, then the criterion for convergence and stability of the numerical scheme of Eq. (11.6.3) is given by

$$\frac{T\,\Delta t}{S[(\Delta x)^2 + (\Delta y)^2]} \leq \frac{1}{2}$$

(11.6.4)

Due to this criterion, fine numerical grid resolution requires very small time steps. Therefore it is often more economical, from the point of view of using computer resources, to apply an implicit finite difference numerical scheme to solve Eq. (11.6.1).

Still assuming a one-dimensional flow in the aquifer, a fully implicit scheme to approximate the various terms of Eq. (11.6.1) is written as

$$S\frac{\partial h}{\partial t} \approx S^m_i\frac{h^{m+1}_i - h^m_i}{\Delta t}$$

(11.6.5a)

$$\frac{\partial}{\partial x}\left(T\frac{\partial h}{\partial x}\right) \approx \frac{1}{\Delta x}\left\{\left[T^m\left(\frac{\partial h}{\partial x}\right)^{m+1}\right]_{i+1/2} - \left[T^m\left(\frac{\partial h}{\partial x}\right)^{m+1}\right]_{i-1/2}\right\}$$

$$= \frac{1}{(\Delta x)^2} [T^m_{i+1/2}(h^{m+1}_{i+1} - h^{m+1}_i) - T^m_{i-1/2}(h^{m+1}_i - h^{m+1}_{i-1/2})]$$

$$(11.6.5b)$$

By introducing Eq. (11.6.5) into Eq. (11.6.1), a set of linear equations is obtained as

$$-h^{m+1}_{i-1} \frac{\Delta t}{S^m_i (\Delta x)^2} T^m_{i-1} + h^{m+1}_i \left[1 + \frac{\Delta t}{S^m_i (\Delta x)^2} (T^m_{i-1} + T^m_{i+1}) \right]$$

$$-h^{m+1}_{i+1} \frac{\Delta t}{S^m_i (\Delta x)^2} T^m_{i+1} = h^m_i + \frac{\Delta t}{S^m_i} q^m_i \qquad (11.6.6)$$

The coefficients of this set of linear equations form a tridiagonal matrix. By definition, all elements of a tridiagonal matrix are zeros, except for those of the major diagonal and the upper and lower diagonals on either side of the main diagonal. If the coefficients of a set of linear equations form a tridiagonal matrix, then the solution of these equations can be very rapidly obtained by the employment of the *Thomas algorithm* found in many introductory texts on numerical solutions of sets of simultaneous equations. This algorithm uses three vectors, one for each of the main diagonals, to solve the problem, rather than a full matrix, thus providing considerable savings in computer memory requirements.

For a two-dimensional flow, a common approach is to apply an alternating direction implicit (ADI) numerical scheme. According to the ADI scheme, the calculation of the piezometric heads at the time step $m + 1$ is performed by an implicit scheme for the spatial derivatives in one of the coordinate directions, say the x-direction. The piezometric heads at the following time step, $m + 2$, are then calculated by an implicit scheme with regard to the y-direction. To start the process, the implicit equations defined in Eq. (11.6.5) are used (along with an additional equation for the y-direction) to calculate the piezometric heads at time step $m + 1$,

$$S \frac{\partial h}{\partial t} \approx S^m_{i,j} \frac{h^{m+1}_{i,j} - h^m_{i,j}}{\Delta t} \qquad (11.6.7a)$$

$$\frac{\partial}{\partial x} \left(T \frac{\partial h}{\partial x} \right) \approx \frac{1}{\Delta x} \left\{ \left[T^m \left(\frac{\partial h}{\partial x} \right)^{m+1} \right]_{i+1/2,j} - \left[T^m \left(\frac{\partial h}{\partial x} \right)^{m+1} \right]_{i-1/2,j} \right\}$$

$$= \frac{1}{(\Delta x)^2} [T^m_{i+1/2,j}(h^{m+1}_{i+1,j} - h^{m+1}_{i,j})$$

$$- T^m_{i-1/2,j}(h^{m+1}_{i,j} - h^{m+1}_{i-1/2,j})] \qquad (11.6.7b)$$

$$\frac{\partial}{\partial x}\left(T\frac{\partial h}{\partial y}\right) \approx \frac{1}{\Delta y}\left\{\left[T\left(\frac{\partial h}{\partial y}\right)\right]^m_{i,j+1/2} - \left[T\left(\frac{\partial h}{\partial y}\right)\right]^m_{i,j-1/2}\right\}$$

$$= \frac{1}{(\Delta y)^2}[T_{i,j+1/2}(h_{i,j+1/2} - h_{i,j})$$

$$- T_{i-1/2,j}(h_{i,j} - h_{i,j-1/2})]^m \qquad (11.6.7c)$$

These equations are introduced into Eq. (11.6.1) to obtain a set of linear equations for each point of the finite difference grid:

$$-h^{m+1}_{i-1,j}\frac{\Delta t}{S^m_{i,j}(\Delta x)^2}T^m_{i-1/2,j} + h^{m+1}_{i,j}\left[1 + \frac{\Delta t}{S^m_{i,j}(\Delta x)^2}(T^m_{i-1/2,j} + T^m_{i+1/2,j})\right]$$

$$-h^{m+1}_{i+1,j}\frac{\Delta t}{S^m_{i,j}(\Delta x)^2}T^m_{i+1/2,j} = h^m_{i,j} + \frac{\Delta t}{S^m_{i,j}}q^m_{i,j} + \frac{\Delta t}{S^m_i(\Delta y)^2}$$

$$\times [T_{i,j+1/2}(h_{i,j+1} - h_{i,j}) - T_{i,j-1/2}(h_{i,j} - h_{i,j-1})]^m \qquad (11.6.8)$$

The coefficients of this set of equations again form a tridiagonal matrix. As just described, the resulting set of equations can be solved using the Thomas algorithm to provide the distribution of piezometric heads in the domain at the $m + 1$ time step.

Following a similar procedure, implicit finite difference approximations are written for the calculation of the piezometric heads at time step $m + 2$,

$$S\frac{\partial h}{\partial t} \approx S^{m+1}_{i,j}\frac{h^{m+2}_{i,j} - h^{m+1}_{i,j}}{\Delta t} \qquad (11.6.9a)$$

$$\frac{\partial}{\partial x}\left(T\frac{\partial h}{\partial x}\right) \approx \frac{1}{\Delta x}\left\{\left[T\left(\frac{\partial h}{\partial x}\right)\right]^{m+1}_{i+1/2,j} - \left[T\left(\frac{\partial h}{\partial x}\right)\right]^{m+1}_{i-1/2,j}\right\}$$

$$= \frac{1}{(\Delta x)^2}[T_{i+1/2,j}(h_{i+1,j} - h_{i,j}) - T_{i-1/2,j}(h_{i,j} - h_{i-1/2,j})]^{m+1}$$

$$(11.6.9b)$$

$$\frac{\partial}{\partial x}\left(T\frac{\partial h}{\partial y}\right) \approx \frac{1}{\Delta y}\left\{\left[T^{m+1}\left(\frac{\partial h}{\partial y}\right)^{m+2}\right]_{i,j+1/2}\right.$$

$$\left. - \left[T^{m+1}\left(\frac{\partial h}{\partial y}\right)^{m+2}\right]_{i,j-1/2}\right\}$$

$$= \frac{1}{(\Delta y)^2}[T^{m+1}_{i,j+1/2}(h^{m+2}_{i,j+1} - h^{m+2}_{i,j}) - T^{m+1}_{i,j-1/2}(h^{m+2}_{i,j} - h^{m+2}_{i,j-1})]$$

$$(11.6.9c)$$

Introducing these into Eq. (11.6.1), a second set of linear equations is obtained for the $m + 2$ time step,

$$-h_{i,j-1}^{m+2} \frac{\Delta t}{S_{i,j}^{m+1}(\Delta y)^2} T_{i,j-1/2}^{m+1} + h_{i,j}^{m+2} \left[1 + \frac{\Delta t}{S_{i,j}^{m+1}(\Delta y)^2} (T_{i,j-1/2}^{m+1} + T_{i,j+1/2}^{m+1}) \right]$$

$$-h_{i,j+1}^{m+2} \frac{\Delta t}{S_{i,j}^{m+1}(\Delta y)^2} T_{i,j+1/2}^{m+1} = h_{i,j}^{m+1} + \frac{\Delta t}{S_{i,j}^{m+1}} q_{i,j}^{m+1} + \frac{\Delta t}{S_{i,j}^{m+1}(\Delta x)^2}$$

$$\times [T_{i+1/2,j}(h_{i+1,j} - h_{i,j}) - T_{i-1/2,j}(h_{i,j} - h_{i-1,j})]^{m+1} \qquad (11.6.10)$$

Again, the coefficients of this set of equations form a tridiagonal matrix, and the Thomas algorithm can be used to provide the distribution of piezo-metric heads in the domain, at the $m + 2$ time step. In successive simulation time steps, the calculations continue to alternate between Eqs. (11.6.8) and (11.6.10).

As long as estimates are needed for quantities of water that can be pumped out of an aquifer, employment of the Dupuit approximation is usually satisfactory. However, if estimates of flow velocity distribution are needed for the evaluation of contaminant advection in the aquifer, then the approximate two-dimensional models originating from Eq. (11.6.1) may need to be replaced by a completely three-dimensional modeling approach. Although flow velocity in the aquifer may be dominated by the horizontal flow, the small vertical velocity, as well as the heterogeneity of the aquifer in the vertical direction, may contribute to the migration of the contaminant in the aquifer. However, if the vertical component of the velocity is extremely small, then transverse dispersion of the contaminant may be the mechanism leading to the contaminant penetration into the entire thickness of the aquifer. Sometimes the aquifer can be considered as a combination of several subaquifers. Then it is common to apply the Dupuit approximation to each subaquifer and to take into account the transfer of water between adjacent subaquifers or aquifer layers. Such models can be termed as quasi-three-dimensional models. Complete three-dimensional models are usually formulated to solve for the distribution of piezometric heads without use of the Dupuit approximation. Sometimes such models suffer from difficulties associated with boundary conditions. A common example, noted previously, is the free surface of the phreatic aquifer, whose location is not known prior to the performance of the numerical simulation. Such a boundary is nonlinear with regard to the determination of the potential function and the velocity components.

11.6.2 The Equation of Contaminant Transport

Calculations of contaminant transport in an aquifer should follow the determination of velocity components at each nodal point of the finite difference grid, as explained in the previous section. In some cases, the contaminant concentration varies considerably in the domain, and it may also affect the density of the water, as in cases of salinity intrusion into the aquifer. In such cases, the flow equation and contaminant transport equation must be solved simultaneously.

The general partial differential equation that is applied to determine the transport of contaminants in an aquifer is given by Eq. (11.3.1). Basic transport features of this equation are determined by the advection [term (2)] and diffusion–dispersion [term (3)] terms. If contaminant diffusion is negligible, then Eq. (11.3.1) can be approximated, in the case of a one-dimensional domain, by the advection equation,

$$\frac{\partial C}{\partial t} + u\frac{\partial C}{\partial x} = 0 \tag{11.6.11}$$

A finite difference approximation for this equation was suggested by *Lax-Wendroff*,

$$C_i^{m+1} = C_i^m - \frac{u\Delta t}{\Delta x}(C_{i+1}^m - C_{i-1}^m) + \left(\frac{u\Delta t}{\Delta x}\right)^2 (C_{i+1}^n - 2C_i^m + C_{i-1}^m) \tag{11.6.12}$$

This expression is second order accurate in the time step, i.e., the accuracy is of order $O(\Delta t)^2$. This scheme is stable, provided that

$$\frac{u\Delta t}{\Delta x} \leq 1 \tag{11.6.13}$$

Alternative numerical schemes have been proposed for the solution of the advection equation. One such scheme, using a combination of explicit and implicit terms that is relatively common, is

$$C_i^{m+1}\left(1 + u_i^{m+1}\frac{\Delta t}{\Delta x}\right) = C_i^m + C_{i-1}^{m+1}\left(u_i^{m+1}\frac{\Delta t}{\Delta x}\right) \tag{11.6.14}$$

where values of the flow velocity are obtained from the flow equation prior to the calculation of the contaminant distribution. This scheme is unconditionally stable, but its accuracy is considerably less than that of the Lax-Wendroff scheme.

If effects of contaminant diffusion in the domain are significant, then all terms of the advection–diffusion equation should be taken into account. For example, in a one-dimensional domain, the advection–diffusion equation is given by

$$\frac{\partial C}{\partial t} + u\frac{\partial C}{\partial x} = D\frac{\partial^2 C}{\partial x^2} \tag{11.6.15}$$

The numerical solution of Eq. (11.6.15) can be based on a combination of a numerical scheme to solve the advection equation and another numerical scheme for simulation of the diffusion equation. This is known as a *split-operator approach*. As an example, an implicit numerical scheme based on the combination of the advection scheme given by Eq. (11.6.14) and a diffusion scheme similar to that of Eq. (11.6.6) can be written as

$$
-C_{i-1}^{m+1}\left[u_i^{m+1}\frac{\Delta t}{\Delta x} + \frac{\Delta t}{(\Delta x)^2}D_i^{m+1}\right] + C_i^{m+1}\left[1 + u_i^{m+1}\frac{\Delta t}{\Delta x} + \frac{2\Delta t}{(\Delta x)^2}D_i^{m+1}\right]
$$
$$
- C_{i+1}^{m+1}\left[\frac{\Delta t}{(\Delta x)^2}D_i^{m+1}\right] = C_i^m \tag{11.6.16}
$$

where values of the flow velocity and the dispersion coefficient, which may depend on the flow velocity, are determined by using the flow equation simulation prior to calculating concentrations. Equation (11.6.16) represents a set of linear algebraic equations, whose coefficients again form a tridiagonal matrix that can be solved using the Thomas algorithm.

PROBLEMS

Solved Problems

Problem 11.1   Water flows in the x-direction through a confined aquifer of thickness B, as shown in Fig. 11.10. The length of the aquifer is L and the hydraulic conductivity is K. The bottom of the aquifer is impermeable, while the top of the aquifer consists of a semipermeable layer of thickness b and hydraulic conductivity α. At the left boundary of the aquifer the piezometric head is h_1. At the right boundary of the aquifer the piezometric head is h_2. The piezometric head in the overlying free surface (phreatic) aquifer is h_0. Apply the Dupuit approximation to derive the expression for the calculation of the piezometric head and flow rate distribution in the aquifer.

Solution

The aquifer flow rate, Q, is given by

$$
Q = qB = -KB\frac{dh}{dx} \tag{1}
$$

where h is the piezometric head of the confined aquifer. The seepage flow from the confined aquifer into the overlying phreatic aquifer is

$$
q_v = \alpha\frac{h - h_0}{b} \tag{2}
$$

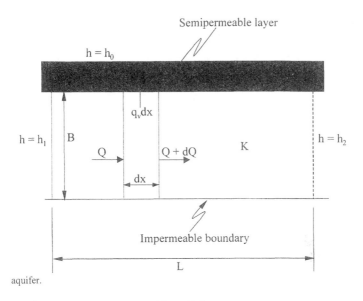

aquifer.

Figure 11.10 Definition sketch, Problem 11.1.

Considering the water flux in the elementary volume of the aquifer bounded by the dashed lines in Fig. 11.10, a mass balance statement can be written as

$$dQ + q_v \, dx = 0 \tag{3}$$

where

$$dQ = \frac{dQ}{dx} = -KB\frac{d^2h}{dx^2} \qquad q_v = \alpha\frac{h - h_0}{b} \tag{4}$$

Introducing Eqs. (1), (2), and (4) into Eq. (3), we obtain

$$BK\frac{d^2h}{dx^2} - \alpha\frac{h - h_0}{b} = 0 \tag{5}$$

Rearranging this result then yields

$$\frac{d^2h}{dx^2} - \beta^2 h = -\beta^2 h_0 \tag{6}$$

where

$$\beta = \sqrt{\frac{\alpha}{bBK}} \tag{7}$$

The solution of Eq. (6) is given by

$$h = C_1 e^{-\beta x} + C_2 e^{\beta x} + h_0 \tag{8}$$

where

$$C_1 = \frac{h_1 - h_2 - (h_1 - h_0)(1 - e^{\beta L})}{1 - e^{\beta L}}$$

$$C_2 = h_1 - h_0 - C_1 \qquad (9)$$

By applying Eqs. (1) and (8), we obtain the desired flow rate,

$$Q = qB = -KB\frac{dh}{dx} = -KB\beta(-C_1 e^{-\beta x} + C_2 e^{\beta x})$$

Problem 11.2   Flow occurs through a phreatic aquifer in the *x*-direction as shown in Fig. 11.11. The length of the aquifer is L and the hydraulic conductivity is K. The bottom of the aquifer is semipermeable, and it separates the aquifer from a deeper confined aquifer with constant head h_0. The thickness of the semipermeable bottom layer is b and its hydraulic conductivity is α. At the left boundary of the phreatic aquifer the piezometric head is h_1. At the right boundary of the aquifer the piezometric head is h_2. Apply the Dupuit approximation to derive the differential equation for the calculation of the piezometric head and in the aquifer.

Solution

The aquifer flow rate, Q, is given by

$$Q = qh = -Kh\frac{dh}{dx} \qquad (1)$$

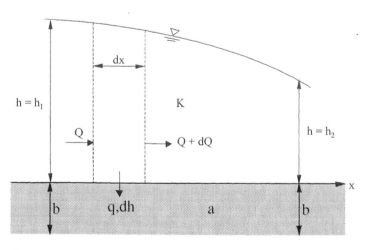

Figure 11.11   Definition sketch, Problem 11.2.

where h is the piezometric head of the phreatic aquifer. The seepage flow from the phreatic aquifer into the underlying confined aquifer is given by

$$q_v = \alpha \frac{h - h_0}{b} \tag{2}$$

Considering the water flux in the elementary volume of the aquifer bounded by the dashed lines in Fig. 11.11, we obtain a mass balance statement as

$$dQ + q_v dx = 0 \tag{3}$$

where

$$dQ = \frac{dQ}{dx} = -K \frac{d}{dx}\left(h \frac{dh}{dx}\right) \qquad q_v = \alpha \frac{h - h_0}{b} \tag{4}$$

Introducing Eqs. (1), (2), and (4) into Eq. (3), we obtain

$$K \frac{d}{dx}\left(h \frac{dh}{dx}\right) - \alpha \frac{h - h_0}{b} = 0 \tag{5}$$

Rearranging this result then leads to

$$\frac{d}{dx}\left(h \frac{dh}{dx}\right) - \left(\frac{\alpha}{Kb}\right)(h - h_0) = 0 \tag{6}$$

This is a one-dimensional nonlinear boundary value problem, subject to the boundary conditions

$$h = h_1 \quad \text{at} \quad x = 0$$
$$h = h_2 \quad \text{at} \quad x = L$$

This type of problem usually requires a numerical method incorporating some sort of linearization.

Problem 11.3 Develop the differential equation and a finite difference numerical implicit scheme for the calculation of piezometric head distribution and flow in a phreatic aquifer. Initially the aquifer flow is subject to the boundary conditions $h = h_1$ at $x = 0$ and $h = h_2$ at $x = L$. The hydraulic conductivity of the aquifer is K. At time $t = 0$, a series of pumping wells started pumping a discharge Q_p per unit width of the aquifer at $x = L/2$.

Solution

The problem is sketched in Fig. 11.12. A mass balance statement is first written for the elementary volume bounded by the dashed lines in Fig. 11.12, resulting in

$$\phi \frac{\partial h}{\partial t} + dQ + q_v dx = 0 \tag{1}$$

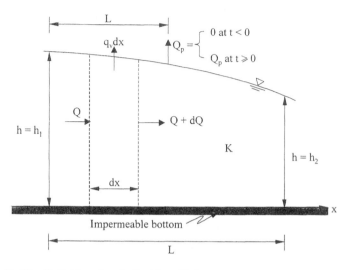

Figure 11.12 Definition sketch, Problem 11.3.

where ϕ is the effective porosity of the aquifer, q_v is the discharge per unit width and unit length pumped out of the aquifer, and dQ is

$$dQ = \frac{\partial Q}{\partial x} = -K \frac{\partial}{\partial x}\left(h\frac{\partial h}{\partial x}\right) \tag{2}$$

We introduce Eq. (2) into Eq. (1) to obtain

$$\phi\frac{\partial h}{\partial t} = K\frac{\partial}{\partial x}\left(h\frac{\partial h}{\partial x}\right) - q_v \tag{5}$$

This is the desired differential equation for the piezometric head distribution. The first-order derivatives of the piezometric head and its gradient may be represented by the following finite difference approximations:

$$\frac{\partial h}{\partial x} = \frac{h_{i+1/2} - h_{i-1/2}}{\Delta x} \qquad \frac{\partial h}{\partial t} = \frac{h_i^{n+1} - h_i^n}{\Delta t}$$

$$\frac{\partial}{\partial}\left(h\frac{\partial h}{\partial x}\right) = \frac{[h(\partial h/\partial x)]_{i+1/2} - [h(\partial h/\partial x)]_{i-1/2}}{\Delta x}$$

$$= \frac{h_{i+1/2}(h_{i+1} - h_i) - h_{i-1/2}(h_i - h_{i-1})}{(\Delta x)^2}$$

where subscripts refer to nodal points and superscripts refer to time steps. All of the spatial gradients are understood to be evaluated at time $(n+1)$, since the problem specifies that an implicit method is to be used. However, in order to deal with the nonlinearities associated with the aquifer transmissivity, the values for h [on the right-hand side of Eq. (5)] are taken at time n. After introducing the finite difference approximations into Eq. (5), the result is

$$\phi \frac{h_i^{n+1} - h_i^n}{\Delta t} = \frac{K}{(\Delta x)^2}[h_{i+1/2}^n(h_{i+1}^{n+1} - h_i^{n+1}) - h_{i-1/2}^n(h_i^{n+1} - h_{i-1}^{n+1})] - q_v$$

This is rearranged to group all like terms in the h values,

$$-h_{i-1}^{n+1}\frac{Kh_{i-1/2}^n\Delta t}{\phi(\Delta x)^2} + h_i^{n+1}\left[1 + \frac{Kh_{i+1/2}^n\Delta t}{\phi(\Delta x)^2} + \frac{Kh_{i-1/2}^n\Delta t}{\phi(\Delta x)^2}\right]$$

$$-h_{i+1}^{n+1}\frac{Kh_{i+1/2}^n\Delta t}{\phi(\Delta x)^2} = h_i^n - q_{v_i}\frac{\Delta t}{\phi}$$

This expression represents a set of linear algebraic equations with a tridiagonal coefficient matrix. Thus the piezometric heads at each nodal point may be evaluated at each time step using the Thomas algorithm. It should be noted that at all nodal points the value of q_v is zero, except for the nodal point representing $x = L/2$. At that particular point, the value of q_v is given by

$$(q_v)_{x=L/2} = \frac{Q_p}{(\Delta x)^2}$$

For the given initial conditions, the piezometric head values are specified at each boundary and, since $q_v = 0$ everywhere for $t < 0$, then under steady state, Eq. (5) and its boundary conditions are written, respectively, as

$$\frac{d}{dx}\left(Kh\frac{dh}{dx}\right) = 0$$

$$h = h_1 \quad \text{at} \quad x = 0$$

$$h = h_2 \quad \text{at} \quad x = L$$

Integrating once provides

$$\frac{K}{2}\frac{dh^2}{dx} = C_1$$

where C_1 is an integration constant. A second integration yields

$$h^2 = \frac{2C_1}{K}x + C_2$$

where C_2 is another integration constant. By applying the boundary conditions, the constants are determined as

$$C_1 = -\frac{(h_1^2 - h_2^2)K}{2L} \qquad C_2 = h_1^2$$

Thus the initial distribution of piezometric head is specified.

Problem 11.4 Consider the steady-state transport of contaminant through a barrier that separates a contaminated area from a freshwater aquifer. Assume that the barrier thickness is L, its width is B, its porosity is ϕ, and its hydraulic conductivity is K. The molecular diffusivity is κ and the Peclet number of the flow through the barrier is small. The two boundaries of the barrier are subject to the following conditions:

$$\text{At} \quad x = 0, h = h_0 \qquad C = C_0$$
$$\text{At} \quad x = L, h = h_E \qquad C = 0$$

where h is the piezometric head and C is the contaminant concentration. Along the barrier a continuous minor discharge q_v per unit area is pumped out of the barrier.

    (a) Develop the differential equations that should be solved to determine the contaminant distribution in the barrier.

    (b) Derive the approximate analytical solution of the differential equations for a small value of q_v.

    (c) Derive the analytical solution for negligible value of q_v.

Solution

(a) Two differential equations should be considered, the flow equation and the contaminant transport equation. These equations are given, respectively, by

$$KB\frac{d^2h}{dx^2} - q_v = 0 \tag{1}$$

$$V\frac{dC}{dx} = \kappa\frac{d^2C}{dx^2} - \frac{q_v}{B}C \tag{2}$$

where V is the local flow velocity through the barrier.

    (b) By integrating the equation of motion (1) twice, we obtain

$$h = h_1 - \frac{h_1 - h_E}{L}x - \frac{q_v x}{2KB}(L - x) \tag{3}$$

The flow velocity through the barrier is given by

$$V = -\frac{K}{\phi}\frac{dh}{dx} = \frac{K}{\phi L}(h_1 - h_E) - \frac{q_v}{2\phi B}(L - 2x) \tag{4}$$

Rearrangement of the equation of contaminant transport yields

$$\frac{d^2C}{dx^2} - \frac{V}{\kappa}\frac{dC}{dx} - \frac{q_v}{B\kappa}C = 0 \tag{5}$$

This expression represents a boundary value problem. Equation (5) can be solved very easily by the employment of a numerical scheme. On the other hand, the analytical solution of Eq. (5) is quite complicated due to the dependence of V on x, as indicated by Eq. (4). However, simplified cases can be solved by simple analytical expressions. Simplification of Eq. (5) depends on the value of the dimensionless quantity,

$$\frac{q_v L^2}{2KB(h_1 - h_E)} \tag{6}$$

If this parameter is very small, then the contribution of q_v to the value of V is minor, and V can be considered as a constant quantity, given approximately by the first term of Eq. (4). Under such conditions, the solution of Eq. (5) is given by

$$C = Ae^{\alpha x} + Be^{\beta x} \tag{7}$$

where A and B are constant coefficients, and α and β are given by

$$\alpha = \frac{V}{2\kappa} + \sqrt{\left(\frac{V}{2\kappa}\right)^2 + \frac{q_v}{BK}} \tag{8a}$$

and

$$\beta = \frac{V}{2\kappa} - \sqrt{\left(\frac{V}{2\kappa}\right)^2 + \frac{q_v}{BK}} \tag{8b}$$

By applying the boundary condition of $C = C_0$ at $x = 0$, and $C = 0$ at $x = L$, we obtain

$$A = C_0 \left[\frac{e^{\beta L}}{e^{\beta L} - e^{\alpha L}} \right] \qquad B = C_0 \left[\frac{e^{\alpha L}}{e^{\alpha L} - e^{\beta L}} \right] \tag{9}$$

(c) If the value of q_v is negligible, then Eq. (1) reduces to

$$\frac{\kappa}{V} \frac{d^2 C}{dx^2} = \frac{dC}{dx} \tag{10}$$

By integrating this expression, we obtain

$$\frac{\kappa}{V} \frac{dC}{dx} = C + A \tag{11}$$

where A is an integration constant. By integrating Eq. (11), we obtain

$$\ln \left[\frac{C + A}{B} \right] = \frac{V}{\kappa} x \tag{12}$$

where B is an integration constant. By applying the boundary conditions, we obtain

$$A = C_0 \left[\frac{1}{\exp\left(-\frac{V}{\kappa} L\right) - 1} \right]$$

$$B = C_0 \left[\frac{\exp\left(-\frac{V}{\kappa}L\right)}{\exp\left(-\frac{V}{\kappa}L\right) - 1} \right] \tag{13}$$

By introducing Eq. (13) into Eq. (12) then gives

$$C = C_0 \left\{ \frac{\exp\left[-\frac{V}{\kappa}(L - x)\right] - 1}{\exp\left(-\frac{V}{\kappa}L\right) - 1} \right\}$$

which is the desired concentration distribution.

Unsolved Problems

Problem 11.5   For the conditions of problem 11.1, assume $h_1 = 40$ m, $h_2 = 30$ m, $L = 5,000$ m, $h_0 = 25$ m, $K = 40$ m/d, $\alpha = 0.01$ m/d, $B = 25$ m, and $b = 5$ m.

(a) Using a spreadsheet, solve for and draw the curves describing the variation of the piezometric head and aquifer discharge along the aquifer.

(b) Draw several curves, describing the ratio between h and h_1 along the aquifer for various values of h_0/h_1.

(c) Draw several curves describing the ratio between the aquifer discharge and the transmissivity KB for various values of α/K.

Problem 11.6   For the conditions of problem 11.2, assume $h_1 = 40$ m, $h_2 = 30$ m, $L = 5,000$ m, $h_0 = 25$ m, $K = 40$ m/d, $\alpha = 0.01$ m/d, and $b = 5$ m.

(a) Develop a numerical scheme incorporating linearization of the basic differential equation and an iterative procedure for the calculation of the piezometric head distribution in the aquifer. As an alternative, apply a perturbation approach with another iterative scheme. In your solution, state clearly which approach you are using.

(b) Apply your numerical scheme and solve the nonlinear boundary value problem for the distribution of piezometric head in the aquifer, as well as the variation of the aquifer flow rate and the rate of seepage along the aquifer.

Problem 11.7   For the conditions of problem 11.3, assume $h_1 = 40$ m, $h_2 = 30$ m, $L = 5,000$ m, $K = 40$ m/d, and $Q_p = 2$ m^2/d.

(a) Calculate the distribution of the piezometric head in the aquifer after two days of pumping.

(b) Calculate the piezometric head distribution under the new steady-state conditions.

Problem 11.8 Referring to problem 11.4, develop the numerical finite difference scheme for the solution of Eq. (5). Consider that $K = 10^{-8}$ m/s, $L = 2$ m, $B = 10$ m, $h_1 = 25$ m, $h_E = 24$ m, $C_0 = 0.1$ kg/m^3, $\phi = 0.15$, and $q_v = 10^{-8}$ m/s.

Problem 11.9 Refer to problem 11.4. Assume that flow through the barrier is subject to steady-state conditions and develop a numerical finite difference implicit scheme for the calculation of the buildup of the contaminant concentration profile in the barrier. Use quantitative values of physical parameters as given in problem 11.8. Initially the barrier is free of contaminant.

Problem 11.10 Consider cases of very low hydraulic conductivity in the barrier using the program developed in problem 11.9. Compare your simulation results with the analytical solution given by Eq. (11.3.12).

Problem 11.11 Referring to problem 11.8, assume that the barrier is very thick, $L \rightarrow \infty$, and all other physical parameters are the same as previously defined. Apply the numerical program developed in problem 11.9 and compare the simulation results with those obtained by using the analytical expressions given by Eqs. (11.3.12), (11.3.15), and (11.3.16). For the latter two equations, assume $a_L = 0.1$ cm.

Problem 11.12 Consider an aquifer subject to uniform flow with velocity $V = 0.1$ m/d. The transverse dispersivity is $a_t = 2$ cm. At the bottom of the freshwater aquifer the salinity is kept constant with salt concentration $C = 40$ g/l. Assume that the aquifer is very thick. Develop a finite difference implicit numerical scheme for the solution of Eq. (11.4.13) and describe the buildup of the transition zone between fresh and saltwater. Compare your results with the boundary layer approximation, which is given by Eq. (11.4.16). (You could also compare your results here with results obtained in problem 10.34.2)

Problem 11.13 Develop a finite difference numerical scheme for the solution of Eq. (11.5.3). This solution should take into account the conservation of mass principle with regard to the NAPL entrapped within the porous medium.

Problem 11.14 Apply your computer program, developed in problem 11.13, to an aquifer of length $L = 60$ m, contaminated by entrapped NAPL. The porosity $\phi = 0.3$, initial saturation of the entrapped NAPL $S_n = 0.2$, flow velocity $V = 1$ m/d, dispersivity $a_L = 0.3$ m, and $k_f a_b = 0.5$ m^2/d. The equilibrium concentration of dissolved NAPL is $C_s = 0.8$ g/l. Simulate the reclamation process and determine how long it should take to achieve a practical reclamation of the aquifer.

SUPPLEMENTAL READING

Bear, J., 1972. *Dynamics of Fluids in Porous Media.* American Elsevier, New York. (Contains a comprehensive review of many important studies concerning environmental flow in porous media, which were done prior to 1970.)

Bedient, P. B., Rifai, S. H., and Newell, C. J., 1994. *Ground Water Contamination.* Prentice Hall, Englewood Cliffs, NJ. (Contains material referring to groundwater contamination and reclamation.)

Clark, M. M., 1996. *Transport Modeling for Environmental Engineers and Scientists.* John Wiley, New York. (Contains several comprehensive chapters referring to transport in porous media.)

Hofman, J. D., 1992. *Numerical Methods for Engineers and Scientists.* McGraw-Hill, NY. (Many books provide the reader with the basic knowledge of numerical methods required by the topics of the present chapter. Such topics incorporate the solution of boundary value problems in one-and two-dimensional domains, as well as parabolic partial differential equations. However, without denying the value of other books on numerical methods, we find that this textbook incorporates a comprehensive presentation of the topics and the different applicable methods of solution.)

Sun, N.-Z., 1996. *Mathematical Modeling of Groundwater Pollution.* Springer-Verlag, New York. (Contains a comprehensive presentation of general topics of contaminant hydrology and numerical methods applications.)

Home Pages on the Internet Concerning Computer Codes

http://www.epa.gov/ada/models.html
http://aapg.geol.lsu.edu/rbwinsto.htm
http://www.ibmcug.co.uk/~bedrock/gsd/
http://hydro.geo.ua.edu
http://www.umanitoba.ca/geo_eng?Groundwater/data.html
http://www.esnt.com
http://www.access.digex.net:80/~scisoft
http://hydrosystems.com/pub/ENVIROMOD
http://www.rrze.uni-erlangen.de/doc/FAU/fakultaet/natIII/geamin/geologic/soft.html
http://www.bossintl.com

12

Exchange Processes at the Air/Water Interface

12.1 INTRODUCTION

One of the major boundaries in most natural surface water bodies is the interface with the atmosphere. In order to describe the distribution of various properties of the water body it is necessary to specify boundary conditions for the transport equations for those properties at this (as well as other) boundaries. The usual kinds of boundary conditions are applicable here, i.e., specification of either the values for the property of interest or its gradients. For the air/water interface it is more common to prescribe fluxes or transport rates for the property of interest. A significant feature of this transport is the wind, which strongly affects property fluxes. Because of limited *fetch*, wind effects in open channel flows are less significant than in lakes and reservoirs, which have much larger surface areas. In this chapter we consider transport of momentum, heat, gases, and volatile organic chemicals across the air/water interface.

12.2 MOMENTUM TRANSPORT

The main driving force for momentum is shear stress exerted as a result of velocity gradients across the air/water interface. The main effects are generation of surface drift currents, waves, setup and *seiche* motions, as illustrated in Fig. 12.1. In lakes and reservoirs wind is a primary driving force for general circulation.

Shear stress at the air/water interface is exerted as a result of differences between the wind speed and direction, and the water surface velocity. Part of this stress works to develop the wave field and part is used for generation of surface drift currents. In the present section the focus is on generation of drift currents and circulation — see Chap. 8 for a discussion of surface water

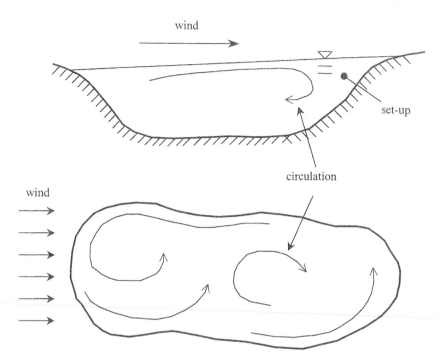

Figure 12.1   Illustration of setup and circulation (side view, top) and horizontal circulation (plan view, bottom) generated by wind over a lake.

waves. For the present discussion, it is assumed that a steady wind is blowing over a water surface and that a fully developed wave field is present, i.e., wind/wave interactions are in equilibrium and there is no further partitioning of the surface stress into wave development.

Figure 12.2 illustrates a possible velocity profile for wind and water, where W_z is the wind speed measured at position z above the water surface and u_d is the surface drift velocity in the water. The mean surface level is at $z = 0$, and z increases upwards (it is convenient to work with z increasing upwards when describing the wind velocity profile; however, when the main concern is with properties of the water body, it may be more convenient to work with z increasing downwards from the surface — see the following section, for example). The shear stress at the surface is given by

$$\tau_s = c_z \rho_a (W_z - u_d)^2 \cong c_z \rho_a W_z^2 \tag{12.2.1}$$

where c_z is a drag coefficient and ρ_a is air density. The approximation of the second part of this equation results from the assumption that $u_d / W_z \ll 1$. The

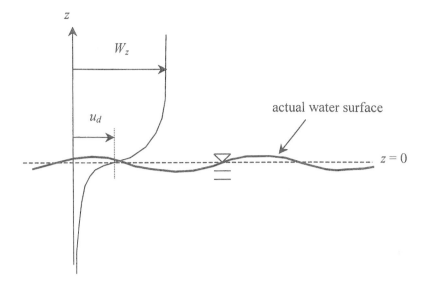

Figure 12.2 Velocity profiles for wind over a water surface.

shear stress also is defined in terms of a friction velocity,

$$\tau_s = \rho_a u_{*a}^2 \tag{12.2.2}$$

where u_{*a} is the friction velocity for the wind profile.

It is commonly assumed that the wind velocity distribution follows a boundary layer logarithmic profile,

$$W_z = \frac{u_{*a}}{\kappa} \ln\left(\frac{z}{z_0}\right) \tag{12.2.3}$$

where κ is the von Karman constant, equal to about 0.4, and z_0 is the virtual origin of the profile (note that $W_z \rightarrow 0$ for $z \rightarrow z_0$). In order for this assumption to be valid, the measurement height (z) must be chosen so that the logarithmic profile is valid. According to Wu (1971), this is satisfied when

$$z = \begin{cases} 10 \text{ cm}, & \text{Re} < 5 \times 10^7 \\ 7.35 \text{ Re}^{2/3} \times 10^{-5} \text{ cm}, & 5 \times 10^7 < \text{Re} < 5 \times 10^{10} \\ 10 \text{ m}, & \text{Re} > 5 \times 10^{10} \end{cases} \tag{12.24}$$

where $\text{Re} = W_z L / \nu$ is a *fetch Reynolds number*, with $L = $ fetch $=$ distance over which wind has blown over the surface of the water and $\nu = $ kinematic viscosity. When Eq. (12.1.4) is satisfied, z is greater than the *significant wave amplitude* and less than about four-tenths of the total boundary layer thickness, so it is well within the range where the logarithmic profile should be valid.

The magnitude of z_0 depends on the roughness of the surface and is assumed to be a linear function of roughness, similar to a turbulent boundary layer in pipe or open channel flow. In a fully developed wave field, the dynamic roughness length scale is estimated by (u_{*a}^2/g), so

$$z_0 = \alpha_c \frac{u_{*a}^2}{g} \tag{12.2.5}$$

where α_c is known as the *Charnock coefficient* and has a value between 0.011 and 0.035, with most reported values falling in the range 0.011 to 0.016. Upon substituting Eqs. (12.2.1), (12.2.2), and (12.2.5) into (12.2.3), an expression for c_z is obtained,

$$\frac{1}{\sqrt{c_z}} = \frac{1}{\kappa} \ln \left(\frac{gz}{\alpha_c c_z W_z^2} \right) \tag{12.2.6}$$

which must be solved iteratively since c_z is found in different terms on both sides of the equation. Normally, c_z has a value on the order of 10^{-3}. Once c_z is found, the surface shear stress is calculated from Eq. (12.2.1).

For drift current, again assuming a fully developed wave field, it is assumed that the surface shear stress in the water is the same as that in the air, $\tau_0 \cong \tau_s$, where

$$\tau_0 = \rho_w u_*^2 \tag{12.2.7}$$

and ρ_w is water density at the surface. (Note that $\tau_0 < \tau_s$ when the wave field is not fully developed, since part of the surface shear is used in developing the waves.) By combining Eqs. (12.2.2) and (12.2.7), we find

$$u_* = u_{*a} \sqrt{\frac{\rho_a}{\rho_w}} \tag{12.2.8}$$

Then, assuming that the general shapes of the velocity profiles in air and water are similar (see Fig. 12.2, where the profile in water is inverted and reversed but has the same general shape), it follows that the drag coefficient should be the same on both sides of the air/water interface. This gives

$$\tau_0 = c_z \rho_w u_d^2 = c_z \rho_a W_z^2 \tag{12.2.9}$$

from which it is found that

$$u_d = W_z \sqrt{\frac{\rho_a}{\rho_w}} \tag{12.2.10}$$

The ratio of air to water density is approximately 10^{-3}, so this last result implies that surface drift velocity is about 3% of wind speed, which may be used as a rough rule of thumb. Also, if c_z is estimated as being about 10^{-3}, then from Eqs. (12.2.1) and (12.2.2), $u*_a \cong 0.03 W_z$, and, combined with Eqs. (12.2.8) and (12.2.10), we find $u* \cong 0.03 u_d$. This gives an alternative relation that may be used to estimate the velocities, at least for the conditions of steady wind, long fetch, and fully developed wave field. A more realistic relationship for lakes and reservoirs, with limited fetch, is $(u*/u_d) \cong 0.1-1$.

Finally, the kinetic energy flux across the surface is

$$KE_F = u_d \tau_s = C \rho_w u_*^3 \tag{12.2.11}$$

where the coefficient C incorporates various assumptions noted above, including the relationship between u_d and $u*$. This last expression is needed to calculate possible mixing induced by wind (see Sec. 13.5).

12.2.1 Seiches

When fetch is limited, the possibility of wind-generated *seiche motions* must be considered. These are in essence wavelike motions with a half-wave length given by the fetch L. Seiche motions are primarily of interest when winds are relatively constant in speed and direction over a long period of time, as sketched in Fig. 12.3. The surface shear stress exerted by the wind causes the water surface to tilt, establishing a pressure gradient to balance this stress. The tilted water surface position is given by $\eta(x)$, and the difference between the tilted surface and the horizontal equilibrium position is referred to as the *wind setup*. If the lake is large enough that the dynamics are affected by the Earth's rotation, then the position of maximum setup moves in the counterclockwise direction (northern hemisphere). For the present discussion we neglect this effect. If the wind suddenly stops, the tilted water surface moves back towards a level condition. As the water reaches this condition, however, it still has momentum and overshoots, resulting in a setup on the

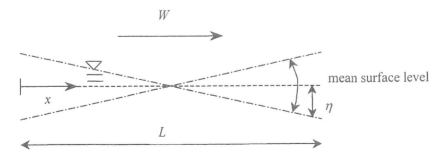

Figure 12.3 Wind-generated seiche motion.

opposite side of the lake. This motion is eventually arrested as the kinetic energy of the moving water is transferred to the potential energy of the setup. The flow then reverses and the surface "rocks" back towards the original setup condition. This oscillatory motion continues until eventually viscous effects cause it to die out. Depending on the lake geometry and relative wind direction, seiche motions can be quite substantial. The frequency of the oscillations is known as the *natural* or *inertial frequency* of the lake.

The tilted water surface profile is found by considering a force balance on a small control volume, as shown in Fig. 12.4. Due to circulation of water resulting from the surface shear, there is some motion along the bottom, which is estimated to generate an additional shear stress approximately equal to 10% of τ_s. Under steady-state conditions, the momentum fluxes into and out of the control volume are equal, and a force balance in the horizontal direction (per unit width) gives

$$1.1\tau_s\, dx = \frac{1}{2}\gamma\left[\left(h + \frac{\partial h}{\partial x}\, dx\right)^2 - h^2\right] \qquad (12.2.12)$$

where hydrostatic pressure is assumed for the two sides, h is depth, and $\gamma = (\rho g)$ is specific gravity. In general, the wind shear acts at an angle to the horizontal, once the water surface tilts. However, this angle is very small, and its cosine is assumed to be approximately 1. Also neglecting the second-order

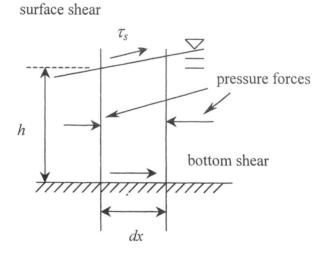

Figure 12.4   Control volume for analysis of water surface profile.

term in dx, the term in brackets in Eq. (12.2.12) is simplified as

$$\frac{\partial h}{\partial x} = \frac{\partial \eta}{\partial x} = \frac{1.1\tau_s}{\gamma h} \tag{12.2.13}$$

If τ_s is assumed constant, or at least is known as a function of x, this equation can be integrated, along with a boundary condition such as $h(0)[\text{or}\eta(0)]$, to calculate the water surface profile $\eta(x)$. A simple approximation for the water surface is a linear profile, using mean water depth H:

$$\eta \cong \frac{1.1\tau_s}{\gamma H}\left(x - \frac{L}{2}\right) \tag{12.2.14}$$

12.3 SOLAR RADIATION AND SURFACE HEAT TRANSFER

12.3.1 Temperature Equation

Before discussing surface heat transfer, it is helpful to review formally the derivation of the *temperature equation*. This is directly related to the thermal energy equation, which was introduced in Sec. 2.9.3. The temperature equation is derived from the general conservation of energy statement, Eq. (2.8.7), which is repeated here for convenience:

$$\frac{dQ}{dt} - \frac{dW_s}{dt} = \frac{\partial}{\partial t}\int_U \rho e\, dU + \int_S \left(\frac{V^2}{2} + gz + u + \frac{p}{\rho}\right)(\rho \vec{V} \cdot \vec{n}\, dS) \tag{12.3.1}$$

where $Q =$ heat added, $W =$ work done by the fluid on its surroundings, $e =$ energy per unit mass, the first integral on the right-hand side is the rate of change of energy in the control volume, and the second integral is the net rate at which energy is transported across the control surface. Note that thermo-dynamic convention is followed here in writing the work term as a positive quantity when the fluid does work on its surroundings.

The heat added can be represented as a surface integral of *heat flux*, $\vec{\varphi}$ (energy transport per unit time and per unit area), which in turn is the sum of radiation ($\vec{\varphi}_r$) and conduction ($\vec{\varphi}_c$) terms,

$$\frac{dQ}{dt} = -\int_S \vec{\varphi} \cdot \vec{n}\, dS = -\int_S (\vec{\varphi}_r + \vec{\varphi}_c) \cdot \vec{n}\, dS$$

$$= -\int_V \nabla \cdot (\vec{\varphi}_r + \vec{\varphi}_c)\, d\forall \tag{12.3.2}$$

where S indicates the control surface, dS is an elemental area of the control surface, $\vec{n}$ is a unit normal vector pointing out of the control volume (refer to

Chap. 2 for further discussion), and the *divergence theorem* has been used to write the surface integral as a volume integral in the last part of Eq. (12.3.2). The negative sign is added since the flux, by convention, is defined as being positive when directed outwards from the control volume.

The work rate term is considered to consist of gravity work and surface work. The gravity work rate is given by a volume integral,

$$\int_V \rho \vec{g} \cdot \vec{V} d\mathsf{V}$$

and the surface work rate, from Eq. (2.8.5), is

$$\int_S p\vec{V} \cdot \vec{n} \, dS - \int_S \tilde{\tau} \cdot \vec{n} \cdot \vec{V} \, dS = \int_S \tilde{S} \cdot \vec{n} \cdot \vec{V} \, dS$$

where $\tilde{\tau}$ is the deviatoric stress tensor and $\tilde{S}$ is the full stress tensor, as introduced in Chap. 2. Rewriting the surface integrals as a volume integral, again using the divergence theorem, the total work rate is

$$-\frac{dW}{dt} = \int_V \left[\rho \vec{g} \cdot \vec{V} + \nabla \cdot (\tilde{S} \cdot \vec{V}) \right] d\mathsf{V}$$

$$= \int_V \left[(\rho \vec{g} + \nabla \cdot \tilde{S}) \cdot \vec{V} + (\tilde{S} \cdot \nabla) \cdot \vec{V} \right] d\mathsf{V} \qquad (12.3.3)$$

The first term in the last expression in Eq. (12.3.3) may be rewritten using the momentum equation (2.7.1),

$$\rho \vec{g} + \nabla \cdot \tilde{S} = \rho \frac{D\vec{V}}{Dt} \qquad (12.3.4)$$

Then, noting also that

$$\frac{D\vec{V}}{Dt} \cdot \vec{V} = \frac{D}{Dt} \left(\frac{V^2}{2} \right) \qquad (12.3.5)$$

we substitute Eqs. (12.3.2) and (12.3.3) into Eq. (12.3.1), rewriting the surface integral in Eq. (12.3.1) as a volume integral, and combine all volume integrals to obtain

$$\frac{\partial}{\partial t} \left[\rho \left(\frac{V^2}{2} + u \right) \right] = -\nabla \cdot \vec{\varphi} + \rho \frac{D}{Dt} \left(\frac{V^2}{2} \right) + (\tilde{S} \cdot \nabla) \cdot \vec{V}$$

$$- \nabla \cdot \left[\rho \left(\frac{V^2}{2} + u \right) \vec{V} \right] \qquad (12.3.6)$$

where u is internal energy, from Eq. (2.8.4).

This last result is further simplified using the continuity equation (2.5.6) and noting that the terms in $(V^2/2)$ cancel, so that

$$\rho \frac{Du}{Dt} = -\nabla \cdot \vec{\varphi} + (\tilde{S} \cdot \nabla) \cdot \vec{V} \tag{12.3.7}$$

The left-hand side of Eq. (12.3.7) is the time rate of change in internal energy of a fluid element, and the right-hand side expresses the heat flux and work done by stresses. Usually the contribution of stresses is negligible in affecting temperature, as shown by the scaling arguments for viscous heating discussed in Sec. 2.9.4. Internal energy in liquids is assumed to be a function of temperature only, i.e., $u = u(T)$, where T = temperature, and from the definition of specific heat (Eq. 2.8.9), a change in energy is related to a change in temperature by

$$du = c\,dT \tag{12.3.8}$$

where $c = c_p = c_v$ since in liquids the specific heats for constant pressure and constant volume are nearly equal. Again considering the heat flux term to consist of conduction and radiation terms, the conduction term is expressed using *Fourier's law* of heat conduction, which states that the conductive heat flux is proportional to the temperature gradient,

$$\vec{\varphi}_c = -\kappa_T \nabla T \tag{12.3.9}$$

where κ_T is *thermal conductivity*.

Making the foregoing simplifications and substitutions, Eq. (12.3.7) becomes

$$\rho c \frac{DT}{Dt} = \rho c \left(\frac{\partial T}{\partial t} + \vec{V} \cdot \nabla T \right) = \nabla \cdot (\kappa_T \nabla T) - \nabla \cdot \vec{\varphi}_r \tag{12.3.10}$$

where constant c has been assumed. This is the *temperature equation*, which has the general form of an advection–diffusion equation expressing conservation of thermal energy, with a main source term due to radiative heat flux. If κ_T is constant, this equation becomes

$$\frac{\partial T}{\partial t} + \vec{V} \cdot \vec{\nabla} T = k_T \nabla^2 T - \frac{1}{\rho c} \nabla \cdot \vec{\varphi}_r \tag{12.3.11}$$

where $k_T = \kappa_T/\rho c$ is *thermal diffusivity*. This result is similar to Eq. (2.9.28), but without the viscous or compression heating terms, which have been neglected here.

In order to solve Eq. (12.3.11), initial and boundary conditions are needed, and possible source terms for radiation must be specified. For natural water bodies, the primary considerations are the heat transfer rate at the air/water interface (which is normally specified as a boundary condition) and the solar heating rate due to absorption of solar radiation (an internal source term). We examine this latter term first.

12.3.2 Solar Radiation Absorption

Solar radiation consists of a large range of wavelengths, that have different absorption properties in water. For light of a given frequency, Beer's law states that the radiation intensity decreases exponentially with depth (z is assumed positive downwards in the following):

$$\Phi'_s(\omega, z) = \Phi_{sn}(\omega)e^{-\eta_s(\omega, z)z} \tag{12.3.12}$$

where $\Phi'_s(\omega, z)$ is the solar radiation intensity at depth z for frequency ω, $\Phi_{sn}(\omega)$ is the net (after reflection) radiation intensity at the water surface for frequency ω, and $\eta_s(\omega, z)$ is the extinction, or absorption coefficient, as a function of ω and z. The total radiation intensity at any depth z is then the sum of $\Phi'_s(\omega, z)$ over all ω. However, in practice insufficient data are available to evaluate such an integral. A simpler approach has been found to give adequate results, in which the range of light frequencies is divided into one or more subranges, with surface radiation intensity and extinction coefficient values defined for each subrange, rather than for each individual frequency. The total is then the sum over all subranges,

$$\Phi'_s(z) = \sum_{i=1}^{n} (\Phi_i)_{sn} e^{-\eta_{si} z} \tag{12.3.13}$$

where n is the number of subranges.

Longer wave radiation (infrared range) tends to be absorbed strongly in water, relative to shorter wave radiation, and has correspondingly higher values for η_s. Thus when a small number of terms is used in the summation of Eq. (12.3.13), an adjustment must be made to account for this stronger absorption near the surface and to provide a better fit for the exponential model. This is usually done by defining a fraction, β_s, of the surface radiation as that portion of the radiation intensity with relatively high extinction coefficients, absorbing nearly completely within a shallow depth near the surface. In general, different values of β_s should be defined for each of the subranges used in Eq. (12.3.13), but a simple common approach is to use a single term in the summation. In that case, a single value is needed, as well as a single value for the extinction coefficient (η_s), to calculate the exponential decay of the remaining fraction of radiation that is not absorbed near the surface. The resulting equation for a one-term model ($n = 1$) is

$$\frac{\Phi'_s(z)}{\Phi_{so}} = (1 - \beta_s)e^{-\eta_s z} \tag{12.3.14}$$

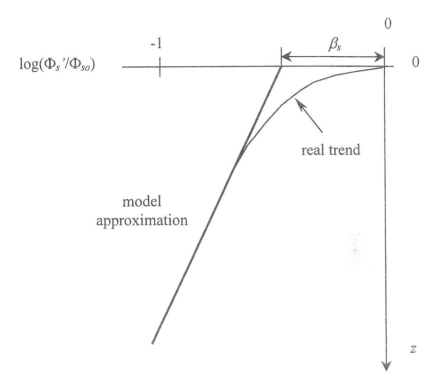

Figure 12.5 Variation of solar radiation intensity with depth, compared with model approximation (Eq. 12.3.14).

where Φ_{so} is now the total radiation intensity at the surface, including all frequencies. This model is sketched in Fig. 12.5, where it can be seen that predictions very close to the surface are not very good. This region is typically only 5 or 10 cm deep at most, and the problem is usually not significant when modeling large water bodies, where variations within this region close to the surface are not of major concern. Also note that the horizontal axis in this figure is a log scale, so the magnitude of the slope of the model curve is given by η_s. The offset at the top (at $z = 0$) is the fraction assumed to be absorbed near the surface (β_s).

In general, η_s is a function of z in Eq. (12.3.14) and may be affected by turbidity gradients in the water, though it is often considered a constant for a particular water body. It is related to the *secchi depth*, which is a simple measurement used to describe water clarity. The *secchi disk* is circular, with alternating black and white sections painted on top. The disk is lowered into the water until it is no longer visible, and the depth at which that occurs is the secchi depth. Larger secchi depths correspond with greater water clarity. Typical values for the parameters in Eq. (12.3.4) are $\beta_s \cong 0.5$ and $\eta_s \cong 0.5 \text{ m}^{-1}$ (higher values correspond with higher turbidity).

12.3.3 Surface Heat Exchange

In addition to solar radiation absorption, surface heat transfer represents a major source or sink of heat in determining the thermal structure of a water body. It is incorporated as a boundary condition in models formulated to solve for vertical temperature distribution, or as an internal source for vertically averaged models. The basic elements of surface heat transfer include solar radiation (Φ_s), atmospheric radiation (Φ_a), back radiation from the water surface (Φ_b), evaporation (Φ_e), and conduction (Φ_c). These elements are sketched in Fig. 12.6 and discussed in each of the subsections below.

Solar Radiation

Ideally, solar radiation intensity is measured directly at a particular site of interest. It consists of direct and diffuse radiation components, usually measured with a *pyrheliometer* or *pyrenometer*. Unfortunately, direct measurements are not very common. At many weather stations only the "percent possible sunshine" or some other measure of sunlight is reported. Solar radiation at the edge of the earth's atmosphere is, however, well known, and values of solar radiation intensity have been tabulated as a function of location and time of year (see, for example, Kreith and Kreider, 1979). These values must be modified according to pollution or water vapor content (clouds, fog, smog, etc.) of the air. Another measure of possible sunshine is the percent of cloud cover C_c, and if the clear-sky radiation intensity is known, the net

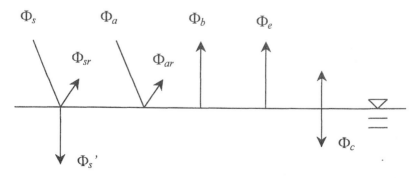

Figure 12.6 Components of surface radiative heat transfer.

solar radiation can be estimated from

$$\Phi_s = \Phi_{sc}(1-0.65C_c^2) \tag{12.3.15}$$

where Φ_{sc} is the clear-sky value and C_c is expressed in decimal units.

The reflected solar radiation, Φ_{sr}, depends on angle of incidence and on water surface roughness (a smooth surface will reflect a higher proportion). The reflected radiation ranges between about 3 and 10% of the incident radiation, and an average of 6% is reasonable. Thus the total net radiation passing into the water surface is estimated as

$$\Phi_{sn} = \Phi_{so} \cong 0.94\Phi_{sc}(1-0.65C^2) \tag{12.3.16}$$

where Φ_{sn} denotes net solar radiation, defined in the previous section for calculating radiation intensity in the water column.

Atmospheric Radiation

Atmospheric radiation consists of long wave black body–type radiation, emitted mostly from water vapor and ozone. It is calculated from the Stefan–Boltzmann law,

$$\Phi_a = \varepsilon\sigma T_{a*}^4 \tag{12.3.17}$$

where ε is emissivity (with values ranging between about 0.7 for clear sky and close to 1 for heavily overcast skies), σ is the Stefan–Boltzmann constant, and T_{a*} is air temperature, on an absolute scale. The *Swinbank formula* suggests that emissivity for a clear sky is proportional to T_{a*}^2 and is written as (incorporating the value for σ)

$$\Phi_{ac} = 1.2 \times 10^{-13}(T_a + 460)^6 \quad \text{(Btu/ft}^2\text{-day)} \tag{12.3.18a}$$

where Φ_{ac} is the clear-sky value for atmospheric radiation and T_a is air temperature in °F, measured 2 m above ground level. In SI units,

$$\Phi_{ac} = 5.35 \times 10^{-13}(T_a + 273)^6 \quad \text{(W/m}^2\text{)} \tag{12.3.18b}$$

where T_a here is in °C.

Similar to the formulation for solar radiation, a cloud cover correction is usually added,

$$\Phi_a = \Phi_{ac}(1 + KC^2) \tag{12.3.19}$$

where K has a value between 0.04 and 0.25, with an average around 0.17. Reflected atmospheric radiation is approximately 3% of incident, so the net atmospheric radiation is

$$\Phi_{an} = 0.97(1.2 \times 10^{-13})(T_a + 460)^6(1 + KC^2)$$
$$\text{(Btu/ft}^2\text{-day)} \tag{12.3.20a}$$

or

$$\Phi_{an} = 0.97(5.35 \times 10^{-13})(T_a + 273)^6 (1 + KC^2) \quad (W/m^2) \quad (12.3.20b)$$

where, as before, T_a is in °F in Eq. (12.3.20a) and in °C in Eq. (12.3.20b).

Back Radiation

Back radiation is longwave radiation from the water surface to the atmosphere. It represents a loss of heat from the water body. The intensity is calculated in a manner similar to atmospheric radiation, with an emissivity of about 0.97, so

$$\Phi_b = 0.97\sigma T_{s*}^4 \tag{12.3.21}$$

where T_{s*} is the absolute water surface temperature. Substituting for σ,

$$\Phi_b \cong 4 \times 10^{-8}T_{s*}^4 \qquad (Btu/ft^2 - day) \tag{12.3.22a}$$

or

$$\Phi_b \cong 1.4 \times 10^{-8}T_{s*}^4 \qquad (W/m^2) \tag{12.3.22b}$$

where T_{s*} is in °R in Eq. (12.3.22a) and in °K for Eq. (12.3.22b).

Evaporation

Evaporation is driven by several transport mechanisms, both molecular and turbulent. First, it should be noted that evaporation refers to the transfer of water molecules from the liquid to the gaseous phase, so *mass* transfer is the primary consideration. Heat transfer occurs due to the latent heat of vaporization associated with this phase change. Molecular diffusive-type transport represents a limiting rate, while *free* and *forced convection* usually play a much more significant role in determining evaporative fluxes. Free convection is due to buoyancy effects related to a heated water surface (relative to the air temperature), and forced convection is related to wind. Both of these transport mechanisms are turbulent in nature.

Evaporative mass transfer is proportional to the difference between saturated vapor pressure (as a function of temperature) and the actual atmospheric vapor pressure and may be expressed as

$$E = \rho_v f'(W)(e_s - e_a) \tag{12.3.23}$$

where E is mass flux, ρ_v is vapor density, e_s is saturated vapor pressure, e_a is atmospheric vapor pressure, and $f'(W)$ is a function of wind speed that must take into account the effects of both free and forced convection. The effect of wind or buoyant (free) convection is to introduce turbulence, which moves vapor away from the interface, thus maintaining a higher difference between

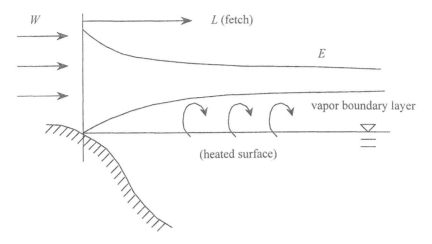

Figure 12.7 Illustration of forced and free convection effects on evaporation.

vapor pressures, with a corresponding increase in evaporation, as illustrated in Fig. 12.7 The evaporation rate driven by wind is highest near the shore, where the vapor boundary layer thickness is smallest, with a correspondingly higher vapor pressure gradient. In this case, the evaporation rate is a function of fetch, L. Free convection, on the other hand, does not depend on L, since it is a function of heating of the water surface, relative to the air. The resulting buoyant convective motions serve to transport vapor away from the interfacial region and increase the gradient driving E.

The heat flux associated with E is

$$\Phi_e = L_v E = f(W)(e_s - e_a) \tag{12.3.24}$$

where L_v is the latent heat of vaporization and $f(W)$ is a *wind speed function* that incorporates ρ_v and L_v into $f'(W)$. Many forms of the wind speed function have been proposed, usually as a simple constant or first- or second-order polynomial in W. An example of a first-order function of W is the *Lake Hefner formula*,

$$f(W) = 17W_2 \tag{12.3.25}$$

where W_2 is the wind speed in mph, measured at a height 2 m above the water surface. This formula was developed to predict evaporation over a medium-size lake and was found to give reasonable estimates for wind speeds between about 3 and 20 mph. This formula, however, does not account for free convection effects.

To develop an expression for free convection, consider first a heated flat plate, held at temperature T_s, which is assumed to be greater than the

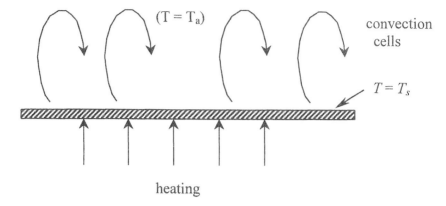

heating

Figure 12.8 Convective heat transfer above a heated flat plate.

air temperature T_a (Fig. 12.8). Due to heating from the plate, the air near the plate is warmed, becomes unstable, and begins to rise (see Chap. 13 for further discussion of convective instability). As the heated air rises, it is replaced by cooler air, thus maintaining a relatively high temperature gradient near the plate surface. The convective motions are turbulent, and the heat transfer rate is assumed to be proportional to the temperature difference $(T_s - T_a)$,

$$H = K_h(T_s - T_a)(\rho_v c_p) \tag{12.3.26}$$

where H is heat flux and K_h is a heat transfer coefficient (with units of velocity). From flat plate heat transfer theory, this coefficient is given by

$$K_h = 0.14 \left[\frac{g\beta k_T^2 (T_s - T_a)}{\nu} \right]^{1/3} \tag{12.3.27}$$

where β is the thermal expansion coefficient and ν is kinematic viscosity, as before.

These results may be used to estimate water vapor mass transfer, assuming that the vapor is transported at the same rate as heat (i.e., *Reynolds' analogy* applies, so that the heat transfer coefficient is the same as a mass transfer coefficient). Similar to the previous development for wind-induced evaporation (Eq. 12.3.13), the mass flux due to convection is written as a function of the difference between saturated vapor density (ρ_{vs}) and the actual vapor density in the air (ρ_v),

$$E = K_m(\rho_{vs} - \rho_v) \tag{12.3.28}$$

where K_m is a mass transfer coefficient equal to K_h. This equation can be written in terms of vapor pressure by introducing the perfect gas law,

$$e_a = \rho_v R_v T_* = \rho_v \frac{R}{M_v} T_* \tag{12.3.29}$$

where R_v is the specific gas constant, R is the universal gas constant, and M_v is the molecular weight of the vapor. Substituting Eqs. (12.3.27) and (12.3.29) into Eq. (12.3.28) and multiplying by latent heat of vaporization to convert mass transport to heat transport, we have

$$\Phi_e = \frac{0.14 L_v M_v}{R T_*} \left[\frac{g \beta k_T^2 (T_s - T_a)}{v} \right]^{1/3} (e_s - e_a)$$

$$= \Gamma (T_s - T_a)^{1/3} (e_s - e_a) \tag{12.3.30}$$

where Γ incorporates all the constants listed in the first half of the equation. Note that the vapor pressure difference is still important for evaporation, as it was before heating was considered, but now an additional factor is present, which is the buoyancy driving force given by the temperature difference.

One further modification is obtained by using *virtual temperatures*, T_v, rather than the actual temperatures. The virtual temperature is defined as the temperature of dry air having the same density as the actual moist air. This provides a more realistic driving force for the convection. The virtual temperature is related to actual temperature by

$$T_{v*} = T_* \left(1 - 0.378 \frac{e_a}{p} \right) \tag{12.3.31}$$

where p is pressure. Using the virtual temperature difference in Eq. (12.3.30),

$$\Phi_e = \Gamma (T_{sv} - T_{av})^{1/3} (e_s - e_a) \tag{12.3.22}$$

and $\Gamma (T_{sv} - T_{av})^{1/3}$ takes the place of $f(W)$ (compare with Eq. 12.3.14).

As noted previously, there are many forms for $f(W)$ suggested in the literature. Other than modeling natural water bodies, a specific engineering application for these calculations is in the design of cooling ponds for disposal of waste heat from power plants. There the effects of free convection are much more important than for a natural water body, and research has focused more on developing expressions for buoyancy effects. For any application, the choice of $f(W)$ depends on the relative importance of forced and free convection effects. Ryan and Harleman (1973) suggested a formula that incorporates both effects,

$$f(W) = 22.4 (T_{sv} - T_{av})^{1/3} + 14 W_2 \tag{12.3.33}$$

$f(W)$

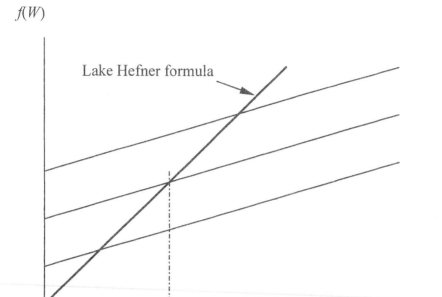

Figure 12.9 Qualitative comparison of Lake Hefner formula and Eq. (12.3.33) (for different temperature differences); the vertical dashed line distinguishes approximately between regions where free convection (low W) and forced convection (high W) dominate the evaporation flux.

where W_2 has the same meaning as in the Lake Hefner formula and temperatures are in °F. A qualitative comparison of formulas for $f(W)$ is shown in Fig. 12.9. For small W it is expected that free convection effects are relatively more important, while for large W, forced convection dominates. A formula such as the Lake Hefner formula performs better for higher W but may underpredict evaporation at low W. Formulas such as Eq. (12.3.33), which depend on temperature, are better able to simulate fluxes for low to medium W but may underpredict evaporation for large W.

Conduction

Conduction is driven by temperature differences between the water and the atmosphere. It can be either a source or a sink of heat for the water body, depending on the relative magnitudes of T_a and T_s. Conduction also is relatively small, compared with the turbulent convection effects associated with evaporation, or the radiative transfer rates from solar, atmospheric, and back radiation. The approach for estimating convective heat flux is similar to that for evaporation, except here the driving force is temperature difference rather than vapor pressure difference,

$$\Phi_c = f^*(W)(T_s - T_a) \tag{12.3.34}$$

where $f^*(W)$ is a wind function for conduction. Conduction is usually related to evaporation. By taking the ratio of conductive to evaporative heat flux,

$$\frac{\Phi_c}{\Phi_e} = \frac{f^*(W)(T_s - T_a)}{f(W)(e_s - e_a)} \tag{12.3.35}$$

Based on the Reynolds analogy (again), heat transfer and mass transfer are accomplished by the same turbulent motions. Thus it may be assumed that $f^*(W)$ is proportional to $f(W)$,

$$f^*(W) = c_b f(W) \tag{12.3.36}$$

where c_b is known as the *Bowen constant*, with a value of 0.255 mm Hg/°F. The product of c_b and the ratio of temperature difference to vapor pressure difference is known as the *Bowen ratio*,

$$R_b = c_b \frac{(T_s - T_a)}{(e_s - e_a)} \tag{12.3.37}$$

so that, substituting into Eq. (12.3.35),

$$\Phi_c = R_b \Phi_e = c_b f(W)(T_s - T_a) \tag{12.3.38}$$

Total Heat Flux, Linearized Approach

The total net heat flux is the sum of each of the above processes (as illustrated in Fig. 12.6),

$$\Phi_n = \Phi_{sn} + \Phi_{an} - \Phi_b - \Phi_e - \Phi_c \tag{12.3.39}$$

where Φ_n is positive when the water body is being heated. The first two terms on the right-hand side represent sources of heat and are functions of meteorology only. Back radiation and evaporation represent losses and are functions of both meteorology and surface water temperature. Conduction may be positive or negative, depending on temperatures. The sum in Eq. (12.3.39) represents the boundary condition for a thermal energy calculation for a water body.

In some studies it may be preferable to use a linearized approach, in which Φ_n is expressed as a linear function of temperature,

$$\Phi_n = -K_n(T_s - T_e) \tag{12.3.40}$$

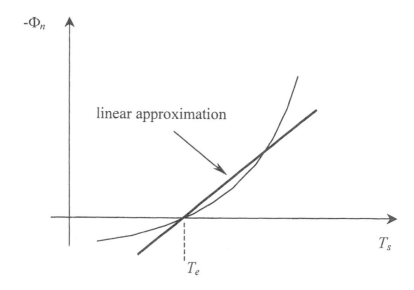

Figure 12.10   Linear approach for calculating Φ_n.

where K_n is a *net heat exchange coefficient* (in general, a function of $T_s - T_e$) and T_e is the *equilibrium temperature*, i.e., the water surface temperature at which $\Phi_n = 0$ (Fig. 12.10). In other words, for a given T_e, K_n is the rate of change of heat flux rate with temperature,

$$K_n = -\frac{d\Phi_n}{dT_s} \tag{12.3.41}$$

This is a useful concept for some analytical solutions but is not of great interest when using numerical modeling approaches in which the various heat flux terms can be calculated directly (otherwise, K_n must be calculated from Eq. (12.3.41), which requires evaluation of the various terms in Φ_n, anyway). On the other hand, Eq. (12.3.40) may be used for quick estimates even in numerical studies, as long as a reasonable value for K_n is known.

12.4   EXCHANGE OF GASES.

Historically, most of the work done on surface gas transfer has involved oxygen, and the present text also will focus on this parameter. Other gases could be treated similarly as in the following development. In fact, as discussed in this and the following sections, bulk mass transfer coefficients are usually assumed to be independent of properties of the specific gas under consideration — they are more a function of physical driving mechanisms such as wind speed. The emphasis on oxygen transport is primarily due to the role *dissolved oxygen* (DO) plays as a water quality parameter. DO levels are critical in determining the general health of a system, the diversity of species, and the number of organisms that can be supported. For example, most fish require a concentration of at least 4 to 5 ppm (or mg/L) DO for survival. A major factor in modeling DO in surface waters is in defining the reaeration rate, or the rate at which oxygen is transported across the air/water interface, since the atmosphere is the primary source of oxygen. A large number of DO models have been developed, with perhaps the best known due to Streeter and Phelps (1925), as described briefly later in this chapter.

In this section it will be convenient to define gas concentrations in terms of mass densities (i.e., mass per unit volume). The oxygen mass flux across the air/water interface is assumed to be proportional to the *DO deficit* in the water body, which is defined as the difference between saturated DO concentration, as a function of temperature, and the actual concentration,

$$J = K_L(C_s - C) \tag{12.4.1}$$

where J is oxygen mass flux, C is DO concentration, C_s is saturated DO concentration, and K_L is a bulk mass transfer coefficient. The value for C_s is temperature dependent, decreasing from 16.4 ppm at $T = 10°C$ to 7.8 ppm at $T = 30°C$. It represents the concentration in water that is in equilibrium with the partial pressure of the gas in the atmosphere. This equilibrium relationship may be expressed by the *Henry's law* constant, defined either in dimensional form,

$$K_H = RT_* \frac{C_{sg}}{C_s} = \frac{P_g}{C_s} \tag{12.4.2}$$

or in dimensionless form,

$$K_H' = \frac{K_H}{RT_*} = \frac{C_{sg}}{C_s} \tag{12.4.3}$$

where C_{sg} is the equilibrium concentration in the gas phase and p_g is the partial pressure of the gas. K_H is generally a function of temperature, meaning that different equilibrium relationships exist between the liquid and gaseous phases for different temperatures. This may have significant consequences for global transport of some materials (Sec. 12.5.1). Thus it may be important to model temperature whenever DO or any other gas is being modeled.

For a well-mixed water body of average depth H, the rate of change of DO due to atmospheric flux is

$$\frac{dC}{dt} = \frac{JA}{\forall} = K_L \frac{A}{\forall}(C_s - C) \tag{12.4.4}$$

where A is water surface area and $\forall$ is volume of the water body. Since $\forall = AH$, this last result is often written in terms of a *bulk reaeration coefficient*, K_2, where

$$K_2 = K_L \frac{A}{\forall} = \frac{K_L}{H} \qquad (12.4.5)$$

A large number of models have been developed to estimate K_L or K_2, some purely empirical and some based on conceptual models. However, nearly all are calibrated and reported in terms of mean hydraulic quantities such as depth, velocity, or friction slope. Many of these models are developed for rivers and are based on the assumption of complete vertical mixing over the depth H. This approach is useful for many river applications but is not appropriate for lakes or stratified rivers or estuaries. In these latter cases it is necessary to calculate flux directly (as in Eq. 12.4.1), as a boundary condition for a DO model that incorporates vertical gradients.

The most common conceptual model for gas mass transfer across the air/water interface is the two-film theory developed by Lewis and Whitman in the earlier part of this century. In this approach it is assumed that there are two laminar films, one on either side of the interface, sandwiched between bulk liquid and gas layers, which are assumed to be turbulent (Fig. 12.11).

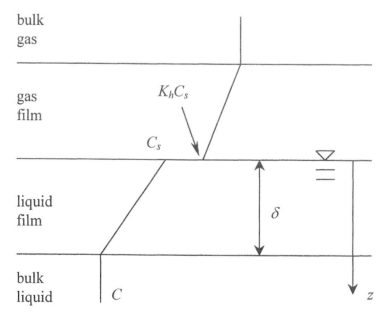

Figure 12.11 Definition sketch for two-film model.

Since the films are laminar, transport is governed by *Fickean diffusion* acting on the gradients in each film layer (see Sec. 10.2). Across the liquid layer, the flux (Eq. 12.4.1) must be equal to the diffusive flux,

$$J = D_m \frac{dC}{dz} \cong D_m \left(\frac{C_s - C}{\delta_1} \right) \tag{12.4.6}$$

where the gradient has been approximated as being constant over the film thickness δ_1, and D_m is the molecular diffusivity of the gas in water. For steady transport, this must be equal to the flux across the gas film layer,

$$J = K_g(C_g - C_{sg}) \cong D_{mg} \left(\frac{C_g - C_{sg}}{\delta_g} \right) \tag{12.4.7}$$

where K_g is a bulk mass transfer coefficient for the gas layer, D_{mg} is the molecular diffusivity for the gas phase, C_g is the gas phase concentration, and δ_g is the gas film thickness (Fig. 12.11).

Using Eqs. (12.4.6) and (12.4.7), and the definition of Henry's constant [here it is convenient to use the dimensionless form, Eq. (12.4.3)], C_s and C_{sg} may be eliminated to obtain

$$J = (C_g - K'_H C) \left(\frac{1}{K_g} + \frac{K'_H}{K_L} \right)^{-1}$$

$$= \left(\frac{C_g}{K'_H} - C \right) \left(\frac{1}{K_L} + \frac{1}{K'_H K_g} \right)^{-1} \tag{12.4.8}$$

or,

$$J = \kappa_g(C_g - K'_H C) = \kappa_L \left(\frac{C_g}{K'_H} - C \right) \tag{12.4.9}$$

where

$$\frac{1}{\kappa_g} = \frac{1}{K_g} + \frac{K'_H}{K_L} \qquad \frac{1}{\kappa_L} = \frac{1}{K_L} + \frac{1}{K'_H K_g} \tag{12.4.10}$$

The values defined in this last equation represent resistances to mass transport across the gas and liquid film layers, respectively. The total resistance is expressed in terms of either one of these quantities and depends on the exchange constants of the individual phases and on K'_H. Table 12.1 summarizes mass transfer coefficients and Henry's law constants for a number of common gases crossing the air/sea interface. Except for SO_2, each of these gases is assumed to be nonreactive, and the relatively large value for K_L for this gas is related to possible reactions for SO_2 that must be considered in the transport process.

Table 12.1 Mass Transfer Coefficients and
Henry's Law Constants for Some Common Gases
for the Air/Sea Interface

Gas	K_g (cm/hr)	K_L (cm/h)	K_H
SO_2	1,600	34,420	0.038
N_2O	1,900	20	1.6
CO	2,400	20	50
CH_4	3,180	20	42
CCl_4	1,030	10.7	1.08
CCl_3F	1,085	11.3	5

Source: Adapted from Liss and Slater (1974).

Because oxygen is only slightly soluble in water, molecular diffusion in the gas film is greater than in the liquid film. Also, the film thickness is likely to be smaller for the gas layer than for the water layer, creating a larger gradient. Therefore diffusive transport through the liquid film is normally considered to represent the limiting process for oxygen mass transport to the bulk water layer, and much of the focus on oxygen transport has been on an evaluation of K_L or K_2. Upon comparing Eqs. (12.4.6) and (12.4.1), it is seen that $K_L = D_m/\delta_l$. Although the film approach for describing oxygen transport across the air/water interface is effectively reduced to a one-film model, the difficulty remains to determine δ_l, though several possible approaches are presented below.

Other conceptual models for surface gas transport include the penetration theory of Higbie (1935), the surface renewal theory of Danckwerts (1951), and the film penetration theory of Dobbins (1956), which combines certain aspects of penetration and surface renewal. Just as with the two-film theory, each of these approaches has its own limitations. For example, the renewal rate must be determined for application of surface renewal theory, and exposure time must be found for penetration theory. A detailed comparison of a number of models with a very large data set was reported by Wilson and Macleod (1974), who demonstrated that there was considerable variability in the estimates for K_2 or K_L. They concluded that the best overall calculation was provided by an empirical relationship developed by Parkhurst and Pomeroy (1972),

$$K_L = 2(1 + 0.17 \, \text{Fr}^2)(SU)^{0.375} \tag{12.4.11}$$

where K_L is in units of (ft/h), S is energy slope, U is mean velocity (ft/s), and Fr is the Froude number. Another formula that performed well was that of O'Connor and Dobbins (1956), which related the surface renewal rate to basic hydraulic properties of the stream:

$$K_L = \frac{D_m^{1/2}(Sg)^{1/4}}{\kappa^{1/2}H^{1/4}} \tag{12.4.12}$$

where κ is the von Karman turbulence constant. The problem with both these expressions is their dependence on mean hydraulic properties, as noted above.

A more conceptually based approach to surface gas transfer can be developed, based on the film layer approach, by specifying the film thickness δ_1. Since turbulence controls transport when $z > \delta_1$, it is necessary to find the location at which turbulence starts to become important, in other words, to define the lower boundary of this film layer. Two possible approaches may be considered, one based on a comparison of molecular and turbulent diffusivities and the other based on an estimate for the smallest turbulence length scale expected in the bulk layer.

First, the vertical profile of turbulent diffusivity may be estimated using the assumption of linear variation of shear stress in open channel flow,

$$\tau = \rho v_t \frac{du}{dz} = \tau_0 \frac{z}{H} \tag{12.4.13}$$

where v_t is the turbulent eddy viscosity and τ_0 is the bottom shear stress. If a logarithmic velocity profile also is assumed (other profiles could be assumed, leading to modifications in the results presented here),

$$\frac{u - u_0}{u_*} = \frac{1}{\kappa} \ln\left(\frac{H - z}{z_0}\right) \tag{12.4.14}$$

where u_0 is the velocity when $z = H - z_0$ (i.e., at the origin of the profile; u_0 is often taken as 0, as was implicitly done in Eq. 12.2.3), and u_* is the friction velocity. Differentiating u with respect to z, and substituting into Eq. (12.4.13), an expression for v_t is obtained,

$$v_t = \kappa z u_* \left(1 - \frac{z}{H}\right) \tag{12.4.15}$$

This value must be compared with the molecular viscosity v and, by setting $v_t = v$, we obtain

$$\delta_1 \cong z = C_1 \frac{v}{\kappa u_*} \tag{12.4.16}$$

where C_1 is a proportionality constant and (z/H) has been neglected, relative to 1. This result is consistent with turbulent boundary layer theory, which defines (v/u_*) as the characteristic length scale for determining laminar sublayer thickness. Gulliver and Stefan (1984) reported on a number of flume experiments that suggested $\delta_1 \cong 10(v/u_*)$, implying a value of $C_1 \cong 4$.

The second approach for estimating δ_1 is based on the assumption that δ_1 is related to (proportional to) the smallest eddies in the flow, i.e., the *Kolmogorov microscale*. As developed in Chap. 5, the microlength scale is given by

$$\eta = \frac{v_{3/4}}{\varepsilon^{1/4}} \qquad (12.4.17)$$

where $\varepsilon = u'^3/l'$ is the kinematic turbulent kinetic energy dissipation rate and u' and l' are characteristic velocity and length scales, respectively, for the turbulent eddies. The problem with this approach is that the characteristic scales must be related to more easily measured quantities, such as mean flow values. For example, if it is assumed that $u'/U = l'/H = C'$ (C' has typical values between 0.1 and 0.5), and $\delta = C''\eta$, where C'' is an order 1 coefficient, then

$$\delta_1 \cong C_2 \left(\frac{v^3 H}{U^3} \right)^{1/4}. \qquad (12.4.18)$$

where $C_2 = C''/C''^{1/2}$.

Substitution of either Eq. (12.4.16) or Eq. (12.4.18) into Eq. (12.4.5), along with the relation that $K_L = D_m/\delta_1$, gives alternative relationships for K_2 or K_L, which have been shown to provide results that are very close to the other models noted above when compared under similar conditions for U and H. Application of these procedures to lakes is, however, difficult, due to the underlying assumptions regarding velocity profiles and distribution of turbulent diffusivities. Most formulas for lake reaerations are empirical, based on wind speed rather than mixing in the water (see also Sec. 12.5).

12.4.1 Dissolved Oxygen in Open Channel Flow

Open channel flow water quality problems are often solved using a one- or two-dimensional framework. In the case of DO modeling in rivers, the classic analysis involves variations in the longitudinal direction only, assuming well-mixed conditions at any cross section. The general transport equation is the one-dimensional advection–dispersion model, which is obtained from Eq. (10.4.10). After multiplying by ρ to write the equation in terms of mass concentration, we obtain

$$\frac{\partial \overline{C}}{\partial t} + U \frac{\partial \overline{C}}{\partial x} = \frac{\partial}{\partial x} \left(E_L \frac{\partial \overline{C}}{\partial x} \right) + R + M \qquad (12.4.19)$$

where $\overline{C}$ is the cross-sectional average concentration, U is the cross-sectional average velocity $= Q/A$, Q is the flow rate, A is cross-sectional area, E_L is the longitudinal dispersion coefficient, which in general may be a function of x and possibly also of time t, R represents the gain or loss of concentration by internal sources and sinks, due to chemical and biological reactions, and M represents the gain or loss of concentration by mass transfer at boundaries of the cross section. Note that this last term, although representing processes occurring at boundaries, appears as a source/sink term in this one-dimensional approach, since all processes over the cross section are averaged. Examples of processes that might be included in this last term, for a DO model, are *surface reaeration* and *benthic* or *sediment oxygen demand*. The primary term included in R is normally biochemical oxygen demand (BOD), which represents the potential for oxygen depletion by biological and chemical processes (e.g., decay of organic wastes). Because BOD may change as DO is used up, any time-dependent model for DO generally requires solution of a coupled set of equations, one for BOD and one for DO.

The various source and sink terms in Eq. (12.4.19) are usually evaluated using first-order kinetics. For example, if M represents reaeration, we may write (see Eq. 12.4.4)

$$M = K_2(\overline{C}_s - \overline{C}) \tag{12.4.20}$$

where $\overline{C}_s$ is the saturated concentration, based on the cross-sectional average temperature. Also, the decay of BOD due to oxygen uptake is

$$\frac{\partial \overline{C}_{BOD}}{\partial t} = -K_1 \overline{C}_{BOD} \tag{12.4.21}$$

where $\overline{C}_{BOD}$ is the cross-sectional average BOD concentration and K_1 is the first-order rate constant, which also applies to the rate of depletion of DO. The solution to Eq. (12.4.21) is

$$\overline{C}_{BOD} = \overline{C}_{BOD_0} e^{-K_1 t} \tag{12.4.22}$$

where $(\overline{C}_{BOD})_0$ is the initial value. This is the same as the steady-state solution for $\overline{C}_{BOD}$, if dispersion is neglected and time t is replaced with travel time (x/U). In other words, for a steady-state problem in which dispersion is negligible, the governing equation is

$$U\frac{\partial \overline{C}_{BOD}}{\partial x} = -K_1 \overline{C}_{BOD} \tag{12.4.23}$$

which has an exponential solution,

$$\overline{C}_{BOD} = \overline{C}_{BOD_0} \exp\left(-\frac{K_1 x}{U}\right) \tag{12.4.24}$$

where $(\overline{C}_{BOD})_0$ here is a boundary value.

A well-known solution for the one-dimensional DO problem in open channel flow was developed by Streeter and Phelps in 1925. They considered steady-state and *plug-flow* conditions (i.e., dispersive transport is neglected — see the following discussion for criteria to decide the appropriateness of these assumptions). A source stream enters a river at $x = 0$, as shown in Fig. 12.12. The upstream flow is Q_r, BOD is $(\overline{C}_{BOD})_r$, and the oxygen deficit $(D = \overline{C}_s - \overline{C})$ is D_r. The corresponding source stream values are Q_i, $(\overline{C}_{BOD})_i$, and D_i, respectively. The source stream and river waters are assumed to mix instantaneously at $x = 0$, resulting in initial values (at $x = 0$),

$$Q_0 = Q_r + Q_i$$

$$\overline{C}_{BOD_0} = \frac{Q_r\overline{C}_{BOD_r} + Q_i\overline{C}_{BOD_i}}{Q_0}$$

$$D_0 = \frac{Q_r D_r + Q_i D_i}{Q_0}$$

The solution for BOD is Eq. (12.4.24). For DO, the governing equation for these conditions is

$$U\frac{\partial \overline{C}}{\partial x} = -K_1\overline{C}_{BOD} + K_2 D \tag{12.4.25}$$

Now, assuming $\overline{C}_s$ is constant in x, D is substituted for $\overline{C}$ to obtain

$$-U\frac{\partial D}{\partial x} = -K_1\overline{C}_{BOD} + K_2 D \tag{12.4.26}$$

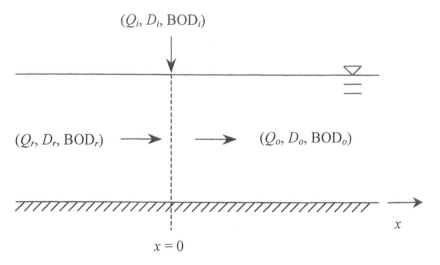

Figure 12.12 Source flow into a river and initial conditions for DO calculation.

Upon substituting Eq. (12.4.24) for $\overline{C}_{BOD}$ and applying the above boundary conditions, the solution is

$$D = \frac{K_1 \overline{C}_{BOD_0}}{K_2 - K_1} \left[\exp\left(-\frac{K_1 x}{U}\right) - \exp\left(-\frac{K_2 x}{U}\right) \right]$$
$$+ D_0 \exp\left(-\frac{K_2 x}{U}\right) \qquad (12.4.27)$$

which is known as the *Streeter–Phelps equation*.

This solution produces the well-known *dissolved oxygen sag curve*, sketched schematically in Fig. 12.13. The critical region of this curve is where D is largest (corresponding to lowest DO values). The location of the highest D is found by differentiating Eq. (12.4.27) with respect to x and setting the derivative equal to 0. The result is (for $D_0 = 0$)

$$x_{crit} = U \frac{\ln(K_2/K_1)}{K_2 - K_1} \qquad (12.4.28)$$

and the corresponding maximum D is

$$D_{max} = \overline{C}_{BOD_0} \frac{K_1}{K_2} \exp\left[-\frac{K_1}{K_2 - K_1} \ln\left(\frac{K_2}{K_1}\right) \right] \qquad (12.4.29)$$

This last result shows the dependence of D_{max} on the relative magnitudes of K_1 and K_2. Figure 12.14 shows the relationship between D_{max} and (K_1/K_2), where L'Hopital's rule has been used for the case $K_1 = K_2$.

One further note on this solution concerns the possibility that calculated values for D might exceed $\overline{C}_s$, which is physically impossible, since it would

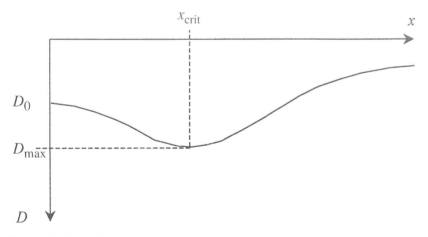

Figure 12.13 DO sag curve (in terms of DO deficit, D).

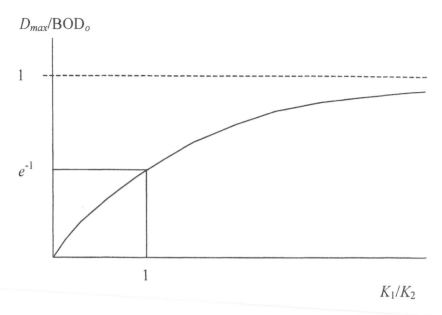

D_{max}/BOD_o

Figure 12.14   Variation of maximum DO deficit with (K_1/K_2).

imply $\overline{C} < 0$. Whenever applying this solution, care should be taken to limit the maximum value of D to $\overline{C}_s$. Typical values for K_1 and K_2 range between about 0.1 and 1 day^{-1}, and with typical river velocities in the range of 0.25 to 1 m/s, values for x_{crit} (location of minimum DO) are found in the range of several tens of kilometers.

Time Scale Analysis

The assumptions of a steady state and nondispersive plug flow can be checked using a time scale analysis. This is done using a nondimensional form of the governing equation (12.4.13). Also, for simplicity, a single first-order reaction rate constant K is used to represent the net effects of all sources and sinks as discussed above, and time scales are defined for the various processes of interest. This approach in general follows the scaling analysis procedures presented in Sec. 2–9.

First, define $T_K = 1/K$ = reaction time scale, where K is the net rate constant; $T_E = L^2/E_L$ = dispersive time scale, related to the time required for mass to be dispersed over a distance L, where L is a characteristic length; $T_a = L/U$ = advection time scale; and T_d = discharge time scale (see Fig. 12.15; T_d approaches infinity for a steady discharge). Defining nondimensional parameters, $t' = t/T_d$ and $x' = x/L$, and assuming constant E_L, the governing

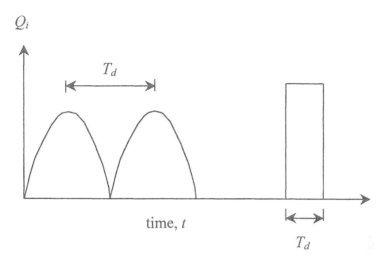

Figure 12.15 Illustration of two possible definitions for discharge time scale T_d, a periodic discharge, and a single constant discharge of limited duration.

equation is found by substitution into Eq. (12.4.19),

$$\frac{1}{T_d}\frac{\partial \overline{C}}{\partial t'} + \frac{U}{L}\frac{\partial \overline{C}}{\partial x'} = \frac{E_L}{L^2}\frac{\partial^2 \overline{C}}{\partial x'^2} + K\overline{C} \qquad (12.4.30)$$

Using the definitions for time scales, Eq. (12.4.30) is rewritten as

$$\frac{1}{T_d}\frac{\partial \overline{C}}{\partial t'} + \frac{1}{T_a}\frac{\partial \overline{C}}{\partial x'} = \frac{1}{T_E}\frac{\partial^2 \overline{C}}{\partial x'^2} + \frac{1}{T_K}\overline{C} \qquad (12.4.31)$$

Now, if T_d is very large, the first term on the left-hand side may be neglected and the problem may be treated as steady state. The relative importance of advection to dispersion is evaluated by comparing the ratio of T_E to T_a, which gives $T_E/T_a = LU/E_L$, which has the form of a Peclet number, Pe $= T_E/T_a$. In other words, a large T_E, relative to T_a, indicates that advection occurs much more quickly than dispersion, so that dispersion is negligible at least over time T_a. For a nonconservative substance, the length scale is often estimated as $L \cong UT_K$, so Pe $= U^2/KE_L$. For typical riverine conditions, Pe is of order 10^2, indicating that the assumption of plug flow is often justified. The usual value used to distinguish plug flow is Pe > 10. Dispersive flow (i.e., neglecting the advective term in the governing equation) is assumed when Pe < 1, and an advective–dispersive regime is defined for intermediate values.

For a conservative material, $K \cong 0$ and $T_K \to \infty$. The scaled equation may then be written as

$$\left(\frac{T_a}{T_d}\right) \frac{\partial \overline{C}}{\partial t'} + (1) \frac{\partial \overline{C}}{\partial x'} = \left(\frac{T_a}{T_E}\right) \frac{\partial^2 \overline{C}}{\partial x'^2} \tag{12.4.32}$$

which facilitates evaluation of the relative importance of the remaining terms. In this case L must be specified by other considerations of the problem (since $K = 0$). The relative importance of the unsteady and dispersive terms is $T_E/T_d = L^2/E_L T_d$, which is determined for site-specific conditions. When $T_E/T_d \ll 1$, the problem may be approximated as being steady, while for $T_E/T_d \gg 1$, dispersion may be neglected. Either of these approximations introduces considerable simplification in solving the governing equation, and use of scaling analyses such as this is very helpful in justifying simplified approaches.

12.5 MEASUREMENT OF GAS MASS TRANSFER COEFFICIENTS

There are several methods available for measuring the air/water exchange of gases. Interest is usually directed at finding appropriate values for the mass transfer coefficients (i.e., K_L or K_g). These methods include laboratory experiments and field studies. Although it is possible to maintain more controlled conditions in the laboratory, it is impossible realistically to reproduce field conditions necessary for accurate evaluation of transport properties. The problem is related mostly to the limited fetch available for wind and development of the boundary layers at the air/water interface. Nonetheless, as noted below, laboratory experiments can be useful for reasonable estimates of the transfer coefficients.

In the field, a method of measurement that has seen increased use in the past decade or so, particularly for lake applications, involves the use of *sulphur hexafluoride* (SF_6), a weakly soluble inert (nonreactive) gas that can be detected at very low concentrations in the environment. In addition, background atmospheric concentrations are extremely low, so that experimental concentrations are easily distinguishable. These conditions make it possible to measure purely physical transport properties. The measured transport coefficients presumably account for variations in film thickness, water surface roughness, wave breaking and spray formation, water and atmospheric turbulence, and any other processes that might affect gas transfer. The main assumption is that these same physical processes apply to other gases, with minor modifications needed for the bulk mass transfer coefficients. These minor modifications are normally made on the basis of the gas *Schmidt number* (see below), which accounts for differences in both temperature and molecular diffusivity.

The basic approach is rather straightforward. A known quantity of SF_6 is distributed evenly in the lake (either throughout the epilimnion or throughout the entire depth if the lake is shallow and well mixed), and aqueous and atmospheric concentration measurements are made over time. The analysis is based on an equation similar to Eq. (12.4.1),

$$J_{SF_6} = k_{SF_6}(C_w - C_0) \tag{12.5.1}$$

where J_{SF6} is the flux of SF_6, k_{SF6} is its bulk mass transfer coefficient, C_w is the concentration in the water immediately below the air/water interface, and C_0 is the gas concentration in water in equilibrium with the atmospheric concentration (i.e., this is equivalent to C_s defined for DO calculations). The rate of change of mass of SF_6 in the lake, based on average concentration measurements, is

$$\frac{dM}{dt} = J_{SF_6}A \tag{12.5.2}$$

where A is the lake surface area. For SF_6 distributed over a depth H (either the mean total lake depth or the depth of the epilimnion), Eq. (12.5.2) is written equivalently in terms of concentration as

$$\frac{dC_w}{dt} = \frac{J_{SF_6}}{H} \tag{12.5.3}$$

If it is assumed that C_0 is much smaller than C_w [according to Gerrard, 1980, C_0 is of order 10^{-16} mol/kg, which is several orders of magnitude smaller than concentrations that might be used in a typical test (Upstill-Goddard et al., 1990)], and that J_{SF6} is approximately constant over a time interval Δt, then Eq. (12.5.3) can be rearranged and integrated to obtain

$$k_{SF_6} = \frac{H}{\Delta t} \ln \frac{C_i}{C_f} \tag{12.5.4}$$

where C_i and C_f are the initial and final concentrations, respectively, over time Δt.

Values of k_{SF6} are related to transfer coefficients for other gases by taking into account differences in molecular diffusivities (recall the earlier discussion on the relationship between transfer coefficient and diffusivity) and temperatures. This is usually done using a function of respective Schmidt numbers (Sc), which accounts for both temperature and diffusivity. The Schmidt number is the ratio of kinematic viscosity of a particular gas to its diffusivity, both of which are functions of temperature. The introduction of viscosity is appropriate because of its relationship with film layer thickness, as was assumed

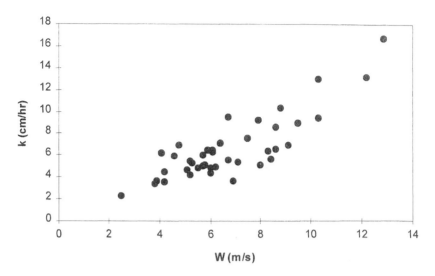

Figure 12.16 Mass transfer coefficients (k_{SF6}), adjusted to 20°C value, as a function of wind speed at 10 m, from measurements at two well-mixed lakes reported by Upstill-Goddard et al. (1990).

to derive Eq. (12.4.16), for example. Following Upstill-Goddard et al. (1990), the appropriate function is

$$\frac{k_{SF_6}}{k_g} = \left(\frac{Sc_{SF_6}}{Sc_g}\right)^{-1/2} \tag{12.5.5}$$

where the subscript g indicates any other gas of interest. Their results for transfer coefficient, normalized to a 20°C value, are plotted in Fig. 12.16 as a function of wind speed, W_{10}, measured at 10 m above the water surface. Although there is some scatter in these data, there is a clear dependence on wind speed, with higher values corresponding to higher W_{10}.

12.5.1 Exchange of Volatile Organic Chemicals

Vapor exchange of organic compounds across the air/water interface is an important process in understanding their fate and transport in the environment. A particular problem is the cyclical transport of these compounds between air and gas phases, due in large part to changes in the value of K_H as a function of temperature and other atmospheric conditions, which contributes to spreading far beyond the original source region of these materials. For example, in warmer climates K_H is generally larger, indicating greater partitioning to the gas phase (see Eq. 12.4.2 or 12.4.3). However, in cooler climates

K_H is smaller, with a corresponding tendency for the material to enter the liquid phase. Thus chemicals produced in countries in warmer climates enter the gas phase and are transported widely through the atmosphere but reenter the water phase in cooler regions. This is a problem of special interest for the Laurentian Great Lakes in North America, which have a very large surface area for gas transport. Organic chemicals no longer produced or even used in this region may be transported to the lakes through this mechanism (a process known as *cold deposition*).

There is particular concern for the transport of various organic chemicals, owing to detrimental environmental and health effects. Polychlorinated biphenyls (PCBs), for instance, degrade only slowly in the environment and tend to bioaccumulate, interfering with bird and mammal reproduction and causing developmental disorders in humans. Some PCBs, as well as other organic compounds, also are suspected carcinogens. Many of these compounds have relatively high values of K_H, and volatilization may be an important environmental transport pathway, as noted above.

The usual approach for calculating volatilization or mass transfer rates is the two-layer model discussed in Sec. 12.4. Equation (12.4.9) is used to calculate the fluxes, with the overall mass transfer coefficients defined in Eq. (12.4.10). The critical variables in this calculation are Henry's law constant K_H and the transport rates for each of the two films, K_L and K_g. Values for K_H can be obtained from solubility and vapor-pressure data or directly from gas-stripping experiments, while K_g and K_L must be measured in laboratory or field experiments. Field experiments based on the use of SF_6 injections were described in the previous section. Laboratory experiments are generally conducted using large wind-wave tanks, with a fetch length of at least several meters so that reasonable boundary layers can be developed, leading to results more representative of field conditions. Experiments conducted by Mackay and Yeun (1983), for instance, utilized a 6 m tank, and results confirmed the validity of the two-layer model for high K_H compounds. However, they also found that experimentally determined values were likely to be higher than field values, based on different surface stress values for a given wind speed (also related to the limited fetch in a laboratory experiment, compared with the field). For environmental conditions, they suggested the following correlations for the mass transfer coefficients:

$$K_L = 34.1 \times 10^{-6}(6.1 + 0.63W_{10})^{.5}W_{10}Sc_L^{-.5} \text{ (m/s)} \tag{12.5.6}$$

and

$$K_g = 46.2 \times 10^{-5}(6.1 + 0.63W_{10})^{.5}W_{10}Sc_g^{-.67} \text{ (m/s)} \tag{12.5.7}$$

where W_{10} is wind speed in m/s, measured at 10 m height, and Sc_L and Sc_g are the Schmidt numbers for the compound in water and air, respectively.

These correlations are averaged values based on data from eleven different organic compounds, not including PCBs. The flux of PCBs also is calculated on the basis of the two-film model. For example, Hornbuckle et al. (1994) looked at the flux of PCBs across the air/water interface in Lake Superior, particularly as a function of seasonal-average conditions. Values for K_L were found to be approximately 0.05 m/day for winter conditions and 0.2 m/day for summer, while K_g values were around 0.01–0.05 m/day and 0.04–0.18 m/day for winter and summer conditions, respectively. The clear difference between winter and summer values is predominantly temperature related and is consistent with the earlier discussion of the dependence of K_H on temperature.

PROBLEMS

Solved Problem

Problem 12.1   Formally derive Eq. (12.4.19) by writing a mass balance for dissolved oxygen concentration, referring to the control volume shown in Fig. 12.17.

Solution

First, let the mean depth be H and note that the cross-hatched area at the top of the control section represents the area over which reaeration takes place. The concept of oxygen mass balance, applied to the control volume shown, may be stated as: (1) [rate of change of mass in the control volume] is equal to (2) [net rate at which mass is transported across the boundaries of the volume] plus

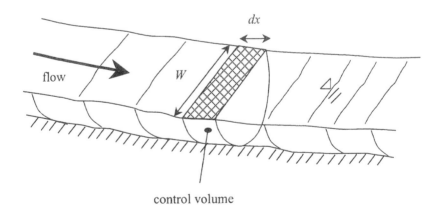

control volume

Figure 12.17   Control volume definition, Problem 12.1.

(3) [rate at which mass grows or decays by biochemical processes]. The mass in the control volume is equal to the product of concentration and volume, so the first term is simply

$$\frac{\partial(\forall \overline{C})}{\partial t} \tag{1}$$

where $\overline{C}$ is the cross-sectional average concentration and $\forall$ is the volume. Transfer of oxygen mass across the boundaries occurs by advection and dispersion (on the left and right faces of the control volume) and by transport across the air/water interface. Dispersion is used instead of diffusion, since the problem is stated in one-dimensional terms and primary interest is in the cross-sectional average concentration. The advective and dispersive transport terms on the left side are

$$\left(U\overline{C} - E_{L}\frac{\partial \overline{C}}{\partial x} \right) A \tag{2}$$

where A is the cross-sectional area on the left side. The advective and dispersive terms on the right side are written using linear Taylor series expansions,

$$\left[\left(U\overline{C} + \frac{\partial(U\overline{C})}{\partial x}dx \right) - E_{L}\frac{\partial \overline{C}}{\partial x} - \frac{\partial}{\partial x}\left(E_{L}\frac{\partial \overline{C}}{\partial x} \right)dx \right]\left(A + \frac{\partial A}{\partial x}dx \right) \tag{3}$$

Following the discussion in Sec. 12.4.1, the main boundary transfer other than advection and dispersion is reaeration. The flux (mass transfer per unit area per unit time) is given as J in Eq. (12.4.1). The rate of change of mass is then the flux multiplied by the surface area,

$$J W\, dx \tag{4}$$

The final term is the internal source/sink term, and this is written simply as $R' =$ rate of mass produced or decayed by biochemical processes per unit time.

We now write the mass balance statement using each of (1) through (4), along with R', noting that (2) is added while (3) is subtracted. This results in

$$\frac{\partial(\forall \overline{C})}{\partial t} = A\,dx\left[-\frac{\partial(U\overline{C})}{\partial x} + \frac{\partial}{\partial x}\left(E_{L}\frac{\partial \overline{C}}{\partial x} \right) \right]$$
$$+ \frac{\partial A}{\partial x}dx^2\left[-\frac{\partial(U\overline{C})}{\partial x} + \frac{\partial}{\partial x}\left(E_{L}\frac{\partial \overline{C}}{\partial x} \right) \right] + J W\,dx + R'$$

We then neglect the nonlinear terms in dx (dx is assumed to be small) and note that $\forall = A\,dx$ is constant and can be divided on both sides of the equation. Finally, let $M = J W\,dx/\forall$ ($= J W/H$, where H is the mean depth) be the rate of change of mass per unit volume per unit time due to reaeration, and $R = R'/\forall$ be the rate of production or decay of mass per unit volume and time. Making these substitutions and rearranging then leads to Eq. (12.4.19).

Unsolved Problems

Problem 12.2 In order to evaluate the possible mixing effect of wind blowing over water it is first necessary to calculate the surface shear stress. Using the data below, with h = distance above water surface and W = wind speed, estimate the surface shear stress. State any assumptions made.

h (m)	0.5	1.0	3.0	5.0	7.0	10.0
W (m/s)	3.6	5.2	7.7	8.8	9.6	10.4

Problem 12.3 In a lake with an average fetch length of 10 km, reform Eq. (12.2.4) in terms of a range of wind speeds, instead of a range of Re. Use $v = 10^{-2}$ cm^2/s. According to this equation, what is the minimum height at which wind measurements should be taken in this lake for a wind speed of 3 m/s?

Problem 12.4
   (a) If the wind speed measured at a height of 10 m above a lake is 7.5 m/s, what is the expected shear stress at the water surface? Use $\alpha_c = 0.013$ and $\rho_a = 1.9$ kg/m^3.
   (b) What are the friction velocities in air and water? Assume $\rho_w = 1000$ kg/m^3.
   (c) Assuming a logarithmic profile, calculate the expected wind speed at $z = 2$m above the water surface.
   (d) What are the surface drift velocity and rate of kinetic energy transfer across the surface?

Problem 12.5 As shown in Fig. 12.18, the depth profile (side view) of a lake is described approximately by a parabola,

$$h = 20 - 3.2x^2 \tag{1}$$

where h is the depth in m and x is the distance from the middle of the lake in km.
   (a) What are the length and average depth of this lake?
   (b) Assuming a constant surface shear stress of 0.2 Pa, use Eq. (12.2.14) to calculate the surface profile [note that x in this equation is defined differently

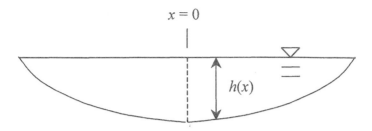

Figure 12.18 Definition sketch, Problem 12.5.

from x in Eq. (1)]. How far from the undisturbed shoreline does the water recede on the upwind end of the lake?

(c) Using the same boundary condition as found in part (b) at the upwind end of the lake, calculate the surface profile according to Eq. (12.2.13). This is a nonlinear equation and may require some iterative solution scheme. Comment on the degree to which the result compares with the result from part (b).

Problem 12.6 Recall that equilibrium temperature T_e is defined as the water temperature at which there is zero net surface heat flux (i.e., $\phi_n = 0$ when $T_s = T_e$). For this condition, and assuming there is no motion in the water, derive an expression for the steady state vertical temperature distribution. Would you expect this distribution to exist in nature? Why or why not? Under what conditions might this profile not be valid? The vertical diffusion coefficient is k_t. Assume an exponential decay in solar radiation intensity. Use $z = 0$ at the surface and positive downwards.

Problem 12.7 It is desired to characterize the clarity of a water body in terms of the extinction (absorption) coefficient value that would be used in an exponential decay solar radiation model. If it is assumed that 50% of the net solar radiation is absorbed very close to the surface, use the data below to estimate the extinction coefficient. State any assumptions made.

z (depth, m)	0	0.02	0.5	1.0	2.5	5.0	10.0
$\varphi_s'(W/m^2)$	250	125	93	70	28	6	0.3

Problem 12.8 Water clarity, or turbidity, is often measured using a Secchi disk, which is a round, flat disk painted in alternating white and black sections that is lowered into the water. The point at which the disk is just no longer discernible to the eye is called the *Secchi depth*. There are various opinions

concerning the relationship between the Secchi depth and the amount of light penetration, but in the Finger Lakes region of upstate New York the Secchi depth has been found to be approximately 1/5 to 1/2 of the depth at which the light intensity reaches 1% of the surface value. If the Secchi depth is measured as 2 m and it is assumed that half the solar intensity is absorbed near the surface, calculate the possible range of extinction coefficient values that would be expected.

Problem 12.9 The average solar radiation intensity reaching the outer earth's atmosphere is approximately 2 cal/cm^2 $-$ min. If, on average, the earth's surface emits 0.25 cal/cm^2 $-$ min (as longwave radiation), why is it that the atmosphere does not experience a steady accumulation of heat of about 1.75 cal/cm^2 $-$ min?

Problem 12.10 It has been reported that the average annual evaporation of the oceans is about 1 m. How much energy is used to produce this evaporation? Assume that sea water has the same latent heat of vaporization as pure water and express your answer in units of W/m^2.

Problem 12.11 Estimate representative seasonal values (fall, winter, spring, summer) for each of the surface heat flux terms, Φ_{sn}, Φ_{an}, Φ_b, Φ_e, and Φ_c for a lake at a site of your choice. Clear sky solar radiation values can usually be found in solar engineering handbooks. Clearly state all assumptions you make.

Problem 12.12 An experiment is conducted to evaluate the reaeration coefficient under a known mixing condition. In the experiment a closed tank of water initially has zero dissolved oxygen. At time $t = 0$, the cover is removed and the surface is exposed to the atmosphere, while the tank is stirred continuously and completely. The tank is cylindrical, with a height of 1 m and a diameter of 0.75 m. If the saturated oxygen concentration is 10 mg/L and the following data are collected for concentration as a function of time, estimate the bulk reaeration coefficient (K_2) value.

t (min)	0	5	10	20	60	120	240	480	1440
C (mg/l)	0	0.12	0.25	0.49	1.39	2.59	4.51	6.99	9.73

At $t = 240$ min, what is the mass flux of oxygen into the tank?

Problem 12.13 Compare values of the bulk transfer coefficient (K_L) calculated according to Eqs. (12.4.11) and (12.4.12), for a river with a bed slope of 1:6000 and a discharge per unit width of 2.8m^2/s. Plot the results for a range of depths from 1 m to 5 m, for constant discharge. Assume the diffusivity of oxygen in water is $10^{-5}\text{cm}^2/\text{s}$.

Problem 12.14 Compare values for δ_1 calculated from Eqs. (12.4.16) and (12.4.18), using $\nu = 10^{-2}\text{cm}^2/\text{s}$, $u* = 3\text{cm/s}$, $H = 2\text{m}$, and $U = 0.7\text{m/s}$. Use $C_1 = 4$ and find the value for C_2 so that the two equations give the same result. Does this seem like a reasonable value for C_2?

Problem 12.15 Rewrite Eq. (12.4.19) by substituting expressions for M and R in terms of appropriate rate constants, where M represents reaeration and R represents decay of DO due to BOD. Show that, under the assumptions used by Streeter and Phelps, this equation reduces to Eq. (12.4.25).

Problem 12.16 In a river the mean velocity is 0.5 m/s, the saturated DO concentration is 9.6 mg/L, $K_1 = 0.2$ day^{-1}, and $K_2 = 0.35$ day^{-1}. An industry discharges an oxygen-consuming waste into the river, with a resulting initial C_{BOD} of 5 mg/L in the river after the waste has been mixed in. If the initial oxygen deficit is zero, calculate the minimum DO concentration and the location at which the minimum concentration should be found, relative to the discharge location. Repeat the calculations for an initial DO deficit of 1 mg/L. Dispersion can be neglected.

Problem 12.17 An industry discharges a waste into a river following a "one hour on, one hour off" schedule. That is, the waste is discharged at a steady rate for one hour, then there is zero discharge, followed by another hour of steady discharge, and so on. There is a water supply intake 50 km downstream of the discharge point. The mean velocity in the river is 0.4 m/s and the longitudinal dispersion coefficient is 100 m^2/s. If the waste is conservative, evaluate whether the concentration response at the intake may be treated as a steady-state problem, or whether the problem can be treated as plug flow or not. If the waste were nonconservative, how large would the decay rate have to be before the assumption of plug flow would not be valid?

SUPPLEMENTAL READING

Atkinson, J. F., Blair, S., Taylor, S. and Ghosh, U., 1995. *Surface aeration. J. of Environmental Engineering* **121**(1):113–118.

Chapra, S. C., 1997. *Surface Water Quality Modeling.* McGraw-Hill, Boston.

Danckwerts, P. V., 1951. Significance of liquid film coefficients in gas absorption. *Industrial Engrg. Chem.* **43**(6):1460–1467.

Dobbins, W. E., 1956. The nature of the oxygen transfer coefficient in aeration systems. *Biological treatment of sewage and industrial waste*, Reinhold Publishing Corp., New York, 141–253.

Gulliver, J. S. and Stefan, H. G., 1984. Prediction of non-reactive water surface gas exchange in streams and lakes. In *Gas transfer at water surfaces*, W. Brutsaert and G. Jirka, eds., D. Reidel Publishing Co., Boston.

Higbie, R., 1935. The rate of absorption of a pure gas into a still liquid during short periods of exposure. *Trans. ASCE J.* **31**:365–389.

Hornbuckle, K. C., Jeremiason, J. D., Sweet, C. W. and Eisenreich, S. J., 1994. Seasonal variation in air-water exchange of polychlorinated biphenyls in Lake Superior. *Environ. Sci. Tech.* **28**:1491–1501.

Kohler, M. A., 1954. Lake and pond evaporation in water loss investigations — Lake Hefner studies. Prof. Paper 269, U.S. Geol. Surv.

Kreith, Frank and Kreider, Jan, F., 1978. *Principles of Solar Engineering*. Hemisphere Publishing Corp., Washington.

Lewis, W. K. and Whitman, W. G., 1924. Principles of gas absorption. *Industrial Engrg. Chem.* **16**(12):1215.

Liss, P. W. and Slater, P. G., 1974. Flux of gases across the air-sea interface. *Nature* **247**:181–184.

Mackay, D. and Yuen, A. T. K., 1983. Mass transfer coefficient correlations for volatilization of organic solutes from water. *Environ. Sci. Technol.* **17**:211–217.

O'Connor, D. J. and Dobbins, W. E., 1956. Mechanics of reaeration in natural streams. *J. Sanit. Eng. Div. ASCE* **82**(6):1–30.

Parkhurst, J. D. and Pomeroy, R. D., 1972. Oxygen absorption in streams. *J. Sanit. Eng. Div. ASCE* **98**(1):101–124.

Rathbun, R. E., 1998. Transport, behavior and fate of volatile organic compounds in streams. U.S. Geological Survey Professional Paper 1589, U.S. Government Printing Office, Washington.

Ryan, P. J. and Harleman, D. R. F., 1973. An analytical and experimental study of transient cooling pond behavior. Tech. Rep. 161, R.M. Parsons Lab., M.I.T., Cambridge, MA.

Streeter, H. W. and Phelps, E. B., 1925. A study of the pollution and natural purification of the Ohio River. Bull. No. 146, U.S. Public Health Service (reprinted 1958).

Swinbank, W. C., 1963. Longwave radiation from clear skies. *Quar. J. Roy. Met. Soc.*, London, 89.

Upstill-Goddard, R., Watson, A., Liss, P. and Liddicoat, M., 1990. Gas transfer velocities in lakes measured with SF6. *Tellus* **42B**:364–377.

Wilson, G. T. and Macleod, N. A., 1974. A critical appraisal of empirical equations and models for the prediction of the coefficient of reaeration of deoxygenated water. *Water Res.* **8**(6):341–366.

Wu, J., 1971. Wind-induced turbulent entrainment across a stable density interface. *J. Fluid Mech.* **61**:275–287.

13

Topics in Stratified Flow

13.1 BUOYANCY AND STABILITY CONSIDERATIONS

Nearly all natural surface water bodies are stratified at least part of the time. This means there are density variations, usually in the vertical direction. Horizontal variations also may exist, but not in steady state unless there are other forces such as Coriolis effects present to balance the resulting pressure differences (see Chap. 9). Density variations exist most commonly because of temperature and/or salinity gradients. Salinity is the main contributor to density in the oceans and in some inland lakes such as the Great Salt Lake in Utah or the Dead Sea in Israel, though these water bodies also may have temperature gradients. In freshwater lakes, temperature stratification is most important. In fact, from a water quality point of view, there is usually great interest in modeling the temperature structure of water bodies. Most biological and chemical reactions depend on temperature, and fish choose habitats based partly on this parameter. As discussed in Chap. 12, gas transfer across the air/water interface also depends on temperature.

A closely related parameter to density is *buoyancy*, defined as

$$b = g\frac{\rho_0 - \rho}{\rho_0} = g'$$ (13.1.1)

where ρ is the density of a fluid parcel and ρ_0 is the reference density. Buoyancy is also known as *reduced gravity*, g', and is defined so that a fluid parcel tends to rise when its buoyancy is positive (i.e., its density is less than the reference value) and a particle with higher buoyancy has a greater tendency to rise. Buoyancy may be treated like other state variables, such as temperature or concentration, and a conservation equation can be defined,

$$\frac{\partial b}{\partial t} + \vec{V} \cdot \vec{\nabla}b = \vec{\nabla} \cdot k_b\vec{\nabla}b + (source/sink)$$ (13.1.2)

where k_b is diffusivity for buoyancy and the source/sink terms are defined depending on which component is contributing to the buoyancy. For example, a solar heating term might generate buoyancy in a temperature-stratified system (Sect. 12.2), while evaporation might cause (negative) buoyancy at the surface, due either to cooling (heat loss due to evaporation) or to increased salinity in the case of saline water. Usually, an *equation of state* linking density or buoyancy to these other properties is needed in addition to the general equations of motion in order to define these systems.

13.1.1   Equation of State

Density is related to temperature and salinity and possibly other properties of a particular system through an equation of state. A well-known example of such an equation is the *perfect gas law*, which relates density, pressure, and temperature through the gas constant. Similarly, a general equation of state may be formulated for water systems as

$$\rho = \rho(T, C, p) \tag{13.1.3}$$

where T = temperature, C = concentration of dissolved species, and p = pressure. Since the main stratifying agent for most natural systems, in terms of dissolved species, is salinity S, this will be used in place of C. Also, except in the deep oceans or the atmosphere, the incompressible assumption implies that density should not be a function of pressure. Thus

$$\rho = \rho(T, S) \tag{13.1.4}$$

A number of expressions have been proposed to define this relationship, usually based on high-order polynomial fits to tabulated values of density as a function of T and S. Relations have been proposed for simple sodium chloride solutions and also to actual seawater solutions. In general the dependence on temperature has been found to be approximately parabolic, with a maximum at 4°C (Fig. 13.1). However, the temperature of the density maximum changes with increasing salinity, decreasing to about 0°C for highly saline systems. To a first-order approximation, density is linearly dependent on salinity over much of the normal range of interest.

The dependence of density on T and S is expressed through values of the thermal and saline expansion coefficients, α and β, respectively, where

$$\alpha = -\frac{1}{\rho_0}\frac{\partial \rho}{\partial T} \tag{13.1.5}$$

and

$$\beta = \frac{1}{\rho_0}\frac{\partial \rho}{\partial S} \tag{13.1.6}$$

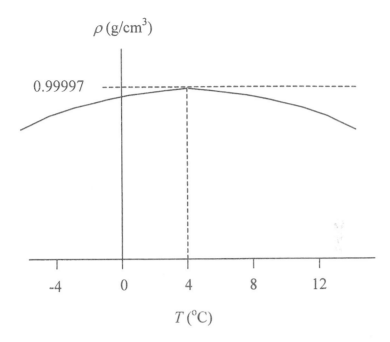

Figure 13.1 Variation of freshwater density with temperature.

The negative sign in the definition for α is meant so that, at least for $T > 4°C$, the coefficient has a positive value. Note that α changes sign for temperatures less than the temperature of the density maximum. In general, both α and β are functions of T and S, but approximate values are $\alpha \cong 2 \times 10^{-4}$ °C^{-1} and $\beta \cong 7.5 \times 10^{-3}S^{-1}$, where S is in weight percent. It is clear from these values that salinity has a much greater effect on density than does temperature.

The equation of state is written to express density in terms of α and β, and deviations of T and S from standard or reference values as

$$\rho = \rho_0(1 - \alpha\Delta T + \beta\Delta S) \tag{13.1.7}$$

where $\Delta T = T - T_0$, $\Delta S = S - S_0$, $\rho = \rho_0$ when $T = T_0$ and $S = S_0$, and T_0 and S_0 are the reference values for temperature and salinity, respectively. Normally, $T_0 = 4°C$ and $S_0 = 0$ (%). Note that S is usually expressed in units of weight percent, parts per thousand, or mg/L. Typical seawater has $S \cong 3.5\%$, 35 ppt (o/oo), or 35,000 ppm. When α and β are taken as constants, the resulting equation is called a *linear equation of state*. A constant value for β is usually a good approximation, but in order to account for the parabolic nature of the temperature dependence, α should be a function of T. A simple expression that gives reasonable results over much of the range of normally occurring values for T and S (except near freezing) is

$$\rho = \rho_0[1 - 0.00663(T - 4)^2 + 7.615S] \tag{13.1.8}$$

where T is in °C and S is in weight percent. For freshwater bodies, with $S = 0$, this equation is reasonable even near freezing. When S is high, however, greater than about 5%, higher order equations should be used to estimate ρ.

13.1.2 Gravitational Stability

When density differences exist in a fluid system, an important consideration is that of stability. In Chap. 10 the concept of convective transport was
introduced, where it was noted that convection is the result of a gravitationally unstable condition. This is simply saying that lighter, more buoyant fluid
should tend to rise and that the system would be stable only when heavier fluid
underlies lighter fluid. Buoyancy instabilities give rise to convective motions,
which tend to mix the fluid system vertically. This is demonstrated mathematically by applying a *perturbation analysis* to a system in which there is no
motion initially ($u = v = w = 0$), with a density stratification $\rho(z)$, as illustrated in Fig. 13.2. A fluid particle is considered with initial position at $z = 0$,
which is an arbitrary point within the fluid.

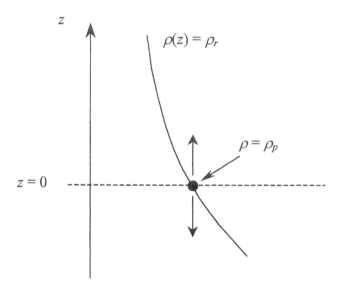

Figure 13.2 Fluid particle at $z = 0$, in fluid with ambient stratification given by
$\rho(z) = \rho_r$ (reference distribution).

The governing equation for this problem is

$$0 = -\frac{1}{\rho}\nabla \cdot p + \vec{g} \qquad \Rightarrow \qquad \frac{\partial p}{\partial z} = -\rho g \tag{13.1.9}$$

This is simply the momentum equation for the vertical direction, which reduces to the hydrostatic result for the case of zero flow. The reference density, ρ_r, must satisfy this equation. An equation of state also should be considered to relate density to other properties of the system, as noted above. The fluid is assumed to be incompressible (note that an incompressible fluid does not have to have the same density everywhere). If the fluid particle of Fig. 13.2 is displaced vertically from its equilibrium location by an amount z, there will be a buoyancy force acting on the particle due to the difference in density between the particle and its new surroundings. The resulting force acting on the particle, per unit volume, is $[-g(\rho_p - \rho_r)]$, where ρ_p is the fluid particle density. Applying Newton's law,

$$\vec{F} = m\vec{a} \qquad \Rightarrow \qquad \frac{F}{\forall} = \frac{m}{\forall}a \qquad \Rightarrow \qquad g(\rho_r - \rho_p) = \rho_p\ddot{z} \tag{13.1.10}$$

where m is the mass of the fluid particle, $\forall$ is its volume, and the vector notation is dropped with the understanding that the force balance is in the vertical (z) direction. The double dots over z in the last part of this equation indicate a second derivative with time (i.e., acceleration).

Taylor series expansions are now defined for ρ_r and ρ_p, in terms of a reference value ρ_0:

$$\rho_r = \rho_0 + z\frac{d\rho_r}{dz} + [O(z^2)] \tag{13.1.11}$$

$$\rho_p = \rho_0 + z\frac{d\rho_p}{dz} + [O(z^2)] \tag{13.1.12}$$

where the last term in both of these expressions indicates that the neglected or truncated terms in the series approximations are of order $O(z^2)$ (i.e., second order). These terms will be neglected in the following. Assuming there is no heat transfer, the relationship between pressure and density is given by the adiabatic relationship, $dp = c^2 dp$, where $c =$ sonic velocity. Making these substitutions, along with the hydrostatic pressure result (13.1.9), into Eq. (13.1.10), we have

$$\frac{d\rho_p}{dz} \cong \frac{d\rho_0}{dz} = \frac{1}{c^2}\frac{dp}{dz} = -\frac{\rho_0 g}{c^2} \tag{13.1.13}$$

and

$$\rho_p\ddot{z} = g\left(\rho_0 + z\frac{d\rho_r}{dz} - \rho_0 + z\frac{\rho_0 g}{c^2}\right)[+O(z^2)] = gz\left(\frac{\rho_0 g}{c^2} + \frac{d\rho_r}{dz}\right) \tag{13.1.14}$$

The *Brunt–Vaisala* or *buoyancy frequency*, N, is defined in terms of the square root of the negative of the term in parentheses in this last expression. Note that the density gradient is assumed to be negative, as it must be for a gravitationally stable water column; N is undefined for an unstable density distribution. However, the sonic velocity c is normally large, and the first term in parentheses on the right-hand side of Eq. (13.1.4) can be neglected under most conditions. This is equivalent to neglecting the small compressibility of the fluid parcel (resulting in a small change in density) due to the change in ambient pressure at the perturbed location. In general, N is defined for most applications by

$$N^2 = -\frac{g}{\rho_0}\frac{d\rho_r}{dz}$$

(13.1.15)

and Eq. (13.1.14) then becomes

$$\rho_p \ddot{z} = -N^2 z \rho_p$$

(13.1.16)

where it has been assumed that $\rho_p \cong \rho_0$. The final differential equation is

$$\ddot{z} + N^2 z = 0$$

(13.1.17)

which is an equation of simple harmonic motion.

Equation (13.1.17) has solutions of the form

$$z \propto e^{iNt} \qquad N^2 > 0$$

(13.1.18a)

$$z \propto e^{|N|t} \qquad N^2 < 0$$

(13.1.18b)

$$z \propto e^0 \qquad N^2 = 0$$

(13.1.18c)

These three solutions correspond to stable, unstable, and neutral conditions, respectively. In other words, $N^2 > 0$ implies a gravitationally stable condition, with density increasing with depth. When $N^2 < 0$, heavier fluid overlays lighter fluid, which is an unstable condition leading to convection (in fact, N is undefined in this situation, as noted above). A neutrally stable condition is one where there is no acceleration since there is no net gravitational force acting on the fluid particle when it is displaced from its original equilibrium position, and particle position remains constant. For the stable situation, substituting the Euler formula,

$$e^{\pm i\phi} = \cos\phi \pm i\sin\phi$$

(13.1.19)

it is seen that oscillatory motions are expected, with amplitude depending on the magnitude of the original displacement of the fluid particle. These oscillations decay over time due to viscous effects, which have not been considered here but could be included in the equation of motion for the fluid particle, Eq. (13.1.17), if desired. For the unstable case, the particle position grows exponentially with time. Thus any perturbation of particle position in an unstable environment will lead to large-scale convective motions.

13.2 INTERNAL WAVES

We now consider wave motions that are possible in a stratified fluid. Internal waves can propagate along the interface between fluid layers of different densities (note that surface waves, as discussed in Chap. 8, propagate along the air/water interface, which is an extreme example of fluids of two different densities) or, more generally, at an angle to the horizontal through a density stratified fluid, with $N^2 > 0$.

Consider a stratified fluid with density and pressure fields given by

$$\rho(\vec{x}, t) = \rho_0 + \rho'(z) + \rho''(x, y, t) \tag{13.2.1}$$

$$p(\vec{x}, t) = p_0 + p'(z) + p''(x, y, t) \tag{13.2.2}$$

where x and y are the horizontal coordinates, subscript 0 indicates a reference value, a single prime denotes a function of z only (vertical position) and the double prime indicates a function of horizontal position and time. Normally, it can be assumed that the magnitude of ρ_0 is much greater than the magnitude of ρ', which in turn is much greater than the magnitude of ρ'' (one or two orders of magnitude difference between each component). Since we are dealing primarily with water, incompressibility dictates that the density following a fluid particle is constant, or

$$\frac{D\rho}{Dt} = \frac{\partial \rho}{\partial t} + \vec{u} \cdot \nabla \rho = 0 \tag{13.2.3}$$

As shown in Chap. 2, this leads to the usual continuity equation for an incompressible fluid,

$$\nabla \cdot \vec{u} = \frac{\partial u_j}{\partial x_j} = 0 \tag{13.2.4}$$

The general momentum equation is (neglecting Coriolis terms),

$$\frac{Du_k}{Dt} = \frac{\partial u_k}{\partial t} + u_j \frac{\partial u_k}{\partial x_j} = -\frac{1}{\rho} \frac{\partial p}{\partial x_k} + g_k + \nu \nabla^2 u_k \tag{13.2.5}$$

For the analysis of stratified fluids it is convenient to consider a reference state of zero motion. The reference density is $\rho_r = (\rho_0 + \rho'(z))$ (see Fig. 13.2). The pressure is given similarly as $p_r = p_0 + p'(z)$. Substituting this definition for

ρ_r into Eq. (13.2.3) results in

$$\frac{D\rho}{Dt} = \frac{D\rho_r}{Dt} + \frac{D\rho''}{Dt} = \frac{\partial \rho_r}{\partial t} + u_j \frac{\partial \rho_r}{\partial x_j} + \frac{D\rho''}{Dt} = 0$$

$$\Rightarrow \frac{D\rho''}{Dt} = -w \frac{d\rho_r}{dz} \tag{13.2.6}$$

since ρ_r is a function of z only. The momentum equation for the reference state is simply the hydrostatic result,

$$0 = -\frac{\partial p_r}{\partial x_k} + \rho_r g_k \tag{13.2.7}$$

Subtracting Eq. (13.2.7) from Eq. (13.2.5) results in

$$\rho \frac{Du_k}{Dt} = -\frac{\partial (p - p_r)}{\partial x_k} + (\rho - \rho_r) g_k + \mu \nabla^2 u_k \tag{13.2.8}$$

Then, using the definitions for p_r and ρ_r, along with the approximation that $(1/\rho) \cong (1/\rho_0)$ (this is the Boussinesq approximation discussed in Sect. 2–7),

$$\frac{Du_k}{Dt} = -\frac{1}{\rho_0} \frac{\partial p''}{\partial x_k} + \frac{\rho''}{\rho_0} g_k + \nu \nabla^2 u_k \tag{13.2.9}$$

This is the governing equation for momentum, though in the following we also will neglect viscous effects (high Reynolds number assumption).

A wave equation is derived by writing the momentum equations separately for the horizontal and vertical directions:

$$\frac{\partial u_h}{\partial t} + \frac{1}{\rho_0} \frac{\partial p''}{\partial x_h} = -u_j \frac{\partial u_h}{\partial x_j} \tag{13.2.10}$$

$$\frac{\partial w}{\partial t} + \frac{\rho''}{\rho_0} g = -u_j \frac{\partial w}{\partial x_j} \tag{13.2.11}$$

where u_h denotes a horizontal velocity component, i.e., h takes values of 1 or 2, for the x or y direction, respectively. The fact that $p'' \neq f(z)$ has also been taken into account in writing Eq. (13.2.11). Taking the derivative of Eq. (13.2.10) with respect to z and subtracting the derivative of Eq. (13.2.11) with respect to x_h gives

$$\frac{\partial}{\partial t} \left(\frac{\partial u_h}{\partial z} - \frac{\partial w}{\partial x_h} \right) - \frac{\partial}{\partial x_h} \left(\frac{\rho''}{\rho_0} g \right) = \frac{\partial}{\partial x_h} \left(u_j \frac{\partial w}{\partial x_j} \right) - \frac{\partial}{\partial z} \left(u_j \frac{\partial u_h}{\partial x_j} \right)$$

$$\tag{13.2.12}$$

This last result is then differentiated with respect to t and the two component equations (for $h = 1$ or 2) are then

$$\frac{\partial^2}{\partial t^2}\left(\frac{\partial u}{\partial z} - \frac{\partial w}{\partial x}\right) - \frac{\partial^2}{\partial t\,\partial x}\left(\frac{\rho''}{\rho_0}g\right) = \frac{\partial^2}{\partial t\,\partial x}\left(u_j\frac{\partial w}{\partial x_j}\right) - \frac{\partial^2}{\partial t\,\partial z}\left(u_j\frac{\partial u}{\partial x_j}\right)$$

(13.2.13)

$$\frac{\partial^2}{\partial t^2}\left(\frac{\partial v}{\partial z} - \frac{\partial w}{\partial y}\right) - \frac{\partial^2}{\partial t\,\partial y}\left(\frac{\rho''}{\rho_0}g\right) = \frac{\partial^2}{\partial t\,\partial y}\left(u_j\frac{\partial w}{\partial x_j}\right) - \frac{\partial^2}{\partial t\,\partial z}\left(u_j\frac{\partial v}{\partial x_j}\right)$$

(13.2.14)

Differentiating the first of these with respect to x and differentiating the second with respect to y and adding the results gives

$$\frac{\partial^2}{\partial t^2}\left(\frac{\partial^2 u}{\partial x\,\partial z} + \frac{\partial^2 v}{\partial y\,\partial z} - \frac{\partial^2 w}{\partial x^2} - \frac{\partial^2 w}{\partial y^2}\right) - \frac{g}{\rho_0}\left[\frac{\partial}{\partial t}(\nabla_h^2\rho'')\right]$$

$$= \frac{\partial}{\partial t}\left[\nabla_h^2\left(u_j\frac{\partial w}{\partial x_j}\right)\right] - \frac{\partial^3}{\partial t\,\partial x_h\partial z}\left(u_j\frac{\partial u_h}{\partial x_j}\right)$$

(13.2.15)

where

$$\nabla_h^2 = \frac{\partial^2}{\partial x^2} + \frac{\partial^2}{\partial y^2}$$

(13.2.16)

is a two-dimensional horizontal Laplacian operator and index notation is used for subscript h. The first term on the left-hand side is simplified using the continuity Eq. (13.2.4):

$$\frac{\partial u}{\partial x} + \frac{\partial v}{\partial y} = -\frac{\partial w}{\partial z} \Rightarrow \frac{\partial^2 u}{\partial x\,\partial z} + \frac{\partial^2 v}{\partial y\,\partial z} = -\frac{\partial^2 w}{\partial z^2}$$

(13.2.17)

The second term on the left-hand side also is rewritten using Eq. (13.2.6):

$$\frac{g}{\rho_0}\left[\frac{\partial}{\partial t}(\nabla_h^2\rho'')\right] = \frac{g}{\rho_0}\nabla_h^2\left(\frac{\partial\rho''}{\partial t}\right) = \frac{g}{\rho_0}\nabla_h^2\left(\frac{D\rho''}{Dt} - u_j\frac{\partial\rho''}{\partial x_j}\right)$$

$$= \frac{g}{\rho_0}\nabla_h^2\left(-w\frac{\partial\rho_r}{\partial z} - u_j\frac{\partial\rho''}{\partial x_j}\right)$$

$$= \frac{g}{\rho_0}\left[-\frac{\partial\rho_r}{\partial z}\nabla_h^2 w - \nabla_h^2\left(u_j\frac{\partial\rho''}{\partial x_j}\right)\right]$$

(13.2.18)

Substituting these last two results into Eq. (13.2.15), rearranging, and using the definition of the buoyancy frequency (13.1.15) gives

$$\frac{\partial^2}{\partial t^2}(\nabla^2 w) + N^2(\nabla_h^2 w) = -\frac{\partial^3}{\partial t\,\partial x_h\partial z}\left(u_j\frac{\partial u_h}{\partial x_j}\right)$$

$$-\nabla_h^2\left[\frac{g}{\rho_0}u_j\frac{\partial\rho''}{\partial x_j} + \frac{\partial}{\partial t}\left(u_j\frac{\partial w}{\partial x_j}\right)\right]$$

(13.2.19)

The right-hand side is generally negligible compared with the other terms in the equation, since it involves all nonlinear terms. A possible exception to this is when there is strong mean shear, in which case the velocity gradients may be important. The simplified final equation is

$$\frac{\partial^2}{\partial t^2}(\nabla^2 w) + N^2(\nabla_h^2 w) \cong 0 \tag{13.2.20}$$

which is a wave equation similar to Eq. (13.1.17).

It is beyond the scope of the present text to provide a full discussion of the solutions to this equation (or the more general Eq. 13.2.19) and the resulting behavior of the fluid motions that it describes. There are many books that cover this material in depth, and the present analysis is restricted to a discussion of some properties of linear internal waves and their relationship to surface waves. First, however, we consider the lowest mode solutions, which correspond to horizontal propagation.

13.2.1 Lowest Mode Solutions

By assuming wavelike disturbances,

$$w = W(z)\exp[i(k_1 x + k_2 y - \sigma t)] \tag{13.2.21}$$

where k_1 and k_2 are wave numbers for the x and y directions, respectively, and σ is frequency, Eq. (13.2.20) becomes

$$\frac{d^2 W}{dz^2} + \left(\frac{N^2 - \sigma^2}{\sigma^2}\right) k_h^2 W = 0 \tag{13.2.22}$$

where $k_h^2 = k_1^2 + k_2^2$. This is a wave equation for W, which expresses wave propagation in the horizontal direction. Boundary conditions must be applied at the surface and at the bottom.

The free surface is defined by $\eta(x, y, t)$, as sketched in Fig. 13.3. At the surface, $z = \eta$, the dynamic boundary condition is

$$p_r(\eta) + p''(x_h, \eta, t) = p_{atm}(= 0) \tag{13.2.23}$$

and the kinematic boundary condition is

$$\frac{D\eta}{Dt} = w \tag{13.2.24}$$

However, application of boundary conditions is problematic at $z = \eta$, since η itself is an unknown, obtained as part of the solution. For a first-order solution, it is more convenient to write the surface boundary conditions at $z = \eta = 0$ and to account for variations between this level and the actual

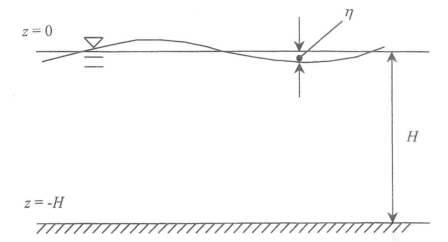

Figure 13.3 Definition sketch for surface level.

surface through linear Taylor series expansions of Eqs. (13.2.23) and (13.2.24), assuming approximately hydrostatic pressure variations and small η. Then, for $z = 0$, the dynamic condition is

$$p_r(0) + p''(x_h, 0, t) - \rho_0 g\eta = p_{atm} \qquad (13.2.25)$$

and the kinematic condition is approximated as

$$\frac{\partial \eta}{\partial t} = w(x_h, 0, t) \qquad (13.2.26)$$

Then, differentiating Eq. (13.2.25) with respect to t and substituting Eq. (13.2.26) gives

$$\frac{\partial p''}{\partial t} - \rho_0 gw = 0 \qquad (13.2.27)$$

Taking the time derivative of the linearized vertical momentum equation (from Eq. 13.2.11), we obtain

$$\frac{\partial^2 w}{\partial t^2} + \frac{g}{\rho_0}\frac{\partial p''}{\partial t} = -\frac{1}{\rho_0}\frac{\partial^2 p''}{\partial t \partial z} \qquad (13.2.28)$$

and the linearized mass conservation equation (from Eq. 13.2.6) is

$$\frac{\partial p''}{\partial t} = -w\frac{d\rho_r}{dz} \qquad (13.2.29)$$

Substituting this last result into Eq. (13.2.28), along with the boundary condition (13.2.27), gives

$$\left(\frac{\partial^2}{\partial t^2} + N^2\right) w = -\frac{1}{\rho_0}\frac{\partial^2 p''}{\partial t\, \partial z} = -g\frac{\partial w}{\partial z} \tag{13.2.30}$$

Then, assuming the same wavelike disturbances as before (Eq. 13.2.21), this last result becomes

$$\sigma^2\frac{dW}{dz} - gk_{\mathrm{h}}^2 W = 0 \tag{13.2.31}$$

which is the boundary condition applied at $z = 0$. At the bottom ($z = -H$), $w = 0$. Thus, for a given frequency σ, an eigenvalue problem is defined for k and W.

From examination of the governing Eq. (13.2.22), if $N^2 < \sigma^2$, then W is monotonic in z ($W \propto \exp(|\lambda|z)$, where $\lambda^2 = (N^2 - \sigma^2)/\sigma^2 < 0$ — recall the earlier discussion of gravitational stability in Sec. 13.1). In this case, the magnitude of the motions decays exponentially with depth, as z becomes more negative. A limiting case is when $N = 0$ (no stratification), where only surface waves can exist. In fact, surface waves are thus seen as a special case of the general internal wave solution. If $\sigma < N_{\max}$, where $N_{\max}$ is the maximum value of the buoyancy frequency in the water column, then there is a range of z over which W and (d^2W/dz^2) must have opposite signs, in order that Eq. (13.2.22) can be satisfied. This is a characteristic of oscillating functions, and it may be concluded that W is oscillating in this region (i.e., wave motions exist in this range of z). Then, for a given $\sigma < N_{\max}$, there are many possible values for k_{h}. The lowest value (lowest mode) corresponds with surface waves, and higher modes correspond with internal waves. For the lowest internal wave mode, the entire thermocline (or pycnocline) moves up and down in unison. As the mode increases, corresponding to larger k_{h}, there is a shortening of the vertical scale of motions, with more zero-crossings. The vertical scale of these motions is estimated by

$$L_z \approx \left(\frac{W\, d^2W}{dz^2}\right)^{1/2} \tag{13.2.32}$$

and the horizontal scale is simply the inverse of the wave number, $L_{\mathrm{H}} \approx k_{\mathrm{h}}^{-1}$, so, using Eq. (13.2.22),

$$\frac{L_{\mathrm{H}}}{L_z} \cong \left|\frac{N^2 - \sigma^2}{\sigma^2}\right|^{1/2} \tag{13.2.33}$$

For σ approaching 0, L_{H} is very large compared with L_z, and the scaling analysis suggests a horizontal drift, uniform in x and y. At the other extreme, for $\sigma \cong N$, $L_z \gg L_H$, and variations in the vertical flow field are small; this is the case for the lowest mode of internal wave.

For this lowest internal mode, consider the situation where the density stratification approaches a step function, as in the two-layer stratification illustrated in Fig. 13.4. Here, $N = 0$ everywhere except at the density interface at $z = -h$, where it is large. The interface has thickness δ, which is small compared with h. In the uniform regions, the governing wave equation (13.2.22) reduces to

$$\frac{d^2W}{dz^2} - k_h^2 W = 0 \tag{13.2.34}$$

and the boundary conditions are $W = 0$ at $z = 0$ and at $z = -H$. The bottom condition is exact, but the surface condition is a first-order approximation, as noted previously. This equation can be integrated within each uniform region, using these boundary conditions, with the constraint that the values for W at the interface (or, just above and just below the interface) must match. In other words,

$$W|_{z=h_+} = W|_{z=h_-} \tag{13.2.35}$$

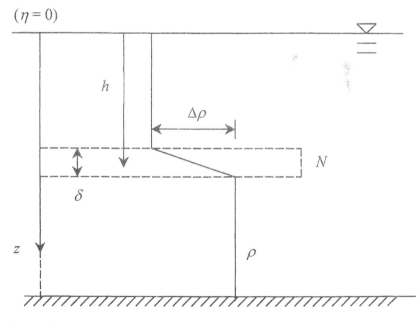

Figure 13.4 Distribution of density and buoyancy frequency in a two-layer stratified system.

The full equation (13.2.22), integrated over the interfacial thickness δ, gives

$$\frac{dW}{dz}\bigg|_{z=-h_+} - \frac{dW}{dz}\bigg|_{z=-h_-} = k_h^2 W\big|_{z=-h}\left[\delta - \frac{g}{\sigma^2}\frac{\Delta\rho}{\rho_0}\right] \tag{13.2.36}$$

where $\Delta\rho$ is the density difference over the interface. These last two results (for W and dW/dz at the interface provide matching conditions that must be satisfied in the overall solution for W.

The solution of (13.2.34), with the stated boundary conditions, is

$$W(z) = \begin{cases} C_1 \sinh(k_h z), & 0 > z > -h \\ C_2 \sinh[k_h(z+h)], & -h > z > -H \end{cases} \tag{13.2.37}$$

and from Eq. (13.2.35),

$$\frac{C_1}{C_2} = -\frac{\sinh[k_h(H-h)]}{\sinh(k_h h)} \tag{13.2.38}$$

while Eq. (13.2.36) implies

$$\sigma^2 = gk_h\frac{\Delta\rho}{\rho_0}\{k_h\delta + \coth(k_h h) + \coth[k_h(H-h)]\}^{-1} \tag{13.2.39}$$

This last result is known as a *dispersion relationship*, which relates k_h and σ. If the interface is relatively thick, or $(k_h\delta) \to \infty$, the wavelengths of the disturbance are short compared with δ, and

$$\sigma^2 \cong \frac{g}{\delta}\frac{\Delta\rho}{\rho_0} \tag{13.2.40}$$

For the other extreme, $(k_h\delta) \ll 1$, which is more commonly the case, and

$$\sigma^2 \cong gk_h\frac{\Delta\rho}{\rho_0}\{\coth(k_h h) + \coth[k_h(H-h)]\}^{-1} \tag{13.2.41}$$

Finally, in the deep ocean, with large $H (H \gg h)$, the dispersion relation can be approximated by

$$\sigma^2 \cong \frac{\Delta\rho}{\rho_0}\frac{gk_h}{\coth(k_h h)} \tag{13.2.42}$$

Surface Manifestation of Lowest Internal Wave Mode

One of the most important effects of the lowest internal wave mode is its relationship to surface waves. This results from the pressure field created by the internal wave. Consider a wave function describing the motions at the interface, given by

$$\zeta = A\exp[i(k_1 x + k_2 y - \sigma t)] \tag{13.2.43}$$

where A is the amplitude of the motions. Using a similar approach as in describing the boundary condition at $z = \eta$ (see text following Eq. 13.2.24), the vertical velocity at the interface is

$$\frac{D\zeta}{Dt} = w|_{z=-h} \tag{13.2.44}$$

Neglecting nonlinear terms and using the same assumption for w as before (Eq. 13.2.21),

$$w|_{z=-h} \cong \frac{\partial \zeta}{\partial t} \Rightarrow W|_{z=-h} = -iA\sigma \tag{13.2.45}$$

Now, assuming that the motion outside the interfacial region is irrotational, a velocity potential φ can be defined so that

$$\nabla^2 \varphi = 0 \qquad \frac{\partial \varphi}{\partial z} = w \tag{13.2.46}$$

Given the assumed form for w, φ is assumed as

$$\varphi = \Phi(z) \exp[i(k_1 x + k_2 y - \sigma t)] \tag{13.2.47}$$

Then, using Eq. (13.2.37) for the region above the interface,

$$\frac{d\Phi}{dz} = W = C_1 \sinh(k_h z) \Rightarrow -C_1 \sinh(k_h h)$$

$$= -iA\sigma \Rightarrow C_1 = \frac{iA\sigma}{\sinh(k_h h)} \tag{13.2.48}$$

so that

$$\Phi = \frac{iA\sigma}{k_h \sinh(k_h h)} \cosh(k_h z) \tag{13.2.49}$$

At the surface, the linearized horizontal momentum equation is (from Eq. 13.2.9, without the viscous term)

$$\frac{1}{\rho_0} \frac{\partial p''}{\partial x} = -\frac{\partial u}{\partial t} = -\frac{\partial^2 \varphi}{\partial t \partial x} \tag{13.2.50}$$

which is evaluated at $z = 0$. Carrying out the differentiation on the right-hand side of this last result and then integrating, we find the pressure perturbation at the surface is

$$p''|_{z=0} = i\rho_0 \sigma \Phi \exp[i(k_1 x + k_2 y - \sigma t)] = \rho_0 g \eta \tag{13.2.51}$$

This, combined with Eq. (13.2.49), gives an expression for surface displacement,

$$\eta = -\frac{A\sigma^2}{g k_h \sinh(k_h h)} \exp[i(k_1 x + k_2 y - \sigma t)] \tag{13.2.52}$$

Upon comparing this result with Eq. (13.2.43), the magnitude of η relative to ζ is seen to be of order $(\Delta\rho/\rho_0) \cong 10^{-3}$ (using Eq. 13.2.41 for σ^2), which suggests that the surface "signature" of the internal wave is relatively small, compared with the amplitude of the internal wave itself.

The *phase speed* of the waves is defined as the ratio of wavelength to period,

$$c = \frac{L}{T} \tag{13.2.53}$$

where $L = 2\pi/k_h$ is the *wavelength* and $T = 2\pi/\sigma$ is the *period* of the waves. Then, using the dispersion relation (13.2.39), the phase speed is written as

$$c = \left[\frac{g}{k_h}\left(\frac{\Delta\rho}{\rho_0}\right)\right]^{1/2} \{\coth(k_h h) + \coth[k_h(H-h)]\}^{-1/2} \tag{13.2.54}$$

which describes the velocity at which the waves propagate.

13.2.2 Small-Scale Internal Waves

The preceding discussion focused on the lowest internal wave mode and its relationship with surface waves. The analysis is now generalized to consider higher mode internal waves, but with the assumption that the vertical scale of motion is small relative to the scale over which N varies, so that at least locally, it may be assumed that N is approximately constant. The equation describing vertical velocity is Eq. (13.2.20), with solutions assumed to be of a form similar to Eq. (13.2.21) but generalized to include a vertical component of the wave number vector,

$$w \propto \exp[i(k_1 x + k_2 y + k_3 z - \sigma t)] \tag{13.2.55}$$

Upon substituting this into Eq. (13.2.20), a dispersion relation is obtained,

$$\sigma^2 = \frac{N^2(k_1^2 + k_2^2)}{(k_1^2 + k_2^2 + k_3^2)} \qquad \Rightarrow \qquad \sigma = \pm N\cos\theta \tag{13.2.56}$$

where θ is the angle between the horizontal and the total wave number vector, as shown in Fig. 13.5, and $\cos\theta = k_h/k(k^2 = k_1^2 + k_2^2 + k_3^2)$. This is the angle of wave propagation, relative to horizontal.

From Eq. (13.2.56), the maximum value for σ occurs when $\theta = 0$, and $|\sigma_{max}| = N$, which corresponds with horizontal wave propagation ($k_h = k$). Vertically propagating waves result for $\sigma = 0(k_h = 0)$. It also is of interest to note that, in general, the velocity field can be represented by

$$\vec{u} = \vec{U}\exp[i(\vec{k}\cdot\vec{x} - \sigma t)] \tag{13.2.57}$$

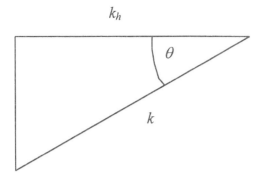

Figure 13.5 Definition sketch for wave propagation angle θ.

(refer to Eq. 13.2.55). For incompressible fluid, the divergence of the velocity is 0, so

$$\nabla \cdot \vec{u} = 0 \qquad \Rightarrow \qquad i\vec{k} \cdot \vec{U} = 0 \qquad (13.2.58)$$

which implies that the wave number vector and the velocity vector are perpendicular to each other. This is in contrast to surface waves, where the velocity and wave number vectors are in the same direction. A further result of Eq. (13.2.58) is

$$\vec{u} \cdot \nabla \vec{u} = \vec{0} \qquad (13.2.59)$$

which shows that the linearized solution is an exact solution to the governing equations, since the neglected nonlinear terms in the linearized solution are in fact zero.

13.3 MIXING

13.3.1 Linear Stability Theory

One of the consequences of internal wave propagation is the transport of energy, and under certain circumstances waves may break and release energy that can be used for mixing, which is a topic of considerable interest in stratified flow modeling. One of the more well-known modes of this type of mixing is a process known as *Kelvin–Helmholtz instability*, which refers to instability and breaking of an interfacial wave. First, however, consider the general stability problem for a system with a steady, two-dimensional, inviscid, mean shear flow $U(z)$. Hydrostatic pressure is assumed, and the mean velocity field, $\vec{u} = [U(z), 0, 0]$ satisfies the exact (nonlinear) two-dimensional governing equations. Perturbations to this mean velocity field are assumed as

$$\vec{u} = (U + u', w') \qquad p = p_0 + p' \qquad \rho = \rho_0 + \rho' \tag{13.3.1}$$

where primes indicate perturbation quantities and p_0 and ρ_0 are the pressure and density, respectively, of the unperturbed system; these are both functions of z only.

The continuity equation is

$$\frac{\partial u}{\partial x} + \frac{\partial w}{\partial z} = 0 = \frac{\partial (U + u')}{\partial x} + \frac{\partial w'}{\partial z}$$

and since the mean flow satisfies continuity, so must the perturbations,

$$\frac{\partial u'}{\partial x} + \frac{\partial w'}{\partial z} = 0 \tag{13.3.2}$$

This is similar to the behavior of the turbulent velocity fluctuations described in Chap. 5. The momentum equations are (neglecting viscous effects)

$$(\rho_0 + \rho') \left[\frac{\partial u'}{\partial t} + (U + u') \frac{\partial (U + u')}{\partial x} + w' \frac{\partial (U + u')}{\partial z} \right]$$
$$= -\frac{\partial (p_0 + p')}{\partial x} = -\frac{\partial p'}{\partial x} \tag{13.3.3}$$

and

$$(\rho_0 + \rho') \left[\frac{\partial w'}{\partial t} + (U + u') \frac{\partial (w')}{\partial x} + w' \frac{\partial (w')}{\partial z} \right]$$
$$= -\frac{\partial (p_0 + p')}{\partial z} - g(\rho_0 + \rho') \tag{13.3.4}$$

Upon carrying out the multiplications on the right-hand sides and linearizing with respect to the perturbation quantities, we obtain

$$\rho_0 \frac{\partial u'}{\partial t} + \rho_0 U \frac{\partial u'}{\partial x} + \rho_0 w' \frac{dU}{dz} \cong -\frac{\partial p'}{\partial x} \tag{13.3.5}$$

and

$$\rho_0 \frac{\partial w'}{\partial t} + \rho_0 U \frac{\partial w'}{\partial x} \cong -\frac{\partial p'}{\partial z} - g(\rho_0 + \rho') \tag{13.3.6}$$

Subtracting the derivative of Eq. (13.3.5) with respect to z from the derivative of Eq. (13.3.6) with respect to x eliminates the pressure terms, and

$$\rho_0 \left\{ \frac{\partial^2 w'}{\partial x \, \partial t} + U \frac{\partial^2 w'}{\partial x^2} - \frac{\partial^2 u'}{\partial z \, \partial t} - \frac{dU}{dz} \frac{\partial u'}{\partial x} - U \frac{\partial^2 u'}{\partial x \, \partial z} - \frac{\partial w'}{\partial z} \frac{dU}{dz} - w' \frac{d^2 U}{dz^2} \right\}$$
$$- \frac{d\rho_0}{dz} \left(\frac{\partial u'}{\partial t} + U \frac{\partial u'}{\partial x} + w' \frac{dU}{dz} \right) = -g \frac{\partial \rho'}{\partial x} \tag{13.3.7}$$

A solution is found by assuming wavelike expressions,

$$\phi'(x, z, t) = \hat{\phi}(z) \exp[ik(x - ct)] \tag{13.3.8}$$

where ϕ denotes any of the perturbation quantities, k is the wave number for the disturbances in the x-direction, and c is the phase speed. The "hat" indicates a function of z only. On substitution into Eq. (13.3.7) and rearranging, the result is

$$\rho_0 \left\{ -k^2(U - c)\hat{w} - \frac{d\hat{w}}{dz}\frac{dU}{dz} - \hat{w}\frac{d^2U}{dz^2} \right\} - \frac{d\rho_0}{dz}\hat{w}\frac{dU}{dz}$$

$$- (ik)\left\{ \rho_0(U - c)\frac{d\hat{u}}{dz} + \rho_0\hat{u}\frac{dU}{dz} + (U - c)\frac{d\rho_0}{dz}\hat{u} - g\hat{\rho} \right\} = 0 \tag{13.3.9}$$

Now, from continuity (Eq. 13.3.2),

$$\frac{\partial w'}{\partial z} = -\frac{\partial u'}{\partial x} \quad \Rightarrow \quad \frac{d\hat{w}}{dz} = -(ik)\hat{u}, \quad \frac{d^2\hat{w}}{dz^2} = -(ik)\frac{d\hat{u}}{dz} \tag{13.3.10}$$

Substituting this into Eq. (13.3.9) and rearranging, we obtain the *Taylor–Goldstein equation*,

$$\frac{d}{dz}\left[\rho_0(U - c)\frac{d\hat{w}}{dz} \right] - \frac{d}{dz}\left(\rho_0\frac{dU}{dz}\hat{w} \right)$$

$$- \left[\frac{g}{(U - c)}\frac{d\rho_0}{dz} + \rho_0 k^2(U - c) \right]\hat{w} = 0 \tag{13.3.11}$$

Applying the Boussinesq approximation, this result may be rewritten as

$$(U - c)^2\frac{d^2\hat{w}}{dz^2} + \left[N^2 - (U - c)\frac{d^2U}{dz^2} - (U - c)^2 k^2 \right]\hat{w} = 0 \tag{13.3.12}$$

and for the case $N = 0$, this is known as the *Rayleigh equation*. This last result forms an eigenvalue problem for c. If the imaginary part of c is greater than 0, the perturbations will grow exponentially and the flow is dynamically unstable.

To develop the result for *Kelvin–Helmholtz instability*, consider a two-layer system, with $z = 0$ defined at the interface between the two layers and with densities and velocities ρ_1 and U_1, and ρ_2 and U_2, in the upper and lower layers, respectively (Fig. 13.6). Following the preceding analysis, vertical velocity w' is assumed to be described by an equation of the form of Eq. (13.3.8). Within each layer, $N = 0$ and U is constant, so Eq. (13.3.12) reduces to

$$\frac{d^2\hat{w}}{dz^2} - k^2\hat{w} = 0 \tag{13.3.13}$$

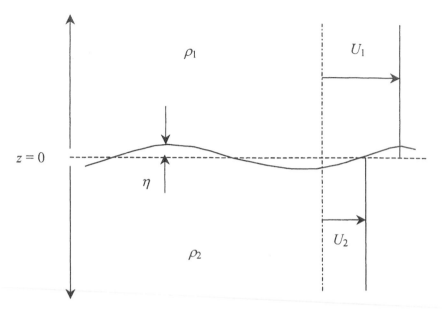

Figure 13.6 Definition sketch for stability analysis of two-layer system.

(note the similarity of this result to Eq. 13.2.34). The solution is an exponential, and in order to keep the velocities bounded as $z \to \pm\infty$, we have

$$\hat{w} = \begin{cases} C_1 e^{-kz} & z > 0 \\ C_2 e^{kz} & z < 0 \end{cases} \tag{13.3.14}$$

The boundary conditions are

$$\frac{D\eta'}{Dt} = \frac{\partial \eta'}{\partial t} + U_1 \frac{\partial \eta'}{\partial x} = w_1'|_{z=0_+} \tag{13.3.15a}$$

and

$$\frac{D\eta'}{Dt} = \frac{\partial \eta'}{\partial t} + U_2 \frac{\partial \eta'}{\partial x} = w_2'|_{z=0_-} \tag{13.3.15b}$$

where η' is the perturbed position of the interface (Fig. 13.6), which is also assumed to be given by a wave equation similar to the velocity. The differential operator on the left-hand side of Eq. (13.3.15) may be written as

$$\left(\frac{\partial}{\partial t} + U \frac{\partial}{\partial x} \right) = ik(U - c) \tag{13.3.16}$$

Then, combining Eqs. (13.3.14) and (13.3.15) gives

$$\hat{w}_1|_{z=0_+} = ik(U_1 - c)\hat{\eta} = C_1 \tag{13.3.17a}$$

$$\hat{w}_2|_{z=0_-} = ik(U_2 - c)\hat{\eta} = C_2 \tag{13.3.17b}$$

Since these must be equal, a relationship between C_1 and C_2 is obtained,

$$C_2 = C_1 \frac{(U_2 - c)}{(U_1 - c)} \tag{13.3.18}$$

A second matching condition is derived by setting the pressures just above and below the interface equal to each other. Using Taylor series expansions,

$$(p_{01} + p_1') + \frac{\partial(p_{01} + p_1')}{\partial z}\eta' \cong (p_{02} + p_2') + \frac{\partial(p_{02} + p_2')}{\partial z}\eta' \tag{13.3.19}$$

Noting that p_0 is the same in the upper and lower layers, and using a hydrostatic approximation, this result is rewritten as

$$p_1' - p_2' \cong (\rho_1 - \rho_2)g\eta' \tag{13.3.20}$$

Differentiating with respect to x,

$$\frac{\partial}{\partial x}(p_1' - p_2') = (\rho_1 - \rho_2)g\frac{\partial\eta'}{\partial x} \tag{13.3.21}$$

Substituting for the pressure derivatives from the momentum equations for the upper and lower layers, this becomes

$$-\rho_1\left(\frac{\partial}{\partial t} + U_1\frac{\partial}{\partial x}\right)u_1' + \rho_2\left(\frac{\partial}{\partial t} + U_2\frac{\partial}{\partial x}\right)u_2' = (\rho_1 - \rho_2)g(ik)\hat{\eta} \tag{13.3.22}$$

This result may be written, using Eq. (13.3.16) as

$$-\rho_1 ik(U_1 - c)u_1' + \rho_2 ik(U_2 - c)u_2' = (\rho_1 - \rho_2)g(ik)\hat{\eta} \tag{13.3.23}$$

Using Eq. (13.3.10), along with Eq. (13.3.14) to evaluate dw'/dz at the interface ($z = 0$), we find

$$u_1' = \frac{i}{k}\frac{dw_1'}{dz} = -iC_1 \qquad u_2' = \frac{i}{k}\frac{dw_1'}{dz} = iC_2 \tag{13.3.24}$$

Substituting these expressions for u_1' and u_2' into Eq. (13.3.23), along with Eq. (13.3.17), assuming $C_1 \neq 0$ and rearranging, we obtain

$$(\rho_1 + \rho_2)c^2 - 2(\rho_1 U_1 + \rho_2 U_2)c + (\rho_1 U_1^2 + \rho_2 U_2^2) - (\rho_2 - \rho_1)\frac{g}{k} = 0 \tag{13.3.25}$$

This last result is a simple quadratic equation for c, which when solved provides a dispersion relationship,

$$c = \frac{\rho_1 U_1 + \rho_2 U_2}{\rho_1 + \rho_2} \pm \left[\frac{g}{k}\left(\frac{\rho_2 - \rho_1}{\rho_1 + \rho_2}\right) - \frac{\rho_1 \rho_2}{(\rho_1 + \rho_2)^2}(U_1 - U_2)^2\right]^{1/2}$$

$$(13.3.26)$$

For the special case of no motion, $U_1 = U_2 = 0$, this reduces to

$$c = \pm\left[\frac{g}{k}\left(\frac{\rho_2 - \rho_1}{\rho_1 + \rho_2}\right)\right]^{1/2}$$

$$(13.3.27)$$

which is a real-valued function. On the other hand, if $\Delta U = (U_1 - U_2)$ is large enough that the term under the radical sign in Eq. (13.3.26) is negative, then c is imaginary and the perturbations are unstable. This last condition may be written as

$$(\Delta U)^2 > \frac{g}{k}\frac{(\rho_1 + \rho_2)(\rho_2 - \rho_1)}{\rho_1 \rho_2} \cong 2\frac{g}{k}\frac{\Delta \rho}{\rho_2}$$

$$(13.3.28)$$

where it has been assumed that $\Delta \rho = \rho_2 - \rho_1$ is small relative to either ρ_1 or ρ_2.

The stability condition is usually written in terms of an *interfacial Richardson number*,

$$\mathrm{Ri_i} = \frac{g}{k\rho_2}\frac{\Delta \rho}{(\Delta U)^2}$$

$$(13.3.29)$$

so that when $\mathrm{Ri_i} < 0.5$, the system is unstable. In other words, for a given ΔU, there exists a k large enough that instability occurs, as an exponentially growing wave. If $\Delta \rho = 0$, any perturbation is always unstable, since Eq. (13.3.28) will be satisfied for any ΔU.

A generalized stability condition was derived by Rayleigh, based on a variation of the Taylor–Goldstein equation (13.3.11),

$$\frac{d}{dz}\left[\rho_0 \frac{d\hat{w}}{dz}\right] - \left\{\frac{d}{dz}\left(\rho_0 \frac{dU}{dz}\right)(U - c)^{-1}\right.$$
$$\left. -\rho_0 N^2(U - c)^{-2} + \rho_0 k^2\right\}\hat{w} = 0$$

$$(13.3.30)$$

Let ψ represent the left-hand side of this equation. Then, by taking the difference, $(\psi\hat{w}^* - \psi^*\hat{w})$, where an asterisk indicates the complex conjugate, and integrating between two positions z_1 and z_2, the result is

$$c_i \int_{z_1}^{z_2} \frac{|\hat{w}|^2}{(U - c_r)^2 + c_i^2}\left\{\frac{d}{dz}\left(\rho_0 \frac{dU}{dz}\right)[(U - c_r)^2 + c_i^2]\right.$$
$$\left. -2\rho_0 N^2(U - c_r)\right\}dz = 0$$

$$(13.3.31)$$

where $c = c_r + ic_i$ has been substituted. If $c_i > 0$ (i.e., the system is unstable), then in order for this last result to apply, the term in brackets in the integrand must change sign at least once in the region between z_1 and z_2 (note that the term preceding the brackets is always positive). This will occur if the velocity profile has a point of inflection, where $d^2U/dz^2 = 0$, and Rayleigh's result may be stated as a necessary condition for instability, that is, $U(z)$ must have an inflection point. This condition also is known as the *Synge theorem*.

To go one step further, a sufficient condition for stability can be developed directly from Eq. (13.3.11). Letting $U' = (U - c)$ and $\hat{w} = FU'$, where F is a function of z, the Taylor–Goldstein equation may be written as

$$\frac{d}{dz}\left[\rho_0 U'\left(U'\frac{dF^*}{dz} + \frac{dU'}{dz}F\right)\right]$$
$$-\frac{d}{dz}\left(\rho_0 U'\frac{dU'}{dz}F\right) + \rho_0(N^2 - k^2 U'^2)F = 0 \qquad (13.3.32)$$

It can be shown that this equation reduces to

$$\frac{d}{dz}\left[\rho_0 U'^2\frac{dF}{dz}\right] + \rho_0(N^2 - k^2 U'^2)F = 0 \qquad (13.3.33)$$

Now assume that $U' \neq 0$ and that F is an unstable solution. For convenience, define another function G, so that $G = FU'^{1/2}$, so Eq. (13.3.33) becomes

$$\frac{d}{dz}\left[\rho_0 U'\frac{dG}{dz}\right] - \left[\frac{1}{2}\frac{d}{dz}\left(\rho_0\frac{dU'}{dz}\right)\right.$$
$$\left. +\rho_0 k^2 U' + \frac{\rho_0}{U'}\left(\frac{1}{4}\left(\frac{dU}{dz}\right)^2 - N^2\right)\right]G = 0 \quad (13.3.34)$$

This equation is then multiplied by G^* and the resulting product is integrated over the region z_1 to z_2. Both the real and the imaginary parts of the result must be zero. In particular, when the imaginary part is zero,

$$c_i \int_{z_1}^{z_2} \rho_0 \left\{\left[\left|\frac{dG}{dz}\right|^2 + k^2|G|^2\right] + \left|\frac{G}{U'}\right|^2\left[N^2 - \frac{1}{4}\left(\frac{dU}{dz}\right)^2\right]\right\} dz = 0$$

$$(13.3.35)$$

Thus, if $c_i > 0$ (unstable), then $(N^2 - 1.4(dU/dz)^2)$ must be less than 0 for some range of z between z_1 and z_2. *Miles' theorem* then states that a sufficient condition for stability is

$$\mathrm{Ri_g} \geq \frac{1}{4} \qquad (13.3.36)$$

where $\mathrm{Ri_g} = N^2/(dU/dz)^2$ is a *gradient Richardson number*.

13.3.2 Convection

Mixing may be generated as a result of instabilities in the flow system, as predicted by the analysis of the foregoing section. Mixing also may be produced by buoyancy instabilities, resulting in convection. This process was introduced briefly in Chap. 12, in connection with surface heat transfer. The stability analysis of Sec. 13.1 showed when convection might occur, but here we discuss what happens when convection does take place. Heat transfer and temperature differences are the usual cause of these motions. First, recall that either free (buoyancy-driven) or forced (flow-driven) convection can occur. An example of free convection is a wall being heated, generating vertical motions as illustrated in Fig. 13.7. A pan of water heated on a stove also exhibits free convection as the water on the bottom becomes heated and rises. An example of forced convection is wind-driven evaporation, as discussed in Sec. 12.3.3.

The analysis of convective motions begins with the general Boussinesq equations (the equations for continuity and momentum, Eqs. (13.2.4) and (13.2.8), respectively, are repeated here for convenience), with an equation for temperature and an equation of state added to complete the mathematical description of the system:

$$\frac{\partial u_j}{\partial x_j} = 0 \tag{13.3.37}$$

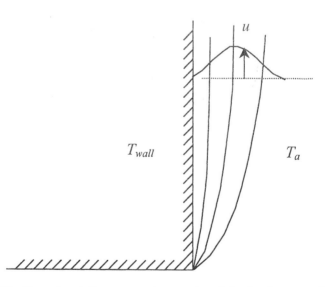

Figure 13.7 Example of free convection, produced by heating a vertical wall ($T_{\text{wall}} > T_a$ = ambient temperature).

$$\frac{\partial u_i}{\partial t} + u_j \frac{\partial u_i}{\partial x_j} = -\frac{1}{\rho_0} \frac{\partial p}{\partial x_i} - \frac{\rho - \rho_0}{\rho_0} g_i + \nu \nabla^2 u_i \qquad (13.3.38)$$

$$\frac{\partial T}{\partial t} + u_j \frac{\partial T}{\partial x_j} = k_T \nabla^2 T \qquad (13.3.39)$$

$$\rho = \rho_0 (1 - \alpha \Delta T_0) \qquad (13.3.40)$$

where g_i is taken as a positive quantity (for downward acceleration of gravity), heat sources are neglected for the present analysis, and α is the thermal expansion coefficient defined in Eq. (13.1.5). By introducing typical scaling quantities U for velocity, L for length, Δp_0 for pressure variation, ΔT_0 for temperature variation, and $\Delta \rho_0$ for density variation, a nondimensional version of the momentum equation is obtained by normalizing all terms with respect to the advective acceleration term,

$$\left[\frac{L}{Ut} \right] + [1] \approx -\left[\frac{\Delta p_0}{\rho_0 U^2} \right] + \left[\frac{\alpha \Delta T_0 gL}{U^2} \right] + \left[\frac{\nu}{UL} \right] \qquad (13.3.41)$$

Nothing that $(\alpha \Delta T_0 = -\Delta \rho / \rho_0)$, the second term on the right-hand side is seen as a variation of the Richardson number defined in Eq. (13.3.29). This term is designated as Ri and represents a ratio of the relative effects of stability, expressed by the buoyancy difference, compared with destabilizing effects of shear. In general, a system with larger Ri is more stable, consistent with previous results (e.g., Eqs. 13.3.29 and 13.3.36). Also, if Ri $\ll$ 1, it may be concluded that buoyancy effects are negligible, compared with inertia, and motions are not significantly affected by temperature (buoyancy) variations. However, when the magnitude of Ri is large, but Ri < 0, buoyancy will supply most or all of the energy for convective motions.

In addition to Ri, another parameter of interest is the Reynolds number, Re, which is the inverse of the last term on the right-hand side of Eq. (13.3.41). When Re is large, there is generally forced convection. The nondimensional pressure term is called either a *Cauchy number* or an *Euler number*, but this term is generally not of major interest in the analysis of convection. One further dimensionless number arising from a nondimensional representation of the temperature equation (13.3.39) is the *Prandtl number*, Pr $= \nu/k_T$. This gives the relative importance of momentum diffusion, compared with heat diffusion. Including the nondimensional temporal term, a simple dimensional analysis suggests solutions of the form

$$\frac{u_i}{U} = f_1 \left(\frac{x_i}{L}, \frac{Ut}{L}, \text{Re}, \text{Pr}, \text{Ri} \right) \qquad (13.3.42a)$$

and

$$\frac{\Delta T}{\Delta T_0} = f_2 \left(\frac{x_i}{L}, \frac{Ut}{L}, \text{Re}, \text{Pr}, \text{Ri} \right) \qquad (13.3.43b)$$

where ΔT is a temperature variation, relative to a background level. For forced convection problems, Ri is small and can be neglected.

For free convection problems, it is convenient to define a buoyancy parameter that does not depend on an imposed external velocity scale, since velocity is not a driving force and in fact is one of the parameters for which a solution is sought. This buoyancy parameter is obtained by combining Ri, Re, and Pr as a *Rayleigh number*,

$$\text{Ra} = \text{Ri } \text{Re}^2 \text{ Pr} = \frac{g\alpha\Delta T L^3}{vk_T} \tag{13.3.43}$$

This parameter encompasses a balance between the buoyancy forces tending to drive the motion and the two diffusive processes that tend to dampen motion; it may be thought of as the product of the ratios of advective to diffusive transport rates for momentum and heat, and suggests a buoyancy velocity scale,

$$u_b = (g\alpha\Delta T_0 L)^{1/2} \tag{13.3.44}$$

Using this instead of U, and substituting Ra for Ri and Re, Eq. (13.3.42) becomes

$$\frac{u_i}{u_b} = f_1\left(\frac{x_i}{L}, \frac{u_b t}{L}, \text{Ra}, \text{Pr}\right) \tag{13.3.45a}$$

and

$$\frac{\Delta T}{\Delta T_0} = f_2\left(\frac{x_i}{L}, \frac{u_b t}{L}, \text{Ra}, \text{Pr}\right) \tag{13.3.45b}$$

These solutions take different forms for different types of problems, depending on geometry, initial and boundary conditions.

Rayleigh–Benard Convection

One of the classical analyses for free convection is based on a perturbation analysis, similar to that described in Sec. 13.1.2, due originally to Benard and extended by Rayleigh. Consider a fluid layer with infinite horizontal dimensions, thickness h, no motion, and a linear (unstable) temperature gradient, as sketched in Fig. 13.8. As the magnitude of the temperature gradient increases, convection will eventually begin as the layer becomes unstable. The convective motions are initially in the form of semistable two-dimensional rolls, which break down into fully turbulent convective motions as Ra is increased. The governing equations are Eqs. (13.3.37)–(13.3.40), with additional perturbation equations,

$$T = T_0 - \frac{\Delta T}{h}z + T' \tag{13.3.46}$$

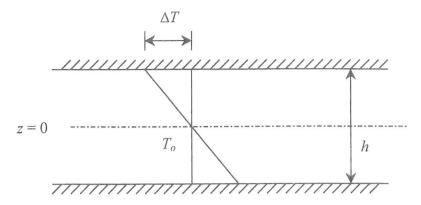

Figure 13.8 System definition sketch for analysis of free convection.

$$p = p_s + p' \tag{13.3.47}$$

$$\frac{\partial p_s}{\partial z} = -g\rho_0 \left(1 + \frac{\alpha \Delta T}{h} z\right) \tag{13.3.48}$$

where p_s is hydrostatic pressure and primes indicate perturbation quantities (i.e., small variations from the initial condition). The temperature equation (13.3.39), upon substituting from Eq. (13.3.46), becomes

$$\frac{\partial T'}{\partial t} + u_j \frac{\partial T'}{\partial x_j} - w \frac{\Delta T}{h} = k_T \nabla^2 T' \tag{13.3.49}$$

where $w = u_3$ is the vertical velocity component. Note that the initial, mean velocity field is one of no motion, so that all velocity components refer to fluctuating terms (primes are omitted for simplicity of notation).

The equations are written in nondimensional form by introducing characteristic scales: h for length, (h^2/k_T) for time, $(vk_T/\alpha gh^3)$ for temperature, and $(k_T^2\rho_0 p'/h^2)$ for pressure. A time scale also is defined as the ratio between the length and velocity scales. The resulting nondimensional equations are (for simplicity, the same notation is used as before, but it should be kept in mind that all variables in the following are dimensionless)

$$\frac{\partial u_j}{\partial x_j} = 0 \tag{13.3.50}$$

$$\frac{\partial u_i}{\partial t} + u_j \frac{\partial u_i}{\partial x_j} = -\frac{\partial p'}{\partial x_i} + \text{Pr} T'(k_i) + \text{Pr} \nabla^2 u_i \tag{13.3.51}$$

$$\frac{\partial T'}{\partial t} + u_j \frac{\partial T'}{\partial x_j} = \nabla^2 T' + \text{Ra}\, w \tag{13.3.52}$$

where $k_i = (0, 0, 1)$ is the vertical unit vector. Following similar procedures as in past derivations (neglect non linear terms in the perturbation quantities, eliminate pressures by cross-differentiating and subtracting the equations for the different momentum components, introduce continuity), a simplified set of equations may be derived as

$$\frac{1}{\text{Pr}} \nabla^2 \left(\frac{\partial w}{\partial t} \right) = \nabla_{\text{h}}^2 T' + \nabla^4 w \tag{13.3.53}$$

$$\frac{\partial T'}{\partial t} = \nabla^2 T' + \text{Ra}\, w \tag{13.3.54}$$

where ∇_{h}^2 is the horizontal Laplacian operator defined in Eq. (13.2.16). Boundary conditions are now needed in order to solve this set of equations.

For free boundaries, $w = T' = 0$ (i.e., vertical velocity kept at 0, temperature maintained constant), while for rigid boundaries, $w = \partial w / \partial z = 0$ (no vertical velocity or momentum transport). Considering free boundary conditions, we use separation of variables and assume solutions of the form

$$w = f_1(t) \cos(k_1 x + k_2 y) \sin(\pi z) \tag{13.3.55a}$$
$$T' = f_2(t) \cos(k_1 x + k_2 y) \sin(\pi z) \tag{13.3.55b}$$

where k_1 and k_2 are horizontal wave number components, $k_1 = A \cos\varphi$ and $k_2 = A \sin\varphi$, and A is the wave number of the disturbance. Note that these solutions satisfy the above free boundary conditions, since $\sin 0 = \sin \pi = 0$ (i.e., free boundaries at $z = 0$ or 1). Substituting these expressions for w and T' into Eqs. (13.3.53) and (13.3.54) and combining to eliminate $f_2(t)$, we obtain

$$\frac{1}{\text{Pr}} (A^2 + \pi^2) \frac{d^2 f_1}{dt^2} + \left(1 + \frac{1}{\text{Pr}} \right) (A^2 + \pi^2)^2 \frac{d f_1}{dt}$$
$$+ \left[(A^2 + \pi^2)^3 - \text{Ra} A^2 \right] f_1 = 0 \tag{13.3.56}$$

This last result is of the form of a forced, damped oscillator. The stability of the system depends on the sign of the term in brackets (multiplying f_1). If this term is negative, the system is unstable, and this defines a critical value for Ra, Ra_{c}, indicating the condition at which instability occurs. In general, Ra is a function of A, and Ra_{c} can be calculated by first setting the bracketed term equal to zero. Let Ra_0 be the value for Ra when this condition is satisfied, so that

$$(A^2 + \pi^2)^3 - \text{Ra}_0 A^2 = 0 \quad \Rightarrow \quad \text{Ra}_0 = \frac{(A^2 + \pi^2)^3}{A^2} \tag{13.3.57}$$

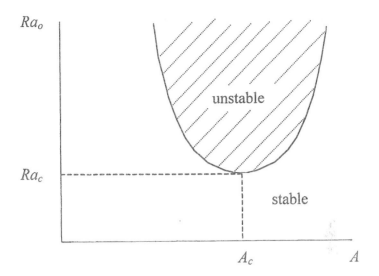

Figure 13.9 Unstable and stable values for Ra_0; the curve denotes the marginal stability condition.

Then, by differentiating Ra_0 with respect to A and setting the result equal to 0, the critical value is

$$A_c = \frac{\pi}{\sqrt{2}} \cong 2.22 \quad \Rightarrow \quad Ra_c = \frac{27\pi^4}{4} \cong 658 \qquad (13.3.57)$$

This result is illustrated in Fig. 13.9. When $Ra < Ra_c$, the system is stable, and convection does not occur. When $Ra > Ra_c$, the system is unstable. The condition when $Ra = Ra_c$ is one of marginal or neutral stability, and wave motions with wave number A_c are set up but do not grow. A similar analysis for rigid boundary conditions gives $A_c \cong 3.12$ and $Ra_c = 1,708$. When one boundary is free and the other rigid, $A_c \cong 2.68$ and $Ra_c \cong 1,101$.

13.4 DOUBLE-DIFFUSIVE CONVECTION

A special case of convective motion occurs when a system is stratified in two components, usually temperature and salinity for environmental applications, and where one of the components is stably stratified and the other component is unstable. Furthermore, the two components must have different diffusivities (note that the diffusivity for temperature is about two orders of magnitude greater than that for salinity). A common example is the upper layer of the ocean, which may be warmer due to solar radiation, and slightly saltier due to evaporation. Thus temperature would provide a stable density stratification, while the density due to the salinity distribution would be unstable. The opposite case, where temperature is unstable and salinity is stable, may occur in shallow saline lakes. This provides a sort of natural solar heat collection system, which was investigated in the 1970s and 1980s for possible commercial use through the use of so-called *salt gradient solar ponds*. These two types of stratification are illustrated in Fig. 13.10. An essential feature of these systems is that instability and convection may occur even when the overall density gradient is stable.

The form of convection resulting from instability depends on which of the components provides the destabilizing force. In Fig. 13.10a, both temperature and salinity decrease with depth, so that temperature (higher diffusivity component) provides the stabilizing contribution to the density gradient. In this case, the instability takes the form of long, thin convection cells known as salt fingers. Stommel et al. (1956) were the first to describe the possibility of salt fingering in the oceans and this is now recognized as an important transport process, with transport rates increased substantially over normal diffusive fluxes. To understand how the fingers are generated, imagine that a fluid particle is displaced downwards from its equilibrium position, where it is in a new environment where both the temperature and the salinity are less. Because of the relatively fast diffusivity of temperature, the particle adjusts quickly to the new temperature, while maintaining its excess salinity, relative to its surroundings. Thus it will experience a net gravitational force and will

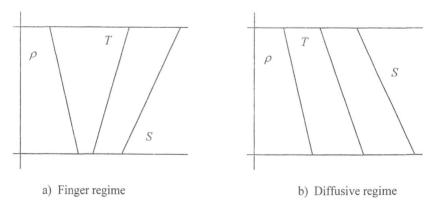

a) Finger regime   b) Diffusive regime

Figure 13.10 Illustration of two types of stratification found in thermohaline double-diffusive systems; in both cases the overall density is stable, although one of the components is unstable.

continue to move downwards. This leads to the development of a "salt fountain," with vertical motion and transport driven essentially by the different diffusivities of the two components. A similar argument applies if the particle is originally displaced upwards, in which case it will continue to rise.

The stratification shown in Fig. 13.10b is known as the diffusive or oscillatory regime. In this case, instabilities are in the form of overstable oscillations. One of the more interesting features observed in experiments is that when a system with a stable salinity gradient is heated from below, a series of well-mixed layers will form, separated by relatively thin interfaces. In other words, unlike a homogeneous system, which would experience convection over the full depth, the stable salinity gradient controls the region over which the convective motions occur. The mechanism for this behavior is again related to the different diffusivities for salt and heat. Basically, while both heat and salt diffuse across an interface, heat diffuses so much more quickly that the diffusive boundary layer becomes unstable and begins to convect. A well-mixed layer then forms, with another interfacial diffusive boundary layer growing ahead of it. With time, these layers have been observed to merge, when the buoyancy difference across adjoining interfaces becomes small. Layer formation also has been observed in fingering systems, where relatively thin salt finger interfaces separate well-mixed layers. In both cases the double-diffusive convective motions provide a much more efficient transport mechanism than molecular diffusion, and as noted before it should be emphasized that these motions occur even though convection would not normally be expected, since the overall density stratification is still stable.

Stability Analysis

The stability of a double-diffusive system can be evaluated using a perturbation analysis that extends the linear stability analysis for Benard convection presented in Sec. 13.3.1, by considering the salinity stratification. Considering stratification by salinity and temperature, the problem is known as the *thermohaline Rayleigh–Jeffreys problem*. The analysis assumes a fluid layer of depth h, bounded above and below by infinite horizontal planes, and with temperature and salinity steps, ΔT and ΔS, respectively, across the layer, with constant gradients. The assumptions of infinite horizontal dimensions implies that horizontal gradients may be neglected. It is easily shown that there are four main nondimensional parameters that describe the system. These are the Rayleigh (Eq. 13.3.43) and Prandtl numbers, as well as a *diffusivity ratio*, τ, and a *stability ratio*, R_ρ, defined as

$$\tau = \frac{k_S}{k_T} \qquad (13.4.1)$$

where k_S is the diffusivity for dissolved salt, and

$$R_\rho = \frac{\beta \Delta S}{\alpha \Delta T} \tag{13.4.2}$$

where α and β are the thermal and saline expansion coefficients defined in Eqs. (13.1.5) and (13.1.6), respectively. In addition to Ra, we will define an equivalent *saline Rayleigh number*, $Ra_S = R_\rho Ra$. It will also be convenient to introduce the Schmidt number, $Sc = \upsilon/k_S$.

The governing equations for this problem are the same as for the Benard convection problem, with the addition of a salinity balance equation,

$$\frac{\partial S}{\partial t} + u_j \frac{\partial S}{\partial x_j} = k_S \nabla^2 S \tag{13.4.3}$$

Also, the equation of state is modified to include salinity effects,

$$\rho = \rho_0 (1 - \alpha \Delta T_0 + \beta \Delta S_0) \tag{13.4.4}$$

An initial quiescent ($u_i = 0$) steady state is assumed, with linear temperature and salinity distributions, as noted previously. This system is then subjected to infinitesimal perturbations u_i', T', S', p' and ρ' (see Eqs. 13.3.46–13.3.48). An equation similar to Eq. (13.3.46) is written for salinity, which when introduced into Eq. (13.4.3) results in

$$\frac{\partial S'}{\partial t} + u_j \frac{\partial S'}{\partial x_j} - w \frac{\Delta S}{h} = k_S \nabla^2 S' \tag{13.4.5}$$

which is analogous to Eq. (13.3.49) for the temperature distribution. As in the preceding section, primes are dropped from the velocity terms for simplicity of notation, and it is understood that the mean velocities are zero.

The linearized equations for the fluctuations are developed from Eqs. (13.3.37)–(13.3.39), along with Eqs. (13.3.49) and (13.4.5),

$$\frac{\partial u_j}{\partial x_j} = 0 \tag{13.4.6}$$

$$\frac{\partial u}{\partial t} = -\frac{1}{\rho_0} \frac{\partial p'}{\partial x} + \upsilon \nabla^2 u \tag{13.4.7}$$

$$\frac{\partial v}{\partial t} = -\frac{1}{\rho_0} \frac{\partial p'}{\partial y} + \upsilon \nabla^2 v \tag{13.4.8}$$

$$\frac{\partial w}{\partial t} = -\frac{1}{\rho_0} \frac{\partial p'}{\partial z} - \frac{\rho'}{\rho_0} g + \upsilon \nabla^2 w \tag{13.4.9}$$

$$\frac{\partial T'}{\partial t} - w \frac{\Delta T}{h} = k_T \nabla^2 T' \tag{13.4.10}$$

$$\frac{\partial S'}{\partial t} - w\frac{\Delta S}{h} = k_S \nabla^2 S' \tag{13.4.11}$$

where (u, v, w) has been substituted for (u_1, u_2, u_3). We now take the derivative of Eq. (13.4.7) with respect to z and subtract the derivative of Eq. (13.4.9) with respect to x. A second equation is obtained by subtracting the derivative of Eq. (13.4.9) with respect to y from the derivative of Eq. (13.4.8) with respect to z. Upon adding the derivative of the first equation with respect to x to the derivative of the second equation with respect to y, we obtain

$$\frac{\partial}{\partial t}\left(\frac{\partial^2 u}{\partial x\,\partial z} - \frac{\partial^2 w}{\partial x^2} + \frac{\partial^2 v}{\partial y\,\partial z} - \frac{\partial^2 w}{\partial y^2}\right) = \frac{g}{\rho_0}\left(\frac{\partial^2 \rho'}{\partial x^2} + \frac{\partial^2 \rho'}{\partial y^2}\right)$$
$$+ \nu\nabla^2\left(\frac{\partial^2 u}{\partial x\,\partial z} - \frac{\partial^2 w}{\partial x^2} + \frac{\partial^2 v}{\partial y\,\partial z} - \frac{\partial^2 w}{\partial y^2}\right) \tag{13.4.12}$$

Making a substitution as in Eq. (13.2.17) for the velocity gradients, this becomes

$$\left(\frac{\partial}{\partial t} - \nu\nabla^2\right)\nabla^2 w = -g\nabla_h^2(-\alpha T' + \beta S') \tag{13.4.13}$$

where Eq. (13.4.4) has been used to substitute for density.

Nondimensional equations are developed by introducing scaling quantities into Eqs. (13.4.3), (13.4.10), and (13.4.11): h for length, h^2/ν for time, k_T/h for velocity, ΔT for temperature, and ΔS for salinity. The resulting non-dimensional equations are

$$\left(\frac{\partial}{\partial t} - \nabla^2\right)\nabla^2 w = \text{Ra}\nabla_h^2 T' - \frac{1}{\tau}\text{Ra}_S\nabla_h^2 S' \tag{13.4.14}$$

$$\left(\text{Pr}\frac{\partial}{\partial t} - \nabla^2\right)T' = w \tag{13.4.15}$$

$$\left(\text{Sc}\frac{\partial}{\partial t} - \nabla^2\right)S' = w \tag{13.4.16}$$

Using a similar approach as for the Benard convection problem, by assuming wavelike solutions, it can be shown that for free, conducting boundaries, the equations representing marginal stability for the system are

$$\text{Ra} - \frac{1}{\tau_S}\text{Ra}_S \leq \frac{27}{4}\pi^4 \tag{13.4.17}$$

and

$$\frac{\text{Sc}^2\text{Ra}}{(\text{Pr} + \text{Sc})(\text{Sc} + 1)} - \frac{1}{\tau}\frac{\text{Pr}^2\text{Ra}_S}{(\text{Pr} + \text{Sc})(\text{Pr} + 1)} \leq \frac{27}{4}\pi^4 \tag{13.4.18}$$

This result is analogous to Eq. (13.3.58) for a temperature-stratified system. In fact, when there is no salinity stratification, $Ra_S = 0$ and Eq. (13.4.21) reduces to Eq. (13.3.58). As before, the factor $(27\pi^4/4)$ arises from consideration of the fastest growing (most unstable) mode.

It is convenient to rewrite (13.4.18) in terms of an "effective" Rayleigh number,

$$Ra_e = Ra - \frac{(Pr + \tau)}{(Pr + 1)}Ra_S \leq \frac{27}{4}\pi^4\left[(\tau + 1)\left(1 + \frac{\tau}{Pr}\right)\right] \qquad (13.4.19)$$

This parameter plays the same role in double-diffusive convection that Ra does in thermal convection. Equations (13.4.17) and (13.4.19) are plotted in Fig. 13.11, along with the line for static gravitational stability, where $Ra = Ra_S$ (i.e., density is constant). The shaded areas represent parameter ranges where the system is gravitationally stable but subject to double-diffusive instabilities.

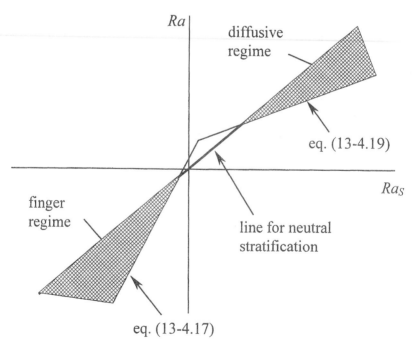

Figure 13.11 Thermohaline stability diagram; lines for marginal stability are plotted along with line of neutral stratification (with $Ra = Ra_S$).

13.5 MIXED-LAYER MODELING

13.5.1 Introduction

A problem of particular interest in water quality modeling is in the mixing of a stratified water body due to the action of surface wind stress, and possibly buoyant convection. Vertical mixing plays an important role in determining the distribution of various water quality parameters, such as temperature, dissolved oxygen, nutrients, and contaminants. Early attempts at modeling wind-induced mixing involved specification of a depth-dependent diffusivity as a function of surface shear stress. Several approaches are possible for estimating these diffusivities, such as the wave orbital velocities model, where the wind-induced surface wave field (see Chap. 8) is needed as input. Convective motions driven by an unstable surface buoyancy flux also may generate mixing. None of these approaches has proved to be totally satisfactory, due to large data requirements for calibration, or to other assumptions needed for their application.

Starting several decades ago, the mixed layer approach has been developed as an alternative to the variable diffusivity approach. This approach is based primarily on the observation that the near-surface water in lakes and oceans tends to be well mixed and bounded below by a well-defined thermocline (or halocline, in the case of salinity stratification), where the density gradient becomes very steep, as illustrated in Fig. 13.12. Also shown in this

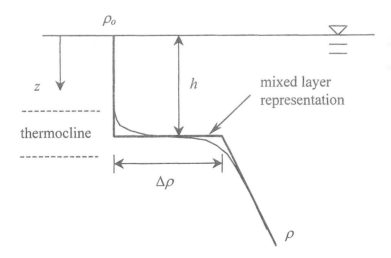

Figure 13.12 Typical density structure in lake or ocean under stratified conditions, along with mixed layer representation.

figure is the density structure assumed in the mixed layer approximation, with a completely mixed upper layer bounded by a sharp density step at its base. The main advantage of this approach is that it is not necessary to specify diffusivities in the upper layer, since there are no gradients on which a diffusivity might act. The problem is in specifying conditions at the upper and lower boundaries of the layer and in determining the mixed layer depth, h. The layer grows by the process of *entrainment* of water from below the interface.

13.5.2 Analysis

In the application of mixed layer models it is usually assumed that the horizontal extent of the system is very large, and that horizontal gradients are small and can be neglected. The analysis then centers on vertical processes. The rate of change of temperature for the upper mixed layer can be calculated from an integrated form of the one-dimensional (vertical) temperature equation, written here to include the solar heating term (Chap. 12),

$$\int_0^h \left\{ \frac{\partial T}{\partial t} + w \frac{\partial T}{\partial z} \right\} dz = \int_0^h \left\{ \frac{\partial}{\partial z} \left(k_T \frac{\partial T}{\partial z} \right) + \eta_s \frac{\Phi_{so}}{\rho c} (1 - \beta_s) e^{-\eta_s z} \right\} dz$$

$$(13.5.1)$$

where c is the specific heat. In carrying out this integration, we assume $w = 0$ at the surface ($z = 0$) and $w = w_e = dh/dt$ (this is known as the *entrainment velocity*) at $z = h$. After integration, Eq. (13.5.1) becomes

$$h \frac{d\overline{T}}{dt} + \overline{T} w_e + (wT)|_h - (wT)|_0 = \left(k_T \frac{\partial T}{\partial z} \right)\bigg|_h - \left(k_T \frac{\partial T}{\partial z} \right)\bigg|_0$$

$$- \left[\frac{\Phi_{so}}{\rho c} (1 - \beta_s) e^{\eta_s z} \right]\bigg|_0^h \quad (13.5.2)$$

where the overbar indicates depth averaging. Note that the sum of the first two terms on the left-hand side, when multiplied by (ρc), is the time rate of change of heat for the upper mixed layer. The last term on the left-hand side is zero because of the boundary condition for w. The advective term at $z = h$ is equal to -$w_e T''$, where T'' is the temperature immediately below the thermocline and the negative sign indicates that the relative movement of water at this temperature is upwards (i.e., in the negative z-direction) into the mixed layer as entrainment occurs. The first term on the right-hand side is the diffusive flux of heat across the interface — this term is usually negligible when entrainment is active. The second term is the heat flux at the surface and is equal to the net heat flux as calculated in Sec. 12.3. The third term gives the difference between solar radiation intensity at the surface and at the thermocline, this difference being absorbed within the layer and serving as a source heating term. Making substitutions and rearranging, Eq. (13.5.2) becomes

$$h\frac{d\overline{T}}{dt} = \left(kT\frac{\partial T}{\partial z}\right)\bigg|_{h} + \frac{1}{\rho c}\{\Phi_n + \Phi_{so}[1 - (1 - \beta_s)e^{-\eta_s h}]\} - w_e(\overline{T} - T'')$$

(13.5.3)

From this result we see that the upper layer temperature may grow according to several different processes: (1) net positive surface heat flux, (2) absorption of solar radiation, and (3) diffusion and entrainment of warmer water from below, if there is a positive temperature step at the base of the layer. However, the temperature step is usually negative, which leads to a lowering of mixed layer temperature due to transport across the interface. The diffusive term is kept for completeness, though it is usually negligible compared with the entrainment flux. This term also is difficult to calculate directly because the interface thickness is not known. Furthermore, the interface thickness is assumed to be vanishingly thin in the mixed layer idealization, so interfacial heat flux would normally be calculated using a bulk transfer coefficient similar to the entrainment term. Diffusion may be thought of as a limiting condition for interfacial heat flux, with the main contribution normally due to the entrainment term. Our main interest here is with this last term in Eq. (13.5.3).

A simple entrainment model is obtained by considering the change in potential energy of the upper layer associated with a change in layer depth, *dh*, that occurs for a given amount of turbulent kinetic energy (TKE) supplied (Fig. 13.13). Possible sources of mixing energy include wind and an unstable

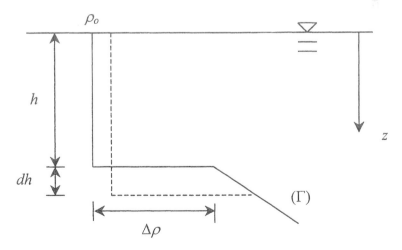

Figure 13.13 Change in mixed layer structure due to entrainment (mixed layer deepens by amount *dh*); Γ is density gradient below interface, assumed constant.

surface buoyancy flux, leading to penetrative convection. The input energy flux due to wind is given in Sec. 12.2 as

$$\frac{dKE}{dt} = Au_d\tau_s = CA\rho_0 u_*^3 \quad \text{(wind)} \tag{13.5.4}$$

where A is surface area, C is a coefficient incorporating a number of proportionality constants, and ρ_0 is the surface layer density. For buoyant-induced mixing (penetrative convection), the surface heat flux gives rise to a buoyancy flux equal to

$$B_F = \frac{\alpha g \Phi_n}{\rho c} \tag{13.5.5}$$

When Φ_n is negative, corresponding to surface cooling, particles of fluid near the surface become accelerated downward due to loss of buoyancy. If it is assumed that the vertical acceleration, a_z, is approximately constant and equal to $g' = g(\Delta\rho/\rho_0)$, where $\Delta\rho$ is the density difference between the fluid particle and ρ_0, and fluid particles are assumed to start from rest, then the velocity at distance h is approximated by $w* \approx a_z t'$, where t' is the time required to reach depth h, and is estimated by $t' \approx h/w*$. Making the appropriate substitutions, it is then easy to see that the velocity scale for penetrative convection can be estimated as

$$w_* \cong (-B_F h)^{1/3} \tag{13.5.6}$$

Note that $w*$ is relevant only when B_F is negative. Alternatively, $(-B_f h)(= w*^3)$ is the kinematic rate at which kinetic energy is added to the upper layer, and some portion of this energy will be available for mixing. A velocity scale for turbulent mixing in the upper layer can then be defined as a combination of the velocity scales relevant for each of these sources. Based on the energy supply rates, a possible scale is

$$\sigma = (w_*^3 + C_v^3 u_*^3)^{1/3} \tag{13.5.7}$$

where C_v is a proportionality coefficient.

The total rate at which kinetic energy is added to the upper layer is then

$$\frac{dKE}{dt} = A\rho_0 \sigma^3 \tag{13.5.8}$$

When an amount of fluid of thickness dh is entrained from below the interface and mixed into the upper layer, the resulting change in potential energy of the upper layer is

$$dPE = \frac{1}{2}gA(\Delta\rho h\,dh + \Gamma h\Delta h^2) \tag{13.5.9}$$

where the density below the interface is given by $\rho = \rho_0 + \Delta\rho + \Gamma(z - h)$ and Γ is the density gradient in the fluid below the interface (Fig. 13.12). Since dh is assumed to be small, the quadratic term in dh is neglected, and the rate at which potential energy changes is approximately

$$\frac{dPE}{dt} \simeq \frac{1}{2}gA\Delta\rho h\frac{dh}{dt} = \frac{1}{2}gA\Delta\rho h w_e \qquad (13.5.10)$$

The ratio of change in potential energy to input kinetic energy is known as a *flux Richardson number* and is obtained from the ratio of Eqs. (13.5.10) and (13.5.8),

$$\text{Rf} = \frac{dPE}{dKE} = \frac{1}{2}gw_e\frac{\Delta\rho}{\rho_0}\frac{h}{\sigma^3} \qquad (13.5.11)$$

This is simplified by introducing the *bulk Richardson number*, similar to Eq. (13.3.29),

$$\text{Ri}_\sigma = g\frac{\Delta\rho}{\rho_0}\frac{h}{\sigma^2} = g'\frac{h}{\sigma^2} \qquad (13.5.12)$$

so that

$$\text{Rf} = \frac{1}{2}\frac{w_e}{\sigma}\text{Ri}_\sigma \qquad (13.5.13)$$

The flux Richardson number is an indication of the efficiency of mixing; a value less than one indicates that some input kinetic energy must be used; either it increases the overall kinetic energy of the upper layer or it is dissipated. As long as the entrainment function, w_e/σ, can be evaluated, this efficiency can be calculated. On the other hand, a simple assumption (common in early mixed layer approaches) is that Rf is a constant. This leads to a simple entrainment law,

$$E = \frac{w_e}{\sigma} \propto \text{Ri}_\sigma^{-1} \qquad (13.5.14)$$

where E is the nondimensional *entrainment rate*. If the mixing is perfectly efficient, where all of the input kinetic energy is converted to potential energy by entrainment, $\text{Rf} = 1$ and $E = 2\text{Ri}_\sigma^{-1}$. Several early mixed layer models used this relation to calculate mixing of the upper layer. However, there are other sinks for the input energy, which must be taken into account in order to evaluate E more accurately. This can be done through an evaluation or parameterization of the TKE budget for the upper layer.

In addition to entrainment, kinetic energy that is added to the layer may increase the kinetic energy level in the layer, it may be dissipated within the layer, and it may generate internal waves. These waves may propagate along the interface and eventually break or, in the case of a stratified lower layer, the waves may radiate away from the interface, thus representing a leakage of energy from the layer. Each of these processes causes Rf to be less than one. Assuming that viscous transport is negligible, and also neglecting all horizontal gradients, the TKE budget is (see Chap. 5)

$$\frac{\partial K}{\partial t} = -\frac{g}{\rho_0}\overline{w'\rho'} - \frac{\partial}{\partial z}\left(\frac{\overline{w'p'}}{\rho_0} + \overline{w'K}\right) - \overline{u'w'}\frac{\partial U}{\partial z} - \varepsilon \qquad (13.5.15)$$

where $K = 1/2\overline{u_i'^2}$ is the TKE per unit mass, overbars denote mean (time-averaged) quantities, and U is the time-averaged velocity in the upper layer. The left-hand side of this equation is the time rate of change of TKE, the first term on the right-hand side is gravity work due to entrainment, the second term is the flux divergence, the third term is the mechanical production of TKE by the working of the Reynolds stresses on the mean shear, and the last term is the viscous dissipation rate of TKE per unit mass. The relative magnitudes for each of these terms must be determined in order to relate w_e to the parameters describing the turbulence.

Since the upper layer is considered to be mixed, it is more convenient to work with an averaged form of Eq. (13.5.15), by integrating over the upper layer depth, h. By doing this, vertical gradients disappear, except as they describe transport at the upper and lower boundaries of the layer. For example, the flux divergence term provides the main source of energy due to surface wind stress or penetrative convection. It also incorporates the net effects of shear production at the lower boundary (representing a source for K in the layer) and loss of energy due to internal wave production (i.e., energy is used to generate waves). Zeman and Tennekes (1977) and Sherman et al. (1977) developed parameterizations for the averaged TKE budget, in order to evaluate the entrainment rate. This parameterization process is similar to treating the equation through dimensional analysis, in order to estimate the relative importance of each term.

First, the time-dependent term is parameterized as

$$\frac{\partial \overline{\overline{K}}}{\partial t} \approx C_t \sigma^2 \frac{w_e}{h} \qquad (13.5.16)$$

where the double overbar indicates a layer-averaged quantity and C_t is a proportionality coefficient. This equation results from the observation that the mean TKE should be proportional to σ^2 and that the time scale of interest is approximately (h/w_e). The buoyancy term is written as the buoyancy flux across the interface,

$$-\frac{g}{\rho_0}\overline{\overline{w'\rho'}} \approx w_e g(\alpha \Delta T) \qquad (13.5.17)$$

The integrated flux divergence term represents the main source of TKE (at the boundaries) and is proportional to σ^3/h. The mean viscous dissipation also is proportional to this quantity (see Chap. 5). As noted previously, another form of energy loss is that due to internal waves that might radiate energy away from the interface, when $\Gamma \neq 0$. This flux will depend on the amplitude and period of the generated waves, which in turn depends on the stratification (Brunt–Vaisala frequency N — see Eq. 13.1.15). The amplitude a can be estimated by equating the kinetic energy of an eddy impinging on the interface with the potential energy associated with the maximum displacement of the interface. Referring to Fig. 13.12, when Γ is small, a may be approximated as $a \propto \rho_0 \sigma^2/g\Delta\rho$. However, if $\Delta\rho$ is small, then $a \propto \sigma/N$. The energy in the waves is proportional to $(aN)^2$, and this energy propagates at a rate N, so that the "leakage" rate of energy is proportional to $a^2 N^3$. Following Sherman et al. (1977) in defining a using an interpolating function between the two limits noted above, the total energy loss rate due to internal wave radiation is

$$\varepsilon' \approx C_{\mathrm{d}} \sigma^2 N \left(\frac{\rho_0 \sigma N}{\rho_0 \sigma N + g\Delta\rho} \right)^2 \tag{13.5.18}$$

The net result of the flux divergence source and viscous dissipation and internal wave losses is written as

$$-\frac{\partial}{\partial z} \overline{\left(\frac{\overline{w'p'}}{\rho_0} + w'K \right)} - \bar{\bar{\varepsilon}} \approx C_{\mathrm{f}} \frac{\sigma^3}{h} - \varepsilon' \tag{13.5.19}$$

where C_{f} is a constant. The shear production term is written in a form similar to the temporal term, but with U instead of σ,

$$-\overline{u'w'} \frac{\partial U}{\partial z} \approx C_{\mathrm{s}} U^2 \frac{w_{\mathrm{e}}}{h} \tag{13.5.20}$$

where C_{s} is another constant.

Substituting the parameterized terms into the averaged TKE budget (13.5.15) and rearranging, an expression for w_{e} is obtained,

$$\frac{w_{\mathrm{e}}}{\sigma} = \frac{C_{\mathrm{f}} - \varepsilon' \dfrac{h}{\sigma^3}}{C_{\mathrm{t}} + \mathrm{Ri}_\sigma - C_{\mathrm{s}} \left(\dfrac{U}{\sigma} \right)^2} \tag{13.5.21}$$

Coefficient values are determined from idealized experiments, and ranges of values found are reported in Table 13.1. Equation (13.5.21) can then be used in a mixed layer formulation to model the growth of the upper layer.

It is interesting to note that a simplified version of Eq. (13.5.21) is the inverse Richardson number relation discussed previously (Eq. 13.5.14).

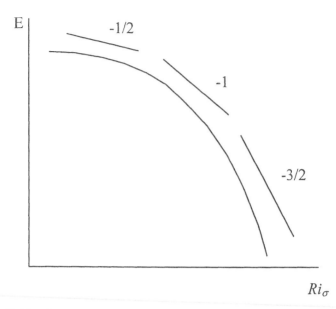

Figure 13.14 Nondimensional entrainment rate as a function of Richardson number (log scales); the straight lines indicate curves with constant slope, corresponding to the exponent used in an inverse Richardson number relationship.

Table 13.1 Coefficient Values for Entrainment Relation

Coefficient	Range
C_v	1.0–2.2
C_f	0.2–1.0
C_d	0–0.04
C_t	0–3.6
C_s	0–1.0

Source: Adapted from Sherman et al., 1978; Zeman and Tennekes, 1977.

Most entrainment relations, in fact, have been expressed in terms of an inverse Richardson number dependency, with the exponent varying between roughly $-1/2$ and $-3/2$, as illustrated in Fig. 13.14. Christodoulou (1987) explained the different ranges (and corresponding exponents) in terms of the relative importance of different mixing mechanisms, and Atkinson (1987) suggested that in the limit of very high Ri_σ, entrainment should eventually cease and interfacial fluxes would be controlled by diffusion. Figure 13.14 shows qualitatively the variation of E with Ri_σ, as observed in a number of experiments reported in the literature.

PROBLEMS

Unsolved Problems

Problem 13.1 Calculate the buoyancy of a fluid particle with zero salinity, at a temperature of $15°C$, relative to a reference density of $\rho_0 = 1000 \text{ kg/m}^3$.

Problem 13.2 What is the relative error in calculating the density of freshwater at $20°C$, using a constant value of $\alpha = 2 \times 10^{-4} °C^{-1}$, compared with density calculated according to Eq. (13.1.8)?

Problem 13.3 If the sonic velocity in water is 1500 m/s, what is the maximum adverse density gradient (i.e., with density increasing upwards) possible so that the water column will not be unstable?

Problem 13.4 How would you include viscous effects directly in the analysis of gravitational stability? Develop the form of the equation similar to Eq. (13.1.4). It is not necessary to solve the equation.

Problem 13.5 Consider a water column with uniform density.
   (a) Find a general solution for the vertical component w, assuming wave-like disturbances as in Eq. (13.2.21) [(in other words, solve for $W(z)$]). Assume deep water and that the waves propagate in the x-direction only.
   (b) Using the approximate boundary condition Eq. (13.2.31), show that the dispersion relationship for this problem is $\sigma^2 = gk_h$.

Problem 13.6 Assuming a relatively thin density interface in a deep ocean that is otherwise well mixed, calculate and plot the relationship between frequency and wave number for an interfacial thickness of 2 m. Use $\Delta\rho/\rho_0 = 0.005$ and assume the interface is centered at a location 10 m below the surface. What is the maximum value for k_h you would use so that the interface would still be considered "thin"? What is the wavelength that corresponds to this maximum value for k_h?

Problem 13.7 Plot the variation of phase speed for a small-scale internal wave as a function of k_h, in an environment where $N = 0.1\ s^{-1}$. Choose several values of k_h/k between 0 and 1 for your plot.

Problem 13.8 Below Eq. (13.2.59) it is stated that "the linearized solution is an exact solution to the governing equations." What are the governing equations for which Eq. (13.2.57) is a solution? What boundary conditions are applicable? Where, in fact, was the linearization assumption made in the derivation?

Problem 13.9 Derive Eq. (13.3.28). Physically, what mechanisms are represented by the interfacial Richardson number defined in Eq. (13.3.29)? What is the effect of a density difference across the interface on the development of instabilities?

Problem 13.10 Estimate the magnitude of the temperature gradient associated with the critical Richardson number defined in Eq. (13.3.57). Compare this result with the gravitational stability criterion developed in Sec. 13.1 (also see problem 13.3). Comment on differences or similarities in the results.

Problem 13.11 Recall that the turbulent kinetic energy (TKE) budget may be written as

$$\frac{\partial K}{\partial t} = -\frac{g}{\rho}\overline{w'\rho'} - \frac{\partial}{\partial z}\overline{(w'K + w'p'/\rho)} - \overline{u'w'}\frac{\partial \bar{u}}{\partial z} - \varepsilon$$

where K = TKE per unit mass. For application to the upper layer in a wind-mixed layer model, a possible parameterization of these terms can be written as

$$[\text{temporal change}] \approx C_T w_e \frac{\sigma^2}{l} \qquad [\text{shear production}] \approx C_s \sigma^2 \frac{\bar{u}}{l}$$

$$[\text{flux divergence - dissipation}] \approx C_F \frac{\sigma^3}{l} \qquad [\text{buoyancy}] \approx C_B u_e g \frac{\Delta\rho}{\rho_0}$$

where w_e is the entrainment velocity, σ and l are turbulent velocity and length scales, respectively, and C_T, C_S, C_T, and C_B are coefficients. Substitute these parameterizations (watch signs) to develop a nondimensional entrainment law for (w_e/σ). Compare this with the "simple" model derived in Eq. (13.5.13),

$$\frac{w_e}{u_*} = 2\frac{\text{Rf}}{\text{Ri}_\sigma} \qquad (\text{you may assume } \sigma = u_*)$$

In other words, under what conditions do these models have the same general form with respect to Ri_σ?

SUPPLEMENTAL READING

Atkinson, J. F. and Munoz, D., 1988. A diffusive limit for entrainment. *J. Hydraul. Res.* **26**:117–130.

Atkinson, J. F., 1988. Note on interfacial mixing in stratified flows. *J. Hydraul. Res.* **26**:27–32.

Bloss, S. and Harleman, D. R. F., 1979. *Effect of wind-mixing on the thermocline formation in lakes and reservoirs.* Tech. Rep. 249, Ralph M. Parsons Lab., M.I.T., Cambridge, MA.

Bo Pedersen, F., 1980. *A monograph on turbulent entrainment and friction in two-layer stratified flow.* Series paper 25, Inst. Hydrodyn. Hyd. Eng., Tech. Univ. Denmark.

Chandrasekhar, S., 1961. *Hydrodynamic and Hydromagnetic Stability.* Clarendon Press, Oxford.

Christodoulou, G. G., 1987. Interfacial mixing in stratified flows. *J. Hydraul. Res.* **24**:77–92.

Craik, A. D. D., 1985. *Wave Interactions and Fluid Flows.* Cambridge University Press, Cambridge.

Fischer, H. B., List, E. J., Koh, R. C. Y., Imberger, J. and Brooks, N. H., 1979. *Mixing in Inland and Coastal Waters.* Academic Press, NY.

Harleman, D. R. F., 1961. Stratified flow. In *Handbook of Fluid Mechanics*, ed. V. L. Streeter, ch. 26, McGraw-Hill, New York.

Lighthill, James, 1978. *Waves in Fluids.* Cambridge University Press, Cambridge.

Philips, O. M., 1977. *Entrainment from Modeling and Prediction of the Upper Layers of the Ocean*, ed. E.B. Kraus, Pergamon Press, Oxford.

Sherman, F. S., Imberger, J. and Corcos, G. M., 1978. Turbulence and Mixing in stably stratified waters. *Ann. Rev. Fluid Mech.* **10**:267–288.

Stommel, H., Arons, A. B. and Blanchard, D., 1956. An oceanographical curiosity: the perpetual salt fountain. *Deep-Sea Res.* **3**:152.

Turner, J. S., 1973. *Buoyancy Effects in Fluids.* Cambridge University Press, Cambridge.

Turner, J. S., 1979. *Buoyancy Effects in Fluids.* Cambridge University Press, Cambridge.

Watson, Andrew, E. P. (ed.), 1976. *The dynamics of stratification and of stratified flow in large lakes.* Proc. Int'l Joint Commission research Advisory Board workshop. International Joint Commission, Windsor, Ontario.

Zeman, O. and Tennekes, H., 1977. Parameterization of the turbulent energy budget at the top of the daytime atmospheric boundary layer. *J. Atmos. Sci.* **34**:111–123.

14
Dynamics of Effluents

14.1 JETS AND PLUMES

There are many problems in both natural and engineered systems that involve flows with large sources of momentum and/or bouyancy. A common example is the outflow of a river or estuary into a large lake or coastal region. The river flow may have significant velocity and thus momentum, and also may be buoyant relative to the receiving water. Estuary flow normally is buoyant due to lower salt content, compared with the ocean or sea into which it flows. Similarly, rivers flowing into lakes or reservoirs may be warmer (positively buoyant) or colder (negatively buoyant) than the lake water. In addition to such surface discharges and dense undercurrents, interflows are also possible; when the receiving water body is stratified the inflow will "find" its density equilibrium level. These three possibilities are illustrated in Fig. 14.1. Similar examples for engineered systems include heated discharge from a power plant (condenser water discharge), which may be mixed with the receiving water or discharged at the surface, and sewage treatment plant effluent discharged into the ocean or other large receiving water body, consisting of a buoyant (less saline) fluid which is commonly discharged through submerged multiport diffusers, as illustrated in Fig. 14.2.

All of the above examples involve situations in which both momentum and buoyancy may be important in describing the dynamics of the flow. Whether momentum or buoyancy is more important in driving the flow determines whether it is called a jet or a plume, respectively. Traditionally, a jet has its primary source of kinetic energy and momentum flux due to a pressure drop through an orifice, while a plume derives its main driving force from buoyancy. In natural systems the relative importance of momentum is often characterized by the value of the source Froude number or in terms of certain length scales, as discussed further below. In particular, when there are density differences between the discharge and the receiving water, a *densimetric Froude number*, Fr_d, is defined, using reduced gravity g' in place of g,

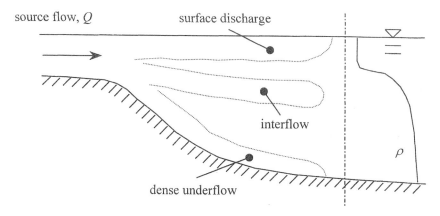

Figure 14.1 Illustration of three flow situations, depending on the density of inflow, relative to the density structure of the receiving water body: surface discharge occurs when the inflow has a density less than or equal to the surface density; underflow occurs when the inflow density is greater than or equal to the bottom density; and interflow occurs when the inflow density is at an intermediate value.

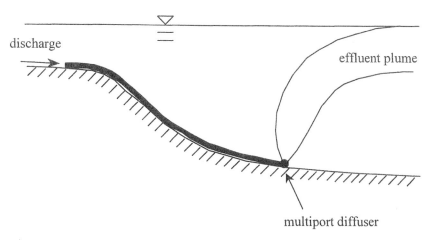

Figure 14.2 Multiport diffuser used to discharge wastewater into the ocean; there is usually intense local mixing near the nozzles of the diffuser, after which the mixed fluid rises as a buoyant plume.

i.e.,

$$\text{Fr}_d = \frac{U}{(g'h)^{1/2}} \tag{14.1.1}$$

This parameter is similar to the inverse square of the bulk Richardson number, introduced in Chap. 13. Flows where both momentum and buoyancy are important are called *buoyant jets* or *forced plumes*, and "large" Fr_d indicates more jetlike behavior, while "low" values indicate more plumelike behavior.

14.1.1 Jets

Before discussing natural discharges, it is helpful first to define basic parameters using the example of a pure (nonbuoyant) axisymmetric jet, as shown in Fig. 14.3. The jet discharges with flow rate Q_0, initial velocity U_0, and diameter D_0 into otherwise quiescent fluid. As the flow moves forward along the x-direction it expands due to turbulent entrainment of the receiving water, with $b(x)$ denoting the radius of the jet at downstream locations. The rate of entrainment decreases as the jet momentum becomes more dilute, but the flow rate increases even as the forward mean velocity diminishes. Normal (Gaussian) distributions about the jet centerline are usually assumed to describe the variation of jet properties such as mean velocity, turbulence intensity, shear stress, or concentration. This suggests the idea of *self-similarity* for the jet profiles, i.e., when properly scaled, all profiles have the same basic shape, independent of position along the centerline. In fact, observations indicate

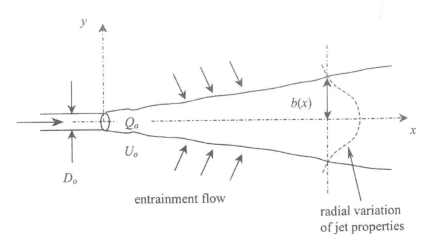

Figure 14.3 Neutral axisymmetric jet discharge; the x axis corresponds with the jet centerline.

this is true only after some distance downstream, when the jet has reached a condition of turbulent stress equilibrium. This state is reached for x greater than about 40 jet diameters downstream (i.e., $x > 40D_0$). For x less than about $6D_0$, there is large-scale vortex generation and entrainment as these vortices engulf the surrounding fluid. As the jet progresses further out, these vortices interact with each other and break down into smaller turbulent eddies.

The *near-field* region of the jet is defined as the region within about $(6-10)D_0$ of the discharge point. An intermediate region then develops, followed by the far-field, self-similar region. In this far-field region, with turbulent stress equilibrium, experiments have shown that the average relative turbulence intensity in the streamwise direction (x) reaches a constant value,

$$\frac{\sqrt{\overline{u'^2}}}{U_m} \cong 0.28 - 0.29 \qquad (14.1.2)$$

where U_m is the maximum (centerline) velocity. The average relative turbulence intensities in the y- and z-directions, also normalized by U_m, are slightly less, approximately $0.23-0.25$. The distributions of these intensities follow a Gaussian profile, as does mean (time-averaged) axial velocity. For example, the mean axial velocity can be written as

$$U = U_m \exp\left[-\left(\frac{r}{b_u}\right)^2\right] \qquad (14.1.3)$$

where r is radial position and b_u is the radial distance from the centerline at which the mean velocity is 37% of the local centerline velocity. In other words, $e^{-1} = 0.37$ when $r = b_u$.

By integrating the time-averaged axial momentum equation over the cross section of the jet, the *specific momentum flux* is defined as

$$M = \int_0^{b(x)} \left(U^2 + \overline{u'^2} + \frac{\overline{p} - \overline{p_\infty}}{\rho_0}\right) 2\pi r\, dr \qquad (14.1.4)$$

where p is pressure and p_∞ is the ambient pressure at some large distance from the jet. This quantity determines to a large extent the basic properties of the jet and its development in the receiving fluid. The integral in Eq. (14.1.4) has been estimated using

$$M \cong \frac{\pi}{2} b_u^2 U_m^2 (1 + \delta_u - \delta_p) \qquad (14.1.5)$$

where δ_u is the contribution to M due to turbulence and δ_p is the contribution due to pressure. Typical values found for these parameters are $\delta_u \cong 0.15$ and $\delta_p \cong 0.10$. Thus most of the contribution to M comes from the mean flow U, and M is often approximated simply by

$$M \cong QU \tag{14.1.6}$$

Also, if shear stresses are neglected (i.e., frictionless flow), then M remains constant with x and

$$M = M_0 = Q_0 U_0 = A_0 U_0^2 \tag{14.1.7}$$

where A_0 is the discharge cross-sectional area. Thus, although the velocity decays, M remains constant because of a corresponding increase in flow rate due to entrainment.

Furthermore, from experimental observations it has been found that b_u may be expressed as a linear function of x, so that $b_u = c_1 x + c_2$, where c_1 and c_2 are constants. Then, by combining Eqs. (14.1.5) and (14.1.7), we find

$$\frac{U_m}{U_0} = \frac{1}{c_1} \frac{2}{\pi} \frac{1}{1 + \delta_u - \delta_p} \frac{\sqrt{A_0}}{x + c_1/c_2} = K \frac{\sqrt{A_0}}{x + c_1/c_2} \tag{14.1.8}$$

where $K \cong 7.46$ if δ_u and δ_p are ignored. This gives a convenient means of calculating the decay of the jet centerline velocity as it spreads into the receiving fluid.

Measurements of a tracer in the jet also are found to be described by similarity profiles,

$$\overline{C} = C_m \exp\left[-\left(\frac{r}{b_\xi}\right)^2\right] \tag{14.1.9}$$

where C is the concentration of any tracer, the overbar indicates a time-averaged value, and C_m is the maximum (centerline) value. The similarity of this profile to the velocity profile (Eq. 14.1.3) is obvious from inspection. Following a similar procedure as with momentum, the total flux of the tracer in the x-direction can be calculated by integrating the time-averaged advection–diffusion equation (Chap. 10) over the cross-sectional area. Again assuming that the contributions to the transport due to turbulence are relatively small compared with the mean flow, and neglecting any sources or sinks of the tracer, an expression analogous to Eq. (14.1.6) results,

$$C_F = QC \tag{14.1.10}$$

where C_F is the *specific flux* of the tracer. A special case of interest is the *specific buoyancy flux*, B, which for a jet may be defined as

$$B = Qb \tag{14.1.11}$$

where $b(= g')$ is the buoyancy or reduced gravity defined previously. As will be shown below, B is defined differently for a *plume*, in which the motions are driven by buoyancy rather than momentum.

With entrainment, the flow rate of the jet increases with distance downstream. This enters directly into any conservation equation applied to the jet. For example, considering the jet control volume of Fig. 14.4, a simple statement of mass balance says the sum of the rate of transport of mass into the control volume across the left-hand face plus the rate of mass transport across the radial boundary by entrainment must equal the rate of mass transport across the right-hand face (assuming no accumulation within the volume). Mathematically, this is written as

$$\int_0^{b(x)} U 2\pi r\, dr + v_e 2\pi \left(b + \frac{1}{2} \frac{\partial b}{\partial x} dx \right) dx = \int_0^{b(x+dx)} U 2\pi r\, dr \quad (14.1.12)$$

where v_e is the *entrainment velocity*, assumed to be equal to the negative of the velocity V at position $r = b$. This definition means that v_e is positive inwards to the jet. Using a Taylor series expansion, the two transport rates in the axial direction can be related by

$$\int_0^{b(x+dx)} U 2\pi r\, dr = \int_0^{b(x)} U 2\pi r\, dr + \frac{\partial}{\partial x} \left[\int_0^{b(x)} U 2\pi r\, dr \right] dx \quad (14.1.13)$$

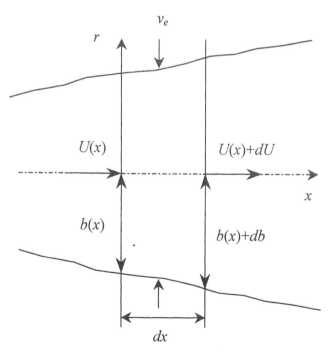

Figure 14.4 Control volume used to define the entrainment function.

Then, substituting Eq. (14.1.13) into Eq. (14.1.12) and rearranging, we have

$$v_e 2\pi \left(b + \frac{1}{2} \frac{\partial b}{\partial x} \, dx \right) \cong v_e 2\pi b = \frac{\partial}{\partial x} \left[\int_0^{b(x)} U 2\pi r \, dr \right] \tag{14.1.14}$$

This is known as the *entrainment function*.

For simplicity of notation, let μ represent the integral in Eq. (14.1.14). Following Taylor, the entrainment is assumed to be proportional to the center-line velocity, and b_u is used to define b, so that Eq. (14.1.14) can be rewritten as

$$\frac{\partial \mu}{\partial x} \longrightarrow \frac{d\mu}{dx} = 2\pi \alpha_e b_u U_m \tag{14.1.15}$$

where α_e is the *entrainment coefficient*. By comparing Eqs. (14.1.14) and (14.1.15) it is seen that the entrainment velocity is estimated as $v_e = \alpha_e U_m$. Alternatively, the entrainment can be specified as a function of M, since M is thought to be the primary controlling parameter for all jet properties. In this case, using the definition in Eq. (14.1.5) and neglecting δ_u and δ_p,

$$\frac{d\mu}{dx} = 2\pi \alpha_e \left(\frac{2M}{\pi} \right)^{1/2} = \alpha_e \sqrt{8\pi M} \tag{14.1.16}$$

Since M is constant, the entrainment also is constant. Integrating Eq. (14.1.16) from zero to some arbitrary x, we find

$$\mu = \alpha_e \sqrt{8\pi M} x \tag{14.1.17}$$

From experiments, the entrainment coefficient has been found to be $\alpha_e \cong 0.05$, so that Eq. (14.1.17) can be written as

$$\mu = 0.25 \sqrt{M} \, x \tag{14.1.18}$$

A *local length scale* can be defined for the jet as

$$\lambda = \frac{\mu}{m^{1/2}} \tag{14.1.19}$$

where m is the local momentum flux, which is equal to M under the no stress assumption. Considering Eq. (14.1.17), this scale is directly proportional to x, which is consistent with the observation that, if viscous forces are negligible, x is the only overall length scale available for the problem. This was implicitly assumed earlier when defining b_u as a function of x (before Eq. 14.1.8). The ratio of λ/x is a measure of the *spreading angle* of the jet, which for $m = M$ is evaluated by substituting Eq. (14.1.17) into Eq. (14.1.19):

$$\frac{\lambda}{x} = \frac{\mu}{xM^{1/2}} = \frac{\alpha_e \sqrt{8\pi M} x}{xM^{1/2}} = \alpha_e \sqrt{8\pi} \tag{14.1.20}$$

This shows that all (round) jets should spread at a constant angle, as long as α_e is constant.

A *local velocity scale* also can be defined:

$$v = \frac{m}{\mu} \tag{14.1.21}$$

where again, $m = M$ for the zero shear stress assumption. Based on this definition and the length scale λ, a local *jet Reynolds number* is

$$\text{Re} = \frac{\lambda v}{\nu} = \frac{m^{1/2}}{\nu} \tag{14.1.22}$$

where ν is kinematic viscosity. Since $m(= M)$ is a constant, Re also is constant. Again, recall that this result applies to a round, neutrally buoyant jet, as illustrated in Fig. 14.3.

14.1.2 Plumes

As previously noted, plumes derive their motion mainly from body forces. Normally, we consider thermal plumes, where buoyancy is due primarily to heat flux. However, saline plumes also are possible. Plumes may be either positively or negatively buoyant, indicating their tendency to rise or sink, respectively, in the receiving fluid. A positively buoyant plume is illustrated in Fig. 14.5. Instead of momentum flux, plumes are described using the value

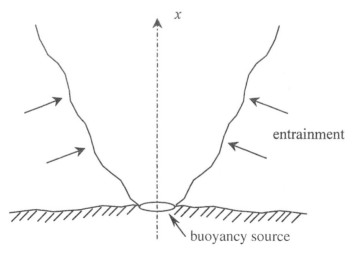

Figure 14.5 Development of a buoyant plume.

of the *specific buoyancy flux,* which is defined (for a thermal plume) as

$$B = \frac{\alpha g H}{\rho c_p} \tag{14.1.23}$$

where α is the thermal expansion coefficient (not to be confused with the entrainment coefficient α_e), H is the heat supply rate, and c_p is specific heat at constant pressure. For plumes, B takes the place of M in determining the general behavior of the flow. The corresponding definition of buoyancy flux for a saline plume is

$$B = \frac{\beta g F_s}{\rho} \tag{14.1.24}$$

where β is the saline expansion coefficient and F_s is the salinity supply rate. Alternatively, if the flux comes from a discharge of magnitude Q, the buoyancy flux is defined by

$$B = g\frac{\rho_a - \rho_0}{\rho_0}Q = g'Q \tag{14.1.25}$$

which is the same as Eq. (14.1.11). Here, ρ_a is the ambient (receiving fluid) density and ρ_0 is the density of the plume discharge. Equilibrium turbulence intensities tend to be somewhat larger in plumes than in jets, due to buoyancy production terms (refer to Chap. 5).

From dimensional considerations, the local momentum flux is related to the buoyancy flux by

$$M = k_m B^{2/3} x^{4/3} \tag{14.1.26}$$

where k_m is a constant. This result is different from that in jets, since M increases with x, whereas in jets M was assumed to remain approximately constant. As with the jet, the entrainment function is proportional to $M^{1/2}$ (refer to Eq. 14.1.16):

$$\frac{d\mu}{dx} = k_e M^{1/2} \tag{14.1.27}$$

where k_e is a constant with measured values in the range 0.25–0.34. This is similar to the constant value in Eqs. (14.1.16) and (14.1.17), where the constant multiplying $M^{1/2}$ is about 0.25. Also, with M proportional to $(b^2 U_m^2)$ and b proportional to x (same assumptions as for jets), combined with Eq. (14.1.26), the maximum centerline velocity is then

$$U_m \propto \left(\frac{B}{x}\right)^{1/3} \tag{14.1.28}$$

14.1.3 Turbulent Buoyant Jets

As defined previously, turbulent buoyant jets are discharges that have characteristics intermediate between the limits of a pure jet and a pure plume. In addition to the densimetric Froude number Fr_d, various length scales are used to describe the characteristics of these flows. This can be seen through a simple dimensional analysis, as follows. In general, combining the possible effects of momentum (jetlike behavior) and buoyancy (plumelike behavior) and considering the results already presented, properties of the discharge such as centerline velocity may be assumed to be functions of x, M, and B, i.e.,

$$U_m = f(x, M, B) \tag{14.1.29}$$

Using standard dimensional analysis (Chap. 1), two dimensionless groupings of variables should result from these parameters, and these are written as

$$\frac{U_m x}{M^{1/2}} = f\left(\frac{xB^{1/2}}{M^{3/4}}\right) \tag{14.1.30}$$

A *buoyancy length scale* is thus defined by

$$L = \frac{M^{3/4}}{B^{1/2}} \qquad \text{(round or three-dimensional discharge)} \tag{14.1.31}$$

or

$$L = \frac{M}{B^{2/3}} \qquad \text{(slot or two-dimensional discharge)} \tag{14.1.32}$$

where M is defined by Eq. (14.1.6) and B by Eq. (14.1.25) in the case of a three-dimensional discharge. For two-dimensional or slot-type discharges, the two-dimensional flow rate q is substituted for Q. The two-dimensional approach is helpful for analysis of many open channel flow situations and also is useful in the analysis of discharges from multiport diffusers used to distribute wastewater in the environment (see Sec. 14.2). As usual, q is simply the total flow rate divided by the width of the slot, $q = Q/W$, where W = width.

The usefulness of L is in determining the region over which the flow exhibits more jetlike properties, relative to plumelike behavior. Large values of L characterize jetlike flows, while more plumelike flows have smaller L. A pure plume has $L = 0$. The relative magnitude of x and L thus determines the type of behavior for the discharge, which is easily seen by substituting Eq. (14.1.31) into Eq. (14.1.30):

$$U_m = \left[f\left(\frac{x}{L}\right)\right] \frac{M^{1/2}}{x} \tag{14.1.33}$$

For large M, L is also large and (x/L) approaches zero. In this case, Eq. (14.1.8) implies U_m should be approximately proportional to $(1/x)$, which, along with the observation of approximately constant M for jets (large L), requires that the function in brackets in Eq. (14.1.33) must approach a constant value for small (x/L). For small M (large x/L), the flow behaves like a plume, and U_m is given by Eq. (14.1.27). This requires that the function in Eq. (14.1.33) must be related to $(x/L)^{2/3}$. To summarize,

$$f\left(\frac{x}{L}\right) \longrightarrow \begin{cases} \text{const} & \dfrac{x}{L} \to 0 \\[2ex] \left(\dfrac{x}{L}\right)^{2/3} & \dfrac{x}{L} \to \infty \end{cases} \tag{14.1.34}$$

Again, it should be kept in mind that the small (x/L) result is consistent with jet behavior and the large (x/L) result applies to plumes. For intermediate values of (x/L) the flow is affected by both momentum and buoyancy.

When there is significant discharge strength, a *discharge length scale* also may be defined,

$$L_Q = \frac{Q}{M^{1/2}} = \sqrt{A_0} \tag{14.1.35}$$

For a two-dimensional discharge, the corresponding discharge length is

$$L_q = \frac{q^2}{M} = D \tag{14.1.37}$$

where D is the flow thickness. The ratio of L_Q (or L_q) to L gives the relative effect of discharge buoyancy and momentum fluxes in the flow. In other words, dividing Eq. (14.1.35) by Eq. (14.1.31) gives

$$\frac{L_Q}{L} = \frac{QB^{1/2}}{M^{5/4}} = \left(\frac{\pi}{4}\right)^{1/4} \left\{\frac{g'b}{U^2}\right\}^{1/2} \tag{14.1.38}$$

where the quantity in brackets on the right-hand side is known as the *jet Richardson number* and represents the ratio of buoyancy force to momentum force. In calculating the ratio in Eq. (14.1.38), the initial discharge values are usually used, though calculations based on local values also may be performed.

14.2 SUBMERGED DISCHARGES AND MULTIPORT DIFFUSER DESIGN

Submerged discharges are often used for disposal of treated sewage and also for dispersal of waste heat generated by power plants. The wastewater is typically buoyant with respect to the fluid into which it is discharged, because it is either warmer or less saline (or both), and one of the main engineering considerations is to achieve a rapid dilution and dispersion of the discharge. As a general rule, the desired dilution for wastewater discharges is on the order of 100 (i.e., the concentrations in the discharge are reduced at least 100-fold). Further reductions in concentrations are expected as a result of natural biological decay. Discharges with rapid dilution are most commonly designed using multiport diffusers, which provide an initial dispersion of the injected wastewater. In the following, we discuss the design for outfalls in an ocean environment, which is the most common situation, and this will serve to exemplify the general characteristics of such discharges.

Figure 14.6 shows the general behavior of sewage discharged through a diffuser on the ocean floor. The discharge is usually directed horizontally or at a small angle through round ports along both sides of the diffuser. If possible, the discharge is oriented into the ambient flow, causing additional mixing and dilution. There is a wide variety of designs used, in terms of port diameter and spacing, with port diameters typically between several inches and about a foot (or, between about 10 and 30 cm) and spacings from several feet up to several tens of feet (approximately 1 to 10 m). The ports may be simple holes cut into the distribution pipe or they may be nozzles at the tips of risers that direct the flow from the pipe.

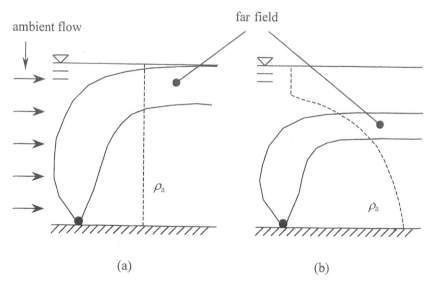

Figure 14.6 Discharge through a diffuser into an ocean environment: (a) uniform ambient density; (b) linearly stratified ambient.

Diffusers are generally located in relatively deep water, to maximize the dilution as the plume spreads upwards. Depending on local conditions, particularly ambient flow direction and strength, diffusers may be located several kilometers offshore, in water that is usually at least 50 m deep. The individual jets usually merge rapidly, so that the diffuser may be considered as a line source (see Chap. 10). There is a *near field* region in which strong mixing takes place, due to the turbulence of the jets themselves and also due to their relative buoyancy. The fluid initially rises, though its relative buoyancy decreases because of entrainment and mixing. In a stratified ambient receiving water, such as is illustrated in Fig. 14.6b, it is possible that a level will be reached at which the density of the rising plume matches the ambient density. In that case, the discharge ceases to rise and spreads laterally at the neutral buoyancy level. Otherwise, the plume rises to the surface and then spreads laterally, as in Fig. 14.6a. In either situation, the *far field* is defined when the plume stops rising and starts to spread laterally. In this region the waste field is transported by ambient currents and mixed by background oceanic turbulence. (Note that ambient currents and turbulence also work on the discharge in the near field, but the dynamics of the flow are much more strongly controlled by the mixing and buoyancy of the discharge in that region.)

Because of different flow dynamics, different modeling approaches are used in the near and far fields. In the near field the important properties to model are the rise height, the initial dilution, the thickness and width of the plume, and the distribution of concentration in the plume. These values are all needed in order to couple the near field to a far field model, which depends primarily on the ambient flow field and the remaining relative buoyancy of the spreading plume. Properties of the diffuser that most closely control the near field characteristics are the jet discharge velocity from each port, the port diameter and spacing, and the buoyancy difference between the discharge and the ambient sea water. In addition, it is necessary to know the conditions in the receiving water, including flow direction and amplitude, and ambient density stratification. Modeling the near field flow is generally difficult due to the complex interactions of these various features. The far field models are somewhat simpler to conceptualize, since they do not include the complicated initial interactions between the discharge and the ambient currents.

The design of the diffuser is accomplished in an iterative fashion, and often physical models are used to evaluate different designs. The diffuser must be located so that the desired level of initial dilution is achieved, and to avoid the transport of wastewater back to shore areas. There should also be an approximately uniform distribution of discharge among the diffuser ports, and all velocities should be sufficiently high that deposition of solids is prevented.

To illustrate the calculation of near field dilution, consider a plane plume in a uniform receiving water, as in Fig. 14.6a. A small control volume is drawn

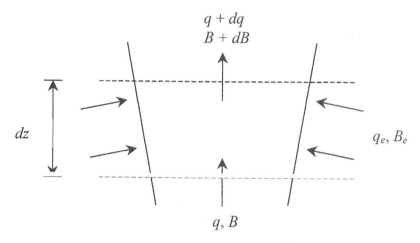

Figure 14.7 Control volume for calculation of dilution in vertically rising two-dimensional plume (B_e is the entrainment buoyancy flux).

in Fig. 14.7, showing the transport of water and buoyancy into and out of the control volume. Since we are considering a plane plume, it is sufficient to look at the two-dimensional, side view. For a steady state, there is zero net transport of mass across the surfaces of the control volume, or

$$q + q_e - \left(q + \frac{dq}{dz} dz \right) = 0 \tag{14.2.1}$$

where q_e is the entrainment flow rate. From the previous discussion, q_e depends on the entrainment velocity, which is proportional to the maximum or center-line velocity of the plume. Also, from Eq. (14.1.28), the maximum velocity is proportional to $(B)^{1/3}$ [the x in the denominator of Eq. (14.1.28) does not appear for a two-dimensional plume, where B is defined using q instead of Q]. Therefore

$$\frac{dq}{dz} \cong B^{1/3} \tag{14.2.2}$$

In general, B should be the local value of buoyancy flux. However, it is much simpler for design purposes to express a final relation in terms of quantities that are better known, so the initial value is used for B. Then, integrating Eq. (14.2.2) over z to evaluate q, and using the definition for dilution, S, as being the local volume flux divided by the initial discharge, we have

$$S \cong c \frac{B^{1/3} z}{q_0} \tag{14.2.2}$$

where q_0 is the initial discharge and the constant c has been found from experiments to have a value of approximately 0.38.

For stratified environments, the calculations are somewhat more complicated, because the buoyancy of the plume changes due to both entrainment and the changing ambient density as the plume rises. In particular, as noted previously, it is possible that the plume reaches a point at which there is no further acceleration due to buoyancy. This maximum rise height, z_m, may be calculated in the case of a linearly stratified receiving water from

$$z_m = 3.6B^{1/3} \left(-\frac{g}{\rho_0} \frac{d\rho_a}{dz} \right)^{-1/2} \tag{14.2.3}$$

where ρ_0 is the initial density of the discharge and $d\rho_a/dz$ is the ambient density gradient, which is negative for a stable stratification. If z_m is calculated to be greater than the water depth, then the plume reaches the surface. The corresponding (minimum) dilution at height z_m is

$$S = 0.24 \frac{B^{1/3} z_m}{q_0} \tag{14.2.4}$$

A typical value for the discharge q_0 is $0.01 \, \text{m}^2/\text{s}$. The total discharge then depends on the length of the distribution pipe. In order to maintain a relatively uniform distribution of flow for each of the ports, either the port diameters or the pipe size is changed along the length, while port spacing is usually kept constant. Generally, the design starts by assuming a discharge for the most downstream port (equal to the discharges for all the ports). A diameter is also chosen, and the pressure in the pipe required to maintain the desired discharge is calculated. A diameter for the first section of pipe upstream of the last port is then chosen, and the change in head is calculated in that section, using a standard Bernoulli-type equation. The pressure at the next port is then known, and the port diameter can be calculated to give the desired discharge. This process then continues until the last (first) port is accounted for and the required pressure at the end of the supply pipe is known. This procedure is generally done in an iterative fashion, until a satisfactory set of pipe and/or port diameters is found.

14.3 SURFACE BUOYANT DISCHARGES

There are a number of different aspects of jet discharges that are of interest for environmental applications. In this section we describe one of the more important types of discharges, that of a buoyant flow that might occur, for example, when a river or estuary discharges to a coastal environment or large lake. For many natural flows the discharge velocity is relatively small, and buoyancy effects are prominent, especially for discharges in ocean environments, where salinity differences generate large buoyancy. There are cases, however, where the flow is more strongly driven by momentum flux. An example of this latter case is the Niagara River flow into Lake Ontario (Fig. 14.8), where temperature differences generate buoyancy. Because Lake Erie, which drains into the Niagara River, is relatively shallow, it tends to heat up much more quickly than the water in Lake Ontario, so that the Niagara River is bouyant for about six months of the year. It also carries a very large flow rate, suggesting that momentum flux is also important. When analyzing flows of this type, it is of interest to be able to describe the spreading behavior of the discharge and the extent of its attachment to the bottom, before buoyancy causes it to lift off and spread over the receiving fluid.

14.3.1 Arrested Wedge Analysis

A particular problem of interest is the description of the density stratification or interface position between the discharge and the receiving water. The

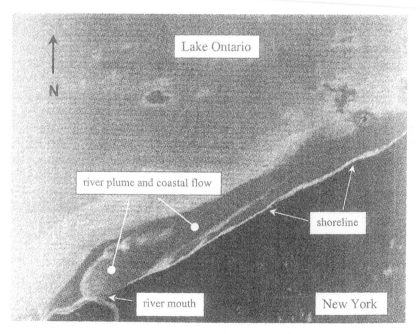

Figure 14.8 Satellite view of Niagara River discharge in Lake Ontario; gray shades indicate different temperatures. The flow typically turns to the right, in response to Coriolis effects and prevailing westerly winds. (Photo courtesy of A. Masse.)

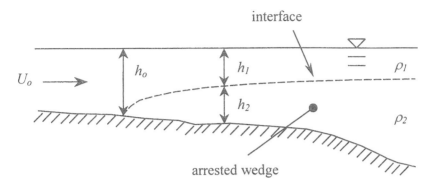

Figure 14.9 Arrested density wedge definition sketch.

analysis considers two-dimensional, steady buoyant flow into a more dense environment ($\rho_2 > \rho_1$), as sketched in Fig. 14.9. At a point sufficiently far upstream, the flow is in contact with the bottom of the channel, which has depth h_0 at that point. The flow velocity is U_0 and the initial value of the densimetric Froude number $\mathrm{Fr_d}$ is assumed to be less than one, where $\mathrm{Fr_d}$ is defined as in Eq. (14.1.1). Reduced gravity is defined by $g' = g(\rho_2 - \rho_1)/\rho_2$, and τ_i is defined as interfacial shear stress. The assumption of $\mathrm{Fr_d} < 1$ is discussed further later in this section.

Downstream of this point, the receiving water has sufficient density that it intrudes upstream, as a "wedge," as shown in Fig. 14.9. Neglecting tidal effects, this wedge is "arrested" at an equilibrium position determined by a balance between hydrostatic pressure differences due to the inequality of densities ($\rho_1 \neq \rho_2$) and interfacial shear stress. For this initial analysis, entrainment across the interface is neglected, and therefore there is no flow in the lower (wedge) layer.

Figure 14.10 shows a control section of the wedge, where $S_s =$ surface slope, $S_i =$ interface slope, and $S_b =$ bottom slope. With h_1 and h_2 denoting the upper and lower layer thicknesses, respectively, geometrical considerations show that

$$\frac{dh_1}{dx} = S_i - S_s \qquad (14.3.1)$$

and

$$\frac{dh_2}{dx} = S_b - S_i \qquad (14.3.2)$$

where $x = 0$ at the position of the farthest upstream extent of the wedge.

For the upper layer, which is labeled as layer 1, note that the (two-dimensional) flow rate, $q_1 = U_1 h_1$, where U_1 is the mean velocity in the upper

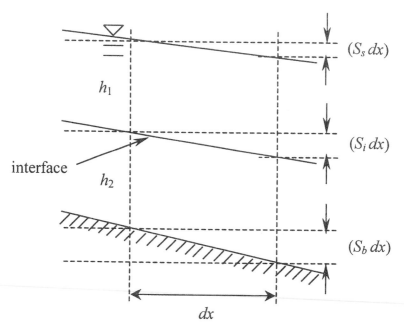

Figure 14.10 Control section of wedge for momentum analysis.

layer, is constant, since there is no entrainment across the interface. Consistent with the two-dimensional approach, forces per unit width are considered in writing a momentum balance applied for the upper layer, in the direction parallel to the bottom. This results in

$$
\rho_1 g \frac{h_1^2}{2} - \frac{\rho_1 g}{2} \left(h_1 + \frac{dh}{dx} dx \right)^2
$$

$$
- \left[\rho_1 g \left(h_1 + \frac{dh}{dx} dx \right) + \frac{\rho_2 g}{2} (S_b - S_i) dx \right] (S_b - S_i) dx
$$

$$
+ \rho_1 g h_1 S_b dx - \tau_i \, dx = \rho_1 q \left[\left(U_1 + \frac{dU_1}{dx} dx \right) - U_1 \right] \tag{14.3.3}
$$

The first two terms on the left-hand side account for pressure forces, the third term is the body force component, and the last term on the left-hand side is the interfacial friction. The right-hand side is the change in momentum between positions x and $(x + dx)$. Although interfacial shear acts in the direction given by S_i, rather than S_b, this adjustment to direction is considered negligible, since $\cos[\tan^{-1}(S_i)] \cong \cos[\tan^{-1}(S_b)] \cong 1$.

After carrying out the multiplication of the terms in Eq. (14.3.3), if we neglect second- and higher order terms in dx and divide by dx, the result is

$$-\rho_1 g h_1 \left[\frac{dh_1}{dx} + (S_b - S_i) \right] + \rho_1 g h_1 S_b - \tau_i = \rho_1 h_1 U_1 \frac{dU_1}{dx}$$

$$\Longrightarrow S_s - \frac{\tau_i}{\rho_1 g h_1} = \frac{U_1}{g} \frac{dU_1}{dx} \qquad (14.3.4)$$

where Eq. (14.3.1) has been used to substitute for (dh_1/dx). Also, by adding Eqs. (14.3.1) and (14.3.2), we find

$$\frac{dh_1}{dx} + \frac{dh_2}{dx} = S_b - S_s \qquad (14.3.5)$$

which when substituted into Eq. (14.3.4) gives

$$\frac{U_1}{g} \frac{dU_1}{dx} + \frac{\tau_i}{\rho_1 g h_1} + \frac{d}{dx}(h_1 + h_2) - S_b = 0 \qquad (14.3.6)$$

This is the equation of motion for the upper layer.

The analysis for the lower layer is similar but simplified because of the no-flow assumption. A momentum balance, again in the direction parallel to the bottom, gives

$$\left(\rho_1 g h_1 + \frac{1}{2} \rho_2 g h_2 \right) + \rho_1 g h_1 (S_b - S_i)\, dx$$

$$- \left\{ \rho_1 g \left(h_1 + \frac{dh_1}{dx} dx \right) + \frac{\rho_2 g}{2} \left(h_2 + \frac{dh_2}{dx} dx \right) \right\} \left(h_2 + \frac{dh_2}{dx} dx \right)$$

$$+ \tau_i\, dx + \rho_2 g h_2 S_b\, dx = 0 \qquad (14.3.7)$$

where each of the terms has a similar interpretation as for the upper layer equation (14.3.3). Following a similar procedure as before, this equation can be rewritten as

$$\rho_1 g \left[-h_1 \frac{dh_2}{dx} + h_1 (S_b - S_i) - h_2 \frac{dh_1}{dx} \right] + \rho_2 g \left(-h_2 \frac{dh_2}{dx} \right) + \tau_i$$

$$+ \rho_2 g h_2 S_b = 0 \Longrightarrow -\frac{\rho_1}{\rho_2} \frac{dh_1}{dx} - \frac{dh_2}{dx} + \frac{\tau_i}{\rho_2 g h_2} + S_b = 0 \qquad (14.3.8)$$

This is further simplified by using

$$\frac{\rho_1}{\rho_2} = 1 - \frac{\rho_2 - \rho_1}{\rho_2} = 1 - \frac{\Delta \rho}{\rho_2} \qquad (14.3.9)$$

so that Eq. (14.3.8) becomes

$$\left(1 - \frac{\Delta\rho}{\rho_2}\right)\frac{dh_1}{dx} + \frac{dh_2}{dx} - \frac{\tau_i}{\rho_2 g h_2} - S_b = 0 \tag{14.3.10}$$

Equations (14.3.6) and (14.3.10) are the basic momentum balances for the upper and lower layers, respectively. They were originally derived by Schijf and Schonfeld in 1953, and there have been a number of modifications and improvements since, notably the introduction of entrainment across the interface. This leads to a nonconstant flow rate in the upper layer and flow in the lower layer, so that bottom friction must be addressed. For the present text, however, the simplified analysis will be continued for one last step, which is to derive the equation describing the interfacial position. This is done by first subtracting the lower layer equation (14.3.10) from the upper layer equation (14.3.6). This gives

$$\frac{U_1}{g}\frac{dU_1}{dx} + \frac{\tau_i}{g}\left(\frac{1}{\rho_1 h_1} + \frac{1}{\rho_2 h_2}\right) + \frac{\Delta\rho}{\rho_2}\frac{dh_1}{dx} = 0 \tag{14.3.11}$$

Then, using the definition of g', with $\mathrm{Fr}_1 = \mathrm{Fr}_d$, based on upper layer properties, and noting that $U_1(dU_1/dx) = 1/2\, d(U_1^2)/dx$, Eq. (14.3.11) is rearranged to obtain

$$\frac{dh_1}{dx} = -\frac{1}{2g'}\frac{dU_1^2}{dx} - \frac{\tau_i}{g'}\left(\frac{1}{\rho_1 h_1} + \frac{1}{\rho_2 h_2}\right) \tag{14.3.12}$$

The derivative on the right-hand side is rewritten using the concept of constant q,

$$\frac{dU_1^2}{dx} = q^2\frac{d}{dx}\left(\frac{1}{h_1^2}\right) = -2\frac{q^2}{h_1^3}\frac{dh_1}{dx} = -2\frac{U_1^2}{h_1}\frac{dh_1}{dx} \tag{14.3.13}$$

The interfacial shear is also written in terms of an interfacial friction factor, f_i, as

$$\tau_i = \frac{f_i}{8}\rho_1 U_1^2 \tag{14.3.14}$$

Upon substituting these last two expressions into Eq. (14.3.12) and rearranging, we obtain

$$\frac{dh_1}{dx} = -\frac{\dfrac{f_i}{8}\mathrm{Fr}_1^2\left(1 + \dfrac{\rho_1 h_1}{\rho_2 h_2}\right)}{1 - \mathrm{Fr}_1^2} \tag{14.3.15}$$

which can be used to calculate the interface position.

In general, Eq. (14.3.15) must be integrated numerically to obtain the interfacial profile, using a known boundary condition, usually stated as a known value for h_1 at some x. There are two possible boundary conditions, either $h_1 = h_0$ at the upstream edge of the wedge, or h_1 is specified at the discharge location and Eq. (14.3.15) is integrated in the upstream direction. The problem with the former approach is that it is not well defined, since the upstream extent of the wedge is not known a priori. At the downstream location, however, a critical condition exists where Fr_1 is close to one (note that this is a singularity point in the solution). This condition provides an effective hydraulic control on the flow and is a natural starting point for the integration. In other words, using an assumed flow rate, q, the initial value for h_1 is chosen as the critical depth. Numerically, a Fr_1 of slightly less than one must be used in Eq. (14.3.15) to calculate the initial value of (dh_1/dx). Then, choosing a dx (or Δx), a value for dh_1 (Δh_1) is calculated. Since the integration is in the upstream direction, $\Delta x < 0$ and $\Delta h_1 > 0$. After the new value for h_1 is found, a corresponding value for U_1 is calculated based on q, resulting in a new value for Fr_1, which is used again in Eq. (14.3.15) to find the next Δh_1. This process is continued until h_1 is the same as h_0, keeping in mind any changes in total depth, depending on channel slope.

An example of this kind of calculation is shown in Fig. 14.11, in which h_2 is plotted as a function of distance upstream from the mouth of the discharge. For these calculations, the total channel depth is taken as 20 m, f_i is 0.02, the salinity difference is 1% by weight, and h_1 at the mouth of the discharge is 5 m. The initial value for Fr_1 is taken as 0.95, resulting in a velocity at that point of 0.6 m/s. According to this calculation, the wedge would extend about 6.5 km upstream. Different wedges would, of course, result from different assumed conditions.

The predicted wedge has singularities at both ends. At the mouth, $Fr_1 \cong 1$ as noted above, and at the upstream end h_2 approaches 0, which again is undefined in the calculation. However, the "nose" shape at the upstream end is characteristic of this type of flow, and the procedure outlined above allows calculations to proceed to a point sufficiently close to $h_2 = 0$ for reasonable estimates of the wedge length to be made. It also should be noted that the general shape of the profile appears to be reasonable, at least as long as $Fr_1 < 1$ (recall that this assumption was made at the beginning of this analysis). If $Fr_1 < 1$, then $dh_1/dx < 0$ according to Eq. (14.3.15). Or, since the positive x-direction has been switched in the integration procedure, $dh_1/dx > 0$, which corresponds with $dh_2/dx < 0$ as shown in Fig. 14.11. As h_1 increases, U_1 must decrease to maintain constant q, resulting in a decreasing value of Fr_1 with increasing distance upstream. In other words, a subcritical flow ($Fr_1 < 1$) is needed in order for the wedge to appear.

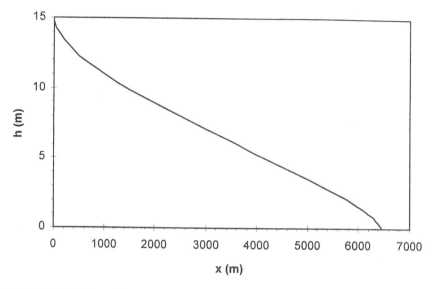

Figure 14.11 Example calculation of interface height for arrested density wedge.

As the flow progresses downstream, starting from the upstream end of the wedge intrusion, Fr_1 increases until it reaches the critical value of one at the mouth. Once past the mouth, the flow is free to spread laterally, and the two-dimensional approach is no longer valid. For supercritical flows, $Fr_1 > 1$, and Eq. (14.3.15) suggests that h_1 should increase with increasing distance downstream. This is just the opposite of the above results. When $Fr_1 > 1$, there should thus be no upstream intrusion of a wedge, and the flow will remain attached to the bottom at least as far as the mouth. Experiments have shown that the critical value of Fr_1, at which there is no intrusion, is somewhat higher than 1, between 1.5 and 2.5. Differences between experimentally observed values and the theoretical value are probably due to friction and other nonideal effects.

14.3.2   Flow on the Slope

When $Fr_1 > 1$, the flow remains attached to the bottom for some distance offshore from the mouth, as shown in Fig. 14.12. As the depth increases, the discharge flow expands to fill the space available, as long as there is sufficient energy and momentum. At high Fr_1, the jet has relatively high momentum, but as depth increases, the flow velocity decreases and Fr_1 decreases. Entrainment occurs from the sides but is prevented at the bottom in this region. Eventually, Fr_1 drops below 1, and the flow cannot remain attached to the bottom as the depth increases further. The point at which the flow ceases to remain in contact with the bottom is called the *detachment point*. A number of experiments have been performed to evaluate the position of this point, which is generally expressed as a function of the Froude number, either locally (at the detachment point) or at the mouth. In general, flows with higher discharge Froude numbers remain attached to the bottom for greater distances off-shore.

The detachment point location is of interest for a number of reasons. First, it locates approximately the transition point from more jetlike flow to more plumelike spreading. Direct interaction between the flow and the sediments also is limited to the attached region. A vertically mixed approach is possible for describing the flow in this area, but a stratified model should be used after the detachment point. In the following discussion, a *near field region* is defined for the flow before it detaches and a *far field region* applies when the flow is spreading mostly by buoyancy. An intermediate field is defined by the transition region between these two extremes (see Fig. 14.12). An approximate expression for the position of the detachment point is obtained by simple scaling analysis, as follows.

Consider the steady-state longitudinal momentum equation,

$$u\frac{\partial u}{\partial x} - fv = -\frac{1}{\rho}\frac{\partial p}{\partial x} + E_h\left(\frac{\partial^2 u}{\partial x^2} + \frac{\partial^2 u}{\partial y^2}\right) + E_z\frac{\partial^2 u}{\partial z^2} \qquad (14.3.16)$$

where f is the Coriolis parameter, E_h is horizontal eddy viscosity (x and y-directions) and E_z is the vertical viscosity. We need to assign typical magnitudes for the various terms in order to estimate their relative importance. In

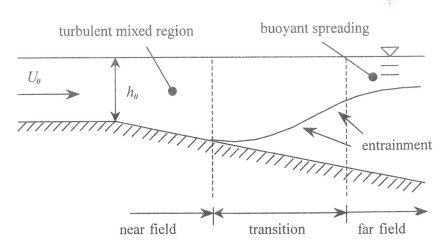

Figure 14.12 Buoyant discharge on a slope, with $Fr_1 > 1$.

the near field there is little velocity in the y-direction, so the Coriolis term is neglected. In addition, the flow is assumed to be well mixed vertically in this region. Thus pressure variations are approximately hydrostatic. The velocity is represented by the mean, vertically averaged velocity U. The gradients are evaluated over a distance L_n = near field length or distance over which the flow remains attached. Thus gradients in the x-direction are represented by $(1/L_n)$. Horizontal diffusion (viscosity) is neglected in the near field, relative to advection. This is equivalent to saying that the *Peclet number* for the flow is large, where Peclet number is defined as $Pe = LU/E$. this assumption also applies for the condition of a quiescent receiving water body. Ambient cross flows, for example, would introduce significant horizontal shear. Finally, the vertical viscosity term is interpreted as the vertical gradient of shear stress (i.e., $\tau = \rho E_z \, dU/dz$), so the last term on the right-hand side of Eq. (14.3.16) is approximated as

$$E_z \frac{\partial^2 u}{\partial z^2} \cong \frac{\partial}{\partial z}\left(\frac{\tau}{\rho}\right) \cong \frac{\tau_0}{\rho H} \tag{14.3.17}$$

where τ_0 is bottom shear stress and H is depth. It is further assumed that τ_0 can be written in terms of a bottom drag coefficient, $\tau_0 = C_D \rho U^2$. Under these assumptions the momentum equation becomes

$$\frac{U^2}{L_n} \cong \frac{1}{\rho}\frac{(\rho g' H)}{L_n} + \frac{C_D U^2}{H} \tag{14.3.18}$$

where the difference in densities between the discharge and receiving waters has been taken into account in evaluating the horizontal pressure gradient. Solving for L_n,

$$L_n = \frac{H}{C_D}\left(1 - \frac{1}{Fr_d^2}\right) \tag{14.3.19}$$

This is consistent with our earlier result, that Fr_d must be greater than one in order for there to be any bottom attachment at all ($Fr_d < 1$ would result in $L_n < 0$, which is clearly not physically reasonable). Also, larger Fr_d gives larger L_n, i.e., the flow remains attached longer with higher Fr_d.

In the intermediate region the bottom friction term drops out as the flow detaches. If lateral velocities are still not too strong, the Coriolis term can still be neglected. The parameterized momentum equation is then the same as in Eq. (14.3.18), but without the drag term. This is easily rearranged to show that $Fr_d \cong 1$, which relates the speed of propagation of the buoyant layer to the relative buoyancy and depth of the layer. From this result it is expected that the local value for Fr_d should be approximately 1 near the detachment point.

For the far field analysis, there is no bottom friction, density differences are still important, so the pressure term must be considered, momentum is relatively small, so the nonlinear advective terms can be neglected, and the Coriolis term must now be included. Under these conditions the flow field is approximately geostrophic (Chap. 9), and a parameterized momentum equation is

$$fU \cong g' \frac{H}{L_f} \tag{14.3.20}$$

where L_f is the far field horizontal length scale. Using the general definition of *Rossby number*, $Ro = U/Lf$, this last result may be rewritten as $Fr_d^2 \cong Ro$ or

$$L_f \cong \frac{r_i}{Fr_d} \tag{14.3.21}$$

where $r_i = (g'H)^{1/2}/f$ is the *internal Rossby radius of deformation*, which is normally interpreted as a lateral distance over which Coriolis and gravity forces (due to deformation of a horizontal surface such as the air water interface) are in balance. Thus the far field length scale is directly related to r_i. It also should be noted that, for constant U and H, increasing buoyancy increases r_i and decreases Fr_d, resulting in larger L_f. As with most scaling analyses, however, the above results should be interpreted as giving first-order estimates only, or guidance for the interpretation of direct observations, or as first steps towards more detailed analyses.

PROBLEMS

Unsolved Problems

Problem 14.1   Assuming a uniform density ocean, what is the minimum depth required to achieve a dilution of 100 for a waste discharge of $0.02 m^2/s$, with $\Delta\rho/\rho = 0.02$?

Problem 14.2   Use a spreadsheet to calculate the interface position of an arrested salinity wedge in an estuary where the mean salinity difference between the freshwater flow and the receiving water is 1% by weight, the interfacial friction factor is 0.015, the total channel depth is 10 m, the freshwater discharge rate (two-dimensional) is 20 m^2/s, and the flow in the bottom layer can be neglected.

Problem 14.3   Under the conditions of problem 14.2, what would be the minimum channel depth required to produce a discharge that would not have an arrested wedge at all?

Problem 14.4   A flow of depth 2.5 m and flow rate 50 m²/s discharges into an environment where the density is 10 kg/m³ greater. The bottom slope is 0.005 on the shelf. If C_D is taken as 0.001, estimate the distance along the shelf at which the detachment point should be found.

SUPPLEMENTAL READING

Atkinson, J. F. 1993. Detachment of Buoyant Surface Jet, Disharged on a Slope. *J. Hydraul. Eng.* **119**(8):878–894.

Fischer, H. B., List, E. J., Koh, R. C. Y., Imberger, J. and Brooks, N. H., 1979. *Mixing in Inland and Coastal Waters*. Academic Press, New York.

Gray, W. G., ed., 1986. *Physics-Based Modeling of Lakes, Reservoirs and Impoundments*. ASCE Publications, New York.

Hutter, K., ed., 1984. *Hydrodynamics of Lakes*. CISM Lecture Series, Springer-Verlag, Wien.

Imberger, J. and Hamblin, P. F., 1982. Dynamics of lakes, reservoirs and cooling ponds. *Ann. Rev. Fluid Mech.* **14**:153–187.

Ippen, and Arthur T., ed., 1966. *Estuary and Coastline Hydrodynamics*. McGraw-Hill, New York.

Schijf, J. B. and Schonfeld, J. C., 1953. Theoretical Consideration of the Motion of Salt and Fresh Water. Proc. Minnesota International Hydraulic Convention, *JAHRA*, 321–333.

Simons, T. J., 1980. Circulation models of lakes and inland seas. *Canadian Bull. Fish Aquatic Sci.*, No. 203.

Turner, J. S., 1973. *Buoyancy Effects in Fluids*. Cambridge University Press, Cambridge.

15
Sediment Transport

15.1 INTRODUCTION

There are a number of problems of hydraulic and environmental interest related to sediment transport. Erosion of solids from hillslopes in overland flow, bed load and suspended load in rivers, and siltation of reservoirs and harbors all are problems that involve movement of solid materials and for which there is considerable engineering interest. It also should be noted that, strictly speaking, any analysis of natural flows, open channel flow for example, should include considerations of sediment transport. For the analyses for open channel flow discussed in Chap. 7 it was implicitly assumed that the channel bed was fixed. In real flows this is not necessarily the case, since sediment material may move along the bed, having an impact on boundary conditions used to derive flow parameters such as shear stress and velocity distribution.

Solids loading to a river or reservoir, with related effects on turbidity, is a water quality parameter of interest in its own right. In addition, many pollutants of interest, particularly those with hydrophobic character, tend to sorb onto the surfaces of particles. This is particularly important for smaller particles, which have a higher ratio of surface area to volume, and for particles with higher organic carbon content, such as might be produced as a result of biological processes.

Although intensively studied for many years, the investigation of sediment transport is still far from complete. Unlike pure water flow, which by itself has a multitude of variables to consider, sediment transport introduces a number of additional features, including settling and resuspension characteristics, drag and lift considerations, and possible interparticle cohesive forces, particularly for smaller particles. Examples of problems in which sediment transport plays a significant role include reservoir and harbor sedimentation, navigation and dredging, channel stability, control of alluvial rivers, and delta formation. These processes all have significant economic impacts, even before pollutants are considered. *In-place pollutants* (i.e., pollutants buried in bottom sediments of rivers and lakes) also pose a serious and continuing potential threat to water quality in many areas.

Solids transported by water can vary significantly in size, from dissolved material such as salt (essentially molecular-size particles), to small pebbles and gravel. In extreme cases much larger material may be moved by a flow, but those cases will not be considered here. Dissolved solids also will not be considered further here, since they are already discussed in Chap. 10. In addition, transport of *colloidal material*, which is defined as suspended solids that do not settle and have typical sizes less than one micron, will not be considered in the present discussion since it is assumed that transport of such material can be treated similarly to dissolved solids. The main concern here is with solid particles having a characteristic size ranging from approximately several microns (clays) to several millimeters (large sand). Table 15.1 provides a listing of normal particle sizes of interest for natural flows.

These particles may be transported as part of the *suspended load*, where turbulence in the water maintains particle positions well above the bottom, or as part of the *bed load*, where particles roll or slide along the bed. In between these two processes is the *saltation load*, where particles "bounce" along the bed, sometimes appearing as part of the suspended load and sometimes as part of the bed load. Figure 15.1 illustrates these different modes of transport.

Often it is difficult to distinguish completely between the different modes of transport, and a complete sediment transport equation should include all in the same function. However, attempts to do this have so far been mostly unsuccessful. Usually, separate equations are applied for each mode of transport. This is often more convenient for many problems, where only one of the modes is of particular interest. For instance, bed load is of interest mainly for relatively fast moving streams and rivers, while suspended load is the main concern in slow moving rivers and in lakes and reservoirs (however, note that bed load *into* a reservoir is a problem of considerable interest). Also, for contaminant transport, suspended load is usually of greater interest, because of the larger relative surface area of smaller particles. Often, the type of transport mode is related to particle size, with larger particles moving closer to the bed, and smaller sizes more likely to be part of the suspended load.

Table 15.1 Particle Sizes in Sediment Transport

Particle type	Approximate size range
Clays, algae	$1-10$ μm
Silts	$10-100$ μm
Sand	$100-1,000$ μm
Gravel, pebbles	$>1,000$ μm

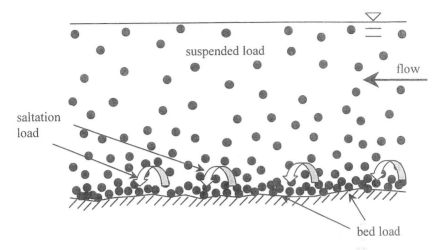

Figure 15.1 Different modes of sediment transport.

15.2 HYDRAULIC PROPERTIES OF SEDIMENTS

As noted above, sediments can be either *cohesive* or *noncohesive*. Much of the classic literature on sediment transport has been concerned with noncohesive material, usually sand and other granular material, since these are the materials of interest for most beds and banks of natural streams. In addition, these materials constitute the bulk of the mass in the bed load, which is usually the main component of the total solid mass transport. Cohesive sediments generally consist of smaller particles (clays and silts) and are more likely to be found in the suspended load. The analysis of cohesive sedimentation is complicated by interparticle bonding forces causing smaller particles to stick together and form larger aggregates, with relatively small bulk density. A problem of particular concern in sediment transport calculations is that of determining the bottom shear stress necessary to cause erosion of the bottom and resuspension of deposited materials. This is generally a site-specific problem, particularly for cohesive sediments, since the cohesive forces are usually not well known. A full analysis requires on-site experimentation to determine values for critical shear stress and other parameters controlling sediment transport.

Perhaps the most basic hydraulic property of sediment is particle size. For noncohesive sediments the size can be measured in several ways:

Equivalent diameter: the diameter of a sphere having the same volume as the particle
Sieve diameter: the minimum seive size opening that passes the particle

Sedimentation diameter: the diameter of a solid sphere having the same density and fall velocity

Surface diameter: the diameter of a sphere having the same surface area.

Of these different measures, seive diameter is the simplest one to measure experimentally, though the equivalent diameter is often used in developing conceptual models of particle and chemical interactions (see below).

Another hydraulic property of interest is the shape of the particle. There are several ways of characterizing this value. Traditionally, the particle shape has been defined in terms either of *sphericity*, which is the ratio of surface area of a sphere of equal volume to that of the particle, or of *roundness*, which is the ratio of the average radius of curvature of the edges of the particle to the radius of the largest circle that can be inscribed within the particle cross section.

In recent years the particle *fractal dimension* also has been used to describe particle geometry, especially for aggregates. The fractal dimension, D_f, is defined by

$$\forall = CL^{D_f} \tag{15.2.1}$$

where $\forall$ is the particle (or aggregate) volume, C is a constant with dimensions of L^{3-D_3}, and L is a characteristic length, either an equivalent diameter as defined above, or, more often, the maximum length of a particle (Fig. 15.2). For planar objects, a similar relationship can be defined in terms of area, rather than volume. In contrast to Euclidean geometry, which has integer values for D_f (i.e., one-, two- or three-dimensional shapes), in fractal geometry D_f can take noninteger values. For planar shapes D_f is generally between 1 and 2, while for three-dimensional shapes it is usually between 2 and 3. A lower value indicates a longer or more spread-out object, and a higher number corresponds with more compact and denser shapes. This can be seen from the images shown in Fig. 15.2. A larger value for L, for a given volume (or area, as pictured here), corresponds with a smaller value needed for D_f in order that Eq. (15.2.1) be valid. Conversely, smaller L requires larger D_f.

It is thought that fractal dimension plays a role in particle interactions and chemical sorption characteristics. For example, for a given mixing condition, particles with smaller D_f are more likely to come into contact with other particles than they would if they had the same mass (or volume) but with larger D_f. Particles, or aggregates, with smaller D_f also tend to be less dense, meaning that they have greater surface area for chemical sorption and other reactions. This is illustrated in Fig. 15.3a, which is a fractal image generated by a process known as *diffusion limited aggregation* (DLA). Experiments with images such as these have shown that they can be described with a fractal dimension (i.e., they are true fractal shapes). These objects clearly are rather

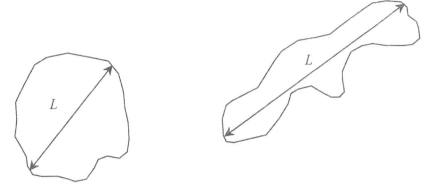

Figure 15.2 Examples of two particles having the same volume (or area, in two dimensions), but different L and thus different D_f. The particle on the left has larger D_f than the one on the right.

diffuse and have large relative surface areas, compared with a more compact, spherical object. It is also evident, looking at Fig. 15.3b, which is an image of a real aggregate, that images produced with the DLA model are a reasonable representation of true particles.

The *specific gravity* or average density of a particle also is important in determining its hydraulic characteristics. Specific gravity is defined as the ratio of the solid weight of a particle to the weight of an equal volume of water under standard conditions. Quartz or silicon-based particles such as sands have a specific gravity of around 2.65, but aggregates of smaller materials typically have much lower specific gravities, sometimes approaching 1.0.

One of the most important hydraulic properties of particles is the *fall velocity* or *settling velocity*, w_s. This is defined as the terminal velocity at which a single particle would move in an unbounded fluid (usually water) that is otherwise still. In terms of equivalent diameter, d, equilibrium between the particle *submerged weight* (i.e., taking buoyancy into account) and fluid drag forces can be written as

$$(\gamma_s - \gamma)\frac{\pi}{6}d^3 = \frac{\pi}{8}C_D\rho d^2 w_s^2 \tag{15.2.2}$$

where γ is the specific weight of the water, γ_s is specific weight of the solid particle, ρ is the water density, and C_D is a dimensionless *drag coefficient*. In general, C_D depends on the *Reynolds number*, $\mathrm{Re} = dw_s/\nu$, where ν is the kinematic viscosity of the fluid, and particle shape. The Reynolds number determines whether the relative motion of the fluid past the particle is laminar

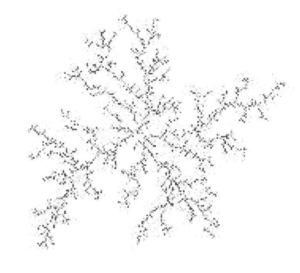

(a)

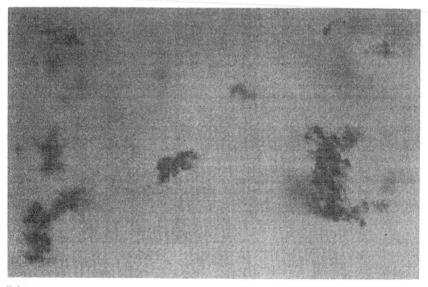

(b)

Figure 15.3 (a) Fractal image produced using diffusion limited aggregation model (using the *Fractint* program), as a representation of real aggregates. (b) Image of real aggregates of smaller clay particles. (Courtesy of R. Chakraborti.)

or turbulent, and a critical value for distinguishing between these two regimes has been found to be approximately 0.1.

Solving for w_s from Eq. (15.2.2), we obtain

$$w_s = \left[\frac{4}{3}\frac{d}{\rho C_D}(\gamma_s - \gamma)\right]^{1/2} = \left[\frac{4}{3}\frac{d}{C_D}g(S-1)\right]^{1/2} \tag{15.2.3}$$

where S is the specific gravity of the particle. For small Re (laminar flow), the *Stokes formula* is often used. This is obtained using the formula for drag coefficient for a sphere in laminar flow, $C_D = 24/\text{Re}$ (Eq. 3.3.5), which when substituted into Eq. (15.2.3) gives

$$w_s = \frac{gd^2}{18\nu}(S-1) \tag{15.2.4}$$

For turbulent flow, the drag coefficient must be determined as a function of Re and particle shape. From experimental evidence, the value for C_D has been found to range from about 0.5 for spheres up to about 3.0 for relatively flat or elongated particles (with small fractal dimension). Sand particles are usually considered to be spherical, but in general direct observations are needed to determine values for C_D. It should also be pointed out that the fall velocity decreases when the particle concentration is very high, due to interparticle effects, such as direct collisions or wake interference. This effect is typically important only for very high concentrations, greater than around 5–10% by dry weight.

15.3 BED LOAD CALCULATIONS

15.3.1 duBoys' Analysis

As previously noted, bed load calculations are primarily concerned with larger, noncohesive sediments. The analysis of sediment in the bed was first described by duBoys in the late 1800s. In this approach, the bed is visualized as a series of layers sliding over each other and with a velocity distribution decreasing linearly with depth. There are n layers and each layer has the same thickness, Δz. If the velocity at the top of the bed layer is V, and the velocity in the first, bottommost layer is 0, then the velocity gradient is $V/(n-1)\Delta z$ and the average velocity of movement in the bed is $V/2(n-1)$. This also implies a velocity difference across each layer,

$$\Delta v = \frac{V}{n-1} \tag{15.3.1}$$

As described in Chap. 7, for uniform open channel flow, a balance between gravitational forces and bottom shear stress leads to

$$\tau_0 = \gamma R_h S_0 \cong \gamma H S_0 \qquad (15.3.2)$$

where τ_0 is bottom or bed shear stress, R_h is hydraulic radius, S_0 is bottom slope, and H is depth. The approximation that $H \cong R_h$ applies for wide, open channel flow. This last expression also may be written in terms of the shear velocity,

$$u_* = \left(\frac{\tau_0}{\rho}\right)^{1/2} = (gHS_0)^{1/2} \qquad (15.3.3)$$

The distribution of shear stress in the bed is such that the assumed linear velocity gradient results. Considering the moving bed (of n layers), overall equilibrium requires that the shear force at the surface, due to τ_0, must be balanced by frictional resistance at the bottom, or, per unit area,

$$\tau_0 = C_f(\gamma_s - \gamma)n\,\Delta z \qquad (15.3.4)$$

where C_f is a friction coefficient and the remainder of the right-hand side is the submerged weight of sediment per unit area. The *critical shear stress*, τ_c, is the value at which bed motion just begins. It is defined by setting $n = 1$ in Eq. (15.3.4),

$$\tau_c = C_f(\gamma_s - \gamma)\,\Delta z \qquad (15.3.5)$$

Values for τ_c have been determined from experimental observations to range from about 0.01 lb/ft^2 (5 dyn/cm^2) for fine sand, with mean diameter around 0.1 mm, to about 0.1 lb/ft^2 (50dyn/cm^2) or higher for coarse sand or gravel, with mean diameter around 5 mm. Comparing Eqs. (15.3.4) and (15.3.5), we find

$$n = \frac{\tau_0}{\tau_c} \qquad (15.3.6)$$

so that the total depth of the moving bed is a function of applied shear stress and can be calculated once Δz is prescribed. When Eq. (15.3.6) results in a noninteger ratio, the closest integer should be taken for n.

The transport of sediment mass, per unit width, passing a given point is a function of the mean velocity of the bed, $V_s = V/2$. It can be expressed as

$$j_s = \gamma_s V_s(n\,\Delta z) = \frac{\gamma_s(n\,\Delta z)V}{2} = \frac{\gamma_s(n\,\Delta z)\,\Delta v}{2}(n-1) \qquad (15.3.7)$$

By substituting from Eq. (15.3.6), this also can be written as

$$j_s = \frac{\gamma_s\,\Delta z\,\Delta v}{2}\left(\frac{\tau_0}{\tau_c}\right)\left(\frac{\tau_0 - \tau_c}{\tau_c}\right) \qquad (15.3.8)$$

As long as the bed shear stress exceeds the critical value, this is an increasing function of τ_0. This equation is normally applicable to bed load calculations in relatively wide and shallow alluvial rivers, where side effects may be neglected. Under these conditions, the *Manning formula* can be used to estimate the depth for a given discharge, so that τ_0 can be calculated. *Total* bed load is then the product of j_s and channel width, B,

$$J_s = B j_s \tag{15.3.9}$$

When the cross section is such that the two-dimensional assumption is no longer valid, the following approach can be used. We first define the minimum depth required to initiate bed motion, $H_c = \tau_c/\gamma S_0$ (see Eq. 15.3.2). Then, bed load is assumed to occur only for that part of the channel with water depth greater than or equal to this value (Fig. 15.4). This range of the channel is then divided into smaller increments, each of width dx, and the above result (Eq. 15.3.8) is applied within each of these sections. The total bed load is then the sum of the load calculated for each of these subsections, multiplied by dx, i.e.,

$$J_s = dx \sum_i j_{si} = dx \sum_i \left[\frac{\gamma_s \, \Delta z \, \Delta v_i}{2} \left(\frac{\tau_{0i}}{\tau_c} \right) \left(\frac{\tau_{0i} - \tau_c}{\tau_c} \right) \right] \tag{15.3.10}$$

where j_{si} is the load, per unit width, in subsection i, and subscript i for the other variables refers to values defined for that particular subsection (see Fig. 15.4). Note that Eq. (15.3.10) assumes there is a single value for the critical shear stress, though varying values (for each subsection) could just as easily be defined.

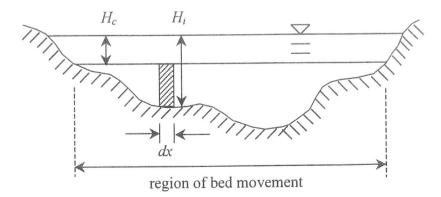

Figure 15.4 Cross section of stream, showing region of bed movement (where H exceeds H_c).

15.3.2 Other Bed Load Formulas

Many other formulas have been developed to calculate bed load, most based on a more detailed analysis of velocity distributions (including turbulent velocity fluctuations) and drag forces exerted on individual particles on the bed. *Sheilds' formula*, for instance, follows this approach and can be written as

$$j_s = \left[\frac{10\gamma^2 \overline{V}}{d(\gamma_s - \gamma)} \right] H S_0 \left(H S_0 - \frac{\tau_c}{\gamma} \right) \tag{15.3.11}$$

where $\overline{V}$ is the average water velocity in the channel. This formula has been shown to provide a reasonable correlation of data from numerous laboratory experiments and has the apparent advantage, not included in duBoys' approach, of incorporating the channel velocity directly. This formula is usually used in a form obtained by substituting the Manning equation and neglecting the critical shear stress term, which is assumed to be small relative to actual shear, so

$$j_s = \left[\frac{10\gamma^2}{d(\gamma_s - \gamma)} \right] \left(\frac{n_m}{1.5} \right)^{3/5} S_0^{17/10} q^{8/5} \tag{15.3.12}$$

where n_m is the Manning roughness coefficient and q is the channel discharge per unit width.

Perhaps the most difficult parameter to measure is the critical shear stress for initiation of particle movement on the bed. Most data for this parameter have been obtained from flume experiments. Figure 15.5 shows the drag and gravity forces acting on a particle. Lift force also may be included and would be an adjustment to the submerged weight. For simplicity the bed is assumed to be horizontal, though a bed slope is easily incorporated. The particle has equivalent diameter d and submerged weight $W = (\pi/6)d^3(\gamma_s - \gamma)$. The drag force is F_D, which is evaluated by considering the moments exerted about the point of contact by the two forces. The resultant force passes through a point of support (P) at angle φ as shown. Assuming that the point P is at a distance $d/2$ from the center of gravity of the particle, the moment arm for W is $(d/2 \sin \varphi)$ and for F_D it is $(d/2 \cos \varphi)$. Then, equating moments, we find

$$F_D = \frac{\pi}{6}d^3(\gamma_s - \gamma) \tan \varphi \tag{15.3.13}$$

It is usually assumed that the angle φ represents the *angle of repose* for the particle. It is further assumed that the critical shear stress for initiation of bed movement is F_D, divided by effective bed shear area. This last parameter is defined as the horizontal area of the single particle, divided by the ratio of the number of particles actually about to move, to the total number of particles

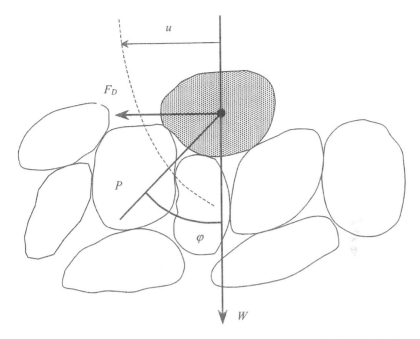

Figure 15.5 Forces acting on a particle on the bed; the velocity distribution is not known exactly but is assumed to have a virtual origin at a location within the bed.

on the bed. This ratio is denoted by p. The critical shear stress is then

$$\tau_c = \frac{(\pi/6)d^3(\gamma_s - \gamma)\tan\varphi}{(\pi d^2/4)/p} = \frac{2}{3}pd(\gamma_s - \gamma)\tan\varphi \tag{15.3.14}$$

Values for p have been estimated to be around 0.25–0.35 (see White, 1940, for example, who found the product $(2/3)p$ to be $\cong 0.18$). Then for sand, with specific gravity 2.65 and assuming $\tan\varphi = 1$, the critical shear is $\tau_c \cong 24d$ (lb/ft^2, when d is in ft and γ is in lb/ft^3).

In fact, significant variations in bed shear stress are expected, related to turbulent velocity fluctuations. Shields observed the random nature of grain movements on the bed and Einstein (1942) first developed a transport relation based on statistical concepts, followed shortly afterwards by Kalinske in 1947. Rouse (1955) later developed a semiempirical formula based on Kalinske's data, in a form similar to Shields' Eq. (15.3.12). Estimates as low as $\tau_c \cong 4d$ have been made for very turbulent flows, where τ_c represents the mean (time-averaged) critical shear stress. In other words, with a mean value of around $4d$, sufficiently large turbulent velocity fluctuations can exist so that the shear reaches $24d$. The random nature of these fluctuations makes it very difficult to determine a specific critical condition for the initiation of grain movement on the bed.

Critical shear stress often is calculated from the *Shields diagram*, which is a nondimensional version of Eq. (15.3.14). By assuming that the initiation of motion depends on τ_c, $(\gamma_s - \gamma)$, d, fluid density, ρ, and viscosity, μ, it can be shown from dimensional analysis that

$$\frac{\tau_c}{d(\gamma_s - \gamma)} = f\left(\frac{U_{*c}d}{\nu}\right) \tag{15.3.15}$$

where ν is kinematic viscosity. The left-hand side is a dimensionless critical shear stress, and the independent variable on the right-hand side has the form of a Reynolds number, based on the particle size and critical shear velocity. On comparing this last result with Eq. (15.3.14), it may be concluded that the function on the right-hand side must incorporate the effects of p and tan φ. This function was evaluated by Shields on the basis of data compiled from a large set of experiments using sand and other materials of similar size, with a range of specific weights. Shields' curve is shown in Fig. 15.6.

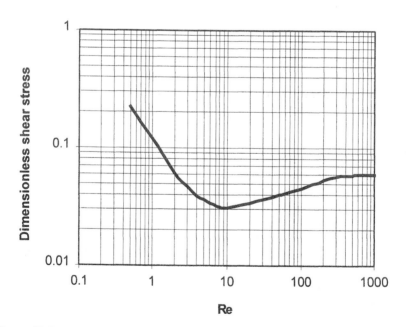

Figure 15.6 Shields' diagram for critical shear stress as a function of grain Reynolds number.

15.4 SUSPENDED SEDIMENT CALCULATIONS

Suspended sediment transport is generally calculated using a slightly modified form of the advection–diffusion equation developed in Chap. 10,

$$\frac{\partial m}{\partial t} + u\frac{\partial m}{\partial x} + v\frac{\partial m}{\partial y} + (w - w_s)\frac{\partial m}{\partial z}$$

$$= \frac{\partial}{\partial x}\left(\varepsilon_x\frac{\partial m}{\partial x}\right) + \frac{\partial}{\partial y}\left(\varepsilon_y\frac{\partial m}{\partial y}\right) + \frac{\partial}{\partial z}\left(\varepsilon_z\frac{\partial m}{\partial z}\right)$$

$$(\pm source/sink\ terms) \tag{15.4.1}$$

where u, v, and w are the fluid velocities in the x-, y-, and z-directions, respectively, z is the vertical coordinate, directed upwards, and ε_x, ε_y, and ε_z are the diffusivities for sediment in the respective coordinate directions. The main difference between this equation and the dissolved mass balance equation from Chap. 10 is the inclusion here of the settling term. The source/sink terms are generally important only when primary production (growth) or predation affects the solids concentration. Deposition and resuspension at the bottom are either incorporated through the bottom boundary condition or, for vertically averaged models, they appear as source/sink terms. Equation (15.4.1) is written in very general form and is usually simplified for most practical applications. Since relatively small particles are normally under consideration (sand and other larger particles are not normally transported as part of the suspended load), it is usually assumed that the same turbulent eddies responsible for transport of momentum also are responsible for transport of the particles. Furthermore, diffusivities often are assumed to be spatially homogeneous, and advective motions are usually considered in one direction only.

The diffusivity is found by assuming that it is the same as for momentum transport (Reynolds analogy for turbulent flows). For open channel flow, assuming a linear variation of shear stress with depth and the usual von Karman logarithmic velocity profile, the variation of turbulent diffusivity is

$$\varepsilon = \kappa u_* z \left(1 - \frac{z}{H}\right) \tag{15.4.2}$$

where κ is the von Karman turbulence constant. The average vertical diffusivity has been shown to be well represented by

$$\bar{\varepsilon} = 0.067\kappa u_* H \tag{15.4.3}$$

and the average transverse diffusivity is given approximately by

$$\bar{\varepsilon}_t = 0.15\kappa u_* H \tag{15.4.4}$$

The nominal value for κ is 0.4, but some observations have indicated that it decreases in the presence of suspended sediment.

As a specific example, consider transport of sediment in a river, with the x-direction defined along the bed. If the problem is considered as one-dimensional (in the x-direction), then any mass loss by deposition or gain by resuspension at the bed is considered as an internal source or sink, since concentration is averaged over the channel cross-sectional area. Assuming that deposition and resuspension are the only source or sink terms, and that diffusivity is spatially constant, Eq. (15.4.1) simplifies to

$$\frac{\partial \overline{m}}{\partial t} + u \frac{\partial \overline{m}}{\partial x} = \varepsilon_x \frac{\partial^2 \overline{m}}{\partial x^2} - \frac{w_s}{H} \overline{m} + \frac{R}{H} \tag{15.4.5}$$

where the overbar here indicates a cross-sectional average value, H is depth, and R is the net resuspension rate, in mass per unit time per unit area, which may in general be a function of x and t (Fig. 15.7). For sandy beds, relations such as those developed in the previous section could be used for R. However, for smaller sediments, resuspension is usually determined empirically as a function of shear stress (or, more directly, as a function of excess shear stress above a critical value for initiation of bed material).

It also should be noted that in writing Eq. (15.4.1) or (15.4.5), it is implied that either one sediment size is being considered, or that w_s is an appropriately defined average settling velocity for a range of sediment sizes and densities. Similarly, R represents either one sediment type or an average. In general, when there is a range of particle sizes, possible interparticle reactions

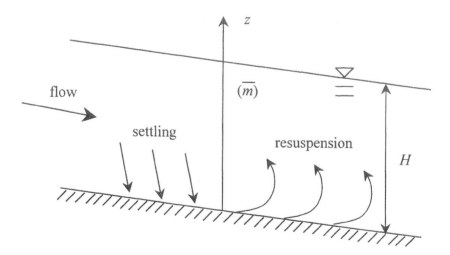

Figure 15.7 One-dimensional sediment transport in open channel flow, with settling (deposition) and resuspension terms.

should be considered (see Sec. 15.5), and separate equations should be written for different size classes, including transformation of particles between classes.

In general, the sediment load passing a given point in the stream is calculated from

$$J_s = \iint\limits_A um\,dy\,dz \tag{15.4.6}$$

where u is the velocity, as a possible function of depth and lateral position. A similar relation for two-dimensional flow may be written, with the integration performed only over the depth H. An equivalent calculation for J_s can be obtained from the product of cross-sectional average values for velocity and concentration. Assuming two-dimensional flow, a common calculation for depth-average velocity is to take the average of velocity readings at $0.2H$ and $0.8H$. Similarly, measurements have shown that an average suspended sediment concentration in river flow can be estimated by the sum of concentration measured at $0.8H$, weighted by a factor of 3/8, and the concentration measured at $0.2H$, weighted by a factor of 5/8, i.e.,

$$J_s \cong \left(\tfrac{1}{2}u_{0.2H} + \tfrac{1}{2}u_{0.8H}\right)\left(\tfrac{5}{8}m_{0.2H} + \tfrac{3}{8}m_{0.8H}\right) \tag{15.4.7}$$

In the vertical direction, the balance between settling and diffusion determines sedimentation behavior. If these terms exactly equal each other, then

$$-w_s\frac{\partial m}{\partial z} = \frac{\partial}{\partial z}\left(\varepsilon_z\frac{\partial m}{\partial z}\right) \tag{15.4.8}$$

From Eq. (15.4.1), it can be seen that this case results for one-dimensional (vertical) transport, steady state, and without any sources or sinks. Integrating once, with w_s assumed to be constant, we obtain

$$-w_s m = \varepsilon_z\frac{\partial m}{\partial z} \tag{15.4.9}$$

where the integration constant is zero because of zero-flux boundary conditions (see below). Thus an equilibrium is established in which the flux due to gravitational settling is balanced by the flux due to turbulent diffusion. This equation can be integrated once more, assuming constant ε_z, (equal to the average value), to write the suspended sediment concentration at any location z, in terms of a known value m_a at $z = a$,

$$m = m_a e^{-(w_s/\varepsilon_z)(z-a)} \tag{15.4.10}$$

This is the profile resulting from an exact balance between settling and diffusion and is known as the *equilibrium profile* (it is also sometimes known as the *Rouse profile*, after Rouse, who originally derived this result in 1938).

15.4.1 Boundary Conditions

For a complete solution to Eq. (15.4.1), initial and boundary conditions must be specified. Initial conditions consist of values for m (or $\overline{m}$) as a function of location, at time $t = 0$. Since the equation is second-order in x, y, or z, two boundary conditions are needed for each coordinate direction considered. If only the x-direction appears in the equation, for example, boundary conditions are needed at the upstream and downstream limits of the model domain in that direction, in terms of either known concentration or known gradient. The choice of specific boundary conditions depends on the particular situation, and general guidelines are similar as for the advection–diffusion equation discussed in Chap. 10.

The boundary conditions in the vertical direction require special attention, however, since they are complicated by the presence of the settling term in the governing equation. For convenience, the following discussion considers a model for the vertical direction only. Also assuming no bulk vertical fluid motion, constant vertical diffusivity and settling velocity, and defining q_m = internal source strength (sediment mass produced, per unit volume per unit time), the governing Eq. (15.4.1) becomes

$$\frac{\partial m}{\partial t} - w_s \frac{\partial m}{\partial z} = \varepsilon_z \frac{\partial^2 m}{\partial z^2} + q_m \qquad (15.4.11)$$

At the surface, $z = H$, there should be zero net flux of sediment, which is obtained by balancing the downward flux due to settling with any upward flux due to diffusion, or

$$w_s m = -\varepsilon_z \frac{\partial m}{\partial z} \qquad \Longrightarrow \qquad w_s m + \varepsilon_z \frac{\partial m}{\partial z} = 0 \qquad (z = H) \quad (15.4.12)$$

It may be noted that this boundary condition results from the same assumptions as were used in deriving the equilibrium profile of Eq. (15.4.10). It implies that the concentration gradient at the surface must be negative (i.e., concentration decreasing upward) or zero, since

$$\frac{\partial m}{\partial z} = -\frac{w_s}{\varepsilon_z} m \qquad (15.4.13)$$

If the concentration is zero, then the concentration gradient also is zero,

$$\frac{\partial m}{\partial z} = 0 \qquad (15.4.14)$$

This last result is consistent with the idea of zero *diffusive flux* across the surface, which also is a possible boundary condition by itself. However, it must be applied with care, since it does not restrict the value of m, and there is a possibility (in a numerical solution, for example) of an artificial settling flux through the surface unless w_s also is set to zero there.

In fact, from a physical point of view, both settling and diffusion stop at the surface, and it is appropriate to set *both* the settling and diffusion fluxes equal to zero. The settling flux should be set to zero by fixing $w_s = 0$, as noted above, and the diffusive flux should be set to zero by fixing the concentration gradient to zero, as in Eq. (15.4.14). Thus Eq. (15.4.12) is still satisfied.

The bottom boundary condition is somewhat more complicated, since there is the possibility that particles settle out of the water column onto the sediment bed or, in the case of resuspension, there is a flux of particles from the bed into the water column. As a first approach, consider a well-mixed water column with a region very close to the bottom in which diffusive-type motions die out and there is settling only, as shown in Fig. 15.8. The net flux of particles at the bottom is then equal to $(-w_s\overline{m})$, where $\overline{m}$ is the depth-averaged concentration. With no gradient or local source terms, integration of Eq. (15.4.11) over the depth yields

$$H\frac{\partial \overline{m}}{\partial t} - (w_s m)\big|_0^H = \left(\varepsilon_z \frac{\partial m}{\partial z}\right)\bigg|_0^H$$

or

$$H\frac{\partial \overline{m}}{\partial t} = \left(\varepsilon_z \frac{\partial m}{\partial z} + w_s m\right)\bigg|_0^H \qquad (15.4.15)$$

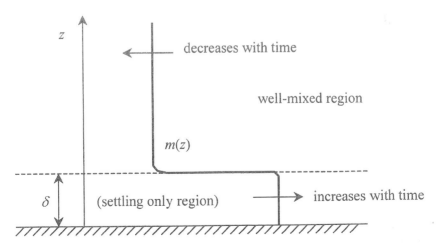

Figure 15.8 Application of bottom boundary condition, used in developing Eq. (15.4.17).

The quantity in brackets on the right-hand side of Eq. (15.4.15) is the *total flux*, which is evaluated at the surface and at the bottom. At the surface, Eq. (15.4.12) indicates that the total flux is zero, and we have already stated that the total flux at the bottom is just the settling term, $(-w_s\overline{m})$, so that Eq. (15.4.15) can be written as

$$\frac{\partial \overline{m}}{\partial t} = -\frac{1}{H} w_s \overline{m} \qquad (15.4.16)$$

The solution is given by

$$\overline{m} = \overline{m}_0 \exp\left(-\frac{w_s}{H} t\right) \qquad (15.4.17)$$

where $\overline{m}_0$ is the initial average concentration. This solution is equivalent to a first-order decay process, with decay rate equal to $(-w_s/H)$.

It should be noted that the solution given by Eq. (15.4.17) applies only under rather restrictive conditions and is not valid, for example, when the water column is not well mixed. Another way to consider this is to look at the total flux expression, evaluated at the bottom. If the total flux is equal to the settling term, then the gradient must be zero, since there is no diffusive contribution to the flux, or

$$w_s m + \varepsilon_z \frac{\partial m}{\partial z} = w_s m \qquad \Longrightarrow \qquad \frac{\partial m}{\partial z} = 0 \qquad (15.4.18)$$

In other words, if the total bottom flux is equal to the settling, based on bottom concentration, then the concentration gradient also must disappear at the bottom. Under conditions of deposition, this is not generally observed (see Sec. 15.4.2).

If the concentration gradient is negative, then clearly the net deposition rate must be less than the rate given by $(w_s m)$. Letting p represent the fraction of $(w_s m)$ that is deposited, then the total flux at the bottom is

$$\left(w_s m + \varepsilon_z \frac{\partial m}{\partial z}\right) = p w_s m$$

or

$$\varepsilon_z \frac{\partial m}{\partial z} + (1 - p) w_s m = 0 \qquad (15.4.19)$$

The parameter p can be thought of as a *probability of deposition*. It represents the percentage of particles coming into contact with the bottom that actually become deposited. A *fully absorbing boundary* has $p = 1$ and a *fully reflecting boundary* has $p = 0$. In general, the value for p is not well known, and it should be considered as a fitting parameter for a given application.

It also should be noted that application of the bottom boundary condition is difficult because the concentration is not well defined right at the boundary. It may be helpful to think of the boundary condition being applied at a location just slightly above the boundary, at $z = \delta$ (Fig. 15.8). This is in fact the assumption already used in developing Eq. (15.4.17).

When resuspension occurs, there is generally strong mixing near the bottom. Usually the resuspension flux is specified externally, so that the total flux expression is set equal to the desired resuspension rate,

$$-\left(w_s m + \varepsilon_z \frac{\partial m}{\partial z}\right) = R \tag{15.4.20}$$

where R is the resuspension rate in terms of mass resuspended per unit area per unit time. This condition may be established by an appropriate choice for the gradient (recall that the gradient is negative, to produce an upward flux). An upward resuspension-type flux also may be established by fixing the bottom concentration at a value higher than in the water just above the bottom, and then letting diffusion move the excess mass upward. Thus

$$C = C_b \qquad (z = 0) \tag{15.4.21}$$

where C_b is the fixed concentration at the bottom. The maximum value for this concentration is the concentration of solids in the bed and would depend on bed porosity. However, it is more realistic to consider the maximum value for C_b to depend on the mixing energy available in the water column, so that C_b decreases with lesser mixing energy available. One advantage of the boundary condition of Eq. (15.4.21) is that it is simple to incorporate. However, it does not provide a well-controlled flux, since the movement of particles upwards depends on the gradient.

One further cautionary note should be considered when deciding on the appropriate boundary conditions for a given problem. Numerical solutions to Eq. (15.4.11) often have problems with maintaining a strict mass balance, when using either fixed gradient or fixed concentration boundary conditions. The problem is more acute at the bottom because the concentrations are generally higher than at the surface. An alternative approach, within a finite difference context, is to write a separate mass balance equation for the numerical grid adjacent to the top and bottom boundaries. For example, Fig. 15.9 shows several finite difference grids near the bottom of a computational domain used to solve the vertical sediment transport equation. A mass balance for grid 1 can be written as

$$\frac{dm_1}{dt} = \frac{1}{\Delta z}\left[w_s m_2 + \varepsilon_z \left(\frac{m_2 - m_1}{\Delta z}\right)\right] + q_1 \tag{15.4.22}$$

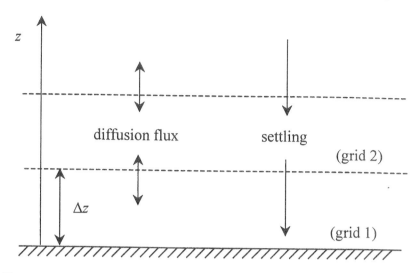

Figure 15.9 Application of no-flux bottom boundary condition, in finite difference approach.

where Δz is the grid spacing, subscript 1 refers to values for the first or bottommost grid, and subscript 2 refers to the next grid above. This equation directly incorporates the fact that both settling and diffusive fluxes are zero at the bottom. The source/sink term q_1 then must be chosen to represent losses due to deposition and sources due to resuspension.

15.4.2 Lake and Reservoir Modeling — Benthic Nepheloid Layer

A particular application of modeling vertical distributions of particulate matter is with large, deep lakes, which often exhibit the formation of a region of higher concentrations near the bottom. This region, called the *benthic nepheloid layer*, may extend several tens of meters above the bottom and have suspended sediment concentrations five to ten times higher than in the overlying water column. These layers were originally observed in marine environments and have also been observed in large freshwater lakes. Nepheloid layers were discovered in the Great Lakes of North America only in the early 1980s. For example, Fig. 15.10 displays vertical profiles recorded at a deep location in the southern part of Lake Michigan in August, 1966. The temperature distribution reflects the formation of a thermocline, with an upper mixed layer. The light transmittance is inversely related to particle concentration. It

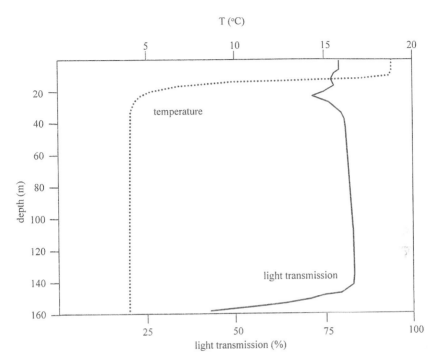

Figure 15.10 Example vertical profiles measured in Lake Michigan.

shows a small peak just below the thermocline, due to primary algal production, and also a decrease near the bottom, indicating a buildup of particulates with increasing depth.

The specific mechanisms leading to formation and maintenance of these layers are not well understood. In deeper parts of the lake, where bottom shears and mixing rates are low, the most probable source of material for the layer is due to primary production closer to the surface (as in Fig. 15.10) and subsequent settling of detritus. These particles tend to be small (<10 μm) and have specific gravity between about 1.03 and 1.1. Their natural Stokes settling velocity is only about 1 m/day, though they tend to move towards the bottom much more quickly than that, due to slow convective motions in the lake. If quasi-steady conditions are assumed, a solution to the vertical sediment transport equation, with zero-flux boundary conditions, is given by Eq. (15.4.10). It is interesting to note that in this case, the governing equation and both boundary conditions are identical. In other words, there is zero net vertical flux at any location in the water column (due to the balance between settling and diffusion), and the only parameter that needs to be specified is the bottom concentration (i.e., $a = 0$ in Eq. 15.4.10).

A simple estimate for the thickness of the layer is obtained using Eq. (15.4.10) and finding the location at which the concentration drops to a given fraction of the bottom value. For example, if we take m to be 1% of the bottom value, with a typical settling velocity of 1 m/day (0.001 cm/s) and a diffusivity of 0.5 cm^2/s, then

$$m = m_o e^{-(w_s/\varepsilon_z)z} \quad \Longrightarrow \quad z = -\frac{\varepsilon_z}{w_s} \ln\left(\frac{m}{m_o}\right) \cong 23 \text{ m} \quad (15.4.23)$$

which is in reasonable qualitative agreement with the thickness indicated in Fig. 15.10. The actual nepheloid layer is of course subject to unsteady conditions, and this calculation should be considered as a rough estimate only.

Other mechanisms that may play a part in nepheloid layer formation include local resuspension due to shear generated from upwelling and downwelling motions and density currents along the bottom, and transport of resuspended material downslope from shallower regions, where particles are resuspended by waves, seiche motion, or other near-shore processes. These mechanisms probably are more important for benthic nepheloid layers in shallower regions of deeper lakes, where such motions would be more common than in deeper parts. Also, it should be noted that the shape of the concentration distribution profile would be different in this case. Due to strong mixing and resuspension flux at the bottom, the concentration should be either uniform near $z = 0$, or exhibit a convex shape, rather than the concave form of the exponential curve. The conditions leading to the previous solution (Eq. 15.4.10) are not valid in this case, due to this strong mixing at the bottom.

Aside from scientific interest concerning formation of benthic nepheloid layers, they also are important in the transport of certain contaminants. In the Great Lakes, for example, there is concern for hydrophobic organic chemicals that tend to sorb onto the surfaces of suspended particles, particularly ones with high organic carbon content. Therefore current large-scale research efforts to examine water quality in the Great Lakes are concerned not only with the possible presence of these layers but also with their effects on the transport and distribution of organic chemical pollutants. Relationships between sediment and contaminant transport are discussed further in Sec. 15.6.

15.5 PARTICLE INTERACTIONS

15.5.1 Size Distribution

As previously noted, there is a wide variety of particle sizes in natural waters. Since particles can freely interact with each other only in a suspended state, the present section deals with smaller particles. In general, there is a range of sizes present in any suspension, and one of the most important physical characteristics of a polydispersed suspension is the size distribution. Size, for example, is a critical parameter in determining settling rates, which must be known for proper design of sedimentation basins in water and wastewater treatment plants. Size also directly influences possible methods of detection, as needed for scientific study of particle interactions and for engineering design of treatment processes. For example, optical microscopes cannot be used to detect particles smaller than the wavelength of visible light (less than about 0.5 μm). The sedimentation rate also is dependent on the square of particle size (Stokes settling — see Eq. 15.2.4), and the total surface area per unit mass depends on particle size, which determines the potential for chemical adsorption (see below). This information is essential for analyzing raw-water particulates, optimizing chemical dosages, measuring particulates in filter effluent and determining filter efficiency, and evaluating particle/pollutant transport mechanisms, among other applications.

It is usually convenient to consider a suspension containing many particles of various sizes as having a continuous size distribution. For continuously distributed particles, three common particle size distribution functions are

$$\frac{dN}{dv} = n_v(v) \quad (\text{number/cm}^3/\mu\text{m}^3) \tag{15.5.1a}$$

$$\frac{dN}{ds} = n_s(s) \quad (\text{number/cm}^3/\mu\text{m}^2) \tag{15.5.1b}$$

$$\frac{dN}{d(d_p)} = n_d(d_p) \quad (\text{number/cm}^3/\mu\text{m}) \tag{15.5.1c}$$

where N is the cumulative size distribution function (number/cm^3), v is particle volume, s is particle surface area and d_p is diameter (subscript "p" is added here to distinguish particle diameter from the differential operator). The total volume of the particle suspension is

$$\forall = \int_0^\infty v n_v(v) \, dv \tag{15.5.2a}$$

Similarly, total surface area is

$$S_p = \int_0^\infty s n_s(s) \, ds \tag{15.5.2b}$$

Combining these with the three parts of Eq. (15.5.1) yields, respectively,

$$\frac{d\forall}{d(\log d_p)} = \frac{2.3\pi}{6} d_p^4 \frac{\Delta N}{\Delta d_p} \tag{15.5.3}$$

$$\frac{dS_p}{d(\log d_p)} = 2.3\pi d_p^3 \frac{\Delta N}{\Delta d_p} \tag{15.5.4}$$

relative frequency

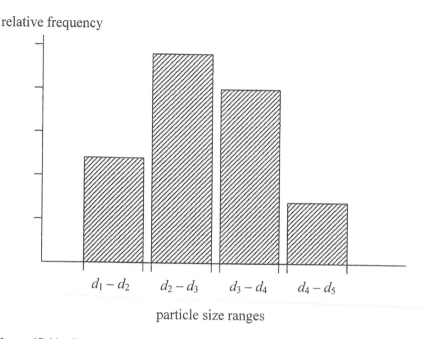

particle size ranges

Figure 15.11 Example of particle size distribution, as relative frequency of particle numbers falling in different diameter ranges.

$$\frac{dN}{d(\log d_{\mathrm{p}})} = 2.3 d_{\mathrm{p}} \frac{\Delta N}{\Delta d_{\mathrm{p}}} \tag{15.5.5}$$

Thus, once $(\Delta N / \Delta d_{\mathrm{p}})$ is known, data can be reported in number, surface, or volume concentrations, depending on the desired application and instrumentation available. This is equivalent to developing a histogram chart of number distribution, as illustrated in Fig. 15.11.

15.5.2 Particle Aggregation — Smoluchowski Model

There are many ways to model aggregation processes, and these can be categorized either as *microscale* or as *macroscale models*. Microscale models describe aggregation at a particle–particle level, considering interaction forces when particles are at close distances. In general, three types of forces are important: (1) hydrodynamic interaction, which prevents particle collisions; (2) van der Waals attractive forces, which promote particle collisions; and (3) electrostatic repulsion, which inhibits particle collision. Once these forces are considered, along with other forces such as gravity and fluid shear stress, the equation of motion of individual particles in a suspension may be solved to predict steady-state or time-dependent positions of every particle in the dispersion. In addition, the interaction of these forces determines the trajectory of particles approaching each other and, eventually, the aggregate geometry. Diffusion Limited Aggregation (DLA) theory, though not considering details of the forces involved, also is an example of a microscale approach. These models have the disadvantage of requiring calculations for every particle in the system, which becomes cumbersome for large numbers of particles.

At the macroscopic level, Eulerian methods have been used to describe the system properties in terms of particle concentrations in space and time, either for steady-state conditions or to reveal time-dependent behavior. Smoluchowski's approach (see below), which describes changes of particle concentration of different sizes in a reactor, best represents macroscale models.

Aggregates form when interparticle cohesive forces become important, as is particularly true with clays and other smaller materials. In general, it is of interest to keep track of the amount of suspended sediment mass in different size classes. In other words, although size distributions are considered to be continuous, it is convenient to model a smaller number of discrete size classes. In order for aggregation to occur, dispersed particles must be brought into contact with one another. In general, this can occur through any of three different transport processes: (1) *Brownian diffusion*, driven by the thermal energy of the fluid; (2) *fluid shear*, which causes velocity differences or gradients in either laminar or turbulent flow fields; and (3) *differential settling*, driven by gravity such that collisions occur due to differences in settling velocity. Once particles are brought very close to or in contact with each other, particle interactions determine whether they become attached, forming a larger aggregate, or remain in a dispersed state. This interaction may be attractive (e.g., van der Waals attraction) or repulsive (e.g., electrical repulsion), depending on the nature of the particle surfaces and on the solution chemistry. In addition, aggregates also may break apart under high shear conditions, generating two or more smaller aggregates.

Transport and attachment can be considered to occur at different scales. This is because direct particle interactions are important only over very short ranges, usually much less than the particle size, so that particles must be very close to each other before there is any significant direct interaction. On the other hand, bulk transport mechanisms may bring particles together from comparatively large distances, to a point where the separation distance is sufficiently small that interaction can take place. For example, in a turbulent flow, particles are largely distributed by the eddies in the integral length scale range. Eddies near the microscale range (but still larger than the particle size) generate the local shear between particles, and finally, attachment is decided by the interactions occurring at distances much smaller than the particle sizes.

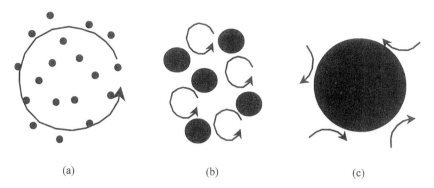

(a)   (b)   (c)

Figure 15.12 Role of eddy size, relative to particle size, in causing particle collisions: (a) eddies are much larger than the particles and tend to transport them all at similar rates, so that collisions are not promoted; (b) eddies and particles are similar in size and collision rate is enhanced; and (c) eddies are much smaller than the particles and do not affect their transport significantly.

The eddies closest in size to the particles or aggregates are the most effective in bringing particles into contact with each other. Much larger eddies transport groups of particles in a sort of bulk or advective-type flow (i.e., relative to the particle size, the eddy motion appears as advective flow and does not contribute to smaller scale mixing), while eddies much smaller than the particles are not effective in moving particles around and can only affect transport conditions between the surface of the particle and the surrounding water. Figure 15.12 illustrates these various cases. There is significant interest in promoting the formation of larger aggregates in treatment plants, for instance, so it is important to understand the mechanisms contributing to good mixing efficiency.

The modern study of aggregation kinetics started with the work of Smoluchowski in the early 1900s. The original Smoluchowski theory accounts for collisions induced by Brownian motion. Generally, Brownian movements are caused by thermal energy and are random in nature. Smoluchowski considered one particle to be a stationary collector and calculated the diffusion rate, caused by Brownian motion, of other particles to this collector. Because particles become attached to the collector and are therefore removed from suspension, a concentration gradient is formed radially from the collector. This gradient can be determined by applying Fick's second law for a spherical collector,

$$\frac{\partial n_i}{\partial t} = \frac{1}{r^2}\frac{\partial}{\partial r}\left(r^2 D_i \frac{\partial n_i}{\partial r}\right)$$

$$(15.5.6)$$

where n_i is the particle number concentration (of size i) in the fluid around the collector, r is the radial distance from the center of the collector, and D_i is the diffusion coefficient for the particles.

An estimate for the *steady state* distribution of particles of size n_i is obtained by integrating Eq. (15.5.6) and applying the boundary conditions: (1) $n_i = 0$ at the surface of the collector, where $r = R_{ij} = $ sum of the radii of the collector, of size j, and an attached particle, of size i, and (2) $n_i = n_{io}$ at $r \to \infty$, implying that at great distances the particle concentration is unaffected by the presence of the collector. A first integration, with the left-hand side set to 0 (i.e., steady state conditions), gives

$$r^2 D_i \frac{dn_i}{dr} = C \qquad (15.5.7)$$

where C is a constant. By separating variables and integrating again, using the specified boundary conditions for n_i, we have

$$n_i \Big|_{R_{ij}}^{\infty} = \frac{C}{D_i} \left(-\frac{1}{r} \right) \Big|_{R_{ij}}^{\infty} \qquad \Rightarrow \qquad n_{io} = \frac{C}{D_i R_{ij}} \qquad (15.5.8)$$

Substituting this last result in terms of C into Eq. (15.5.7) and integrating from R_{ij} to some arbitrary point r, we obtain

$$\frac{dn_i}{dr} = \frac{n_{io} R_{ij}}{r^2} \qquad \frac{n_i}{n_{io}} = 1 - \frac{R_{ij}}{r} \qquad (15.5.9)$$

which is the steady-state distribution for n_i.

In addition, the rate at which particles diffuse to the stationary collector may be calculated by *Fick's first law*,

$$J_i = 4\pi R_{ij}^2 D_i \frac{dn_i}{dr} \qquad (15.5.10)$$

where J_i is the transport rate of particles (number per unit time per collector) moving to a collector of surface area $4\pi R_{ij}^2$. Note that a spherical collector has been assumed — this is a major simplifying assumption in the development of the Smoluchowski model. By considering that the collector also is subject to diffusion, with diffusivity D_j, then the total effective diffusivity causing particles of sizes n_i and n_j to collide is $D_{ij} = D_i + D_j$. Then, with n_j collectors, the total rate of growth of aggregates formed by joining particles i and j is

$$J_{ij} = 4\pi R_{ij}^2 D_{ij} n_i n_j \qquad (15.5.11)$$

For aggregate $k = i + j$ (i.e., a particle of size class k, formed by the joining of particles from size classes i and j), the basic *Smoluchowski equation* may be written as

$$\frac{dn_k}{dt} = \frac{1}{2} \sum_{i+j=k} 4\pi R_{ij}^2 D_{ij} n_i n_j - n_k \sum_{i=1}^{\infty} 4\pi R_{ik}^2 D_{ik} n_i \tag{15.5.12}$$

where the first term on the right-hand side represents the production of k particles by the collision of i and j particles, and the second term represents a sink for particle k, as it joins other particles to form larger aggregates. The one-half coefficient in the first term is to avoid counting the same particle twice, i.e., $i + j = j + i$.

Equation (15.5.12) forms the basis of the Smoluchowski approach for aggregation studies. It should be noted that in this basic form, it is assumed that each collision results in an attachment (aggregation). However, sometimes attachment does not occur, due to particle interaction forces or other factors, and a collision efficiency factor needs to be added (see below). Particles also may break apart, or disaggregate, forming smaller aggregates, but this is not considered in Eq. (15.5.12). In addition, D_{ij} is not limited only to Brownian diffusion. For shear or turbulent flows an appropriate D_{ij} can be calculated according to specific mixing conditions, though the functional form of the equation remains the same.

Equation (15.5.12) has been applied in many studies involving particle aggregation. Usually, as shown above, the most common mechanisms included in these models are gain (aggregation), i.e., $i + j = k$, and loss of particles of class k from its collision with all classes of particles. However, in addition to aggregation, mechanisms affecting particle number and size include settling (loss) and production terms, and noncohesive collisions, when particles collide but do not stick together to form larger aggregates. Also, disaggregation following a collision, or in response to high fluid shear, is not included in Eq. (15.5.12).

With these other features in mind, a more general form of Smoluchowski's equation is

$$\frac{dn_k}{dt} = \frac{1}{2} \sum_{i+j=k} \alpha_{ij} \beta_{ij} n_i n_j - n_k \sum_{j=1}^{\infty} \alpha_{jk} \beta_{jk} n_j - B_k n_k + \sum_{j=k+1}^{\infty} \gamma_{ij} B_j n_j$$

$$- n_k \sum_{j=1}^{\infty} C_{jk} \beta_{jk} n_j \pm W_k - \frac{1}{t_s} n_k \tag{15.5.13}$$

where α_{ij} and β_{ij} are the collision efficiency and frequency, respectively, B is the break-up coefficient, γ_{ij} is the probability that particle k is formed after the disaggregation of a particle of class j, C_{ij} is the probability of disaggregation of particle k after collision with particle i, W is a production/destruction term, and t_s is the settling time scale, which is the ratio between the depth of the water and the settling velocity. The first two terms on the right-hand side correspond directly with the terms on the right-hand side of Eq. (15.5.12). In Eq. (15.5.13), however, collision efficiency has been added, which expresses the probability that aggregation occurs following a collision, and collision frequency has been used. It is easily seen that collision frequency is directly related to particle size and diffusivities. The third term on the right is loss due to the breakup of particle k due to fluid shear, the fourth term is the production of particle k by the breakup of larger particles, the fifth term is the loss of particle k resulting from a collision with another aggregate, the sixth term is an internal source or sink, and the final term represents loss by settling. In many cases, not all of the terms in Eq. (15.5.13) are important. However, collision frequency and efficiency are of interest in almost all applications.

Collision Frequency

As previously noted, there are three general mechanisms that bring particles into contact with each other, including shear (orthokinetic interactions), Brownian motion (perikinetic motion), and differential settling. The simplest case of orthokinetic aggregation considers particle collisions in a uniform shear flow. Neglecting interaction forces introduced as particles approach each other, rectilinear particle trajectories may be assumed. By considering the total flux of particles approaching each other, the collision rate β between two particles of size (radius) R_i and R_j can be expressed as

$$\beta_{ij} = \frac{4}{3}G(R_i + R_j)^3 \tag{15.5.14}$$

where G is the mean velocity gradient, which is normally defined in terms of the total power used to generate mixing (this can be the input power to a mixer in a stirred tank, for example) and fluid viscosity,

$$G = \sqrt{\frac{P}{\mu}} \tag{15.5.15}$$

where P is input power and μ is viscosity. For Brownian motion and differential settling, collision frequency is described, respectively, by

$$\beta_{ij} = \frac{2KT}{3\mu} \frac{(R_i + R_j)^2}{R_i R_j} \tag{15.5.16}$$

and

$$\beta_{ij} = \left(\frac{2\pi g}{9\mu}\right)(\rho_s - \rho)(R_i + R_j)^3(R_i - R_j) \tag{15.5.17}$$

where K is the *Boltzmann constant* and T is *absolute temperature*. The overall collision frequency is a combination of these three mechanisms and is usually calculated as a geometric sum of equations (15.5.14), (15.5.16), and (15.5.17).

Collision Efficiency

Collision efficiency α represents the probability that particles will aggregate after they collide. It is a function of both chemical and physical conditions. In general, there are three types of phenomena that become important when particles come into close proximity. First, water in between the particles must be moved out of the way. This process is called *hydrodynamic interaction* and it tends to prevent particle collision. Second, van der Waals attractive forces exist between any two particles and become significant at small separation distances; these forces tend to promote particle collisions and attachment. Third, if the particles have charged surfaces, electrostatic repulsion also inhibits particle collision. It is difficult to evaluate these mechanisms for all possible cases, and although significant research has been conducted, the best we can say at this time is that the value for α usually lies somewhere between 10^{-4} and 10^{-1} for fresh waters (Weilemann et al., 1989). There is some evidence to suggest its values are higher in salt water by a factor of two to three. Given this wide range, it is difficult to choose a value *a priori* for α for any given application.

    For two particles with the same surface charge in an electrolyte, repulsion force becomes important as they approach each other. For a negatively charged surface submersed in a neutral solution, a high concentration of positive charges forms at the surface and is called the *Stern layer*. Beyond this layer, but before the bulk solution, another layer is formed in which the positive charges gradually diffuse as radial distance increases. This is the source of repulsion potential. On the other hand, there is an attraction force (potential) due to van der Waals forces, which also becomes important when particles are very close to each other. By considering both repulsion and attraction forces, the interparticle potential may be calculated as the net sum of the attraction and repulsion potentials.

15.6 PARTICLE-ASSOCIATED CONTAMINANT TRANSPORT

Many contaminants of interest tend to sorb strongly onto particles, so that a knowledge of the fate and transport of suspended sediments provides much of the story for describing the fate and transport of the pollutant. Interest here is primarily with smaller particles, which have higher surface area (per unit mass), are more likely to be suspended, and also may have relatively high organic carbon content. Heavy metals and hydrophobic organic compounds (HOCs) such as PCBs are two classes of contaminants of particular interest in environmental engineering problems.

The fate of HOCs in aquatic environments is highly dependent on sorptive behavior. Partitioning models, in which equilibrium between the water phase (dissolved) and solid phase (particulate) concentrations is assumed, are often adequate to describe transport phenomena when solid–water contact times are relatively long. For HOCs having equilibrium water phase concentrations less than 10^{-5} M or one-half of the species water solubility, sorption relationships (called *isotherms*) to natural sediments are linear, so that the concentrations in the two phases can be related as

$$r = K_p C_d \tag{15.6.1}$$

where r = concentration in the solid phase (mass compound per unit mass of solid), C_d = concentration in the water phase (mass compound per unit volume of solution), and K_p = *equilibrium partition coefficient*. This partitioning process has been shown to be reversible and independent of sorbent concentration.

Numerous sorption studies for soils and sediments have shown that the partitioning process occurs primarily between the water phase and the organic carbon of the solid phase. The partition coefficient is therefore often expressed as

$$K_p = K_{oc} f_{oc} \tag{15.6.2}$$

where f_{oc} = organic carbon content of the solids and K_{oc} quantifies the partitioning tendency of the compound to organic carbon. This relationship is valid provided the organic carbon content exceeds about 0.1%. Various studies also have found K_{oc} to be a function of the *octanol–water partition coefficient*, K_{ow}, of the same compound, having the form

$$\log K_{oc} = a \log K_{ow} + b \tag{15.6.3}$$

in which a and b are data-fitted constants. The advantage of this result is that the octanol–water partition coefficient value is known for many compounds. This formulation is easily incorporated into transport and fate models and is valid provided the equilibrium assumption (Eq. 15.6.1) holds.

There are many instances, however, for which the flow retention time scales are similar to or less than those for sorption/desorption transfer, therefore invalidating the equilibrium assumption. A common example of this is with the resuspension of contaminated sediments due to storm runoff or other resuspension events. In other words, sorbed contaminants on bottom sediments are usually assumed to be in equilibrium with pore water concentrations, following relationships similar to Eq. (15.6.1). When these sediments are eroded from the bottom and mixed into the overlying water column, they are placed in an environment where the dissolved concentration is presumably much less than the equilibrium value. In these cases the particles are subject to lateral and vertical transport (by advection/diffusion processes), and we can estimate an average resuspension time simply as the height above the bottom to which the particles are raised (h), divided by the net mean settling velocity (v_s),

$$t_r = \frac{h}{v_s} \tag{15.6.4}$$

In many cases v_s can be approximated by w_s, the Stokes settling velocity. However, turbulence and probable high particle concentrations in the water column, along with the bottom boundary condition (for deposition), may affect the net settling rate. While in suspension, the chemical undergoes kinetic desorption, at a rate given by the inverse of a time scale to reach equilibrium partitioning, t_e. The ratio (t_r/t_e) indicates the degree to which the equilibrium partitioning assumption would be valid. When this ratio is large, the equilibrium assumption is reasonable, but when it is small, a kinetic desorption calculation should be performed.

Results from studies in a variety of soil/sediment–water systems have shown that sorption of HOCs onto soils and sediments consists of a rapid initial uptake followed by a slow final approach to equilibration. Similarly, desorption consists of a rapid initial release followed by slow equilibration. This is referred to as a two-step sorption process, and several modeling approaches have been developed to incorporate this phenomenon. The one that seems to relate most closely to the physics of the process, and also is relatively easy to implement, is based on defining two first-order rate constants. It is assumed that the initial release of chemical involves material primarily near the surface of the particles, and therefore is more easily accessible, while the later stage is related to the release of chemical from within the pore spaces of the particle. Results presented by Wu and Gschwend (1986) imply that kinetic effects are particularly acute for compounds having large K_{ow} and small molecular diffusivities. Sorbed PCBs, having $K_{ow} = 10^5$ to 10^8, would therefore require long times to equilibrate in response to changes in water phase concentrations, and equilibration times of up to 280 days have been estimated for typical sediments. This is considerably longer than anticipated resuspension times, which may be on the order of several hours to several days at most.

Often, from a management point of view, it is of interest to determine the rate of natural loss of in-place contaminants (in bottom sediments) from a region by modeling sediment transport processes. In particular, when contaminated sediment is entrained during a high flow event, the question is to what extent the particles are transported downstream and to what extent the contaminant is desorbed while the particles are in suspension. An equilibrium assumption may suggest that more contaminant is lost than would be the case under more realistic kinetic desorption assumptions. In other words, under the equilibrium assumption a greater amount of contaminant would enter the dissolved phase before the particles resettle and would be transported away from the region of interest.

15.6.1 Partitioning

Equilibrium Approach

Total concentration of a contaminant is written as the sum of the particulate (i.e., attached to particles) and the dissolved phase concentrations,

$$C_t = C_p + C_d \tag{15.6.5}$$

where all values are in units of mass contaminant per unit volume of water; C_p is the particulate phase concentration, and C_d is the dissolved phase concentration. The equilibrium partitioning coefficient is defined as in Eq. (15.6.1),

$$K_p = \frac{r}{C_d} \tag{15.6.6}$$

As previously noted, this is the ratio of particulate and dissolved phase concentrations that would exist when the two phases are in equilibrium. If m is the sediment concentration (mass sediment per unit volume of water), then $C_p = mr$, so that the partition coefficient also may be defined as

$$K_p = \frac{C_p}{mC_d} \tag{15.6.7}$$

From these expressions, relations between the two phases can be obtained as

$$C_d = \frac{C_t}{1 + mK_p} \tag{15.6.8}$$

and

$$C_p = \frac{C_t mK_p}{1 + mK_p} \tag{15.6.9}$$

Thus the distribution of dissolved and sorbed concentrations is known, as long as the total concentration, sediment concentration, and partition coefficient values are specified.

Kinetic Approach

Several models have been developed to simulate sorption/desorption kinetics, primarily for HOCs and metals. Early works treated sorption as a simple first-order chemical reaction. Unfortunately, these formulations include rate constants and other parameters that must be determined experimentally for each sediment–pollutant combination. More recent efforts have attempted to develop mechanistically based models that predict sorption kinetics from known or easily measurable properties of the sediments and pollutants. Often these models are based on the assumption that the aggregates are spherical (recall the earlier discussion of the Smoluchowski equation). Most models assume that natural aggregates are made up of fine mineral grains and organic matter. With these assumptions, it has been shown that a single effective diffusivity parameter could be used to quantify sorption kinetics in a diffusion-limited model, where the effective diffusivity is a function of the molecular diffusivity, partition coefficient, intra-aggregate porosity, and solid density. Other models have been based on the idea that partitioning occurs between the water phase and the natural organic matter that coats the soil particles, consistent with the earlier discussion of the influence of organic carbon on the particles (see Eq. 15.6.2). This approach includes diffusion of the sorbate from the bulk liquid through an aqueous diffusion layer, partitioning at the water–organic interface, and diffusion into the organic coating of the soil. Both of these models replicate the rapid uptake/release and subsequent slow equilibration observed in sorption experiments, as previously noted.

These models generally assume that the particles making up sediments can be viewed as porous spherical aggregates of finer grains in which the macroscopic sorption of chemicals occurs microscopically by either sorption to the natural organic matter on the solid matrices or diffusion into the intraparticle pore water. The radial diffusion of HOCs into or out of the aggregate is retarded by local equilibrium partitioning between intraparticle pore water and organic matter associated with the solids. For a spherical aggregate of radius R, the mathematical expression of these transport processes can be stated as

$$\frac{\partial}{\partial t}\{[(1-n)\rho_s K_p + n]C_a\} = \frac{1}{r^3}\frac{\partial}{\partial r}\left(r^3 D_m T \frac{\partial C_a}{\partial r}\right) \tag{15.6.10}$$

where C_a = local volumetric water phase concentration in the aggregate, r = radial distance, t = time, n = interaggregate porosity, ρ_s = mass density of the solid particles comprising the aggregate, T = interaggregate tortuosity, and D_m = molecular diffusivity of sorbate molecules in water. This equation is valid for $0 < r < R$.

When fluid turbulence is of low intensity, an exterior boundary layer of thickness L limits sorptive exchange between the bulk water phase and the sediment aggregate. Transport through this layer is by radial diffusion alone and is governed by

$$\frac{\partial C_{bl}}{\partial t} = \frac{1}{r^2}\frac{\partial}{\partial r}\left(r^2 D_m \frac{\partial C_{bl}}{\partial r}\right) \tag{15.6.11}$$

where C_{bl} is the dissolved phase concentration in the boundary layer. This equation applies for $R < r < R + L$.

Boundary conditions associated with Eqs. (15.6.10) and (15.6.11) include zero flux at the center of the aggregate and the specification of concentration at the diffusion layer–bulk water interface, i.e.,

$$\frac{\partial C}{\partial t}(0, t) = 0 \qquad (15.6.12)$$

and

$$C(R + L, t) = C_d \qquad (15.6.13)$$

where C_d is the bulk water phase concentration. At the aggregate–diffusion layer interface, mass conservation requires that there be no accumulation of the diffusing substance. This requirement is expressed as

$$C_{bl} = C_a \qquad (15.6.14)$$

and

$$n D_m T \frac{\partial C_a}{\partial r} = D_m \frac{\partial C_D}{\partial r} \qquad (15.6.15)$$

at $r = R$.

As noted previously, two-phase partitioning behavior also is modeled directly by considering two separate pools of material, one being relatively easily desorbed and the other diffusing or desorbing more slowly. The simplest approach is then to define separate sets of sorbing and desorbing rates for the two pools, using simple first-order expressions. For example, the desorption flux for one of the components is

$$\frac{\partial C_{p1}}{\partial t} = K_{1d(C_d - C_p1)} \qquad (15.6.16)$$

where subscript "1" refers to either the quickly or the slowly desorbing pool, and K_{1d} is the desorption rate or bulk mass transfer coefficient for component 1. A similar equation can be written for component 2. Equations also can be written for the adsorbtion process. For either component, noting that the desorption flux is primarily proportional to $K_d C_p$, and adsorption flux is proportional to $K_s C_d$, where K_s is a sorption mass transfer coefficient, at steady state (i.e., equilibrium) these fluxes must exactly balance, so

$$K_d C_p \approx K_s C_d \implies m K_p = \frac{K_s}{K_d} \qquad (15.6.17)$$

This demonstrates the relationship between the partitioning coefficient and the two process rates (at steady state) and provides an extra constraint when formulating these models.

PROBLEMS

Solved Problems

Problem 15.1   Show that the duBoys formula, Eq. (15.3.8), can be written as

$$j_s = \psi H S_0 \left(H S_0 - \frac{\tau_c}{\gamma} \right) \tag{1}$$

where H is the water depth, g is the specific weight of the water, and

$$\psi = \frac{\gamma_s \Delta z \, \Delta v}{2} \left(\frac{\gamma}{\tau_c} \right)^2 \tag{2}$$

is called a *sediment characteristic*.

Solution

Starting with Eq. (15.3.8), insert Eq. (15.3.2) for the case of wide flow to substitute for τ_0. This results in

$$\begin{aligned}
j_s &= \frac{\gamma_s \Delta z \, \Delta v}{2} \left[\left(\frac{\gamma H S_0}{\tau_c} \right)^2 - \left(\frac{\gamma H S_0}{\tau_c} \right) \right] \\
&= \frac{\gamma_s \Delta z \, \Delta v}{2} \left(\frac{\gamma}{\tau_c} \right)^2 \left[(H S_0)^2 - H S_0 \frac{\tau_c}{\gamma} \right]
\end{aligned}$$

Using Eq. (2) for ψ and rearranging then gives the result of Eq. (1). Values for ψ have been tabulated for different kinds of sediment, most commonly in English units. Since ψ is a sediment characteristic, it has been found to be a function of sediment size.

Problem 15.2   It can be shown, for cases of flows in which the two-dimensional assumption cannot be applied, that the total bed load transport rate may be written as

$$J_s = \psi S_0^2 A (2\bar{z} + H_c) \tag{1}$$

where ψ is defined in problem 15.1, H_c is the minimum depth required for bed load movement, A is the area of the channel below depth H_c, and $\bar{z}$ is the depth, relative to H_c, to the centroid of A, as sketched in Fig. 15.13. Use Eq. (1) to calculate the bed load in a river with an approximately parabolic cross section and a bed slope of $1:8000$. The river width is 200 m when the flow depth is 3 m. The mean effective bed particle diameter is 2.0 mm. The total water depth is $H = 5$ m. Use $\tau_c = 2.48$ Pa and $\psi = 10.4 \times 10^6$ N/(m^3-s).

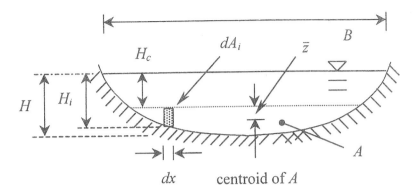

Figure 15.13 Definition sketch, Problem 15.2.

Solution

The derivation of Eq. (1) is left as an exercise for the student. It involves dividing the river cross section, below depth H_i, into small area elements dA and applying an integral form of Eq. (15.3.10). To solve the present problem, we need to first find H_c. This is done using Eq. (15.3.2), with the given value for τ_c:

$$H_c = \frac{\tau_c}{\gamma S_0} = \frac{2.48(8000)}{9810} = 2.02 \text{ m}$$

Then, from geometry, the distance between depth H_c and the centroid of the area below H_c is $(2/5)(H - H_c) = 0.4(2.98) = 1.19$ m. Next we need to find the area below H_c. This is done using the general expression for a parabola, in terms of the width B,

$$B = cH^{1/2}$$

where c is a constant that is determined by the given information that $B = 200$ m when $H = 3$ m, $c = 200/3^{(1/2)} = 115.5(\text{m}^{1/2})$. The width at depth H_c is then $B_c = 115.5(2.98)^{1/2} = 194.4$ m. The area of the parabola below H_c is then

$$A = \tfrac{2}{3}(2.98)(194.8) = 396.1 \text{ m}^2$$

Finally, using Eq. (1),

$$J_s = \psi S_0^2 A(2\bar{z} + H_c) = (10.4 \times 10^6)\left(\frac{1}{8000}\right)^2$$
$$(396.1)(2.38 + 2.02) = 283 \text{ N/s}$$

Problem 15.3 Consider the vertical distribution of suspended sediments in a lake. Assume that the particle settling velocity w_s is constant. There is no mean fluid motion, but turbulent mixing is incorporated through a diffusivity ε, which also can be assumed constant.

    (a) Write the governing equation describing the distribution of sediments in the water column. Assume that there are no sources or sinks.

    (b) What boundary condition would you use at the surface, to express the fact that there is no movement of particles across that boundary?

    (c) Suppose that at the bottom the concentration is m_0. Solve for the steady-state concentration distribution, in terms of m_0.

Solution

    (a) From Eq. (15.4.1), simplifying for the conditions of constant settling speed and diffusivity, no horizontal gradients and no fluid motions and no sources or sinks, we have

$$\frac{\partial m}{\partial t} - w_s \frac{\partial m}{\partial z} = \varepsilon \frac{\partial^2 m}{\partial z^2} \tag{1}$$

    (b) Equation (15.4.12) expresses the condition of zero net flux across the surface:

$$w_s m + \varepsilon \frac{\partial m}{\partial z} = 0$$

    (c) For the steady state, the equation reduces to a second-order ordinary differential equation which, when integrated once, gives

$$w_s m + \varepsilon \frac{\partial m}{\partial z} = C$$

where C is a constant. From the boundary condition of part (b), $C = 0$. Integrating once again, using the boundary condition that $m = m_0$ at $z = 0$, gives

$$\frac{m}{m_0} = \exp\left(-\frac{w_s z}{\varepsilon}\right)$$

Thus the steady-state distribution is an exponential decay with increasing height above the bottom. Note that this result is the same as Eq. (15.4.10), with $a = 0$.

Unsolved Problems

Problem 15.4 For each of the following particles, calculate the Stokes settling speed and check to see whether the Stokes law is in fact valid:

    (a) Sand, with specific gravity 2.65 and equivalent diameter 1 mm

(a)   (b)

Figure 15.14   Two arrangements of particles in an aggregate, Problem 15.5.

(b)   Clay, with specific gravity 2.5 and equivalent diameter 10 μm
(c)   Algae, with specific gravity 1.03 and equivalent diameter 6 μm

Problem 15.5   A two-dimensional version of Eq. (15.2.1) can be defined in terms of area and a two-dimensional fractal dimension, D_2,

$$A = C_A L^{D_2}$$

where C_A is a constant having units of L^{2-D_2}. Compare the values of D_2 calculated for two different shapes formed by an equal number n of circles of diameter d, as shown in Fig. 15.14. Use L as the longest length for each case [i.e., $L = nd$ for case (a) and $L = (2n)^{1/2}d/2$ for case (b)].

Problem 15.6   Consider that an arbitrary particle aggregate can be approximated by a number n of smaller spheres of diameter d, as shown in Fig. 15.15. Assuming that the surface area of the aggregate also can be approximated by the sum of the surface areas of the spheres, what is the relationship between the equivalent diameter D_e and the surface diameter D_s, in terms of n?

Problem 15.7   Calculate the bed load in a wide river with bed slope 1 : 6000 and a normal uniform depth of 4.5 m. The effective particle size is 0.5 mm, critical shear stress is 2.03 Pa (0.0215 psf), and sediment characteristic (Problem 15.1) is 29.4×10^6 N/(m^3-s)[187,000 lb/(ft^3-s)].

Problem 15.8   Derive Eq. (1) in problem 15.2 by first writing an expression for the contribution dJ_s to the total load, that occurs in area dA (see Fig. 15.13).

Problem 15.9   Calculate the bed load for a river with the same sediment characteristics as in problem 15.7, but the river cross section is approximately

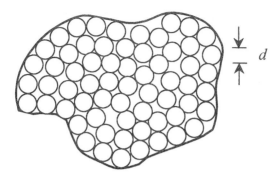

Figure 15.15 Definition sketch, Problem 15.6.

parabolic. The normal uniform depth is 6 m, the river width is 100 m at a depth of 1 m, and the bed slope is $1 : 8000$.

Problem 15.10 A river has an initial suspended sediment concentration m_o, at location $x = 0$, where x is the longitudinal coordinate. The river mean velocity is 0.7 m/s. Assuming that there is no resuspension and that the particles have an average settling speed of 10 m/day, calculate the distance downstream at which the average suspended concentration is one-half of m_o. Compare the result when longitudinal dispersion is neglected and when it takes a value of 200 m²/s. In both cases it may be assumed that the particles are well mixed in the vertical direction.

Problem 15.11 Considering again the conditions specified in problem 15.10, recalculate the distances for the two assumed values for longitudinal dispersivity when there is no vertical mixing.

Problem 15.12 How would the boundary condition of Eq. (15.4.22) be written so that it directly incorporates a probability of deposition p? Assume that resuspension is not taking place.

Problem 15.13 Write an implicit finite difference numerical solution to Eq. (15.4.11). Use a zero-flux boundary condition at the top and a probability of deposition at the bottom, as developed in problem 15.12. Assuming a water column of depth 20 m and with an initial uniform concentration of 10 mg/L, use the solution to calculate the times needed for half of the suspended mass to settle when $p = 0.2$, 0.5, and 0.8.

Problem 15.14 Using Eqs. (15.5.9) and (15.5.10), show that the flux of particles towards a stationary spherical collector is independent of r under steady conditions.

Problem 15.15 In a lake the equilibrium partitioning coefficient for a certain hydrophobic chemical is determined to be 10^{-5} L/mg. The suspended solids concentration is 10 mg/L and the solid phase contaminant concentration is 0.5 µg/g. What is the total concentration of chemical in the lake? What percentage of the total concentration is in each of the dissolved and particulate phases?

Problem 15.16 In a lake the equilibrium partitioning coefficient for a certain chemical is found to be 10^{-4} L/mg. The total concentration is 1 µg/L and the suspended sediment concentration is 7 mg/L. Find the distribution of the chemical between the dissolved and particulate phases.

Problem 15.17 Using the equation for the boundary layer around an aggregate, Eq. (15.6.11), calculate the steady-state concentration distribution for a constant value of the diffusivity D_m, in terms of the parameters defined in Sec. 5.6.1.

Problem 15.18 Consider a chemical sorbed onto a particle that is initially settled onto the bed of a river but becomes resuspended into the water column during a storm event. The chemical dissolved and particulate phases are in equilibrium in the bed environment, but the water column has virtually zero dissolved concentration, so the chemical tends to desorb from the particle once it has been suspended. Formulate an expression for the ratio of chemical desorbed, using a nonequilibrium approach, to the amount desorbed using an equilibrium approach, in terms of the parameters defined in Sec. 15.6.1. For simplicity, assume only one component of desorbing concentration, with one associated desorption coefficient, and also assume that sorption is negligible.

SUPPLEMENTAL READING

Note: There are many books and articles dealing with sedimentation processes. A quick look in any engineering library would likely reveal a number of books. Instead of listing these, we give the following list to provide additional reading sources for some of the specialized topics discussed in the chapter. It is far from an exhaustive list, however.

Ambrose, R. B., Wool, T. A., Connolly, J. P. and Schanz, R. W., 1988. WASP4, A Hydrodynamic and Water Quality Model — Model Theory, User's Manual and

Programmer's Guide. Environmental Research Laboratory, Office of Research and Development, U.S. Environmental Protection Agency, EPA/600/3-87/039.

Apmann, R. P. and Rumer, R. R., 1970. Diffusion of sediment in developing flow. *J. Hydraul. Div.* ASCE **96**:109–123.

Barfield, B. J., Smerdon, E. T. and Hiler, E. A., 1969. Prediction of sediment profiles in open channel flow by turbulent diffusion theory. *Water Resour. Res.* **5**:291–299.

Bayazit, M. 1972. Random walk model for motion of a solid particle in turbulent open-channel flow. *J. Hydraul. Res.* **1**:1–14.

Christensen, B. A., 1987. Transport of suspended sediment in very rough channels, in *Coastal Sediments*, ed. N. Kraus. ASCE, New York, 41–49.

Coates, J. T. and Elzerman, A. W., 1986. Desorption kinetics for selected PCB congeners from river sediments. *J. Contam. Hydrol.* **1**:191–210.

Connolly, John P., Armstron, Neal E. and Miksad, Richard W., 1983. Adsorption of hydrophobic pollutants in estuaries. *J. Environ. Eng.* **109**(1):17–35.

Crank, J. 1975. *The Mathematics of Diffusion.* 2d ed. Clarendon Press, Oxford.

Csanady, G. T., 1980. *Turbulent Diffusion in the Environment.* Reidel, Boston.

DePinto, J. V., Theis, T. L., Young, T. C., Vanetti, D., Waltman, M. and Leach, S., 1989. Exposure and biological effects of in-place pollutants: Sediment exposure potential and particle–contaminant interactions. Report No. 90-3, Department of Civil and Environmental Engineering, Clarkson University, Potsdam, NY, December.

Dobbins, W. E., 1944. Effect of turbulence on sedimentation. *Trans. ASCE* **108**:626–678.

Einstein, H. A. and Chien, N., 1954. Second Approximation to the Solution of the Suspended Load Theory. Series 47, Issue 2, Inst. Eng. Res., Univ. of California, Berkeley, CA, June.

Einstein, H. A. and Chien, N., 1955. Effects of Heavy Sediment Concentration Near the Bed on Velocity and Sediment Distribution. MRD Series 8, Inst. Eng. Res., Univ. of California, Berkeley, CA, August.

Elata, C. and Ippen, A. T., 1961. The Dynamics of Open Channel Flow with Suspensions of Neutrally Buoyant Particles. Tech. Rep. No. 45, Hydrodyn. Lab., Mass. Inst. Tech., Cambridge, MA, January.

Finlayson, B. A., 1980. *Nonlinear Analysis in Chemical Engineering.* McGraw-Hill, New York.

Formica, S., Baron, J. A., Thibodeaux, L. J. and Valsaraj, K. T., 1988. PCB transport into lake sediments. Conceptual model and laboratory simulation. *Environ. Sci. Technol.* **22**:1435–1440.

Gschwend, P. M. and Wu, S. C., 1985. On the constancy of sediment–water partition coefficients of hydrophobic organic pollutants. *Environ. Sci. Technol.* **19**:90–96.

Hino, M., 1963. Turbulent flow with suspended particles. *J. Hydraul. Div.* ASCE **89**:161–185.

Huellmantel, L. L., Reidy, J. E. and Rathbun, J. E., 1991. Horizontal and vertical distributions of contaminated sediment in the Buffalo River, New York. 34th Conference Inter. Assoc. Great Lakes Research, Buffalo, NY, June 2–6.

Jaffe, P. R. and Tuck, D. M., 1988. Theoretical study of partitioning of organic solutes in soil. Proc. Ground Water Geochemistry Conf. Assoc. Ground Water Scientists and Engineers, Denver, CO, February.

Julien, Pierre Y., 1995. *Erosion and Sedimentation.* Cambridge University Press, Cambridge, UK.

Karickhoff, S. W., Brown, D. S. and Scott, T. A., 1979. Sorption of hydrophobic pollutants on natural sediments. *Water Research* **23**:241–248.

Karickhoff, S. W., 1980. Sorption kinetics of hydrophobic pollutants in natural sediments, in *Contaminants and Sediments*, Vol. 2, R. A. Baker, ed. Ann Arbor Science, Ann Arbor, MI. pp. 193–205.

Karickhoff, S. W., 1984. Organic pollutant sorption in aquatic systems. *J. Hydraul. Eng.* **110**:707–735.

Li, R. M. and Shen, H. W., 1975. Solid particle settlement in open-channel flow. *J. Hydraul. Div.* ASCE **101**:917–931.

Lean, G. H., 1971. The settling velocity of particles in channel flow. Inter. Symp. Stochastic Hydraul., C. L. Chiu, ed., Pittsburg, PA, 339–351.

Murray, S. P., 1970. Settling velocities and vertical diffusion of particles in turbulent water. *J. Geophys. Res.* **75**:1647–1654.

Nomicos, G., 1956. Effects of sediment load on the velocity field and friction factor of turbulent flow in an open channel. Ph.D. thesis, Cal. Inst. Tech., Pasadena, CA.

O'Brien, M. P., 1933. Review of the theory of turbulent flow and its relation to sediment transportation. *Trans. Amer. Geophys. Union,* 487–491.

Oddson, J. K., Letey, J., and Weeks, L. V. 1970. Predicted distribution of organic chemicals in solution and adsorbed as a function of position and time for various chemical and soil properties. *Soil Sci. Soc. Am. J.* **34**:412–417.

Rao, P. S. C., Davidson, J. M., Berkheiser, V. E., Ou, L. T., Street, J. J., Wheeler, W. B. and Yuan, T. L., 1982. Retention and transformation of selected pesticides and phosphorus in soil–water systems: A critical review. U.S. Environmental Protection Agency, EPA/600/3-83-060.

Rouse, H., 1938. Experiments on the mechanics of sediment suspension, Proc. 5th Int. *Cong. Appl. Mech.*, Vol. 55, John Wiley, New York, pp. 550–554.

Schwarzenbach, R. P. and Westall, J., 1981. Transport of nonpolar organic compounds from surface water to groundwater — Laboratory sorption studies. *Environ. Sci. Tech.* **15**:1360–1367.

Smith, J., 1977. Modeling of sediment transport on continental shelves, in *The Sea*, Vol. 16:539–577.

Smith, V. E., Rathbun, J. E. and Filkins, J. C., 1991. Mapping contaminated sediments in the great lakes areas of concern (AOC). 34th Conference Inter. Assoc. Great Lakes Research, Buffalo, NY, June 2–6.

van Genuchten, M. T., Davidson, J. M. and Wierenga, P. J., 1974. An evaluation of kinetic and equilibrium equations for the prediction of pesticide movement through porous media. *Soil Sci. Soc. Am. J.* **38**:29–35.

van Genuchten, M. T. and Cleary, R. W. 1979. Movement of solutes in soil: Computer-simulated and laboratory results, in *Soil Chemistry*, Vol. B, Physio-Chemical Methods, G. H. Bolt, ed. Elsevier, New York, pp. 349–386.

van Leussen, W., 1986. Laboratory experiments on the settling velocity of mud flocs, in *River Sedimentation*, S. Y. Wang, H. W. Shen and L. Z. Ding, eds., Univ. Miss., 1803–1812.

Vanoni, V. A., 1946. Transportation of suspended sediment by water. *Trans. ASCE* **111**:67–133.

Vanoni, Vito A., ed., 1975. *Sedimentation Engineering*. ASCE Manuals and Reports on Engineering Practice No. 54. ASCE, New York.

Wang, S. Y., 1981. Variation of von Karman constant in sediment-laden flow. *J. Hydraul. Div.* ASCE, **107**:407–417.

Weilemann, U., O'Melia, C. R. and Stumm, W., 1989. Particle transport in lakes: Models and measurement. *Limnol. Oceanog.* **34**(1):1–18.

Wu, S. and Gschwend, P. M., 1986. Sorption kinetics of hydrophobic organic compounds to natural sediments and soils. *Environ. Sci. Technol.* **20**:717–725.

Yalin, M. S. and Finlayson, G. D., 1972. On the velocity distribution of the flow carrying sediment in suspension, in *Sedimentation*, ed. H. Shen, Col. State Univ., Fort Collins, CO, 8.1–8.18.

Young, T. C., DePinto, J. V. and Kipp, T. W., 1987. Adsorption and desorption of Zn, Cu and Cr by sediments from the River Raisin (Michigan). *J. Great Lakes Res.* **13**(3):554–561.

16

Remediation Issues

16.1 INTRODUCTION

As awareness of environmental problems has evolved over the years, there has been an increasing focus not only on reducing ongoing sources of pollution but also on remediation of sites that are already contaminated. A clear example of this is with the so-called Superfund program set up in the United States during the 1980s to clean up large hazardous waste sites causing serious contamination of groundwater resources. Another example is with efforts in the Great Lakes basin of North America to reverse the trend of eutrophication that had been growing steadily in the middle part of the last century. Reduction of nutrient loadings such as phosphorus and nitrogen have reduced drastically the productivity in the lakes, and now the concern is with cleaning residual sites of contamination, in large part associated with contaminated sediments.

In general, the initial phase of remediation involves controlling contaminant disposal into the particular environment under consideration. This may mean complete cessation of all source loadings or at least reduction of loads to a suitable level to allow recovery of the system, either by natural or by engineered processes. According to the type of environment and the time and length scales of the problem, the remediation may include containment and treatment. Most phases of environmental remediation require an understanding of fluid flow and transport phenomena by advection and diffusion, as discussed in previous chapters of this text. Remediation strategies also may incorporate the use of chemical agents and biodegradation of contaminants. Methodologies that have been developed and used for many decades for water and wastewater treatment can often be adapted for environmental remediation.

16.2 SOIL AND AQUIFER REMEDIATION

16.2.1 General Aspects

In cases of soil and aquifer contamination, the source of contaminant is usually located above ground, or in a shallow depth below the ground surface. The contaminant penetrates first to the *vadose zone*, and later it reaches the groundwater. Therefore the issue of aquifer and groundwater contamination is usually associated with site contamination. On the other hand, if the contaminant penetration is limited to the vadose zone, then site contamination is not necessarily accompanied with aquifer and groundwater contamination. Figure 16.1 shows a schematic description of site contamination, which originates from a typical landfill.

In addition to landfills, other sources of groundwater contamination include spills of soluble substances, which become completely sorbed in the vadose zone and then are gradually released by percolating runoff water. An important category of potentially spilled materials includes a variety of hydrocarbons, such as oils and fuels, and these are collectively known as *nonaqueous phase liquids* (NAPLs). When a NAPL has a density less than that of water, it is referred to as a *light nonaqueous phase liquid*, or *LNAPL*. When NAPL is denser than water, it is called *dense nonaqueous phase liquid* *(DNAPL)*. When LNAPLs are released at the soil surface, they percolate through the vadose zone and eventually float on top of the groundwater table and the capillary zone, while gradually releasing dissolved hydrocarbon into the flowing groundwater. DNAPLs sink through the water layer and rest on the bottom of the aquifer, except for material that may be adsorbed onto soils,

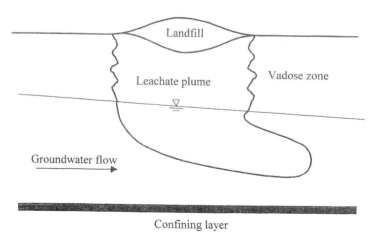

Figure 16.1 A typical site contamination originating from a landfill.

either in the vadose zone or in the water layer itself. LNAPLs also may sorb onto soils.

Assuming that any ongoing source of contamination can be removed, there are different alternatives to consider in deciding how to remediate a given site. Sometimes removal of the contaminated vadose zone is desirable and feasible. In cases of relatively small oil spills it is quite common to excavate the soil contaminated by the spill, to avoid contact between the oil spill and groundwater. However, removal of the contaminated soil introduces an additional problem of disposal of the removed soil. If the amount of contaminated soil is not too large, then incineration may be appropriate. For large quantities of contaminated soil, a more common reclamation method involves soil excavation and deposition in a bioreactor, where biodegradation of the contaminant can be achieved in a comparatively short time period. This requires controlling the appropriate supply of moisture, oxygen, and nutrients for the enhancement of the microorganism development and growth.

16.2.2 Containment of the Contaminated Site

If the contaminated site cannot be excavated economically or technically, then it may be appropriate to contain it and to apply technologies of in-situ remediation. Containment of the contaminated site is obtained by surrounding the contaminated site by *cutoff walls*, or vertical barriers, as shown in Fig. 16.2.

For the analysis of the vertical barrier performance, it is possible to adopt a one-dimensional conceptual model as shown in Fig. 16.3. The barrier consists of a porous medium with very low permeability and it separates the

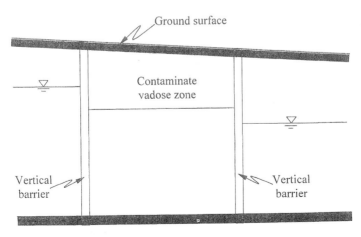

Figure 16.2 Containment of the contaminated site by vertical barriers.

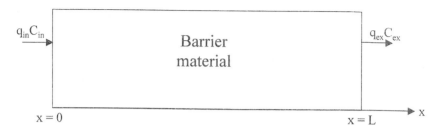

Figure 16.3 Conceptual model of flow with a vertical barrier.

contaminated groundwater from fresh groundwater. According to Eq. (11.3.1), the one-dimensional differential equation of contaminant transport through the barrier is

$$R\frac{\partial C}{\partial t} + v\frac{\partial C}{\partial x} = D\frac{\partial^2 C}{\partial x^2} - \lambda C \tag{16.2.1}$$

where R is the retardation factor, C is the contaminant concentration, t is the time, v is the interstitial fluid velocity, x is the horizontal coordinate, D is the dispersion coefficient, and λ is the decay coefficient for the contaminant.

The contaminant is transported by advection and diffusion through the barrier. At the upstream boundary, namely at $x = 0$, the contaminant concentration is assumed to be C_{en}, which may be time dependent. For example, C_{en} may gradually decrease if the contained area is subject to remediation treatment. However, for a conservative calculation, we may assume that C_{en} is constant. At a downstream cross section the contaminant concentration is C_{ex}. The value of C_{ex} increases with time due to the contaminant flux through the barrier. The increasing value of C_{ex} has no effect on the advective contaminant flux through the barrier, but it may lead to a decreasing diffusive flux, due to a smaller concentration gradient. Therefore, again for a conservative calculation, we consider that $C_{ex} = 0$, and its value is kept constant. Such an assumption has no effect on the advective contaminant flux through the barrier, but it maintains the maximum possible diffusive flux of the contaminant. The instantaneous contaminant flux F, at any cross section of the barrier, is given by

$$F = \phi v C - \phi D\frac{\partial C}{\partial x} \tag{16.2.2}$$

where ϕ is the porosity, v is the interstitial flow velocity, C is the contaminant concentration, D is the dispersion coefficient (including molecular diffusion and mechanical dispersion), and x is the longitudinal coordinate.

We may refer to differences between values of F at the entrance cross section, where $x = 0$, and values of F at the exit cross section, where $x =$

L. At the entrance cross section, the contaminant is subject to advection, represented by the first term on the right-hand side of Eq. (16.2.2), as well as dispersion, which is represented by the second term on the right-hand side of Eq. (16.2.2). At the exit cross section, due to the assumption of $C_{ex} = 0$, the advective contaminant flux vanishes, and the contaminant is transported solely by dispersion.

Even under unsteady-state conditions, the advective flux is assumed to be kept constant at the entrance cross section of the barrier. On the other hand, the dispersive flux gradually decreases, as noted above. Initially it is very large, when the contaminant concentration gradient is large. On the other hand, at the exit cross section the dispersive flux gradually increases from an initial value of zero. Therefore calculation of steady-state conditions may provide an estimate of the maximum contaminant flux that can be expected at the exit cross section of the barrier. Of course, from a practical view point, it is also appropriate to provide an estimate of the time period needed to develop the maximum steady-state flux. For a conservative contaminant, under steady state conditions the flux F is constant at every cross section of the barrier. Under such conditions, Eq. (16.2.2) is obtained by direct integration of Eq. (16.2.1). A further integration of Eq. (16.2.2) then gives

$$\ln\left[K\left(\frac{F}{\phi v} - C\right)\right] = \frac{vx}{D} \tag{16.2.3}$$

where K is an integration constant. Applying the boundary conditions of $C = C_{en}$ at $x = 0$, and $C = 0$ at $x = L$, then shows that

$$K = \frac{1}{F/\phi v - C_{en}} \qquad F = \frac{\phi v C_{en}}{1 - \exp(-vL/D)} \tag{16.2.4}$$

The Peclet number of the barrier is defined by

$$Pe_b = \frac{vL}{D} \tag{16.2.5}$$

If Pe_b is high, then Eq. (16.2.4) can be approximated by

$$F \approx \phi v C_{en} \tag{16.2.6}$$

If Pe_b is very small, then Eq. (16.2.4) can be approximated by

$$F \approx \phi D \frac{C_{en}}{L} \tag{16.2.7}$$

By introducing the expressions of K and F (Eq. 16.2.4) into Eq. (16.2.3), the contaminant distribution in the barrier is found as

$$C = C_{en}\left\{\frac{1 - \exp[-(v/D)(L - x)]}{1 - \exp(-Pe_b)}\right\} \tag{16.2.8}$$

For large values of Pe_b, this expression reduces to

$$C = C_{en} \left\{ 1 - \exp \left[-Pe_b \left(1 - \frac{L}{x} \right) \right] \right\} \qquad (16.2.9)$$

For small values of Pe_b, we use the series expansion of the exponential terms of Eq. (16.2.8) to obtain

$$C = C_{en} \left(1 - \frac{x}{L} \right) \qquad (16.2.10)$$

Also, if Pe_b is large, then steady-state conditions of contaminant transport through the barrier are established after an approximate time period of

$$T \approx \frac{L}{v} \qquad (16.2.11)$$

If Pe_b is small, then the time period required for the establishment of steady-state conditions can be estimated using analytical solutions of the diffusion equation.

16.2.3   Pump-and-Treat of Contaminated Groundwater

Following containment of a contaminated site, appropriate treatment technologies are generally needed to bring the site to full reclamation. Groundwater of the contaminated site can be pumped into a treatment plant and later reinjected into the aquifer. Sometimes the treated groundwater can be used directly, mainly for irrigation purposes. A variety of treatment methods are classified as *in-situ treatment* methods. These can sometimes be applied without physical barriers. A common approach is to apply hydrodynamic isolation approaches, rather than physical barriers, to contain the contaminated portion of the aquifer. Hydrodynamic isolation applies various types of injection and extraction well combinations that do not allow the migration of groundwater from the contaminated site to neighboring aquifers. In Fig. 16.4, schematics of two common options of hydrodynamic isolation are shown.

Calculation of flow conditions in the two examples of Fig. 16.4 can be done using potential flow theory and well hydraulics, as detailed in Chap. 11. Each of these examples is associated with the separation of the aquifer flow into two regions. One region incorporates mainly the fresh groundwater. The other region incorporates a comparatively small portion of the fresh groundwater flow and also the flow of contaminated groundwater. A well-defined line of separation represents the interface between these two regions. The schematic of Fig. 16.4 shows two examples of pumping of contaminated groundwater for its possible treatment by conventional methods of waste

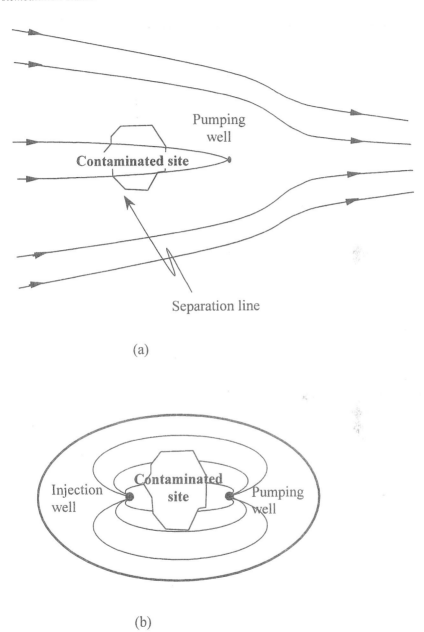

(a)

(b)

Figure 16.4 Hydrodynamic isolation of a contaminated site: (a) isolation by a single pumping well; and (b) isolation by a combination of a pumping well and an injection well.

treatment. Such an approach is called *pump-and-treat*. Hydrodynamic isolation, incorporated with pump-and-treat, can be a comparatively inexpensive method, which leads to gradual reclamation of the contaminated portion of the aquifer.

To obtain high efficiency of the systems shown in Fig. 16.4, it is important to avoid conditions of significant dispersion and mixing between the fresh and contaminated groundwater. However, problems arise if contamination of the aquifer is associated with significant sorption–desorption capacity onto the soil of the aquifer. For example, in cases of soil contaminated with LNAPL, then in the case described by Fig. 16.4a, groundwater is continuously contaminated by the residual adsorbed or entrapped material. It should be noted that in cases of NAPL entrapment, different agents to enhance the remediation, such as surfactants and nutrients for microbial activity, can be added to the water injected into the aquifer. However, such materials should be chosen so as not to cause other types of aquifer pollution.

If the contaminated site of Fig. 16.4b is rich with adsorbed or entrapped contaminant, then injected water is subject to contamination prior to its pumping by the pumping well. Figure 16.5 shows a schematic of a pump-and-treat system, in which the aquifer is contaminated by NAPL. Figure 16.5a illustrates a problem of contamination by LNAPL where, due to seasonal and annual fluctuations of the groundwater, some quantities of the LNAPL are entrapped within the top layers of the aquifer. The flow induced by the pump-and-treat system is associated with dissolution and solubilization of the entrapped NAPL, as well as with penetration of the dissolved constituents into the deeper portions of the aquifer. In the case described by Fig. 16.5b, DNAPL is entrapped throughout the entire thickness of the aquifer. Induced groundwater flow of the pump-and-treat system is associated with the dissolution of the entrapped DNAPL.

Calculations of the performance of the pump-and-treat system shown in Fig. 16.5 can be done using a conceptual one-dimensional flow model. Under such conditions, the process of NAPL dissolution and mass transfer from the entrapped NAPL *ganglia* to the flowing aqueous phase can be calculated using the approach presented in Sec. 11.5. The pump-and-treat system of Fig. 16.5a then appears to be inefficient, as most of the induced groundwater flow cannot be in contact with the entrapped LNAPL. Furthermore, the induced groundwater flow enhances transverse dispersion, which leads to penetration of dissolved constituents into deeper layers of the aquifer.

As an alternative, the pump-and-treat system of Fig. 16.6 is based on the use of a single pumping well. The discharge of the well causes a *drawdown* of the groundwater table and an associated *cone of depression*. The cone of depression contains the lens of LNAPL and avoids the uncontrolled migration

Strip of injection wells

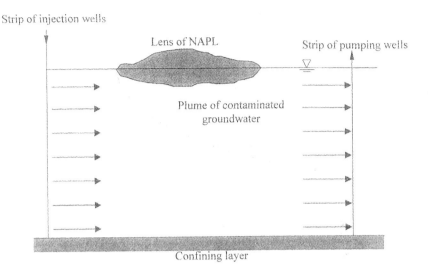

Strip of pumping wells

Lens of NAPL

Plume of contaminated
groundwater

Confining layer

Strip of injection wells

Strip of pumping wells

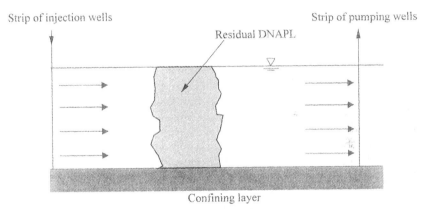

Residual DNAPL

Confining layer

Figure 16.5 Examples of pump-and-treat systems: (a) aquifer contamination by LNAPL; and (b) aquifer contamination by DNAPL.

of NAPL. The floating lens of LNAPL flows towards the pumping well in the region of the cone of depression. Various techniques can then be applied to collect the floating LNAPL in that region, by various types of membranes and floating pumps.

Following the pumping of the contaminated groundwater, it must be treated. The appropriate treatment of the extracted groundwater depends on the type of contaminant. In cases of inorganic contaminants, precipitation is an attractive treatment method. Precipitation is governed by the pH value, which

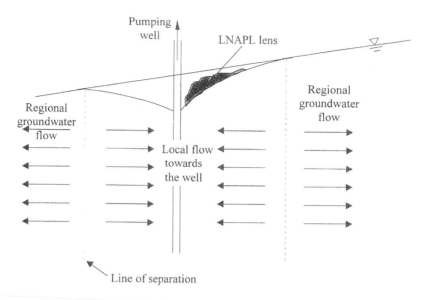

Figure 16.6   Containment of the LNAPL lens by a single pumping well.

may be adjusted by adding lime to the treatment stream. Sometimes aeration of metals creates salts with faster precipitation. Dissolved organic materials can be removed by air stripping. Organic compounds, which have low volatility, cannot be removed efficiently by air stripping. Instead, they can be sorbed onto activated carbon. Other compounds can be treated by biological methods similar to the treatment of domestic wastes.

16.2.4   In-Situ Remediation

In various cases, remediation by pump-and-treat is not feasible or is not the optimal method. For example, when a volatile organic compound is spilled into the unsaturated zone, it partitions between the liquid and vapor state. The vapors may migrate through the vadose zone and accumulate in underground structures like basements, where they pose a threat of fire or explosion. In such cases *soil-vapor extraction* (SVE) methods can provide an appropriate measure of in-situ remediation. According to these methods, wells are installed in the vadose zone and are used to pump air and vapor. Other SVE systems may incorporate air injection wells and air-vapor extracting wells. Such systems can also be used if the contaminant volatility is comparatively low. In such cases the injected humid air enhances the bioactivity and in-situ bioremediation. These kinds of systems are sometimes referred to as *air-sparging* systems.

16.3 BIOREMEDIATION

As indicated in the previous sections, biological activity may comprise a significant part of the remediation procedure. On-site bioremediation is based on excavation of the contaminated soil, and its placement in bioreactors, into which air, water, and nutrients are injected to promote the biological activity. Many engineers and practitioners claim that in most cases the best approach for bioremediation is to enhance the development of the local indigenous microorganism population. Others claim that specially acclimated microorganisms may produce better results. As indicated in the previous section, biological activity can enhance the pump-and-treat approach if the injection wells provide water rich in air and nutrients, in addition to possible surfactants (emulsifiers), which enhance the solubilization and dissolution of entrapped NAPL. Biological activity also can play a significant role in air sparging and other soil vapor extraction procedures.

16.3.1 Basic Concepts and Definitions

Bioremediation is based on the biochemical reactions leading to degradation of the contaminant. Usually, bioremediation is considered for application in cases of degradation of entrapped NAPL, since removal of the NAPL is very difficult. In general, the organic compound is subject to an *oxidation* reaction, in which it loses electrons. An electron acceptor, which is subject to a reduction reaction, participates in the oxidation reaction and it gains electrons. If the electron acceptor is oxygen, then the oxidation of an organic compound is called *aerobic heterotrophic respiration*. If oxygen is not available (*anoxic*, or anaerobic conditions), then anaerobic microorganisms use an alternate electron acceptor.

Several definitions are useful for the classification of microorganisms involved in bioremediation. With regard to concentration of organic carbon in the environment, *obligotrophic* organisms are most active in cases of low concentration of organic carbon. *Eutrophic* organisms are active in cases of high concentration of organic carbon. Regarding the particular nutritional basis of the organisms, *chemotrophic* organisms capture energy from the oxidation of organic or inorganic materials, *autotrophic* organisms synthesize their cell carbon from CO_2, and *heterotrophic* organisms require a source of organic carbon.

The organic compound subject to degradation may be a *primary substrate* for the microorganism, provided that it is a source of energy and carbon. In cases where the compound to be degraded is not a primary substrate, provision of a primary substrate may be needed. Then, the degrading compound may be considered as a *secondary substrate*. A secondary substrate usually cannot provide sufficient energy for sustaining the microorganisms, but its degradation, in the presence of the primary substrate, can provide some important compounds for microorganism growth.

Biodegradation of a contaminant organic compound can be feasible, provided that the following six basic requirements are satisfied:

1. *Presence of the appropriate organisms.* In general, enhancement and development of indigenous microorganisms that are capable of degrading the organic contaminant is recommended, although specialized microorganisms acclimated to a particular contaminant or environmental condition may be required.

2. *Primary substrate.* This is the energy source for the microorganisms. The organic contaminant can be either the primary or secondary substrate. The primary substrate is transformed into inorganic carbon, energy, and electrons.

3. *Carbon source.* About 50% of the microorganism dry weight is composed of carbon. Organic chemicals serve as sources of energy and carbon.

4. *Electron acceptor.* For the oxidation–reduction process, an electron acceptor is required. Typical electron acceptors are oxygen, nitrate, and sulfate.

5. *Nutrients.* Nutrients required for the growth of the microorganisms include nitrogen, phosphorous, calcium, magnesium, iron, and trace elements. These elements are needed for growth of the microorganism cell.

6. *Acceptable environmental conditions.* Such conditions include humidity, temperature, pH, salinity, hydrostatic pressure, radiation, and absence of toxic materials.

16.3.2 Kinetics of Biodegradation

The simplest model usually used for the general expression of rate of growth of any population is

$$\frac{dx}{dt} = \mu x \qquad (16.3.1)$$

where x is the size of the population and μ is the *growth rate coefficient* for the population. The growth of microorganisms in a limited environment such as an aquifer is schematically shown in Fig. 16.7. This figure indicates that the rate of growth in such an environment is subject to changes with the size of the population. One of the most useful models used to describe microorganism population growth in a closed environment is the *logistic model*, which is

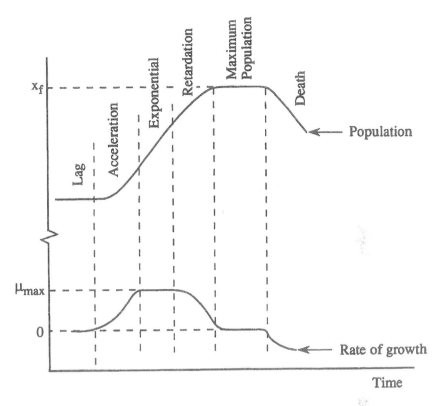

Figure 16.7 Growth curve for microorganisms in a limited environment.

given by

$$\frac{dx}{dt} = \mu_{max}x\left(1 - \frac{x}{x_f}\right) \tag{16.3.2}$$

where μ_{max} is the maximum growth rate and x_f is the maximum population size.

Phases of Microorganism Population Growth

According to Fig. 16.7 and Eq. (16.3.2), the following phases of population growth can be defined in a limited environment:

1. *Lag phase.* In this phase the microorganism population size is extremely small and there is no discernible increase in population size.

2. *Acceleration phase.* This phase shows the beginning of a gradual increase in the population rate of growth.
3. *Exponential phase.* In this phase the rate of growth obtains its maximum value. According to Eq. (16.3.1), the microorganism population size is significantly smaller than its maximum value, and the population size is subject to exponential growth.
4. *Retardation growth.* In this phase the rate of growth of the population starts to decline, as the maximum population is approached.
5. *Maximum population phase.* In this phase the microorganism population is metabolically active, but the population size is kept constant. For this phase, $x = x_f$.
6. *Death phase.* In this phase the population size decreases due to lack of substrate, the accumulation of toxins, or other phenomena. It should be noted that the logistic model of Eq. (16.3.2) cannot describe the death phase, as it considers only effects of the size of the population on its rate of growth.

Various types of kinetic expressions may be useful for the prediction of rates of biodegradation of organic contaminants in aquifers. Such expressions refer to the effect of substrate availability, as well as other limiting factors, on the rate of growth of the microorganisms. A common approach is the *Monod*, or *Michaelis–Menton* kinetic expression,

$$\mu = \mu_{max} \frac{C_s}{K_s + C_s} \tag{16.3.3}$$

where C_s is the concentration of the growth-limiting substrate. The coefficient K_c is called the half-saturation constant and is defined as the growth-limiting substrate concentration that allows the microorganisms to grow at half the maximum specific growth rate. A low value of K_c indicates that the microorganism is capable of growing rapidly in an environment with low concentration of the growth-limiting substrate. If several growth-limiting substrates should be considered, then the concentrations and half-saturation constants of all growth-limiting substrates should be incorporated in Eq. (16.3.3) by products of terms similar to that of Eq. (16.3.3). Usually, besides the organic substrate, it is appropriate at least to consider oxygen as another growth-limiting substrate.

In the limiting case of $C_s \gg K_c$, Eq. (16.3.3) yields

$$\mu = \mu_{max} \tag{16.3.4}$$

When this occurs, the reaction is called a *zero-order reaction.* Alternatively, in the limiting case of $C_s \ll K_c$, Eq. (16.3.3) yields

$$\mu = \frac{\mu_{max}}{K_c} C_s \tag{16.3.5}$$

This is called a *first-order reaction*, with the first-order rate constant given by $\mu_{\max}/K_c$.

If we let M represent the microbial mass per unit of groundwater volume and Y the amount of organism mass formed for each unit of substrate mass utilized (this is called the *yield coefficient*), then the change in substrate concentration can be expressed as

$$\frac{dC_s}{dt} = \frac{\mu_{\max}MC_s}{Y(K_c + C_s)} \tag{16.3.6}$$

The ratio between $\mu_{\max}$ and Y represents the maximum *contaminant utilization rate* per unit mass of microorganisms.

Equation (16.3.6) should be accompanied by the equation of biological mass transport, growth and decay,

$$\frac{dM}{dt} = \mu_{\max}MY\frac{C_s}{K_c + C_s} - bM \tag{16.3.7}$$

where b is a first-order decay coefficient, representing cell death.

16.3.3 Modeling of Biodegradation

The foregoing relations give expressions for the effect of organisms on the decrease of the organic substrate concentration and the effect of the decrease of the substrate mass on the growth of the microorganism mass, under conditions of no limiting supply for all nutrients. However, in a real aquifer, there is usually a limiting supply of at least one of the growth nutrients, and the organic substrate, as well as the limiting growth nutrient and microorganisms also, are subject to transport by advection and dispersion. Using the expressions for Monod kinetics, all of these effects may be combined as

$$R_c\frac{\partial C}{\partial t} + \nabla \cdot (vC) = \nabla \cdot (D\nabla C) - \frac{M_t\mu_{\max}}{Y}\frac{C}{K_c + C}\frac{G}{K_G + G} \tag{16.3.8}$$

$$\frac{\partial G}{\partial t} + \nabla \cdot (vG) = \nabla \cdot (D\nabla G) - \frac{M_t\mu_{\max}}{Y}F\frac{C}{K_c + C}\frac{G}{K_G + G} \tag{16.3.9}$$

$$R_m\frac{\partial M_s}{\partial t} + \nabla \cdot (vM_s) = \nabla \cdot (D\nabla M_s) - M_t\mu_{\max}R_mY\frac{C}{K_c + C}$$

$$\times \frac{G}{K_G + G} + K_cYC_0 - bR_mM_s \tag{16.3.10}$$

where C is the contaminant concentration, G is the concentration of the limiting growth nutrient (usually oxygen), D is the dispersion coefficient, v is the interstitial groundwater velocity, R_c is the retardation coefficient for the contaminant, M_s is the concentration of microorganisms in solution, M_t is the total concentration of microorganisms, R_m is the microbial retardation factor ($M_t = R_m M_s$), Y is the microbial yield coefficient, K_c is the half-saturation constant, K_G is the half-saturation constant of the limiting growth nutrient, C_O is the concentration of the natural organic carbon, F is the ratio of limiting growth nutrient to hydrocarbon consumed, and b is the microbial decay rate.

The set of Eqs. (16.3.8)–(16.3.10) considers that the organic contaminant and oxygen are the growth-limiting substrates. Prior to using these three differential equations, flow conditions in the domain should be evaluated using the appropriate flow equations. The solution of the set of Eqs. (16.3.9)–(16.3.11) requires a complicated numerical scheme. Furthermore, in various cases it is sufficient to use kinetics models simpler than the Monod expressions. For example, the first-order kinetics model is seen as a simplification of Monod kinetics and results in an exponential decay rate. With this model, the concentration of the organic contaminant is given by

$$C = C_{in} \exp(-kt) \tag{16.3.11}$$

where C_{in} is the initial concentration of the contaminant and k is the rate of decrease of the contaminant concentration.

Another simplified model is the *instantaneous reaction model*, which assumes that microbial biodegradation kinetics are very fast in comparison with transport of oxygen. Thus all biochemical reactions of biodegradation, such as growth of microorganisms, utilization of oxygen, and utilization of the organic contaminant, are considered as reactions between the organic contaminant and oxygen. The basic assumption of the instantaneous model implies that the rate of utilization of the organic contaminant and oxygen by the microorganisms is very high, and the time period required to mineralize the contaminant is very small. For this case, biodegradation can be calculated according to

$$\Delta C_R = -\frac{G}{F} \tag{16.3.12}$$

where ΔC_R is the change in contaminant concentration owing to the biodegradation process, G is the concentration of oxygen, and F is the ratio of oxygen to consumed organic contaminant.

Numerical models currently in use usually apply either some form of the Monod kinetics, or a form of a simplified kinetics. However, besides the basic model given by the set of differential equations above (16.3.8–16.3.10), some additional conceptual models have been applied to describe the process of degradation of an organic contaminant by microbial activity. Two such conceptual models are discussed below; (a) the biofilm model and (b) the microcolony model.

16.3.4 The Biofilm Model

Figure 16.8 represents a schematic of the idealized conceptual biofilm model. This model assumes the development of a biofilm, comprising a homogeneous matrix of bacteria, whose extracellular polymers bind them together and to the solid surface. The biofilm is of uniform cell density X_f and locally uniform thickness L_f. Groundwater, flowing in the x-direction, carries the substrates. Substrates are transported from the water to the biofilm in the z-direction through a mass diffusion layer of thickness L.

Following the arrival of the substrates to the biofilm, they are subject to (a) molecular diffusion through the biofilm and (b) utilization by the bacteria according to some relation, such as Monod kinetics. If steady-state conditions are established in the domain shown in Fig. 16.8, then

$$D_f \frac{d^2 C_f}{dz^2} = \frac{k X_f C_f}{K_s + C_f} \qquad (16.3.13)$$

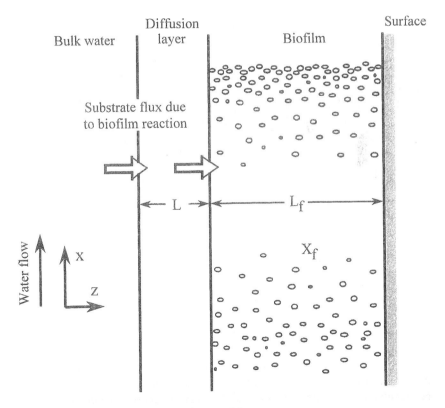

Figure 16.8 Schematic of the biofilm model.

where C_f is the concentration of the growth rate limiting substrate at a point within the biofilm, D_f is the molecular diffusivity of the substrate within the biofilm, k is the maximum specific rate of substrate utilization by the microorganisms, K_s is the Monod half maximum rate concentration, and z is the coordinate normal to the biofilm surface.

Under steady-state conditions, the substrate flux arriving at the biofilm surface is given by

$$J = -D\frac{dC_s}{dz} = D\frac{C_s - C_{surf}}{L} \tag{16.3.15}$$

where J is the flux of the growth rate limiting substrate crossing the diffusion layer, D is the diffusivity of the diffusion layer, and C_{surf} is the substrate concentration at the surface of the biofilm.

The processes described by Eqs. (16.3.13) and (16.3.14) should be matched at the biofilm surface. Figure 16.9 shows the profile of the substrate distribution in the diffusion layer and in the biofilm. Depending on the thickness of the biofilm, it is possible that layers of the biofilm close to the solid surface are subject to a very low supply of the substrate.

16.3.5 Microcolony Modeling

According to the microcolony model, the microorganisms develop in microcolonies, represented as disks of uniform radius and thickness, which attach to the solid surfaces of the porous medium. Through a diffusion layer which surrounds the colony, the substrate and oxygen are transported to the colony.

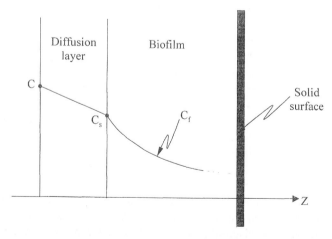

Figure 16.9 Distribution of the substrate in the diffusion layer and in the biofilm.

Microcolony modeling for two and three-dimensional domains have been developed. However, presently, most commercial computer codes apply some simplified approaches, which require smaller resources of computation.

16.4   REMEDIATION OF SURFACE WATERS

Remediation of surface waters involves some of the same issues as for groundwater. For instance, the first step in cleaning up a contaminated site is to determine whether there is a continuing source of contamination. This is sometimes not easy to do, especially in cases where sediments are contaminated. Unlike groundwater, surface waters usually are able to recover naturally on shorter time scales, particularly when there is a significant flow of water through the system. Because surface waters are usually (hydraulically) connected with other water bodies, and because there may be multiple uses for a particular water body of concern, remediation alternatives must be carefully thought out before any action is taken. It is normal to use water quality modeling to evaluate the effects of different remediation options when making management decisions. Models that attempt to simulate entire ecosystems also have been used. These models range in complexity, both in terms of spatial and temporal resolution and also in terms of the sophistication of their chemical and/or biological components. However, they all incorporate some description of transport (i.e., the advection–diffusion equation introduced in Chap. 10) and chemical and biological reactions that control the distribution of specific *state variables* of interest.

The usual remediation options for surface waters include cessation or limitation of one or more sources, artificially altering the physical or chemical environment, and physically removing in-place sources of contaminants. Contaminants may enter a system through both *point* and *nonpoint* sources. A point source is one that occurs at a specific location, such as a sewage treatment plant outfall, while a nonpoint source is distributed. An example of a nonpoint source is direct runoff into a river from an agricultural field that may have been treated with pesticides. Examples of remediation by altering the environment include adding lime to lakes that have become highly acidic due to acid rain, and mixing the water column to eliminate stratification. This latter approach has been followed, for instance, in the Charles River of Boston, Massachusetts, where bubblers inject air at the bottom of the river to mix the benthic waters upward and also to add oxygen and prevent anaerobic conditions from developing. In-place contaminants are usually associated with bottom sediments, which have collected pollutants during many years of industrial activity and disposal into surface waters. Even after direct discharges have been stopped, contaminated sediments may continue to act as a source of pollution for the overlying water for many years, introducing contaminants each time resuspension occurs. This is a significant problem in the Great Lakes basin of North America, and choices must be made to decide whether it is better to leave the sediments as they are, to be eventually covered over by "clean" material by natural sedimentation, or to remove the contaminated solids by dredging. When dredging is used, other problems must be addressed, concerning proper handling and disposal of the dredged material.

As a simple example of application of a water quality model to address a remediation question, consider a lake with one inlet and one outlet, as sketched in Fig. 16.10. An industry just upstream of the inlet discharges a waste stream into the river, so that the inflow concentration of a particular chemical of concern is C_{in}. For this example, the lake is considered to be fully mixed, with initial concentration C_0. The chemical decays with first-order decay rate k, and it is assumed that it does not settle or volatilize. The differential equation describing the concentration of the chemical in the lake is

$$\frac{dC}{dt} = \frac{Q}{\forall}(C_{in} - C) - kC \qquad (16.4.1)$$

where Q is the flow rate (inflow and outflow rates are assumed to be equal, so that the volume remains constant) and $\forall$ is the volume of the lake. The ratio of $Q/\forall$ is called the *retention rate*, and its inverse is called the *retention time*,

$$t_r = \frac{\forall}{Q} \qquad (16.4.2)$$

This is a measure of the average length of time a parcel of fluid remains within the lake.

Now suppose that it is desired to examine the response of the lake concentration to a step change in inflow concentration. In particular, suppose it is desired to know how long it would take for the concentration to fall to a level C_1, if C_{in} is reduced to zero. Setting $C_{in} = 0$ in Eq. (16.4.1) and

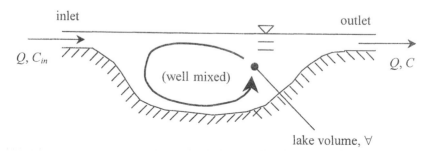

Figure 16.10 Schematic for problem in which a lake has one inlet and one outlet.

integrating, along with the initial condition that $C = C_0$ at $t = 0$, gives

$$C = C_0 \exp\left[-\left(\frac{Q}{\forall} + k\right)t\right] \tag{16.4.3}$$

Or, setting $C = C_1$ and solving for t, we have

$$t = -\left(\frac{Q}{\forall} + k\right)^{-1} \ln\frac{C_1}{C_0} \tag{16.4.4}$$

As should be expected, the required time is shortened for faster reaction rates and for higher through-flow rates (smaller t_r), which increases the rate of dilution.

The above problem is an example of a *one-box* or *fully mixed reactor model*, since only the mean concentration in the lake was considered. A slightly more complicated example may be imagined in which there is interaction between a particular water body of concern and another water body, as shown in Fig. 16.11. This figure shows a bay in a large lake or coastal region. There

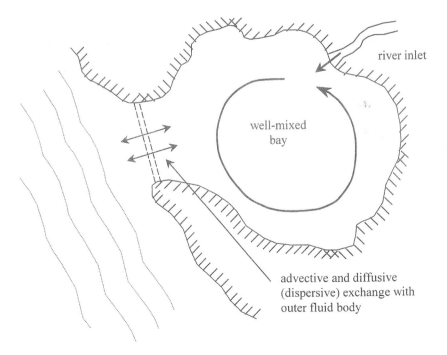

Figure 16.11 Schematic for problem set up for chemical interactions between a bay and a large lake or coastal area.

is a river discharging to the bay, and there may also be groundwater and direct runoff sources. The boundary between the bay and the rest of the lake is relatively long, and the flow patterns along it may be quite complicated. For a steady state, of course, there must be a net flow out of the bay, equal to the river flow and any distributed inflows. However, along the boundary with the rest of the lake there could be regions where there is flow into the bay, and other areas where flow is outward. Even if there were no inflow to the bay from point or non-point sources, this complicated flow pattern at the boundary results in some transport of materials between the bay and the rest of the lake.

In keeping with a box-type approach, one concentration is used to describe the chemical of concern in the bay. Without solving a detailed hydro-dynamic model to specify the flow field exactly, all of the transport across the boundary with the rest of the lake that is not associated with advective flows is lumped into a dispersion-type term. As in Chap. 10, dispersion flux is written in terms of the product of a dispersion coefficient and the concentration gradient, or

$$F_d = -E\frac{dC}{dx} \cong -E\frac{C - C_\infty}{L} \tag{16.4.5}$$

where F_d is the flux due to dispersion, E is the dispersion coefficient, C is the concentration in the bay, C_∞ is the concentration in the lake, and L is a characteristic distance between the regions where the concentration is C and where it is C_∞ (see Fig. 16.11). A reasonable estimate for L is half the distance across the bay, in the offshore direction. The equation describing changes in concentration in the bay may then be written as

$$\frac{dC}{dt} = \frac{1}{\forall}\left[\left(\sum Q_{in}C_{in} - C\sum Q_{in}\right) - E_b(C - C_\infty)\right] - kC \tag{16.4.6}$$

where the summations are taken over all point and nonpoint sources of flow into the bay (along the shoreline), k is again a first-order decay rate, and $E_b = EA/L$, where A is the area of the boundary between the bay and the rest of the lake. The parameter E_b is sometimes called the *dispersivity*, although it should be noted that it does not have the same units as the dispersion coefficient. Often, E_b is calibrated to site data and it may be considered as a fitting parameter.

Most water quality models of large water bodies use box-type or segment-type approaches as illustrated in the foregoing examples, rather than solving the advection–diffusion equation directly. This is partly because of computer resources and the desire for computational speed and also because more detailed spatial and temporal resolution is rarely justified in the model, since data are generally sparse at best. In other words, there may be only a few measurements available to characterize a large body of water, so that it is difficult fully to calibrate a given model, and it is not very meaningful to generate model output for spatial and time resolutions that cannot be checked with data. Water quality models also may involve a relatively large number of reaction equations, to describe the evolution of a number of chemical and/or biological species in the system, and they may be run for very long simulation times, perhaps as much as 50 to 100 years. Thus the choice is usually made to sacrifice some temporal and spatial resolution, in order to incorporate more chemical and biological features and to enable calculations to be performed for reasonable run times.

For smaller systems such as a river reach, it may be possible to develop models based on direct solutions to the advection–diffusion equation, although again, the question of data availability should be considered. For any system, development of a water quality model should be undertaken while considering the management questions that are being asked, as well as the data availability and the characteristics of the system itself. Equations (16.4.1) and (16.4.6) are simple examples of models that might be used to evaluate remediation options, and they can serve as the starting point for more involved models. In addition to a simple first-order reaction, interactions with sediments may be considered (Sec. 15.6), as well as reactions at the air water interface (Secs. 12.4 and 12.5). The level of detail in the hydrodynamic description of the system also may be increased significantly, including both the velocity field and the temperature distribution, with associated effects of stratification (Chap. 13).

No matter what level of detail is used, it is clear that development of models necessary for evaluation of remediation options involves application of many of the ideas and principles described in this text. The specific problem to be addressed will dictate which material needs to be incorporated. Because of the complexity of most systems, computer models will continue to be used for the foreseeable future, and it is expected that successively more detailed and inclusive models will be developed, until eventually we will be able to model entire ecosystems. With advances in computer technologies, the main limitation for such development will most likely be appropriate data acquisition, for both model development (process representation) and calibration/verification.

PROBLEMS

Solved Problems

Problem 16.1   Consider a barrier of length L, hydraulic conductivity K, diffusion–dispersion coefficient D, retardation factor R, and porosity ϕ. The barrier separates a freshwater aquifer from a portion of the aquifer contaminated by a radioactive material with a decay coefficient λ. Initially the barrier is saturated with freshwater. The head difference between the entrance and exit of the barrier is Δh. Develop a finite difference implicit numerical scheme for the solution of Eq. (16.2.1), by which the migration of the radioactive contaminant through the barrier can be simulated.

Solution

The aquifer flow velocity, V, is given by

$$V = -\frac{K}{\phi}\frac{dh}{dx} = \frac{K}{\phi}\Delta h \tag{1}$$

where h is the local piezometric head of the barrier and x is the longitudinal direction. We apply a backward finite difference approximation for the advection term and a central difference for the diffusion term of Eq. (16.2.1) to obtain

$$R\frac{C_i^{n+1} - C_i^n}{\Delta t} + V\frac{C_i^{n+1} - C_{i-1}^{n+1}}{\Delta x}$$
$$= D\frac{C_{i+1}^{n+1} - 2C_i^{n+1} + C_{i-1}^{n+1}}{(\Delta x)^2} - \lambda C_i^{n+1} \tag{2}$$

Rearranging to group similar terms for the concentration values then gives

$$-C_{i-1}^{n+1}\left[\frac{D\Delta t}{(\Delta x)^2} + \frac{V\Delta t}{\Delta x}\right] + C_i^{n+1}\left[R + \lambda\Delta t + \frac{2D\Delta t}{(\Delta x)^2} + \frac{V\Delta t}{\Delta x}\right]$$
$$-C_{i+1}^{n+1}\left[\frac{2D\Delta t}{(\Delta x)^2}\right] = RC_i^n \tag{3}$$

The numerical scheme represented by Eq. (3) is unconditionally stable. Equation (3) represents a set of linear algebraic equations, whose coefficients form a tridiagonal matrix. The solution of these equations can be performed using the Thomas algorithm.

Problem 16.2   Consider an aquifer contaminated by entrapped benzene. At a given location, downstream of the benzene entrapment, the dissolved benzene concentration is 11 mg/L. Biodegradation takes place in the aquifer, and it is assumed that 9 mg/L of oxygen is available for utilization by the microorganisms, over a time period of 12 days.

(a) Assuming that 3 mg/L of oxygen is required to biodegrade 1 mg/L of the contaminant, apply the instantaneous reaction expression to provide an estimate of anticipated reduction in benzene concentration due to the presence of 9 mg/L of oxygen.

(b) Repeat the calculation of (a), but apply the Monod kinetic expression, while assuming an oxygen half-saturation constant of 0.1 mg/L, a benzene half saturation constant of 22.6 mg/L, a maximum utilization rate of 9.3/day-mg, and a microorganism population of 0.05 mg/L.

(c) Repeat the calculation of part (a), but apply a first-order decay expression, while assuming a half-life for benzene of 5 days.

Solution

(a) Since 9 mg/L of oxygen is available, and 3 mg/L are used for every 1 mg/L degraded, the reduction in benzene, R_B, is given by

$$R_B = \frac{9}{3} = 3 \text{ mg/L}$$

and the resulting benzene concentration is

$$C_F = 11 - 3 = 8 \text{ mg/L}$$

(b) According to Monod kinetics, Eq. (16.3.3), the benzene reduction is

$$R_B = 9.3 \times \frac{11}{11 + 22.6} \times \frac{9}{9 + 0.1} \times 0.052 \times 12 = 1.83 \text{ mg/L}$$

The resulting benzene concentration is

$$C_F = 11 - 1.83 = 9.17 \text{ mg/L}$$

(c) For a first-order decay expression, we first need to find the decay coefficient of benzene. This is done using the given information for the half-life, as follows:

$$\frac{dC}{dt} = -kt \qquad \Longrightarrow \qquad \frac{C}{C_0} = e^{-kt}$$

where k is the first-order decay rate. Knowing that $C/C_0 = 0.5$ when $t = 5$ days then allows a solution for k, as

$$k = -\frac{\ln(0.5)}{5} = \frac{0.693}{5} = 0.1386 \text{ day}^{-1}$$

The resulting benzene concentration is then found by substituting back into the first-order decay equation, for $t = 12$ days:

$$C_F = 11 \exp(-1.66) = 2.08 \text{ mg/L}$$

Thus relatively different predictions are obtained using these three different models.

Problem 16.3 Consider a small lake with one inflow and one outflow, as sketched in Fig. 16.10. The inflow and outflow rates are not necessarily equal, nor are they constant. The average inflow and outflow rates on a monthly basis, along with the average monthly inflow concentration of a contaminant of interest, are shown in the first three columns of the spreadsheet below. Values for Q are in m^3/s and C is in mg/L. If the contaminant decays with rate 0.05 day^{-1} and the concentration in the lake is 35 mg/L on January 1, calculate the concentration at the end of the year. The lake volume on January 1 is 10^8 m^3. Assume fully mixed conditions and no other sources or sinks.

Solution

First, it should be noted that Eq. (16.4.1) cannot be used directly, since inflow and outflow rates are not equal, nor are they constant. Instead, we write a modified version of this equation that still expresses a basic mass balance statement for the contaminant that can be applied for each month, during which the flow rates and inflow concentrations are assumed to be constant. Setting the rate of change of contaminant mass equal to the difference between what flows in and what flows out, and accounting for decay, we have

$$\frac{dC_i}{dt} = \frac{1}{\forall_i}[(Q_{in}C_{in})_i - Q_iC_i] - kC_i \tag{1}$$

where the subscript i indicates the time period (month). Since the concentration changes throughout the month, it is not clear which value should be used for the outflow and decay terms on the right-hand side of Eq. (1). However, a reasonable value is the average at the beginning and end of the month. Of course, the value at the end of the month is unknown, so this must be calculated in an iterative manner. The same idea holds also for the volume, except that the volume in each time period is known, so that the average can be used directly.

There are a number of ways to solve this problem; here, we use a spreadsheet, as shown below. It is convenient to rearrange Eq. (1) as

$$\Delta C_i = \Delta t_i \left\{ \frac{(Q_{in}C_{in})_i}{\forall_i} - \overline{C}_i \left(\frac{Q_{out}}{\forall} - k \right)_i \right\} \tag{2}$$

where $\forall_i$ is the average volume in period i and $\overline{C}_i$ is the average concentration. Note that Δt also has a subscript, to account for the different numbers of days in different months. In the spreadsheet table below, "Del V" is the change in volume during the month, which is calculated from the difference between inflow and outflow rates. Avg. V is the average volume, taken as the volume at the beginning of the month plus half of Del V. Ctry is the guessed value for the new average concentration, which is input and must be iterated until there is convergence for the final value for C. dC1 is the change in concentration calculated according to Eq. (2), and C1 is the final concentration, which is the sum of the value of C from the previous time step and dC1. Cavg is the average of C1 from the previous and the present time steps and should be equal to Ctry. Basically, values of Ctry are varied until Cavg is equal to Ctry, at which point the iterations are completed and the calculation moves to the next time step. The final answers are then in the C1 column.

	Qin	Qout	Cin	Del V	Avg. V	Ctry	dC1	C1	Cavg
Jan.	200	195	15	13392000	106696000	12.4699	−5.06051	9.939489	12.46974
Feb.	180	178	10	4838400	109115200	8.1382	−3.60256	6.33693	8.13821
Mar.	350	345	25	13392000	115811200	18.652	24.63048	30.96741	18.65217
April	500	480	35	51840000	141731200	31.1099	0.28534	31.25275	31.11008
May	350	362	32	−32140800	125660800	26.7382	−9.02969	22.22306	26.73791
June	170	175	50	−12960000	119180800	31.1125	19.77962	42.00268	32.11287
July	100	110	45	−26784000	105788800	31.245	−21.5151	20.48763	31.24515
Aug.	25	35	30	−26784000	92396800	13.7396	−13.4953	6.992327	13.73998
Sept.	40	40	45	0	92396800	13.951	13.9141	20.90643	13.94938
Oct.	100	98	40	5356800	95075200	24.4815	7.150793	28.05722	24.48183
Nov.	140	136	27	10368000	100259200	21.9268	−12.2607	15.79655	21.92689
Dec.	155	153	18	5356800	102937600	13.8345	−3.92389	11.87266	13.83461

Unsolved Problems

Problem 16.4   Consider the steady-state distribution of a radioactive contaminant in a barrier. Assume that at the barrier entrance the contaminant concentration is C_0. At the exit of the barrier, the contaminant distribution vanishes. The retardation factor is R, the flow velocity is V, the dispersion coefficient is D, and the decay rate is λ.

(a) Derive the analytical solution of Eq. (16.2.1) for steady-state conditions.

(b) Use the solution from part (a) to determine the contaminant distribution along the barrier.

(c) Determine the conditions needed to assure that there is zero contaminant flux across the exit of the barrier.

Problem 16.5   For the conditions specified in problem 16.4, assume $R = 1.5$, $\Delta h = 1$ m, $L = 3$ m, $K = 10^{-3}$ m/d, $D = 10^{-8}$ m^2/s, and the half-life of the radioactive contaminant is $t_{1/2} = 10$ yr. Also assume that at the barrier entrance the contaminant concentration is $C_0 = 2$ ppm. At the exit of the barrier, as before, the contaminant concentration vanishes.

(a) Apply the numerical scheme of problem 16.1 to simulate the buildup of the contaminant concentration profile, until steady-state conditions are established. Plot several distributions showing the approach to the steady state.

(b) Determine the length of the barrier that is required to avoid any contaminant flux at the barrier exit. Compare your result with the analytical solution obtained in problem 16.4.

Problem 16.6   Consider an infinite aquifer with uniform flow velocity U and porosity ϕ. A portion of the aquifer with width, b, measured in the direction perpendicular to the aquifer flow, is contaminated by entrapped NAPL. Apply the potential flow superposition of a uniform flow and a negative source, to suggest an appropriate location for a pumping well that collects the contaminated flowing groundwater. Consider the pumping well discharge versus the distance between the pumping well and the contaminated region. Provide an estimated time interval, over which the pumping well still supplies freshwater.

Problem 16.7   Consider the same situation as described in problem 16.6, except that two wells are used, an injection well and a pumping well. Again, apply potential flow theory to suggest possible locations for the wells and estimate the time over which the pumping well still supplies fresh water.

Problem 16.8   In problem 16.3 a spreadsheet approach was demonstrated for solving the problem of calculating the time variation of concentration in a lake with variable inflow and outflow conditions. However, it may be noticed that in March, the concentration at the end of the month is calculated to be greater than either the concentration at the beginning of the month or the inflow concentration. What is the problem? Develop a modified approach that resolves this problem.

Problem 16.9   Consider a small lake with one stream flowing in and one stream flowing out, as sketched in Fig. 16.10. The inflow concentration of a contaminant of interest and flow rate records, along the with the outflows, are shown in the table. Concentrations are in mg/L and flow rates are in m^3/s. If the initial lake volume is 5×10^7 m^3 and the concentration on January 1 is 10 mg/L, calculate the concentration at the end of the year. The contaminant decays with first-order rate 0.02 day^{-1}.

	Jan.	Feb.	Mar.	April	May	June	July	Aug.	Sept.	Oct.	Nov.	Dec.
C_{in}	10	8	15	22	20	23	18	15	12	8	7	8
Q_{in}	105	110	135	140	120	88	55	32	47	73	86	98
Q_{out}	103	109	128	133	128	96	63	38	45	67	84	95

Problem 16.10   Consider a lake with one inlet and one outlet and no other significant sources or sinks of flow or of contaminant. The annual average

inflow and outflow rates are 500 m^3/s, and the annual average lake volume is 10^{10} m^3.

(a) What is the mean retention time in the lake, in days?

(b) If the contaminant concentration in the inflow is 100 mg/L and the first-order decay rate is 0.5 day^{-1}, what is the steady-state concentration in the lake? Assume well-mixed conditions.

(c) Assuming that the lake is at steady state with regard to the chemical concentration, calculate the time required for the lake to reach 10% of the steady-state concentration if the inflow concentration is reduced to zero.

(d) How long would it take for the lake to reach a new steady-state condition if, starting at the original steady-state condition of part (b), the inflow concentration is reduced to 10% of the original steady-state concentration?

Problem 16.11   It is proposed to decrease the turbidity of a small pond by filtering the very fine sediment that is causing the turbidity (the sediment is sufficiently fine that it has a negligible settling velocity). It is assumed that the pond water is well mixed. The water is to be pumped out, passed through a filter, and returned to the pond. With each pass through the filter, 80% of the sediment is removed. Calculate the times required to reduce the sediment concentration to 25% of the initial value using pumping rates corresponding to retention times of (a) 1 day, (b) 2 days, (c) 5 days, and (d) 10 days.

Problem 16.12   In Eq. (16.4.6), explain the implication of taking the second summation over all the inflows to the bay. In other words, what is the implicit assumption that has been used?

Problem 16.13   Under steady-state conditions of a well-mixed bay in a large lake or coastal area, with no inflows to the bay and no net flow across the bay/lake interface, what can you say about the relative magnitudes of concentration of any tracer, in the bay and outside the bay?

SUPPLEMENTARY READING

Bedient, P. B., Rifai, S. H. and Newell, C. J., 1994. *Ground Water Contamination*. Prentice Hall, Englewood Cliffs, NJ. (Contains material referring to ground-water contamination and reclamation. Of special interest are portions concerning bioremediation.)

Chapra, S. C. and Reckow, K., 1983. *Engineering Approaches for Lake Management, V. 2:Mechanistic Modeling*. Butterworth.

Chapra, S. C., 1996. *Surface Water Quality Modeling*. McGraw-Hill. (This is an excellent treatment of issues related to modeling contaminants in surface waters.)

Fischer, H. B., List, E. J., Koh, R. C. Y., Imberger, J. and Brooks, N. H., 1979. *Mixing in Inland and Coastal Waters*. Academic Press, New York. (Contains a comprehensive presentation of issues associated with contaminant mixing and transport in surface water flows.)

Hofman, J. D., 1992. *Numerical Methods for Engineers and Scientists*. McGraw-Hill, New York. (Contains a comprehensive coverage of numerical methods associated with remediation and reclamation. Such topics incorporate the solution of boundary value problems in one: and two-dimensional domains, as well as parabolic partial differential equations.)

Sun N. Z., 1996. *Mathematical Modeling of Groundwater Pollution*. Springer-Verlag, New York. (Contains a comprehensive presentation of general topics of contaminant hydrology and numerical methods applications.)

Thomann, R. V. and Mueller, J. S., 1987. *Principles of Surface Water Quality Modeling and Control*. Harper and Row.

Index